D1259374

Changing Perspectives
in the History of Science

CHANGING PERSPECTIVES IN THE HISTORY OF SCIENCE

Essays in Honour of Joseph Needham

edited by
Mikuláš Teich and Robert Young

HEINEMANN
LONDON

Heinemann Educational Books Ltd

LONDON EDINBURGH MELBOURNE TORONTO
AUCKLAND JOHANNESBURG SINGAPORE
IBADAN NAIROBI HONG KONG NEW DELHI
KUALA LUMPUR

ISBN 0 435 54925 1

© Heinemann Educational Books 1973
First published 1973

9
126.8
C45

WITHDRAWN

E. M. CUDAHY
LOYOLA
UNIVERSITY
MEMORIAL LIBRARY

Published by Heinemann Educational Books Ltd
48 Charles Street, London W1X 8AH

Printed in Great Britain by
Morrison & Gibb Ltd, London and Edinburgh

CONTENTS

NOTES ON CONTRIBUTORS

Maria Luisa Righini Bonelli is Director of the Institute and Museum of the History of Science in Florence and *Professore Libero Docente* in the history of science. She is the author of about 130 publications.

Gerd Buchdahl is Reader in History and Philosophy of Science and Fellow of Darwin College at Cambridge University. He has published widely in numerous journals in the fields of logic and philosophy of science, the critical history of scientific ideas, and the relations between philosophy of science and the history of philosophy. He is the author of *Metaphysics and the Philosophy of Science. The Classical Origins: Descartes to Kant* (1969).

Bernard Cohen is Professor of History of Science at Harvard University. His research interests have mainly concentrated on seventeenth and eighteenth-century physics. He recently edited (with A. Koyré) Newton's *Principia* (1969) and published *Introduction to Newton's Principia* (1971).

Allen George Debus is Professor of the History of Science and Director of the Morris Fishbein Center for the Study of the History of Science at the University of Chicago. His research has centred on sixteenth and seventeenth-century science and medicine. He is the author of the *English Paracelsians* (1965) and *Science and Education in the Seventeenth Century: The Webster-Ward Debate* (1970).

Aage Gerhardt Drachmann was librarian at the Copenhagen University Library (1915–1956). His special subject is the study and interpretation of texts on technology composed in Classical Antiquity. Among his chief works are *Ktesibios, Philon and Heron* (1948) and *The Mechanical Technology of Greek and Roman Antiquity* (1963).

André G. Haudricourt is *Directeur de recherche* at the Central national de la recherche scientifique in Paris. In his investigations of linguistic and ethnological problems he utilizes his former agronomical and botanical training. His publications comprise (in collaboration with Louis Hédin) *L'homme et les plantes cultivées* (1943) and (in collaboration with Mariel Jean Brunhes-Delamarre) *L'homme et la charrue à travers le monde* (1955).

Mary Brenda Hesse is Reader in Philosophy of Science and Fellow of University College at Cambridge University. She is the author of a number of articles, especially on aspects of the history and logic of physical science. Her books include *Forces and Fields: The Concept of Action at a Distance in the History of Physics* (1961) and *Models and Analogies in Science* (1963).

Henry Holorenshaw was originally destined for medicine, but he eventually forsook science for history. He published *The Levellers and the English Revolution* (1939), which also appeared in Russian and Italian.

Lu Gwei-Djen is a Fellow of Lucy Cavendish College and a member of the Cambridge University Oriental Studies Faculty. In 1957 she joined Joseph Needham to collaborate on the 'Science and Civilisation in China' project specialising on the history of Chinese medicine and biology.

Neil McKendrick is University Lecturer in Modern English Economic History and Fellow, College Lecturer and Director of Studies at Gonville and Caius College at Cambridge University. He is the author of many studies and articles and is currently engaged in the evaluation of the rôle of Josiah Wedgwood in the development of eighteenth-century business and manufacture.

Walter Pagel is Emeritus Professor of the History of Medicine of Heidelberg University after working as Consulting Pathologist to several hospitals in Britain. His many publications cover the fields of medical pathology and the history of medical and biological thought in the sixteenth and seventeenth century, including *Paracelsus: Introduction to Philosophical Medicine in the Era of the Renaissance* (1958) and *William Harvey's Biological Ideas* (1967).

Roy Porter is Fellow, Director of Studies and College Lecturer in History at Churchill College, Cambridge. He is completing his doctoral thesis on the rise of the science of geology in Britain in the seventeenth and eighteenth centuries.

Derek de Solla Price is Avalon Professor of the History of Science at Yale University. He collaborated with Joseph Needham and Wang Ling on the study of Chinese medieval clockwork and has also worked on the statistical analysis of modern science policy. He is the author of *Science since Babylon* (1962) and *Little Science, Big Science* (1963).

Abdur Rahman is currently in charge of the Research Survey and Planning Division of the Council of Scientific and Industrial Research in Delhi. His main interests have been the history of science, operational research, science planning and social dimensions of science. He is the author of a number of papers and books in these fields.

Pyarally Mohammed Rattansi is Professor of History of Science at University College, London. He has published numerous papers on the social interpretation of sixteenth and seventeenth century natural philosophy and is currently completing a study of Newton's alchemical writings.

Jerome Raymond Ravetz is Senior Lecturer and Head of the Division of the History and Philosophy of Science at the University of Leeds. He has been lately concentrating his attention on the epistemology, sociology and ethics of science and its applications in the technical and human context. He is the author of *Astronomy and Cosmology in the Achievement of Nicolaus Copernicus* (1965) and *Scientific Knowledge and Its Social Problems* (1971).

Cyril Stanley Smith is Professor Emeritus of Massachussetts Institute of Technology. He has researched and written on the science of metals and the history of metallurgy. He has been instrumental in the English publication of the classical texts of Theophilus, Biringuccio, Ercker and Réaumur. He edited a collection of shorter works as *Sources for the History of Science of Steel 1532–1786* (1968). He dealt with the development of metallography in *A History of Metallography* (1st ed. 1960).

Mikuláš Teich is Visiting Scholar at Gonville and Caius College, Cambridge. His publications include work on the history of chemistry and bio-medical sciences, social and philosophical aspects of the development of science, historical relations of science, technology and the economy, and on the history of scientific organization.

Charles Webster is Reader in the History of Medicine at the University of Oxford and Fellow of Corpus Christi College, Oxford. He has published original papers especially on seventeenth-century scientific and medical thought. He edited and wrote the introduction to a selection of educational writings, *Samuel Hartlib and the Advancement of Learning* (1970).

Robert Young is in charge of the Wellcome Unit for the History of Medicine at Cambridge University and Fellow of King's College. His research is concerned with historical, conceptual and social aspects of the bio-medical and human sciences, especially in the nineteenth and twentieth centuries. He is author of *Mind, Brain and Adaptation in the Nineteenth Century* (1970) and other studies on the application of science to the study of man.

INTRODUCTION

This volume has two purposes. The first is to bring together a number of approaches to the history of science which have emerged in recent years and to make available essays by leading scholars in the field which reflect these approaches. The second is to honour Joseph Needham, a scholar whose work has both reflected and inspired interest, research topics, and geographical and historiographic viewpoints which extend and deepen the discipline. The editors are very gratified that an invitation to a number of scholars who wished to pay their respects to Joseph Needham on his seventieth birthday (1970) has produced such a rich set of perspectives on the history of science. We would not claim that the result is comprehensive or synthetic, but it is representative in important ways.

In particular, as its title implies, the volume reflects a fundamental feature of the history of science as a continuously developing discipline—that is, that its scope and texture are rich enough to be constantly changing; not in a single direction but in several, and from various points of view which are relevant to any given historical problem. The cumulative effect of the essays is to demonstrate how the discipline contains a diversity of perspectives which are firmly linked with detailed and specific research. The particularity of a given problem comes to be seen as integral to its content, to the perspective of the scholar, and to the historiographic problems which it reflects. Some of the authors are looking beneath the findings and theories of science to the underlying 'philosophy of nature'; others are searching more widely into the social context of science; and others are opening up new areas of investigation and exploring disciplines in science, pseudo-science, or technology which have not hitherto been carefully considered by scholars in the history of science. Thus, one feature of the current situation is that particular arguments are being integrated with a more intricate appreciation of both the fine texture of, and multiple perspectives on, the relevant research materials.

In a scholarly field there is at any one time an awareness on the part of certain scholars of 'the state of the discipline'. From time to time volumes appear which attempt to present and reflect the cross-currents and attitudes which are prevalent and others which are emerging. We see this volume as following on from two such attempts. The first is *Critical Problems in the History of Science* (Wisconsin, 1959), edited by Professor Marshall Clagett.

It contained papers with commentaries which were given by a group of distinguished scholars attached mainly to American Universities. The stated aim of the conference was to stimulate the study and the teaching of the history of science.

The development of the history of science and technology after the Second World War was not fortuitous, and that conference reflected its maturing condition. It would be misleading to claim that the history of science existed as a coherent discipline before the war. George Sarton, Charles Singer, and a few other individuals were relatively isolated. There was no established profession, no continuing research tradition in the field, and few institutional expressions. There were a limited number of posts, because there was little consciousness of the relevance of the subject. But it would be equally misleading to fail to notice the perils and hopes brought about by scientific and technical developments during the war. Deeply worried scientists and non-scientists began to feel a pressing need to analyse critically the past of science in order to understand its present and to prepare for its future.

It was the attempt to formulate some of the 'critical problems' in the history of science according to the highest standards of scholarship which makes the Wisconsin volume historically significant. It effectively signalled the coming of age of the history of science as a discipline whose contribution to knowledge of the past could no longer easily be ignored by other branches of learning. There were good reasons for the participants' concentration on the Scientific Revolution of the sixteenth and seventeenth centuries. That the birth of modern science has been associated with the breakthrough in mechanics and astronomy very few disputed. But the antecedents of modern science—and later developments—were discussed in far less distinct terms. Evidence had been accumulating that the European Middle Ages were not —as had often been proclaimed—completely or uniformly 'Dark Ages', unconnected with succeeding periods. Possible relations between the artisan and the scholar had been explored. Although this approach was by and large rejected by the historians of science as having no significant bearing on the understanding of the origins of the Scientific Revolution, its historical background and its philosophical and theological dimensions were coming to be seen as a more complex affair than was formerly believed.

The second important volume was *Scientific Change*, edited by A. C. Crombie (Heinemann, 1963), which contained contributions to a symposium organized by the Oxford University Committee for the History and Philosophy of Science in July 1961. In scope and in number of participants, the symposium at Oxford superseded that at Madison. Indeed, the Oxford symposium had been conceived at the Madison meeting as an expansion of that gathering. A representative international group of distinguished historians and philosophers of science was joined by general, social and economic historians and working scientists. Their purpose was to discuss the issue of the structure of scientific change and to throw light on the intellectual, social, and technological conditions for scientific discovery and technical invention, from antiquity to the present. Compared with its predecessor at Madison, the Oxford symposium reflected a more ready acceptance of the idea that the development of the physical and biological sciences was influenced by both 'internal' and 'external' factors, although the consideration of their mutual relations was left very much in the air.

In looking back from the vantage point of a decade later, it emerges more clearly that the Oxford symposium was less concerned with specific problems in historical research than with the consideration of mature ways of *looking at* problems. It was, for example, at that meeting that Thomas Kuhn first acquainted the international community of scholars with his views on 'paradigms' and their rôle in scientific change. (His much-debated historiographic monograph, *The Structure of Scientific Revolutions*, was still in press.) This meeting was also one of the first at which scholars of varied backgrounds had an opportunity to comment on and discuss with Joseph Needham his novel ideas on the history of Chinese science. It was perhaps during this discussion that the sharpest division of opinion occurred among the participants in the symposium.

It is probably an oversimplification—but a useful one—to suggest that the Wisconsin symposium was more concerned with problems and the Oxford one with approaches and solutions. It is clear, however, that since the early 1960s, the awareness has deepened among a number of scholars of the need to integrate the concerns expressed in those volumes with their own research. We believe that the present collection draws attention to some new ideas and new trends in the history of science and technology, especially the increasingly interdisciplinary nature of the subject.

The authors also feel that the person who has most brilliantly contributed to and reflected the changing perspectives in our discipline is Joseph Needham. His research—both in its domain and its conception—has shaped the interests of many scholars, while his writings constitute an unparalleled example of the fruits of fresh approaches to the subject. We feel particularly fortunate that in preparing this volume of essays in his honour, we were able to persuade Henry Holorenshaw to write an incisive biographical essay about Joseph Needham. Holorenshaw has not published in the field of history for several decades, but he was willing to take up his pen again for this special occasion, to write about an intimate of whom he says, 'As I have known him so well for more than sixty years I might be able to explain how it all fits in.'

Where Needham's influence has become most discernible is in the area of *comparative* history of science and technology. As indicated by some of the contributions to this volume, scholars have become increasingly sensitive to the need to take a fresh look at the transmission and exchange of scientific and technical knowledge. As a precondition of this approach, it is necessary to locate the occurrence of particular discoveries in space and time, to consider whether or not they were made independently more than once, and to investigate the conditions of diffusion and the possible modifications of the original ideas or practical contrivances. An example of this kind of history can be found in Cyril Stanley Smith's inquiry into the history of metallurgy affecting two areas in the Far East. Professor Smith doubts that the Chinese independently invented the use of bronze and suggests that they may have acquired it from the Thais. At the same time he derives the change in style in China from the changes in moulding techniques. These, in turn, he traces to Chinese knowledge of the specific properties of the loess, that characteristic yellow material covering vast areas of China.

It is only natural to expect that a study of India on the lines of Needham's work on China could produce equally important results. A. Rahman raises part of the curtain behind which most of the riches of ancient and medieval Indian science remain hidden. Since India does not possess a table of dynasties, the chronology is uncertain. This, and the fact that the sources are composed in Sanskrit, Persian, and Arabic, does not make it easy to work in this area.

In the study of the Greek world, accessible knowledge about science is relatively extensive, while it is less available about technology. Dr. A. G. Drachmann has taken up an earlier suggestion of Needham's and considers the evidence for the claim that the crank may have been an invention of Archimedes. Because of the fundamental importance of this technical device, the tracing of its development assumes a special role in the history of technology. Drachmann's paper also raises another issue: the complex conceptual problems involved in discerning whether or not an artifact, which appears to illustrate an important technical principle, was *seen* at the time as an example of a conception whose general applicability had been grasped. Historians who deal with such issues as this—and the one to which Smith and Rahman address themselves—are expanding the boundaries of the discipline.

It is the custom to write about 'Egyptian', 'Babylonian', 'Chinese', 'Indian', 'Greek' and 'Arabic' science and technology, and on the whole there is a consensus among scholars about the meaning of those terms. A similar consensus also extends, for instance, to descriptions of Greek mathematics as 'largely geometrical' and of Chinese mathematics as 'arithmetical and algebraic'. But when it comes to more recent periods of history, the employment of national adjectives coupled with broad conceptual categories raises questions which (if we except the controversies on 'priorities') historians of science and technology have so far failed to examine seriously. The English scientific attitude is often characterized as 'empiricist', the French as 'analytical', the German as 'philosophical'. Are these merely sweeping generalizations, or do they contain a kernel of truth? Is 'natural philosophy' as much English as '*Naturphilosophie*' is German? Clearly, comparative history without national histories of science and technology becomes meaningless, and that is not the least of Joseph Needham's messages: 'In fact, of course, no ancient or medieval science and technology can be separated from its ethnic stamp, and though that of the post-Renaissance period is truly universal, it is no better understandable historically without a knowledge of the milieu in which it came to birth.'[1]

Joseph Needham is not generally thought of as an historian of the Scientific Revolution of the sixteenth and seventeenth centuries. Yet his studies on the history of embryology (along with his work on the English Revolution), and his reflections on the question of why modern science and technology did not develop in China, make him a participant in the debate which has gathered momentum since the early 1960s. Put in its simplest terms, this debate centres around the question of whether the rise of modern science can be reduced to advances in the knowledge of the mechanical movement of terrestrial and celestial bodies or whether the net has to be cast wider and deeper to include other branches and levels of the interpretation of nature—for instance chemistry, physiology, theology, and the 'philosophy of nature'. This volume contains several contributions which bear on this general issue, one which raises both historiographic and philosophical questions about the nature, scope and definition of science and of rationality. Those who have advocated a more flexible conception of the developments in the period have insisted on the integrity of their detailed studies and have pursued them in their own terms. As a result of these preoccupations, they have been relatively less concerned to connect the results of their studies with the mainstream of interpretation of the period. The essays of Debus, Pagel, Rattansi and Webster fall into this category, while those of Hesse and Buchdahl—and in a different period that of Young—attempt to re-assess the resulting conceptions of science and its history.

Allen G. Debus is concerned to explore in depth the world of the Paracelsians. In his article Debus reaffirms his opposition to any claim that their endeavours and those of the iatrochemists 'had played anything less than a

---

[1] J. Needham, *Science and Civilisation in China*, Vol. 4, Part II (Cambridge, 1965), p. xlvi.

major role' in helping to prepare the ground for the mechanical world picture. Looking further, no student in this area can afford to ignore the work of Walter Pagel, who before the Second World War was a brother in arms with Joseph Needham in attempting to establish the study of the history of science in Cambridge.[2] Pagel's work has significantly influenced that of Debus, Rattansi and Webster. Although his is a relatively short contribution, it provides further proof of his insight into the complexity of scientific development as he considers the relationships of continuity and discontinuity between the work of van Helmont and that of his predecessors. From this account an essentially dialectical relationship emerges in which 'the chemical results . . . are the by-products of his religious and vitalistic philosophy and can only be open to our understanding when so viewed'. Pagel also conveys one of the major trends in the scholarship of the last decade—the determination to make oneself, in so far as possible, 'contemporary to the savants studied. In other words what the scientist looking back into the history of his subject has to cut out of context the historian has to replace. He must do so unafraid of complicating something which appeared to be simple.'

In examining this complexity on its own terms, students of hermeticism, alchemy and the philosophical and theological assumptions underlying the work of natural philosophers, have found that it is very difficult indeed to find ways of phasing out the old and demarcating it from the new. Debus points out how late in the seventeenth century the influence of the Paracelsians and iatrochemists was felt and adds that their approach was less, not more, mathematical than their predecessors and contemporaries. It might be thought that this is mere quibbling about dates, but it has at least two aspects which are more profound. The first is that the unilinear conception of scientific progress is under sustained and heavy attack. The second is that a number of scholars are challenging the traditional historiography of the period of the seventeenth century and are arguing for the persistence of philosophical and theological themes at the heart of the natural philosophy of the period. It will be seen that the assessment of hermeticism and alchemy is central to this debate. While Debus and Pagel offer important evidence on their rôle, two of the historiographic papers in the volume—those of M. B. Hesse and P. M. Rattansi—crystallize the fundamental issues in a debate which many believe has come in the last decade to the centre of the historiography of the development of modern science. Hesse is a philosopher of science who has always related her philosophical work to original research in the history of science, while Rattansi is one of the most outspoken advocates of a fully relativist and contextualist historiography. It is no

[2] Cf. J. Needham and W. Pagel (eds.), *Background to Modern Science* (Cambridge, 1938).

exaggeration to say that the fundamental disagreement between them on the concept of rationality in the period, has implications for the whole conception of science in the modern period. In some ways it is also proving a testing ground for the belief that the history of science and the philosophy of science can remain closely related. Moreover, the controversies about the historiography of the period are showing that the history of science is part of general history.

Two other papers in the volume indicate aspects of this debate in the field. Like Mary Hesse, Gerd Buchdahl believes in the integration of the history and the philosophy of science. His contribution is concerned with the history and the philosophy of science as related to an underlying philosophy of nature. Where the primarily historical studies of Debus, Pagel and Rattansi often appear to challenge the integrity of the subject by showing the important non-mechanical (and even anti-mechanical) aspects of the thought of the period, Buchdahl is probing the deep philosophical structure of conceptions at the centre of the methodological and conceptual framework of modern science. Instead of tidying up the past to suit present fashions in the philosophy of science, he is exploring the richness and complexity of the views of nature in the period and is attempting to provide an analytical framework for studying the ambiguities and intangibles in science. One way of indicating the difficulties in the field is to say that Buchdahl's work is in the same world as Hesse's both philosophically and historically; it is resonant with Rattansi's in its indication of the depth and complexity of the problem; yet the philosophical approach and conclusions of Hesse do not seem amenable to integration with the historical relativism of Rattansi. Hesse concludes that the suggestion of a confluence of hermeticism and mechanism in the 'melting pot of the new science is a mistake', while Rattansi maintains that 'a knowledge of Renaissance Neo-Platonism and Hermeticism, and its vicissitudes through the sixteenth and seventeenth centuries is essential' for the understanding of Newton's work.

The problem is made more intricate by the affinity of Rattansi's approach to that of Charles Webster, and it would appear that there is no reconciling Webster's approach with the mainstream of the history and philosophy of science. Webster is uncompromising in insisting on seeing the issues in the period in their own terms and, more importantly, according to *contemporary priorities*. Consequently, natural philosophy is not mentioned until nearly halfway through his essay. Webster's study of the theological and educational debates in the period makes a highly critical comment on the official historiography of the seventeenth century in what it does *not* say. He places science very much *inside* the religious, anti-élitist and radical thought of the group he is considering. He says of William Dell—a predecessor of Needham as

Master of Caius College—that he 'was not a direct participant in the scientific movement of the puritan revolution; he neither experimented nor contributed to theoretical natural philosophy. His priorities and preoccupations were religious, like the majority of his contemporaries. The quest for salvation and preparation for the millennial kingdom dictated attitudes to all aspects of social behaviour, from economic life to philosophy.' He goes on to say that the enthusiasm of Dell and his associates for experimental science provides a perspective on the motives of an important section of the community for advocating science. The point to be made about Webster's contribution is that one group of historians of science would see it as an important contribution to the understanding of science and its relations *in* the discipline of the history of science, while others would relegate it (whatever its undoubted merits) to social history or the history of education. Whether it is seen as inside or outside, it raises the problem which Young and Teich also discuss of 'internalist' versus 'externalist' historiography.

As long as history of science was relatively centred on locating critical problems and debating the process of change, many of these tensions were masked. As the perspectives of the discipline widen and deepen, and especially as scholars seek more fully to investigate science in its myriad relations, there is an apparent tendency towards fragmentation and lack of communication in the field. The papers in this volume provide a cross-section of evidence that the problem has become central in the last decade.

Another dimension of this fragmentation is the lack of communication between those with historico-philosophical preoccupations and those who have become increasingly concerned with the social relations of science and seek to study its content and methodology as part of a social system. This tendency has grown up in recent years, and the contribution of J. R. Ravetz exemplifies it.

We believe that the diversity in the field—even including its fragmentation and problems of communication—is part of the constructive aspect of the development of knowledge. Indeed, it is a consequence of the multiple, overlapping and changing perspectives in the field that even the introduction to this volume cannot be written in a unilinear way. We pointed out earlier the rôle of alchemy in the historiography of the Scientific Revolution, and from there the argument has developed into the social dimension. But, of course, there is also a *comparative* aspect to the study of alchemy, and another of our contributors, Dr. Lu Gwei-Djen, writes from that perspective. In his introduction to a collective volume on the history of biochemistry (which he also edited), Needham had a good deal to say about the Chinese, Indian, Arabic and European backgrounds to the chemistry of life: 'There are many ways in which the work of Paracelsus . . . was hardly less revolutionary'[3] than that of Copernicus, Keplen and Galileo. He writes of van Helmont that he 'has perhaps as good a title as anyone else to be called the father and founder of all modern biochemistry'.[4] Needham finds that the world-view of the Paracelsian iatrochemists 'derived from Neo-Platonic, Gnostic and Hermetic sources, and . . . was strikingly similar to certain Chinese world-views involving the macrocosm-microcosm doctrine, the principle of action at a distance, the unification of the spiritual and corporeal worlds, the idea of the inter-connexion of all things by universal sympathies and antipathies, and a tendency to numerology as opposed to real mathematics and quanti-fication'.[5] Although a common ground existed, Chinese iatrochemistry developed certain ideas and practices associated with the concept of the 'inner elixir' with no equivalent in its European counterpart. It is fitting that Needham's oldest and closest Chinese collaborator, Dr. Lu, should explain in her contribution some of the fascinating features of this phase of Chinese physiological chemistry.

Needham and Lu are at home in the ideographic language of the Chinese, and most historians of science have to rely on their translations and inter-pretations for understanding the science of that culture-area. Three of our contributors are bringing the problems of pictographic representation and nomenclature closer to home. In his discussion of botanical nomenclature, André Haudricourt shows that naming is in itself an historical process which is intimately bound up with national history: the meaning and reference of such terms are far from clear. In a brief contribution, I. B. Cohen takes this point further and borrows techniques from the history of art to help to illuminate the meaning of a Harveian frontispiece. This approach has been applied with similarly interesting results to Vesalius', and Burnet's and Lyell's frontispieces, and has been discussed at length in a book by Frances Yates.[6] Finally, Derek Price has taken on a daunting task in exploring the relationship between geometrical and scientific talismans and symbols on the one hand, and the concepts to which they relate on the other. The analyses of Haudricourt, Cohen and Price indicate new lines of interpretation for further widening the domain of science in its multiple relations.

---

[3] The *Chemistry of Life; Eight Lectures in the History of Biochemistry* (Edited, with an introduction by J. Needham) (Cambridge, 1970). p. xix.

[4] Ibid., p. x.

[5] Ibid., p. xx.

[6] C. D. O'Malley, *Andreas Vesalius of Brussels, 1514–1564* (California, 1964), pp. 139–144; M. J. S. Rudwick, *The Meaning of Fossils; Episodes in the History of Geology* (Macdonald, in press); 'The Strategy of Lyell's *Principles of Geology*', *Isis* **61** (1969), p. 16; 'Lyell on Etna and the Antiquity of the Earth', in C. J. Schneer (ed.), *Toward a History of Geology* (M.I.T., 1969), p. 304; F. A. Yates, *Theatre of the World* (London, 1969).

The studies of Price and Cohen also point to the relativity of fundamental metaphysical and cosmological notions in different traditions. Buchdahl has brought this issue closer to the modern scientific world view by probing the underlying approach to the explanation of gravity. Maria Luisa Bonelli has taken this line of enquiry further and tests the limits of the history and philosophy of science in another direction in her discussion of the nature of time. She suggests, contrary to the currently accepted view, that time is not an empirical but a rational notion.

It will be seen that a rich and complex set of interrelated perspectives are being brought to bear on problems in the history of science. We have hitherto emphasized the *national, comparative, philosophical, alchemical* and *hermetical* and *conceptual ones.* But a number of essays in this volume point to further dimensions—*economic, political* and *ideological.* Compared with the established links between the history of science and the philosophy of science (however broadly or narrowly defined), there has been very little contact between the history of science and economic history. Of course economic historians have been aware of the contributions of technology to economic development. However, these aspects have usually been registered rather than analysed by economic historians. During the last two decades there has been a noticeable change in the study of economic history with increasing emphasis on economic growth, particularly during the Industrial Revolution. Economic historians have found it necessary to inquire more closely into the historical rôle of technology in economic change. There has been considerable scepticism about the significance of science itself in this process, and it has been argued that the relevant technology was not, on the whole, developed in the tradition of science but grew up within a craft and inventors' tradition relatively out of touch with science. This scepticism is only now coming under scrutiny, and the rôle of scientific rationality and particular scientific developments are coming to be juxtaposed with economic change and technology.

Economic historians are therefore gradually finding that they are co-participants in the same debates as historians of science and technology, philosophers of science, sociologists of science, and others who are basically interested in the same kind of questions but are understandably approaching them from the vantage-point of their own specific field. The issue of what is science and what are the causal relations between science and industry comes under scrutiny in Neil McKendrick's study of one of the leading figures in eighteenth-century industrial England, Josiah Wedgwood. The inclusion of this study in the volume examplifies the movement away from the restricted vision of specialists who had not fully examined Wedgwood's position as both a central entrepreneurial figure in the highly technological pottery industry and a leading member of one of the most celebrated scientific and philosophical circles in the period—The Lunar Society.

Compared with economic historians who have been aware of the existence of scientific and technological developments, the vast majority of historians of science have hitherto exhibited very little interest in economic history. The small minority who have addressed themselves to such questions have been well out of the mainstream and have been influenced by Marxist views. This lack of interest sprang from fear of entanglement in the tenets of the prevailing economic determinism in Marxist circles. This orthodoxy is now on the wane, and Marxist questions are being asked anew by historians of science, as Roy Porter shows in his pioneering examination of the history of geology against the background of the Industrial Revolution. He has pointed the way to a number of new perspectives in addressing himself to the relatively neglected field of the early history of geology and has reviewed its development in the light of a simple economic determinist hypothesis. His long parade of negative conclusions is cast in a new light when he alters the *level* of socio-economic analysis: 'To understand science we must see it in the context of, and constitutive of social structure, social change, and social consciousness. But it is simple-minded always to expect to find science *responding*, in any immediate way, to social conditions. For a mere glance will show that because men of science in East and West have always comprised some sort of mandarinate or clerisy, whether as an élite set apart by wealth, talent, privilege or by state patronage, etc., they have often been cushioned from some of the more obvious and potent social movements of their day, often isolated from economic and technical pressures from below, by systems of cultural mediations.' Porter has, by example, shown how illuminating the detailed study of these mediations can be.

The essays of McKendrick and Porter represent an important new development since the Madison and Oxford volumes, in which there was no serious discussion of the interrelations of science, technology, and the economy. As we said above, comparative history is not possible without national history, and these two essays provide an important beginning. However, it is already clear from Porter's account that the relationships among geology, mining and government on the Continent were different from those in Britain. Economic historians and historians of technology are currently very interested in the problem of the diffusion of technology from Britain (in the first instance) throughout Europe and elsewhere, and the inter-play between the factors of science, technology and government—both nationally and comparatively—should be a most promising development in the future.

It is well established that geological developments in the period after the one discussed by Porter, provided essential groundwork for the development of the modern theory of biological evolution. The essay of Robert Young complements Porter's in a number of ways. It is explicitly concerned with the intersection of historiographic and ideological perspectives as they arise in the interpretation of the nineteenth-century British debate on man's place in nature. Like Porter, Young is attempting to carve a new path between the internalist history of science and the available theories for understanding science as a part of society. He systematically explores a Marxist epigraph drawn from the writings of Joseph Needham as it applies to historians of science and to the subjects of their research. His aim is to stimulate debate on the requirements of a radical historiography. He attempts to show that it is impossible to separate the internal history of biology from concomittant debates on nature, God, Man and Society. He is concerned with a period of transition from social and theological rationalizations which were based on a vew of Britain as a harmonious, pastoral society to ones which were based on a conflict-ridden, urban and industrial one. In challenging the distinction between the substance and the context of science, he is also considering the issues raised in the contributions of Hesse and Rattansi, but he does so from an explicitly ideological point of view and addresses his argument to a different field and a later period.

Where Porter and Young have written about disciplines the histories of which are only now coming to the centre of attention of scholars in the field, the contribution of Mikuláš Teich is concerned with a science—biochemistry—which is of relatively recent origin, has grown spectacularly, and has far-reaching connections with medicine, industry and agriculture. There was no substantive reference to biochemistry in *Critical Problems in the History of Science* or *Scientific Change*, but the continuing growth of the chemical study of living matter is calling forth a systematic analysis of its historical development, thus adding a new perspective in the history of science. Teich's study begins with the cell theory of Schwann and shows that for a long time cell morphology and biochemistry developed more or less independently. The only bridge between them was the concept of 'protoplasm' as the highly-organized material basis of life. The growth of cytology and the disappearance of the protoplasmic 'giant' molecules formed essentially two aspects of the same complex historical process, which bridged the gap and eventually gave rise to the linking of the dynamical and structural approaches in the study of the chemical organization of living matter. In doing that, Teich also places Needham's ideas of the 1930s on the micromorphology of the cell in an historical context. Although seemingly an 'internalist' study, Teich makes it clear that a full account of the development of biochemistry—and indeed of any science—requires a rich dialectic between 'internal' and 'external' approaches.

The volume concludes with a bibliography of Joseph Needham (excluding papers on biochemistry and experimental embryology and morphology) which scholars in a wide spectrum of fields will undoubtedly find convenient.

At the beginning of this introduction we suggested that the volume reflects a diversity of perspectives which are finally linked with solid research. Having listed the contents and alluded to some of the themes which they contain, we would like to conclude by drawing attention to what we feel to be an even more salient aspect of the development of the discipline. That is, the essays defy neat categorization: one could—and should—group them in different ways according to the particular aspects of the available perspectives, emphases, contrasts, and specialist knowledge which are being brought under scrutiny. Whereas the editors of the two earlier volumes which reflected 'the state of the discipline' could attempt to achieve a single clear focus, we find it much more difficult. Whatever interpretation might be put on this feature of the volume, we feel that it reflects an increasing awareness of the requirements for reproducing the full complexity and depth of the historical realities. Particular problems in the history of science are coming to be seen in their relations. Not just the subjects, not just their relations, but detailed research, interpreted—as an integral part of that research—in the *totality of its relations*.

M.T.
R.Y.

The editors wish to express their thanks to Dr. Dorothy Mary Moyle Needham, F.R.S., for her unfailing helpfulness and gentle guidance.

I

THE MAKING OF AN HONORARY
TAOIST

Henry Holorenshaw

All scholars, it has been said, have some blind spot. Historians of science and of thought, for instance, have been known to be allergic to the social and economic background of the development of man's knowledge of Nature, or alternatively if they are willing to give full weight to this they may be very disinclined to recognize religious motives operating perhaps in the formation of scientific ideas. Some have been loth to consider possible aesthetic factors in the world-systems evolved in different civilizations; others have been strongly hostile to metaphysics. Most common, perhaps, has been a firm disinclination to consider the part played in the world history of science or medicine by the non-European civilizations; and it has too often been assumed that since modern science arose in European culture alone no other cultures were worth paying much attention to, even if the difficulties of approach through the Arabic, Sanskrit or Chinese languages could be overcome. But the work of the man for whom this volume of friendly essays has been produced, Joseph Needham of Cambridge, seems to have been more free of these blind spots than most. Hence it came about that I was asked by the Editors, as one who knows him better than most people, to try and sketch the historical background of the historian himself. How did it happen that a biochemist turned into a historian and sinologist? And why did he once call himself an 'honorary Taoist'? This enquiry is naturally difficult, but I shall attempt the task.

There are a number of paradoxes which make the person we are writing about difficult to understand. Many have been puzzled by his seeming synthesis of what are frequently considered irreconcilable contradictions. First, of the 'younger scientists' of the twenties and thirties (such as Julian Huxley, J. B. S. Haldane, C. H. Waddington, Lancelot Hogben, Desmond Bernal, Hyman Levy and others), he alone started out as a man of definite commitment to liturgical religion and alone has held to it throughout his life. This is what one might call the antithesis of science and religion. Secondly, this commitment to religion was somehow not at all incompatible with, indeed it even implied, an equal commitment to socialism, and this not strictly delimited in character either, but including great sympathy with its Marxist forms, and with the efforts of the socialist countries after 1917 to bring about a communist order of society. This is what one might call the antithesis of religion and politics. Thirdly, people who are very much rooted in the religion and society of their own culture often show little comprehension or sympathy for the religious and social forms of civilizations other than their own. Yet this one completely 'fell in love' (as he says himself) with Chinese civilization, finding it of inestimable value not only for its own sake but in the critical appraisal of his own. Moreover he found it possible to enter over the years into intimate and enduring friendships with Chinese, Ceylonese, Indian and other Asian people. This is what might be called the antithesis of East and West. If in fact he did succeed in synthesizing all these things in a life as it was lived, there does seem some interest in trying to elucidate from what he has told his friends the origins of these various ideas, convictions and attitudes; the analysis might be helpful to some other people. This paper is not intended as a biography in any formal sense, but rather a kind of spiritual anatomy, a study of the make-up of one particular person. To the minds of those who only slightly know him, many of the beliefs and positions which he has adopted have sometimes seemed, as I know, strangely contradictory, but this is because they have not seen the coherent pattern which is really there. As I have known him so well for more than sixty years I might be able to explain how it all fits in.

First then, about his upbringing. He came of a family which was neither wealthy nor poor; his father was a physician in general practice who had taught histology at the University of Aberdeen and was to end as a Harley Street specialist (one of the first) in anaesthesia. His mother was a gifted, though feckless, musician and composer, very much with the 'artistic temperament', who published a great many songs, some of which are still sung today. From his father, with whom he consciously most sympathized, he derived the scientific mind; from his mother a certain largeness and generosity in action and initiative. Unfortunately the two did not get on well, and as he himself says, he grew up, as it were, in the midst of a battle-field; or, in what might be perhaps a better image, ferrying between two pieces of land separated by an arm of the sea. Those who like to explain everything in psychological terms might consider that the efforts of the child to link these continents set his mind in a posture of permanent bridge-building, searching always as it were for the union of things separated, of science and religion, of biochemistry and morphology, of religion and socialism, and of East and West.

As a boy he was naturally strongly influenced by his father's attitudes in matters of religion and philosophy. He also benefited by the fine library which his father had collected, as will later appear again. The elder Joseph Needham had been to begin with a figure in the Oxford Movement of Anglo-Catholicism, but later was more attracted to the mysticism of the Society of Friends, and finally to a philosophical theology which took him and his son Sunday after Sunday to sit at the feet of E. W. Barnes, F.R.S., the mathematician, then Master of the Temple in London, later Bishop of Birmingham.

Thus it came about quite naturally that when the younger man went up to Caius College at Cambridge to read medicine, he became a member of the Confraternity of the Holy Trinity (S.T.C.) and Secretary of the Cambridge branch of the Guild of St. Luke, an Anglican society (now long defunct, I think) for medical students. But here again the element of philosophy and comparative religion strongly entered in. He has often said that he learnt more from some of the evening meetings of the Guild of St. Luke than from all his regular lectures. For example, Edward Browne of Pembroke spoke on Persian and Arabic medicine, while F. C. Burkitt of Trinity discoursed on the Manichaean religion, with its service-books recovered from the sands of the Gobi—such scholars as these could hardly be stopped when in full flow, and it was they who first inspired him with a sense of the excitement and romance of humanist scholarship, especially when combined with the history of the natural sciences. And so it came about that he chose for his College prize books the writings of Lancelot Andrewes, Jeremy Taylor, Angelus Silesius, Herbert of Cherbury and Miguel de Molinos, as well as the *Fioretti* of St. Francis; a cause of some astonishment to his Tutor, who was accustomed to other requests from medical students.

He had not particularly enjoyed his time at Oundle School in Northamptonshire, but he feels that he gained there many valuable things. From the headmaster, F. W. Sanderson (a friend of H. G. Wells), he got a deep impression of the value of spacious conceptions in history and life. Although on the Classical Side, he did a great deal of history, and much enjoyed Sanderson's scripture classes where the Old Testament was treated primarily as history and archaeology. Sanderson's insistence on the making of historical charts has also been, he says, of enormous advantage to him throughout his life. One of his earliest publications was a large wall-chart to illustrate the history of physiology and biochemistry. Moreover, in those days, during the First World War, every boy in the school, whether on the classical side or not, had to spend much time in the workshops, and in this way he acquired a fundamental store of engineering knowledge gained among lathes and milling-machines and in the foundry, a precious advantage later on, both in China and when writing on the history of technology.

After graduating, he did not go to a London hospital to do the clinical part of his medical work, but rather stayed in Cambridge to carry on research in biochemistry, for which he was given the Benn Levy Studentship. At that time he was much attracted to the community aspect of religious life and thus became a lay brother of the Oratory of the Good Shepherd for a couple of years. But this turned out not to be a permanent vocation, and he married Dorothy Mary Moyle (also a biochemist) in the same year that he took his doctorate and became a Fellow of Caius (1924). In the end he never did qualify in medicine, for this election implied the feasibility of a career in pure science, and such was indeed the pattern of things for him until in the end the history of science asserted pre-emptive claims.

It would perhaps be hard to over-rate the influence on him of Bishop Barnes, modified though it was by that of his Anglo-Catholic undergraduate friends later. It meant that fear and the irrational never became indissolubly connected in his mind with religion; it never had for him that creepiness or spookiness, nor yet the childish terrorization, which alienated so many of his intellectual contemporaries. Barnes' influence also made it impossible for him to dismiss those rational arguments in favour of religion which are brushed aside by many who are repelled from it for other reasons. All this is not to say that fear never came in—on the contrary, I happen to know that anxiety-neurosis symptoms have been familiar to him throughout his life—but that was a different sort of thing altogether; and the fact that he grew up at a time when everyone was studying Freud, Adler and Jung, meant that he could overcome them unaided. There are still further things that he traces to the influence of Bishop Barnes. For example, the ideal of a spiritual ardour untinged with obsession or fanaticism, and the aim of being without illusions, yet untouched by bitterness or cynicism. Two great men that he knew personally have always exemplified this for him, Charles Sherrington the physiologist, writer of *Man on his Nature*; and Ludwik Rajchman, the medical international civil servant, once of the League of Nations Health Service and then of WHO and UNICEF. And there were others among the men of religion who influenced him much, mostly 'modernists' in one way or another, such as J. W. Hunkin, his Dean at Caius (afterwards Bishop of Truro, and now acclaimed as one of the fathers of the New English Bible), and also his Oratorian brothers, especially Wilfrid Knox and Eric Milner-White. To all such men he feels he owes his emancipation from the conventional attitudes so often found among religious people. In this way his mind was not only open to the personal experience of other religions, but able also to accept the great need for a re-thinking of Christian doctrine and practice in the light of scientific knowledge—for example, the necessity for a thorough reformation in the Church's attitude to sexual questions, race relations and social justice. Enlightenment, one might say, has always been his end in view.

All through this time, of course, he had many intellectual struggles, but he reached the conviction that life consists in several irreducible forms or modes of experience. One could distinguish the philosophical or metaphysical form, the scientific form, the historical form, the aesthetic form and the religious form, each being irreducible to any of the others, but all being interpretable by each other though sometimes in flatly contradictory ways. This conclusion was supported by many thinkers, but particularly R. G. Collingwood in his book *Speculum Mentis*. Vaihinger's *The Philosophy of As If* also exerted influence on him, but if there was one book more than any other which moulded his attitude to religion, it was Rudolf Otto's *The Idea of the Holy*. From this it became overwhelmingly clear that religion did not reside in any dogmas, doctrines or specific rites but rather in the numinous sense, applied at first of course by primitive peoples to all kinds of objects unworthy of it, but in the higher religions firmly attached to ethics and manifested in speech and rite. One of the struggles which he most clearly remembers was in Christian liturgiology itself, where he was captivated early by its wonderful symbolism yet repelled by much of the phraseology, till he came to realize that the words were a form of poetry, not to be dissected by the scientific scalpel or criticized by the methods of the linguistic philosopher. Among the books of his father's library in which he had been able to soak himself were Schlegel's *History of Philosophy*, read at the age of about ten, but also there had been George Herbert's poems, Sir Thomas Browne's *Religio Medici*, and the *John Inglesant* of W. H. Shorthouse, with other books of that Anglican ethos to which he has never lost his early attachment.

I spoke just now of liberation from the bondage of conventional ideas. On race and class I shall have more to say presently, but here may be the place for a word on sex. Needham has often told me how greatly influenced he was by William Blake, D. H. Lawrence, Edward Carpenter and Havelock Ellis, writers some of whom, it may be, are now out of fashion, but only perhaps because their work has been accomplished. Medieval and traditional Christendom, he came to feel, had far too much been distorted by those Gnostic and Manichaean heresies which attributed all evil to matter, and took all sex as sin. Modern science has made it impossible for us today, he says, to think as the early Christians did, ignorant of the physiology of reproduction and lacking all control over the generative processes. Modern psychology has made it clear that the only truly civilized society will be one in which there is the greatest tolerance for individual deviations and unusual life patterns—always so it be under the sign of the *agape tou plesiou* of the Gospels, the love of one's neighbour. And the Church, in its formulae and practice of moral theology, must absorb into itself this new knowledge and sensibility.

> Anni nova novitas
> novas leges afferens
> seque verat veritas
> vetustatem aufferens. . . .

Without these liberating influences, how could Joseph Needham have found such sympathy for those who venerate the *lingam* and the *yoni*, or felt the numinous quality in the symbolism of Krishna's sport with the Gopis, or above all responded in China to the great Taoist affirmation, *i Yin i Yang wei Tao*? Though he never met Lawrence, Carpenter or Ellis, there were others whom he recollects with gratitude as helping along this path—John Moyle, Dorothy's father, a Quaker of deep humanity and insight, and Margery Stephenson, one of the founders of chemical microbiology, their colleague in the laboratory, whose scorn of cant and humbug still lives in their minds.

While still at the Bachelors' Table in Caius, where all the talk was of the new psychology, he persuaded some of the great of his acquaintance (such as Dean Inge, William Brown the psychologist, Malinowsky the anthropologist, Eddington the astronomer, and C. C. J. Webb the theologian) to join together in a book of essays entitled *Science, Religion, and Reality*, later twice reprinted. Also between 1925 and 1941 he collected a number of essays and addresses on science, philosophy, religion and socialism in books such as *The Sceptical Biologist*; *The Great Amphibium*; *Time, the Refreshing River*; and *History is on Our Side*. The introductory essay in the third of these volumes, specially written for it, was a personal stocktaking or self-criticism with an autobiographical content which is relevant here. It recognized ethics as something above and beyond all the forms of experience, and it accepted political action as the implementation of ethics. Ethics and politics were 'the cement necessary for the unification of the divergent forms'. Thus although the primacy of politics was a powerful Marxist element in his thinking, the possibility of uniting the forms of experience only in the individual's life and action was a clearly existentialist one. Looking back, therefore, one is inclined to think that without being directly influenced by any of them (so far as I know), his world-outlook was distinctly akin in various ways to those of Kierkegaard, Jaspers, Gabriel Marcel, Emanuel Mounier and perhaps Sartre. The first of the four titles above derived obviously from Robert Boyle (with a modified meaning), the second took its Brownean title from the very incompatibility of the different forms of experience, while the third was a quotation from W. H. Auden, whose group of poets he greatly admired.

All these writings were of course by-products, the results of evening reading or time for extraneous study snatched while waiting for the completion of a distillation or an incubation. The Cambridge Biochemical Laboratory was his home in the most real of senses from 1920 to 1942, first as a student, then as Demonstrator, lastly as Sir William Dunn Reader, all under the aegis of that beloved *fundator et primus abbas* of modern biochemistry in England, Sir Frederick Gowland Hopkins. His first research was, as we know, on the metabolism of the cycloses, and it was from this by a curious chance that he was led into his major speciality, the biochemistry of embryonic development. Always an avid bibliographer, he came across a dissertation by a young German named Klein who had found just before the First World War that the hen's egg at the beginning of development contains no inositol, yet by the time of hatching plenty has been synthesized. This inspired him, he says, with a vision of the developing egg as a most wonderful factory of changes and syntheses, so that he set out to chart as much as possible of them, warmly supported by Hopkins; work which led to the appearance in 1931 of the three-volume treatise *Chemical Embryology*. In a parallel way Hopkins encouraged the early phases of the life-long activity of Dorothy on the biochemistry of muscle contraction. I have heard it said that *Chemical Embryology* was in parts too much of a compilation, but that is probably inevitable in the works of all who embark on the painting of broad canvasses. Others felt that it had defined what was really a new branch of science; and whatever else may be said it certainly set the stage for the next phase, the first movement directed to finding out, not only what chemical changes went on *pari passu* with morphogenesis, but how they were directly involved in it.

Now it was just in 1931 that a breakthrough occurred in the borderline field of biochemistry and experimental embryology, when the school of Hans Spemann discovered that the primary inductor or 'organizer' of amphibian and avian embryos, the morphogenetic hormone that stimulates the formation of the neural tube and vertebrate axis, is stable to the temperature of boiling water. This fundamental discovery, with all its promise of the possible identification of the molecules which act as morphogenetic hormones, attracted Joseph Needham at once into a new line of work, pursued during the ensuing decade with the collaboration of C. H. Waddington and others. This led to the publication in 1942 of *Biochemistry and Morphogenesis*, a general survey of the field afterwards reprinted, and even now useful as a statement of the many problems still unsolved as they presented themselves at that time. One sometimes hears the criticism that the work of the Cambridge school in those days was based on ideas too 'simplicistic' in nature, and it is certainly true that the model of their thought was endocrinological rather than genetic, but the real difficulties which impeded the work were intrinsic. The elucidation of the chemical aspects of organizer phenomena did indeed turn out much more difficult than was at first anticipated, but today much progress is being made, especially by the schools of investigators in Finland and in Japan. Moreover, after a period of very intensive study of the genetic code and the way in which it works in unicellular micro-organisms there is now noticeable a return to the great problems of morphogenesis seen in the development of the metazoa.

The devotion of Joseph and Dorothy to their great teacher continued unabated to the end, and in 1937 he edited (with David Green) a *Festschrift* for him called *Perspectives in Biochemistry*; while later on he also prepared (with Ernest Baldwin) a commemorative volume entitled *Hopkins and Bio-chemistry* (1949). These books were of particular significance from the theoretical point of view because Hopkins had always taken a special interest in the problems for the philosophy of science presented by biochemistry, and in this all his students participated.

In connection with the books of essays referred to above, mention was made of socialism, and the development of this in Needham's world-outlook calls for explanation. Already in 1917 he had welcomed, with his closest friend, Frank Chambers, as schoolboys, the Bolshevik revolution in Russia. What exactly had influenced his mind in the direction of socialism at that early time, so that he often argued in favour of it with his rather Victorian and old-fashioned father, he cannot now clearly recall, but it must almost certainly have been the writings of H. G. Wells and Bernard Shaw, whom he would always eagerly read, avoiding the books of Trollope, Jane Austen and Dickens pressed upon him by his parents. During his undergraduate time, however, he was not particularly interested in politics, and as the student population in Cambridge at that time was rather homogeneous in class character, he was not brought up sharply against the way the world looks from the working-class point of view. The friendship of Louis Rapkine was therefore cardinal in his development. When he and Dorothy went to work at the Marine Biological Station of Roscoff in Brittany in 1925, they were deeply impressed by this young biologist of Jewish-Lithuanian origin, the son of a shoemaker whose family had fled to Canada to avoid the pogroms, and one whose greatest hero was Spinoza, though extremely well-read in all the Marxist classics. It was through this contact that they began to read them too.

About this time also, or a little later, Joseph and Dorothy first became acquainted with the church of Thaxted in Essex, forming an association which has lasted till the present time. For many years they kept it quiet, not wishing to be mistakenly identified with the 'church-and-king-ism' prevalent in Cambridge at that period. The vicar in those days, Conrad Noel, an inspired parish priest and a prophetic preacher, was in fact another outstanding influence. Thaxted had long been notable for three things: a very thorough-going Christian socialism, a great musical tradition (Gustav Holst lived there and wrote much music for the church) and thirdly a liturgical beauty learnedly yet informally based on medieval English precedents. Thus for forty years the clerks of Thaxted, first under Conrad Noel and then under Jack Putterill, have formed a group of friends drawn from all walks of life devoted to the coming of the Kingdom on earth, the celebration of the holy liturgy, and one another.

Here this doctrine of the Kingdom of God was of particular importance. Joseph Needham formed at that time the conviction, never afterwards abandoned, that it should be regarded as a realm of justice and comradeship on earth, to be brought about by the efforts of men throughout the centuries, not primarily as some mystical body existing already, or some spiritual state to be expected somewhere else in the future. Gradually this became linked up in his mind with a conviction of the essential unity of cosmological, organic, and social evolution, in which the idea of human progress, with all due reservations, would find its place. Parallel with this was the conviction that the Christian must take Marxism extremely seriously, such doctrines as historical materialism and the class struggle being perhaps recognition of the ways in which God has worked during the evolution of society. This general approach was the subject of his Herbert Spencer Lecture at Oxford about 1936, and it was the reason why *History is on Our Side* bore the subtitle *Essays in Political Religion and Scientific Faith*. The necessity of faith in the irrational seemed more and more pointless when everything that we knew of evolution and history indicated a continuous process or plan of salvation which humanity was working out, and this was the reason why later on he came to feel so much sympathy not only with the emergent evolutionists such as Lloyd Morgan or Samuel Alexander, but even more with the religious form of the same thing, represented by Teilhard de Chardin. That great Jesuit he never met in China though both spent so much time there, but only in Paris after the war, when they found themselves very close in outlook, different though their French and English lines of development had been. In considering this conception of progress, Needham always looked at it with the eyes of one familiar with the scattered points on many scientific, especially biological, graphs—scattered they may be, yet a continuous trend there also is which can be depicted by a rising curve, the general course of which over-rides all individual and particular backslidings.

One might say therefore that his idea of progress was essentially statistical, and dependent as it were on a calculus of aberrant observations.

In the thirties too he joined with the theologian Charles Raven and the Marxist Unitarian John Lewis in editing a book of essays *Christianity and the Social Revolution*, still today not superseded. All through that time, moreover, he represented the Association of Scientific Workers on the Cambridge Trades Council, and was one of those who tried hard to awaken the people to the dangers which faced them at the time of the Spanish Civil War and the rise of Nazi-Fascism. He was also active in the Cambridge branch of the Socialist League, a ginger group of the Labour Party led by Stafford Cripps. Moreover he found time to write a preface for my little book on *The Levellers and the English Revolution*. Naturally all these activities did not make him popular with the establishment. But if the renovation of society was slow in coming, one could always in the meantime do one's best to be 'no respecter of persons', and seek to treat all those with whom one came into contact as equals. Afterwards, when he came to know of that splendid saying of Confucius: 'Behave to every man as one receiving a great guest', he saw in it the principle of what he had tried to practise long before.

It was now not long before the great turning point in his life from 1937 onwards, but before describing this one ought to say something about his great predilection for the philosophy of organism. It was entirely in character for him to be searching for something which was neither mechanical materialism nor spiritualist vitalism. This 'middle way' had close connections with dialectical materialism; but it was also something which arose immediately and directly out of that theoretical biology which he had to think about because of his attempts to build a bridge between biochemistry and morphology. The chemical biochemist may like to work with isolated enzyme reactions; the physiologist may like to study the behaviour of isolated organ-systems or aspects of behaviour, while the zoologist and botanist may enjoy the infinite complexities of taxonomy and systematics; but if a biochemist was to try to seek a synthesis with experimental embryology and morphology as in fact Needham did, the philosophy of the organism, the interactions of its parts and the nature of its over-all controls, was inescapable. This was why he was so interested in A. N. Whitehead's philosophy, enjoying the friendship of J. H. Woodger and joining with him in the deliberations over a number of years of a small group which called itself the Theoretical Biology Club. The same interest gave rise about this time to a book of Yale lectures, *Order and Life*, which before the era of electron-microscopy emphasized the rôle of micro-structure in the living cell; this has now been reprinted in paperback on account of its pioneering character. Such studies in the philosophy of organism proved invaluable to him later on when he came to investigate medieval Chinese philosophy, and found that most of it is very much of that complexion. The Chinese scholasticism of the twelfth century A.D., Neo-Confucianism, is a case in point, and all this he was able to expound in the second volume of *Science and Civilisation in China*.

Then came the great divide. It happened because of the appearance in Cambridge of several Chinese biochemists to work for their doctorates. There was Wang Ying-Lai who studied under the great David Keilin at the Molteno Institute; there was Shen Shih-Chang who collaborated directly with him in the Biochemical Laboratory; but the greatest influence was due to Lu Gwei-Djen who joined in collaborative work with Dorothy. The finger of Fate—or of the Tao—might be seen in the fact that while she was at the Lester Institute in Shanghai trying to make up her mind where she would go in England to do her doctorate work, she noticed in an advertisement that Joseph was Treasurer of the Cornford-Maclaurin Fund, founded to help the relatives of the members of the International Brigade in Spain and called after the names of two young Cambridge men who had given their lives in that anti-Fascist war. This was one of the reasons which decided her on the Cambridge Biochemical Laboratory, and looking back now, it seems an essential thread in the network.

The influence of these Chinese friends upon him was quite extraordinarily powerful, almost puzzlingly so. It was as if he received from them some kind of liberation for which he had always been looking. In later years he has told us how dazzled he had been by another of the books in his father's library, Rawlinson's old treatise on the Ancient Egyptians. Perhaps it was that which fixed so firmly in his mind at an early age that all the apparent absolutes of the traditions of Christendom were not absolutes at all, but formulations of relative value, keyed to the particular forms of one civilization only. In any case it was clear that talking with his Chinese colleagues about their cultural background, the traditions of Chinese language and literature, he found something equal and opposite to all that in which he himself had been brought up, and something for that very reason of compelling fascination. Somewhat astonished by this enthusiasm, Lu Gwei-Djen helped him to take the first steps in the written language by correcting his early efforts in letters, marvelling that anyone should suddenly find a Chinese dictionary the most exciting thing in the world, and spend innumerable evenings constructing a fresh one for his own edification. And indeed it would be true to say that he never had any formal education in Chinese, apart from a few seminars which he attended when the London School of Oriental Studies was evacuated to Cambridge after the Second World War had begun. He did also have great kindness from the then Professor of Chinese at Cambridge, the Czech, Gustav Haloun, who, realizing that no ordinary linguistic teaching would meet the case, got him to sit with him for a couple of hours each week, and went over with him translations of the difficult philosophical and economic *Kuan Tzu* text (4th or 5th century B.C.) which he was himself preparing for publication.

One of Dorothy's favourite stories concerns a holiday that a group of friends took one year at Ringstead Mill in Norfolk. Just as they were going to set out for a walk it transpired that Joseph had a bad headache and therefore preferred not to go. He would rest. He would lie down on the sofa. When they returned he was still lying down, shielding his eyes from the light, but Dorothy, sympathetic as usual but also scientifically sceptical, went over and felt the chair by the table on which lay the Chinese dictionary and his notebooks. And indeed its warmth betrayed its recent use. This little anecdote may serve as an indication of an element of his character which has sometimes made life rather difficult for those around him. 'Persistence is my middle name' was a statement always applicable, but obstinacy and determination, even approaching a certain ruthlessness in pursuing objectives, is not always easy to live with. However, this applied primarily to specific tasks—in relations with colleagues or in administrative matters, open-mindedness and diplomatic tact were, I guess, what he gradually learnt to practise.

At the beginning of the Second World War they all dispersed, Wang Ying-Lai returning to China (where he afterwards became Director of the National Institute of Biochemistry at Shanghai), Shen Shih-Chang to Yale (which he never afterwards left), and Lu Gwei-Djen to research first in California, then at Columbia. But Joseph's fate was sealed when in 1942 he was invited by the British Government to go to China as a representative of the Royal Society under the conditions of blockade imposed by the Japanese. The plan at first was just to give some lectures to encourage the Chinese scientists, while at the same time E. R. Dodds, the Oxford Greek philosopher, went out to represent the British Academy and also give some lectures. The original idea of this mission had been due to one of the greatest scholars in the British diplomatic service, the Japanologist Sir George Sansom. It soon became clear, however, that while on the humanistic side not much was possible, on the scientific and technological side there was everything to be done; practical help, indeed, to be given to the scientists and engineers who were trying to carry on their work under the unbelievably difficult conditions of the time. Consequently with the active help and understanding of Sir Horace Seymour, the Ambassador, he stayed 'for the duration', setting up the Sino-British Science Co-operation Office (Chung-Ying K'o-Hsüeh Ho-Tso Kuan), backed in London by the British Council for all the peaceful aspects of science and by the Ministry of Production for the war aspects. Eventually this useful organization grew to a team of about six British and ten Chinese scientists and engineers, having its own head-quarters in Chungking, a fleet of vehicles, and an allocation of tonnage on the transport planes of the R.A.F. which commuted 'over the hump' from India.

He himself always says that this period in China was a providential oppor-tunity for getting to understand the culture and the civilization free from the conventions and restrictions of the old types of people; the business man, the Old China Hand, the missionary or the professional diplomat. He says also that it gave him a unique opportunity of sitting at the feet of Chinese mathematicians, medical doctors, chemists, engineers and physicists, many of whom had a great interest in the history of science and technology in their own civilization. In this way he was able to acquire an understanding of the relevant literature, a knowledge of what books should be bought, and what studied, in fact everything he needed to implement the plan he had formed for a history of science and technology in Chinese culture; something which had never before been envisaged, still less portrayed to the world as a whole. I have not so far said anything about this interest in the history of science, but in fact it was of long preceding date. In his first large work, already mentioned, *Chemical Embryology*, begun when he was only twenty-nine and finished a couple of years later, he had felt it necessary to preface the three volumes by a history of embryology down to the nineteenth century, and this, afterwards separately published, still remains almost the only work of its kind. During those pre-war years he also produced occasional articles on the history of science, and this interest gained him and Dorothy the friendship of Charles and Dorothea Singer, with whom they often stayed at Kilmarth in Cornwall, and to whom they look back with affection and respect as great influences in their lives. Now in 1942, before setting out for China, he conceived the definite intention of writing a work of some kind on the history of science and technology in the Chinese culture-area, and he did indeed after returning to Cambridge in 1950 set on foot the series of volumes *Science and Civilisation in China*, which is still in course of publication.

The fact is that he got on extraordinarily well with the scholars and scientists of China, who would open up, as it were, and talk to him perfectly freely. His relations with them were on a basis of absolute intellectual equality, yet even then not exactly so, for he had the modesty to realize how little he knew of the incredible riches of Chinese language and literature, thought and history. Here was someone who had come to help, yes, 'bringing charcoal when it was snowing' as the old proverb said, but also one who had come to learn much more than to teach. Here he followed the behaviour of his old friend, E. R. Hughes of Oxford, one of those ex-missionaries turned sinologist whose relations with Chinese scholars were always the epitome of traditional courtesy; and these things were all the more extra-ordinary because it was just at the time when the Burma Road had been closed, many British defeats were taking place, and the British were particularly unpopular in China. But that made no difference. Perhaps this was the reason why before he had been long in China he was made a Foreign Member of Academia Sinica (the National Academy), and at the conclusion of the war, towards the end of the united front period, received a decoration, the Order of the Brilliant Star. Some of the best Chinese scholars, such as Kuo Mo-Jo (the present President of the Academy), Tung Tso-Pin and Li Shu-Hua, presented him with epigrams written in calligraphy on scrolls, and these he still greatly treasures.

It is clear that nothing could have exceeded the impact which Lu Gwei-Djen's country and her people had upon him from the latter part of 1942 onwards. In that dreadful time they were far too busy, distressed and disorganized to pay attention to wandering foreigners, and for this reason he was able to penetrate everywhere into the life of villages and towns (roughing it of course a good deal in the process) and bend his solitary steps into Confucian, Buddhist and Taoist temples often deserted, able therefore to savour to the full the great beauty of the traditional architecture in its setting of age-old trees and forgotten gardens. And he was also free to experience the life in Chinese homes and market-places, and to know at first hand the miseries of a society in collapse, in the dark night before the dawn.

About a year before the end of the war Lu Gwei-Djen returned to China to join the group as Nutritional Adviser. She found that Dorothy, who was already in it as Chemical Adviser, had also been captivated by the culture of her country and supported enthusiastically Joseph's idea of writing a definitive work on the history of science and technology in China. By this time Dorothy too had been elected a Fellow of the Royal Society, and for a long time they were the only couple in this position. After she had left for Cambridge to resume scientific work when the war ended, Gwei-Djen accompanied him to the cultural capital, Peking, and other cities, so that he was able to get to know her family and her own city, Nanking. Then she saw him off en route for London and Paris where he had been called to set up the Division of Natural Sciences in the United Nations Educational, Scientific and Cultural Organisation (UNESCO). This development was a natural outgrowth of the Sino-British Science Co-operation Office, and indeed for a long time embodied Field Science Co-operation Offices based upon its model. Moreover, all through the war he had himself been very active in the planning of this specialized agency of the United Nations, urging in a series of memoranda issued from Chungking the great importance of including science in what had originally been planned for education and culture only.

Flying wartime visits to Washington, New Delhi, Canberra, London and Moscow had also given opportunity for pressing this urgently upon the statesmen and political leaders of the time. In all these activities he acknowledges a debt of real inspiration to that great international medical and scientific civil servant already mentioned, Ludwik Rajchman.

In those days he was, so to say, favoured with a following wind. But it should not be thought that he has not been willing to stand up against storms of abuse if that should prove necessary. In 1952 he was a member of the International Commission for the Investigation of Charges of Bacteriological Warfare in North China and Korea. For the affirmative findings, in which he concurred, he had to suffer great public unpopularity at the time, though the Report is now 'required reading' for those who study this highly unpleasant and dangerous subject. His main motive in joining that expedition, he says, was to come to the aid of his many friends among the Chinese scientists at a time when they needed it.

Just two years later there appeared the first volume of *Science and Civilisation in China*. Although he had made up his mind before the war to write a book on the history of science, technology and medicine in Chinese culture, it was not until after the war that he realized what a tremendous field it was. He had vowed never to write another treatise in several volumes, but as the work went on, it became clear that one could not anticipate how much space would have to be devoted to particular sciences and technologies. In the end it was evident that it would have to be a work in many volumes. The third part of the fourth volume (the sixth part out of a planned total of ten or eleven) and the eighth and ninth parts are now published. The best proof of the appreciation of *Science and Civilisation in China* by the Chinese themselves is not that Academia Sinica orders regularly 100 copies of each volume for distribution among its Institutes, but rather that the entire work is being translated into Chinese—both in Peking and in T'aiwan. No doubt the chief reason for this is the comparative method which is used throughout it.

He had from the beginning realized that collaboration with Chinese scholars would be absolutely essential, and succeeded (in the face of considerable financial difficulties, since no generous support for the project was at first forthcoming) in gaining as co-worker Wang Ling (Wang Ching-Ning) whom he had known as a neophyte historian of Academia Sinica at Lichuang in Szechuan. When nine years later Wang Ling departed to a chair at Canberra, Joseph persuaded Lu Gwei-Djen to join him on a permanent basis after a spell of similar length in the Natural Sciences Division of UNESCO. It is a happy circumstance that she is one of the contributors to the present volume. Other friends and collaborators come and go, notably Ho Ping-Yü from Malaya and Lo Jung-Pang from California; and probably by the end of the project there will have been as many as a dozen figuring on the title-pages of the successive volumes.

Not everyone of course has been pleased with *Science and Civilisation in China*, though there are few who do not find in it a mine of information not previously available in the West. After the second volume came out, Arthur Wright, the American Buddhologist, maintained in a review that the work was fundamentally unsound for the following reasons. The authors believe, he said in effect, (1) that human social evolution has brought about a gradual but real increase in man's knowledge of Nature and control of the external world, (2) that this science is an ultimate value and with its application forms today a unity into which the comparable contributions of different civilizations (not isolated from each other as incompatible and mutually incomprehensible organisms) all have flowed and flow, (3) that along with this progressive process human society is moving towards forms of ever greater unity, complexity and organization. In the preface of the fourth volume Joseph Needham remarked: 'We recognized these invalidating theses as indeed our own, and if we had a door like that of Wittenberg long ago, we would not hesitate to nail them to it'. Naturally all those with personal world-outlooks different from his will continue to be out of sympathy with the informing principles of the work, useful though they may find it in detail. He himself, I know, is conscious of many imperfections in the volumes, some doubtless due to inaccessibility of textual or archaeological evidence, and all of which he would repair if it were possible.

There are other criticisms perhaps more interesting. Some have wondered, especially in matters concerned with the history of scientific thought, whether he has not read into the old Chinese texts too much of his own organicism, or in particular cases of scientific discovery or technical invention, too much of our modern knowledge; here time alone, and further research, will show. Certain it is that many have been convinced by his interpretations. Of one of the volumes William Empson wrote: 'The Chinese insisted on an organic approach to Nature, and this (whether or not "scientific humanism" is a suitable term) will be found the only philosophy tolerable to man in the world he is discovering.' And the eminent sinologist Derk Bodde saluted 'the fascinating and sometimes stirring picture of the intellectual background of Chinese science, presented with insight, clarity, originality and grasp of complex detail'. Fresh light has been cast on many things, such for example as Taoist and Neo-Confucian philosophy, medieval navigation, mechanical and hydraulic engineering, ancient meteorology and ancient alchemy.

Yet another objection sometimes voiced is that he has been too partisan now and then in upholding the claims of Chinese civilization vis-à-vis the Arabs or the West; this I think he would not entirely deny, undoubtedly having felt that a little over-statement of the case would be no bad thing to awaken people's minds to much that had in the past been woefully ignored and unrecognized. Sometimes evidence of transmission of admittedly earlier Chinese discoveries and inventions is lacking, but it is a principle of his that the burden of proof should lie with those who wish to maintain the thesis of independent invention. 'We are witnessing', wrote F. I. G. Rawlins of the British Museum Laboratory, 'the emergence of something unique in the history of thought, namely a detailed investigation into, and an appraisal of, a civilization advanced in its own right, exhibiting qualities peculiar to itself, and inadequately appreciated, for the most part, in Western culture.' And Arnold Toynbee struck a similar note when he said: 'The practical importance of the work is as great as its intellectual interest. It is a Western act of "recognition" on a plane higher than the diplomatic.'

During the course of this work Needham had occasion from time to time to treat certain subjects in a way more detailed than was possible in the volumes of *Science and Civilisation in China*. Separate monographs therefore resulted. *Heavenly Clockwork* (with Wang Ling and Derek Price) was a study of the great astronomical clocks of medieval China, that missing link in horological history; *The Development of Iron and Steel Technology in China* (for the Newcomen Society) told an epic story of siderurgical discovery. Other papers have been collected together in a volume entitled *Clerks and Craftsmen in China and the West*; while two further books of collected addresses have also seen the light: *Within the Four Seas; the Dialogue of East and West*, and *The Grand Titration; Science and Society in East and West*. These are in similar style to his essays of earlier biochemical years already mentioned.

Thus Joseph Needham continues to carry out what one feels is for him a quasi-religious vocation, that of rendering justice at last, as well as sympathy and understanding, to a great people whose contributions to human development have been grotesquely underrated. Yet, perhaps because of the impact of old Rawlinson upon him when a boy, he has been indeed a Christian with a difference. I often thought that it was one of the most characteristic yet unexpected things about him that in spite of all that background the one book which he chose to take with him on many dangerous plane journeys during the war was Lucretius' great materialist poem *De Rerum Natura*, and in the aircraft he would occupy his time in trying to improve on W. E. Leonard's English translation of it. What this symbolized was, once again, the relativity of all beliefs. It was as if he knew that nothing that man could ever say could fully capture and enshrine the mystery of the universe, and indeed he has often remarked that of all the works of the patristic period, that of St. John Chrysostom, *On the Inconceivability of God*, was what had most impressed him. Here was the best authority for the belief that even the greatest sayings of men are but babblings. Hence it was natural that though himself inalienably attached to the Christian tradition, he could sympathize as a scientist with those who had been most determinedly atheist. Perhaps it was a kind of fundamental intellectual humility; at any rate it served him in very good stead when coming into contact with the traditions of Chinese civilization.

He often maintains that if he had not been thoroughly incorporated into the living tradition of the religion of his own civilization as a practiser, he would never have been able to enter into sympathetic understanding with Confucians, Buddhists and Taoists. The first essential here was the realization that the numinous must be dissociated from the theology of a creator God, because fundamentally such a conception is antipathetic to Confucianism and clearly absent from Taoist philosophy. If one was prepared to sympathize with Democritus and Epicurus, one could certainly not complain at the Confucian aversion from all gods and spirits, recognizing at the same time that no more numinous place has ever existed than the Confucian temple, with its pure association of the sense of the holy with a this-worldly ethical system. Similarly the Taoist veneration of the immanent Tao was a long way from the transcendence of Christian theology—yet he found himself in tune with it also. Of the 'three doctrines' (*san chiao*) Buddhism was the one for which he felt least sympathy, disliking its strong denial of the possible redemption of this world; yet in later years, when he came into contact with such phenomena as the Association of Socialist Bhikkus in Ceylon, he realized that the *karuna* (compassion) aspect of Buddhism could fully balance the *śunyatā* (emptiness) element, and lead to something akin to a belief in the coming of the Kingdom of God on Earth.

On Taoism a few more words need saying. Ancient Taoist philosophy, with its many parallels to the great Christian paradoxes, its emphasis on the victory of weakness over strength, of Aphrodite over Ares, its belief in spontaneity and naturalness, and its Nature-mysticism, had irresistible attraction for Needham. Medieval Taoist religion, still living in the abbeys and temples of the countryside when he first went to China, also showed him something of its great liturgical treasures, now being more and more revealed by current researches in comparative religion. Only afterwards, when he plunged into the history of discoveries and inventions in Chinese culture, did he come to realize and to expound the leading part which Taoist thought and craftsmanship had played through the centuries in the development of natural knowledge and technological control. This may explain why much later on, being asked on one occasion whether he regarded himself as primarily a scientist or a historian, he remarked that he thought his best designation would be that of 'honorary Taoist'.

With so wide a background it is understandable that he is deeply interested in the oecumenical movement within Christendom at the present time. Of his Anglicanism I have already said something, but further light can be thrown on his outlook by the statement he likes so often to make that he is both Orthodox and very unorthodox. In fact his love for the Greek side of Christendom goes back to the early twenties when he used to attend the liturgy at the Cathedral of the Holy Wisdom in London, and he treasures from those days a missal with the Greek and English on opposite pages. If you ask him why he finds Orthodoxy so attractive, the answer is that it takes one back (if one is interested at all in Christian ways) not only beyond and behind the exaggerations and excesses of both sides in the Reformation period, but also behind the time of the scholastic philosophers themselves, who he feels were largely on the wrong track in trying to encompass mystical religion within the strait-jacket of Aristotelian philosophy. Some people dislike or are offended by what they call 'externals', but for him the practical actions so replete with symbolism, which liturgiology studies, are themselves the carriers of meaning and prayer as much as anything purely mental. And evidently this is equally true of worship in all the great religions of Asia and China.

Broadly speaking, his attitude to the world religions can perhaps best be described as an anthropological one. He feels himself to be a Christian because his ancestors and forefathers were, and if the real mystery of the universe cannot be comprehended in any words, the only natural words for him to use are the words of his own country and people as they have developed down the ages. He feels that it is now quite impossible that China and India will ever accept Christianity, so often presented to them in the past with all the trappings of racial pride and superiority. He feels that it is up to Christians now to look for and find their own values under the forms and within the histories of those other civilizations, and not to expect to find them in forms familiar to Westerners; hence a great effort of mutual understanding and mutual explanation is needed for the welfare and peace of the future world. And thus he believes that the work on the history of science and technology in Chinese culture is and will be a contribution towards this international understanding.

To sum it all up, we have delineated here a mind that might be called, in a sense, eclectic, syncretistic. How one feels about it depends inevitably on whether or not one approves of all-embracingness. There are other ideals, there are those prepared to stake everything on a single throw—the Trappist monk, the mathematical logician, the pious Muslim, the dedicated agnostic scientist, the *pratyeka* Buddha, the Marxist revolutionary. . . . That sort of life might be narrow, exclusive, intolerant, even bigoted, but at least it could be called single-minded, and many may feel that that is a supreme virtue which our subject has altogether missed. 'Don't dissipate your energies, my boy', his father often used to tell him; his only salvation was that he happened to be endowed with an abundance of them, and the catholicity of his sympathies did not prevent the carrying through of large-scale scholarly projects. But where there is a great breadth one does not often find an equal depth, and Needham has often told me how much he has admired those minds of intellectual penetration he has known, able as he is not to grasp the abstract propositions of analytical mathematics, philosophy or genetics. The concreteness of physiology and chemistry was what he grew up in, and by a singular appropriateness he responded as best he could to the ancient alchemical aim of accomplishing, as it were, 'the marriage of Fire and Water'.

I believe I would not be going wrong in thinking that the deepest springs of his being are connected with his conviction of the unity of all peoples in the human race. Many others have given more and suffered much more in the cause of inter-racial friendship, love and understanding than he has, but it seems that he has had a specific contribution to make: a warm appreciation of the transcendence of all racial barriers by science and technology, carried out in terms of the detailed hammering home by chapter and verse of the fact that modern science is not all science, and as he himself likes so often to say, 'Wisdom was not born with Europeans'. All races and peoples have contributed to this great endeavour of mankind. In his case at least, it would seem that neither religion nor science nor philosophy have been divisive forces, but rather unifying ones. The union was of course not intellectual but existential, constructed within a single human being open by nature to all the forms of experience. Hence that many-mindedness which I set out to examine. In spite of all limitations, he has found true what is said in the Confucian *Analects,* that 'for him who respects the dignity of man, and practises what love and courtesy require, for him all men within the four seas are brothers'.

II

❀

BRONZE TECHNOLOGY IN THE EAST:

A METALLURGICAL STUDY OF EARLY THAI BRONZES, WITH SOME SPECULATIONS ON THE CULTURAL TRANSMISSION OF TECHNOLOGY*

Cyril Stanley Smith

I have two motives in offering the following small study to Joseph Needham. First, few people have enjoyed more than he the technical skill of the cultures in the Far East or have contributed more to the study of the conditions under which techniques are transplanted between cultures. Second, since most of the papers in this volume will probably be based ultimately upon verbal sources, it seems not inappropriate to add a reading of a different kind of record—that of the internal microstructure of some artifacts which carries a vivid picture of the human activity in the workshops that produced them.

There are few if any better examples of the interrelationship between the subtleties of an aesthetically satisfactory design and technique than those provided by the development of the decoration of Chinese cast bronze ceremonial vessels from early Shang to late Chou Dynasties. The changes in form and style of detail grow directly out of changes in moulding technique.[1, 2] For a time it seemed as if the early Shang vessels sprang almost

* The author is deeply grateful to Dr. Solheim and the Government of Thailand for making the objects available for study.

The preparation of the specimens and photographs are all the work of James F. Howard, Jr. This examination is part of the general investigation of ancient metallurgical techniques being carried out at Massachusetts Institute of Technology with support by grants from the Sloan Fund for Basic Research, The National Endowment for the Humanities, and Mrs. Dominique de Menil.

[1] Noel Barnard, *Bronze Casting and Bronze Alloys in Ancient China*, Monumenta Serica Monograph XIV (Canberra and Nagoya, 1961). Idem. 'Chou China: a review of the third volume of Cheng Te K'un's Archaeology in China', *Monumenta Serica*, 1965, **25**, pp. 307–459.

[2] Rutherford J. Gettens, *The Freer Bronzes Volume II Technical Studies* (Washington, 1969).

without precedent from a non-metal-using culture, though recent excavations[3] have shown that it was not a completely new industry but an explosive growth in scale. There is little evidence that convincingly points to the introduction of metallurgy into China from other countries, though, of course, by early Shang times cast bronzes had existed for about 1,500 years in the Middle East and casting of some sort was a thousand years older.

Anthropologists have debated at length the problem of the origin and spread of techniques, examining mainly the rôles of independent invention, the diffusion of techniques by personal contact between skilled men, and the diffusion of an idea or information that stimulates a development in a new society. It is, of course, all a case of structural change. As one who has spent a lifetime studying changes of microstructure and crystalline structure in solids, I cannot help but point out the almost exact analogy between the changes of metallurgical phases and changes of a social nature. Any structure is an association of units, and it is stabilized by interaction between its parts, the units themselves internally benefiting from the association. Though the assembly resists change, change becomes more probable when interaction within the structure becomes less satisfying to the units that compose it. Small changes can occur slowly without topological disruption by the production and elimination of gradients within the structure, but the most significant changes occur discontinuously through the formation, somehow, of a nucleus of a new structure, or its introduction from outside. The internal creation of a nucleus is rare because it involves a disruption of many locally satisfactory interactions and until it is large it can experience nothing but the adverse effects of change. But if once supercritical, then it can grow because those units at the boundary can freely choose whichever of the adjacent structures they prefer. The boundary is inevitably a region of strain, disorder and fluidity, but it involves only a small fraction of all the units at any one time and its movement produces the conversion. The solidification of ice, the magnetization of iron, the hardening of steel, and the writing of a poem involve essentially the same patterns of interaction between individuals in a hierarchical assembly with feedback as those that are involved in revolution, religious conversion, the growth and decay of nations, or the rise and fall of the hemline.

It is important to keep in mind that in the appearance of any new form—whether it is a solid, a scientific theory, a political structure, or a new technique—there are at very least two essential ingredients. *Everything* depends upon the nature of the interaction between the thing itself and its environment; the appearance of the new structure requires both a nucleus

[3] Chang Kwang-chih, *The Archaeology of Ancient China.* (Second edition, New Haven and London, 1968) *passim*, especially pp. 234–9.

and an environment which feeds it and is changed in so doing. The units of the new structure may exist in the old environment, but until they associate to give new formative fields nothing can happen—but, conversely, nothing *will* happen unless the new pattern proves itself, by interaction, to be in some way more stable, desirable, or useful, than the old. In some cases, the nucleus may lie around for long times with nothing happening because conditions of growth do not exist. In others, an overripe environment in the absence of a nucleus may remain unchanged for centuries.

Many scholars, particularly Chinese ones, believe that bronze metallurgy in China was an independent invention, a view strongly supported by the lack of any archaeological evidence of artifacts intermediate in space and time that would mark the route between the established metallurgy in the Middle East and the Anyang region from which bronze casting spreads in ever widening circles in China.[4] Personally, I believe that the idea of making metals from rocks is not easily arrived at and a second independent nucleation of metallurgy in China (or the New World either) seems highly improbable. This by no means denies that there is something uniquely Chinese in the technique and beauty of the superb bronzes of Shang and Chou China, but this seems to lie in the making of moulds and the decoration of them, not necessarily in the origin of metallurgy itself.

Just before the Shang Dynasty conditions were ripe for the explosive development of a uniquely Chinese technique once the idea was planted. The bronze casters built on the advanced ceramic technology that existed, and exploited the mould-making properties of the local loess. The nature of their material, the tradition of carved decoration on ceramics, and many subtle aspects of Chinese aesthetic taste which a mere metallurgist cannot comment upon, all joined to enable the rapid development of a superb form. Seams from mould joints led to bold flanges, split moulds gave access to concave surfaces which, combined with the properties of the mould material, inspired a special form of design based on the difference between moulded detail, carved detail, details coming from local superficial applications in relief, local intaglio details impressed with stamps, and premoulded inserted pieces. Every bit of this is Chinese. But how about the first knowledge that metals exist in stones, that metal can be shaped in mere mud, and that alloys make better castings than pure metals? Though recent excavations make it clear that Anyang did not appear suddenly in all its glory, metallurgical precedents are both rather rudimentary in form and limited in time. In the Middle East these stages took millennia to ripen. There was a native copper

---

[4] Noel Barnard, 'The special character of metallurgy in ancient China', in *Application of Science in the Examination of Works of Art*, ed. William J. Young (Boston, 1967), pp. 184–204.

stage and, distinct from this, the use of certain minerals (mainly metallic oxides) in decorating ceramics. Blue-glazed faience appeared at about the time that smelted metal did in a very active period of empirical search for beautiful effects. Then there was a long period in which arsenic-alloys were used and mould-making techniques developed before that superlative alloy bronze was finally discovered. Arsenic-copper alloys were easy to discover because of the mineralogical relationship of the two elements, but it was different in the case of tin, for a conscious search had to be undertaken, supposedly guided by the concept that interesting things occasionally happen when heavy or coloured minerals are heated with each other or with metallic copper. But this is a late stage. It is the lack of a comparable exploratory period in China that makes one doubt the independent invention of bronze. It seems far more likely that the imported idea of bronze combined with a Chinese desire to have decorative vessels, a knowledge of the versatility of the properties of fire, clay and loess, and a local geological history that had resulted in visible deposits of copper and tin ores. The idea had the effect of a spark in gunpowder: but sparks are rare and they invariably come from somewhere else. Chinese bronze was not analogous to the slow spontaneous combustion that occurs in the absence of ignition from outside.

Once the idea of metallurgy took hold in China, its elaboration occurred in a way that owed little or nothing to the transfer of a developed technology from elsewhere. No group of artisans had migrated with their tools on their backs and their skills in their fingers to start a technical colony. Such men are not usually aggressive voyagers. Most travellers are more commercially than technically inclined; however, an occasional traveller driven by intellectual curiosity to see the social and geographic environment of distant places would certainly know the general principles of metallurgy if he had seen it in his homeland or elsewhere. Even though he could not carry the technical details, he could convey the idea that heavy coloured minerals when properly heated give rise to pretty and useful substances that can be shaped by hammering or by pouring molten into formed cavities in earth or stone, and that the admixture of two of these minerals gives interesting results. This idea does not need to carry along any of the social structure that originally nurtured it, but in the rare event that a new environment allows it to take hold at all, it can grow anew by interaction with local institutions.

It is admittedly a sheer act of belief, but everything I have learned about materials with my own hands and about the history of man and his works as preserved in museums and libraries leads me to think that it is vastly more probable that there would have been such an intellectual traveller than that the discovery of basic metallurgy would have been independently made in two regions in neither of which there existed any particular reason for knowing that it could be done. The discovery of smelting was slow to occur even in the Middle East where a lively tradition of the use of coloured minerals and high temperatures in the decoration of ceramics already existed.

The remainder of this paper describes a laboratory examination of some early bronzes from Thailand, a region separated from China by difficult though not impassable mountains and dating several centuries before bronze founding began in China. The technical style is as different from that of the later Chinese as it is from the slightly earlier ones far to the West, but all three depend upon the same basic properties of reduction, alloying and casting that would be transmittable as an idea.

Anthropologists versed in the nature of cultural transmission will properly regard the above as a naïve trespass of a metallurgist upon their field. What follows is metallurgy.

The Non Nok Tha Bronzes

All the objects that are described below were kindly provided by Dr. Wilhelm G. Solheim II of the University of Hawaii, and come from the 1966 excavations at Non Nok Tha that were part of a four-year archaeological programme in Thailand jointly sponsored by the Fine Arts Department of the Thai Government and the University of Hawaii. The site was discovered by Chester Gorman in 1964; in 1965 Ernestene Green dug seven test pits; and in 1966 Dr. Wilhelm G. Solheim II, Hamilton Parker, and Donn T. Bayard, all from the University of Hawaii, excavated a portion of the site, and uncovered the bronzes that constitute the material for the present study. Since the recoveries brought major surprises and raised new questions, Donn Bayard returned again to the site in 1968. A full report is being prepared by the excavators. In the meantime Dr. Solheim has furnished the following description of the site and identification of the objects to amplify the account he published in 1968[5]:

Non Nok Tha is a low mound lying some 500 meters south of the village of Ban Nadi in the northwestern Phu Wiang district, Khon Kaen Province,

---

[5] Wilhelm G. Solheim, II, 'Early Bronze in Northeastern Thailand', *Current Anthropology*, 1968, **9**, pp. 59–62. *Note added in proof:* See also D. T. Bayard, *Non Nok Tha: The 1968 Excavation. Procedure, Stratigraphy, and a Summary of the Evidence* (University of Otago Studies in Prehistoric Anthropology, No. 4, Dunedin, New Zealand, 1971); *idem* 'An early indigenous Bronze technology in North-East Thailand . . .', *Proceedings of the 28th International Congress of Orientalists* (In press, 1972; and *idem*, 'Early Thai Bronze Analysis and New Dates', *Science*, 1972, **176**, pp. 1411–2; also papers by D. T. Bayard, W. G. Solheim and R. Pittioni to be published in Volume 13 of *Asian Perspectives* (Univ. of Hawaii).

Thailand. The latitude is 16° 47′ 57″ N, longitude 102° 18′ 17″ E. The mound is located some 2·5 kilometers north of the foot of a large, low sandstone mountain after which the district is named. The sandstone formation rises some 500 meters above the level of a flat or gently rolling plain that extends north, east and southeast to the Phong and Choen Rivers. This plain is largely riceland alternating with patches of scrub forest on the higher grounds. The mound itself has approximate dimensions 100 m north-south by 150 m east-west. Its surface lies between 80 and 150 cm above the average level of the surrounding rice fields, which in places have been cut into the mound. The general elevation of the fields is from 190 to 195 meters above sea level according to Army Map Service maps. At the present time, the mound is divided by lines of banana trees into four plots owned by four families of Ban Nadi, who use the land for the cultivation of cotton, mulberry, jute, bananas, papayas, and red pepper.

In the portion of the site excavated in 1966, 14 archaeological levels were uncovered representing periods during which the site was used for habitation or as a cemetery or both. The second to lowest layer yielded a radiocarbon date of *ca.* 3500 B.C., but there was one more layer underneath for which no C[14] date could be obtained.

All the bronzes submitted for analysis were recovered in the 1966 excavations. Following are the exact provenences for each specimen:

NP 549   bronze axe, recovered from baulk between squares C6 and C7, associated with burial No. 69, layer 19 (level III). The C[14] date for this level is *ca.* 2300 B.C. [The axe is shown *in situ* in Fig. IIb of reference 5.]

NP 553   one of five bronze bracelets recovered from square D5, layer 19 (level III).

NP 551–1   one of sixteen bronze bracelets associated with burial No. 31, recovered from square E3, layer 17 (level V). The C[14] date for this level is *ca.* 1100 B.C. [Shown *in situ* in Fig. IIa of reference 5.]

NP 294   bronze nodules, recovered from baulk between squares C4 and C5, layer 17 (level V).

NP 521   clay crucible, recovered from baulk between squares C4 and C5, layer 17 (level V).

NP 331B   wire ring fragments. These are not from Non Nok Tha but from Pimai, another site in Northeast Thailand. From a secondary burial in an earthenware pot. Layer 8, probably mid first millennium A.D.

Qualitative spectrochemical analyses have been carried out on all of these objects for the excavators by the renowned analyst Richard Pittioni of Vienna. His unpublished report shows that they all contain copper as major component, with tin and lead in major amounts, and a small amount of silver: they were all normal leaded bronzes. There were also present traces of arsenic, and iron, and often manganese, nickel, antimony, and bismuth, but since the spectrographer's arc struck the corroded outer surface of the samples (the fused spot caused by the arc can be seen about 1 cm. from the lower left corner of the arc in Fig. 1) it is unsafe to draw conclusions from their

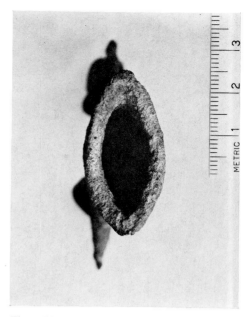

Figure II.1 (MIT 289–1) Bronze socketed axe from Non Nok Tha, Thailand. General view, ×0.96.

Figure II.2 (MIT 289–2) View of axe, looking directly into cored socket. ×1.2.

Figure II.3 (MIT 289–4) Detail of axe, showing line due to mould offset. ×4.1.

Figure II.4 (MIT 289B–1) Transverse section cut from axe at point 1·0 cm from top left corner of Figure 1. ×10.

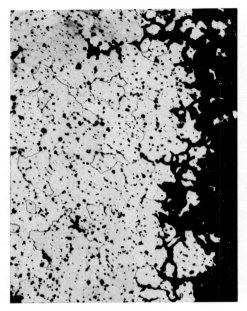

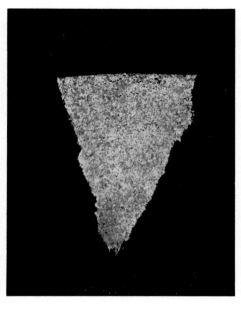

Figure II.5 (MIT 289B–2) Microstructure of specimen shown in Figure 4. Potassium bichromate etch. × 100.

Figure II.6 (MIT 289A–1) Transverse section of cutting edge at centre of axe. Etched. × 18.

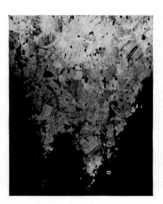

Figure II.7 (MIT 289A–9) Same as Figure II.6. Microstructure of edge at magnification of × 200. At the right is the corroded natural surface of the axe. See text for a discussion of the structural details.

Figure II.8 (MIT 289A–4) Same as Figure II.7, area about 5 mm from edge. × 200.

Figure II.9 (MIT 289A–7) Same as Figure II.7 at lower magnification, × 50. Note the transverse intergranular cracks near the cut surface at the top.

Figure II.10 (MIT 290–2)   General view of two bronze bracelets of squarish section. 6·4 cm outside diameter.

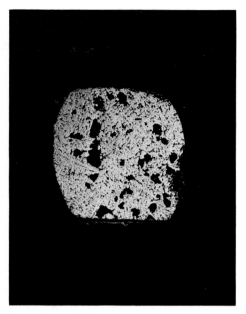

Figure II.11 (MIT 290A–1. Cross section of bracelet shown in Figure II.10. Etched. × 10. The large irregular black spots are holes in the metal due to gas.

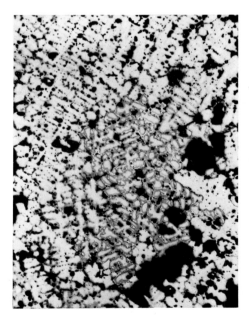

Figure II.12 (MIT 290A–14)   Same as Figure II.11. Microstructure at magnification of 50. Etched.

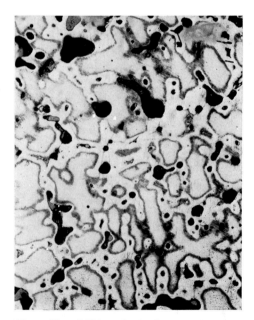

Figure II.13 (MIT 290A–7)   Same as Figure II.12. At magnification of 200.

Figure II.14 (MIT 291–2) Bronze bracelet of D-shaped section. 7·3 cm outside diameter.

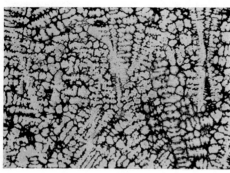

Figure II.15 (MIT 291A–15) Micro-structure of bracelet shown in Figure II.14. Etched. × 50.

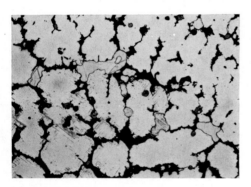

Figure II.16 (MIT 291A–11) Same as Figure II.15 at magnification of 200. Note intergranular and transgranular corrosion, and the pools of copper (very light grey, sometimes containing twins) replacing original cavities in the metal.

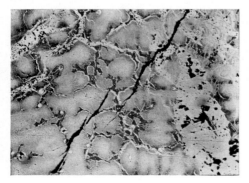

Figure II.17 (MIT 291A–20) Same specimen as Figure II.16. Mineral corrosion products preserving pseudo-morphs of the structure of the original bronze. × 200.

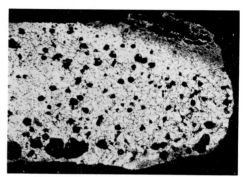

Figure II.18 (MIT 293–2) Two bronze nodules, slightly enlarged. × 1.3.

Figure II.19 (MIT 293A–1) Section of smaller nodule in Figure II.18, showing cast bronze structure and accretion of environmental earth cemented by corrosion products. Unetched. × 10.

Figure II.20 (MIT 293B–1) Section of larger nodule in Figure II.18. Unetched. × 10.

Figure II.21 (MIT 295–2) Crucible from Non Nok Tha. General view, slightly reduced. The few stains visible on the inside are mainly slag and dross but a few corroded metallic droplets are present.

Figure II.22 (MIT 295–1) Another view of the crucible, showing the spout.

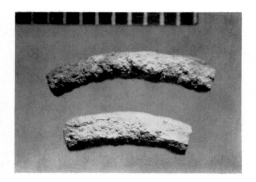

Figure II.23 (MIT 292–1) Fragments of
bronze ring. × 7.2.

Figure II.24 (MIT 292B–9) Microstructure of
polished section of ring fragment, which is
entirely mineralized by corrosion. × 100.

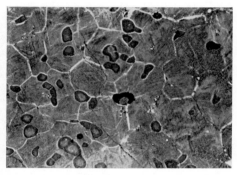

Figure II.25 (MIT 292B–12) Another area of
the specimen shown in Figure II.24. × 200.
The structure is a clear pseudomorph of a
quenched alpha-beta bronze structure.

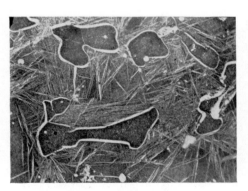

Figure II.26 (MIT 292A–8) Structure of
second ring fragment (Figure II.24). × 500.

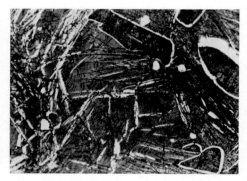

Figure II.27 (MIT 292A–14) Same specimen
as Figure II.26. Another area showing faint
pseudomorph of a eutectoid structure.

The magnification stated in the captions apply to the original photomicrographs which have been
variously reduced in size by the photoengraver. The reductions were as follows: Figures 3, 4, 5, 6,
12 and 13 reduced by factor 0.8. Figures 7, 8, 9, 11, 15, 16, 17, 19, 20, 23, 24, 25, 26 and 27 reduced
by factor 0.6.

minor variations, for they represent the environment of burial more than the original metal.

Metallographic examination was carried out at Massachusetts Institute of Technology, the specimen preparation and photography being the work of James F. Howard, Jr. The microstructure throws considerable light on the techniques of fabrication, for the shape and distribution of the micro-constituents in a bronze, as in any metal, is greatly affected by both thermal and mechanical treatment.

Bronze axe (NP 549, MIT 289)

The general appearance of this axe (Fig. II.1 to 3) suggests that it had been cast in a two-piece (bivalve) mould with a core to shape the socket hole. There is a visible offset on both sides of the socket, clearly due to lateral displacement of two mould halves at their junction plane. This offset is not present on the cored surface at the inside of the socket. The mould must have been identical in construction with the two-piece sandstone mould shown in Fig. IIIa in Solheim's report.[5]

A small specimen cut transversely from a point about 1 cm. from the top shows the offset more clearly (Fig. II.4) and the microstructure at this location (Fig. II.5) confirms that this part of the axe had not been deformed since casting. The structure shows that the tin content is not high: there is no dendritic segregation ('coreing') visible in the grains and there is none of the eutectoid constituent that is commonly found in castings with a tin content of over about 8 per cent, though it is not an equilibrium constituent until over 13 per cent. The shape of the grains is unmistakably that of a cast alloy, and their small size is attributable to the unusually large number of non-metallic inclusions dispersed rather uniformly throughout the metal. These inclusions are probably tin oxide and indicate simply that the bronze had not been melted under the best conditions. There are also inclusions tentatively identified as copper sulphide, and there is a fair amount of lead: both of these would have been in solution in the liquid bronze and took their present form by microsegregation during the solidification.

The cutting edge of the axe has a totally different microstructure (Figs. II.6 to 9). The surface examined was that of a transverse section close to the centre of the edge of the blade. The non-metallic inclusions are still there, essentially unchanged except for being slightly distorted, and the lead is mostly distributed as small particles at the new grain corners. The grains are smaller, more uniform, and of the more clear-cut shape appropriate to metal that has recrystallized after it has been well worked, either when hot, or when cold followed by subsequent annealing. There are also present in-

[5] Ibid.

numerable twins which are a certain sign of deformation and heating. It is unlikely that the axe could have been hot worked to any great extent, for bronzes containing lead are hot-short, but there are some internal intergranular cracks in the middle of the section (visible near the top of Figs. II.6 and 9) that suggest it. The grain size is quite uniform throughout this entire section, and the axe must have been well worked, the deformation extending deeply into the body of the axe. It seems therefore that both the thinning of the metal into the cutting edge and its splaying laterally into its present form came about by extensive though local hammering on a casting of a rather different shape. The initial casting was probably a simple tapered socketed piece, and the beautifully splayed cutting edge, now a line about 9 cm. long, was originally a rounded rectangle or an oval measuring about 2·5 × 1 cm. It could have been shaped with less hammering from an eared shape such as would come from the mould shown in Fig. IIIa of Solheim's report,[5] but unless the edge was reduced to at least half its original thickness it would not be expected to have the uniform structure that is observed.

The thin straight deformation bands crossing the grains in the microstructure near the edge (Figs. II.7 and 8) are a result of some cold deformation after the last anneal. This may have occurred in use, or may have been done as a final shaping or hardening operation. Cold working was not extensive, for the grains are very little distorted. The hardness of the metal in the cast part of the socket (Fig. II.4) is between 97 and 108, average 103 (these are Vickers hardness numbers, using 50 gramme load). The slightly cold worked metal at the cutting edge (Fig. II.6) had a hardness of 195–199, dropping to 150 at 1·4 mm. from the edge and remaining in the range 146–150 up to the end of the specimen at 3·2 mm.

Altogether this axe is a result of a fairly advanced metal working operation, fully exploiting the casting and working propensities of bronze. Though the metal was somewhat oxidized in melting, the casting in a bivalve mould with a core is good. The present shape of the cutting edge is a result of extensive deformation of an originally thicker part of the casting, and its uniform recrystallized microstructure shows knowledge of both mechanical deformation and of the effects of heat.

Bracelets (NP 553, 551–1; MIT 290, 291)
The bracelets submitted for analysis were all of simple uniform cross section. Five of them, all from level V, were D-shaped, i.e. segments of circles, one from this level was circular. Two from the earlier level III were nearly barrel-shaped in section, with two flat parallel sides and two neatly rounded ones. The geometric appearance of all strongly suggested that they had been

[5] Ibid.

made of a hammered or drawn wire bent and joined into a ring. The micro-structure, however, showed unmistakably that they had been cast directly to shape and showed no evidence of working of any kind.

Fig. 10 shows the general appearance of two of the bracelets of barrel-shaped section (NP 553) cemented together by corrosion. The internal diameter was 5·5 cm., the section 0·42 × 0·56 cm., and the weight of two of them 53·4 gm. Figures 11 to 13 show the microstructure of a section of one of them. Corrosion had penetrated deeply into the metal, but the structure, with dendritic coreing and interdendritic porosity with patches of lead and of eutectoid, is unmistakably that of a bronze of medium tin content in the cast condition. The average hardness was 82, varying between 72 and 96.

The D-shaped bracelet selected from the batch NP 551 from level V is shown in Figs. II.14 to 17. It weighed 49·7 gm., had an internal diameter of 6·8 cm. and in section was a segment of a circle with chord 0·9, height 0·24 cm. Despite its more recent date the metal was more heavily corroded than the other, and the microscope reveals that the tin-rich parts of the alloy have been almost completely replaced with corrosion products throughout the entire section. However, the copper-rich part of the dendrites in the body of the metal remains almost untouched, and it is certain that the alloy had been cast and not subsequently deformed in any way. Fig. II.16 shows a region where there is the crystallographic cleavage-plane corrosion which occurs in bronzes that have been slightly strained. This photomicrograph is also interesting for showing a few areas in which original cavities in the alloy have been filled in with nearly pure copper which has been electrolytically redeposited in the course of corrosion.[6] Fig. 17 is a photomicrograph of a thin part of the section where the metal has been completely mineralized by corrosion: the original microstructure of the metal can still be clearly seen, for the minerals are pseudomorphs of the metallic constituents they replace.

It would be necessary to free the objects of corrosion product to obtain further evidence as to manufacturing technique. No evidence of any joints was seen, or seams that would indicate joints in the mould. The barrel section could have been set equally into two halves of a split mould, but no trace of joint flashing was seen and if it had ever been present it had been removed in a final dressing operation. The parallel flat surfaces looked as if they had been finished by rubbing on a flat abrasive stone. The D section would have been difficult to draw from a split stone mould. Though neither certain nor necessary, it is conceivable that the bracelets had been moulded by the *cire-perdue* process. The barrel section could have begun as an extended or rolled rod of wax, bent and joined into a circle, while the D-shaped

[6] For a discussion of the phenomenon of redeposition of copper during the corrosion of bronze, see Gettens, above, pp. 134–7.

section may have been formed over a preformed cylindrical core of clay; then the whole was embedded in clay, the wax melted out, and the metal poured into the space formerly occupied by it.

Bronze nodules (NP 294, MIT 293)

These nodules, from level V, were irregularly-shaped bronze nodules of various shapes and sizes and weighing between 1 and 50 grammes. They had every appearance of being simply irregular drops of metal spilled from a crucible or running out of a mould and solidifying on an earthen floor, unrestrained in shape by a mould or other surface. Two of them (Fig. II.18) were sectioned and examined microscopically. There was a very thick layer of laminated corrosion-product-cemented debris on one of them (Fig. II.19), partly broken away. Under the microscope both samples showed the typical undistorted structure of a cast bronze (Fig. II.20). They were corroded intergranularly throughout, and there are many large blowholes indicating gassy metal. One would not expect to find material of this kind except at the actual site of a metalworking shop.

Clay crucible (NP 521, MIT 295)

This crucible, also from level V, is shown in Figs. II.21 and 22. It measured about 8·5 cm. diameter in the narrowest dimension, 9·6 cm. in the diameter including the spout, and was 4·8 cm. high. It would hold about 800 grammes of bronze. Its own weight was 165 grammes. A few small drops of metal and a little dross were found clinging to the inside surface. The end of the spout had been broken off, and the clay body inside is black, showing that it has not been heated very long under oxidizing conditions, either during manufacture or use.

In all respects, this crucible is similar to the crucibles that are found in early metalworking sites in both the Middle East and in Europe. In common with these, the outside of the crucible shows no special effect of fire, and heat had almost certainly been applied by covering it with fuel on top (probably using bellows or a blow-pipe) not by laying the crucible in or on a fire in the way that is now customary.

Wire ring fragments (NP 331–B, MIT 292)

These rings are not from the same site as the other samples examined. They are from Pimai in Northeast Thailand and date approximately to the middle of the first millennium A.D. They are made of wire about 1·5 to 2·0 mm. diameter. Only fragments remain (Fig. II.23), but their curvature shows that they came from rings about 1·8 cm. in diameter. The microscope showed that corrosion had completely destroyed the metal, but in many areas the resulting corrosion products (copper and tin minerals of different colours) had preserved exactly the structure of the initial bronze (Figs. II.24 to 27). The structure was identical in transverse and longitudinal sections and there were no inclusions to provide evidence that the wire had been mechanically elongated, though it probably had been and the main structural anisotropy eliminated by heat. The astonishing thing is that the structure is not that of a usual bronze with 5–10 per cent tin but is that of a very high tin alloy— about 20 per cent tin, almost a speculum metal. This alloy is far too brittle to be worked cold but is extremely plastic when hot (550–750° C.). It becomes brittle again if slowly cooled to room temperature, but if quenched from above 520° C. it becomes relatively hard and not fragile. It is known in the West only in the form of cymbals, either imported or made by crafts-men from the Middle East who have carried their techniques with them even to the New World. Archaeologically, it has been reported in a Koryu Dynasty bowl from Korea (see E. Voce[7]). The famous Kerman bronze bowls have the same microstructure, and so does an eighteenth(?)-century die for metal-stamping collected by Hans Wulff in Iran in 1966 and examined by the writer. This is a rather unusual alloy for an object such as a ring, and its use is perhaps more common than has been hitherto suspected.

Although the rings came from a cremation site, the structure corresponds to quenching, not slow cooling or low temperature heating and it is unlikely that they had been in a fire.

The micrographs leave no doubt whatever that the rings were made of an alloy of the high-tin kind. The microstructure shows the large crystals of alpha and beta phase and the fine acicular markings denoting a martensitic transformation in the quenched beta phase often shown in metallurgical literature and are identical in nature with the photomicrograph of the Korean bronze published by Voce. Corrosion certainly has distorted the structure,

[7] E. Voce, 'Examination of specimens from the Pitt-Rivers Museum', in H. H. Coghlan, *Notes on the Prehistoric Metallurgy of Copper and Bronze in the Old World* (Oxford, 1951), pp. 105–11. *Note.* For an extensive discussion of structural evolution in many different kinds of systems, see L. L. Whyte, A. Wilson and D. Wilson (eds.), *Hierarchical Structures* (New York, 1969). The relation of microstructures to the processing history is discussed in most modern textbooks of metallurgy and summarized for the archaeologist in C. S. Smith, 'The Interpretation of Microstructures of Metallic Artifacts', in *Application of Science in the Examination of Works of Art*, ed. William J. Young (Boston, 1967), pp. 20–52. Some discussion of the way in which the intimate properties of matter affect the use to which they are put, with some emphasis on Japanese swords and Chinese ceramics and bronzes, is in C. S. Smith, 'Metallurgical Footnotes to the History of Art', *Proceedings of the American Philosophical Society*, 1972, 116, pp. 97–135. On the decorative arts as stimulae to technological discovery see C. S. Smith, 'Art, Technology and Science: Notes on their Historical Interaction', *Technology and Culture*, 1970, 11, 493–549; reprinted in Duane Roller (ed.), *Perspectives in the History of Science and Technology* (Norman, Oklahoma, 1971), pp. 129–65.

for in Fig. II.27 needles are seen to cross an area of lamellar eutectoid that indicates a zone of local not very rapid cooling. (Some of this structure may, however, be a result of the mineralization process accompanying corrosion.)

General Comments

The metallographic examination reported above is in agreement with the spectrographic analyses showing the presence of tin in the objects. The intergranular corrosion could only have resulted from long exposure but does not permit any estimate of actual antiquity. As is usually the case, the tin-rich micro-constituents had more corroded than the copper-rich ones. The alloys are all typical copper-tin alloys and seem to have been well made without any indication of high impurities or alloying elements such as arsenic that in other parts of the world appear at the earliest stages of metallurgy. Though simply shaped, the objects are well fashioned and represent a stage of metallurgy that is by no means primitive. The axe from level III is particularly impressive. Very approximately the technical skill could be equated to that in the Middle East early in the third millennium B.C. although bronze was not widespread at that time. It should be noted that the Thai axe had been cast in a split stone mould and the edge, which shows the effects of both heating and cold work, subsequently hammered to shape. Its technique is quite different from that later used by the Chinese bronze founders in making their ceremonial vessels.

III

☙

THE CRANK IN GRAECO-ROMAN ANTIQUITY

A. G. Drachmann

Dr. Joseph Needham, in his work *Science and Civilisation in China*,[1] Vol. 4, part 2, has reviewed all the different mechanical elements in order to elucidate their history in East and West. On p. 112 he speaks of the crank and remarks in a long footnote *c* that there is considerable doubt about its existence during Antiquity. But in note *d* he writes:

> My friend Dr. A. G. Drachmann, however, feels that certain passages in Oribasius (fl. +362) and even Archimedes, seem to require the assumption of a crank. I hope he will set forth the evidence for this before long.

What better subject, then, could I choose for my contribution to the volume in honour of Dr. Needham?

First of all I have to point out that the English word 'crank' covers two mechanical elements which, though they are related, are nevertheless so different, that in my own language we have two different words for them, which is why I always have to keep them apart.

The principle is the same: a wheel or an axle is turned by means of a pin placed eccentrically parallel to the axle.

In Danish we have two words: *Haandsving* for a crank turned by hand, and *Krumtap* for a crank turned by a connecting rod.

In the latter, the machine crank, there is a connecting rod, one end of which grips the pin, while the other end is connected with a crosshead sliding in straight guides. In this way the revolution is changed to a to-and-fro motion, as when a windmotor drives a pump, or a to-and-fro motion is changed into a revolution, as when a steam cylinder turns a wheel. In the latter case the motion of the connecting rod is directed against the centre of

[1] Joseph Needham, *Science and Civilisation in China*, Vol. 4, Part 2 (Cambridge, 1965).

LOYOLA UNIVERSITY LIBRARY

revolution twice in every revolution, and so there must be either a fly-wheel to help it over the two dead centres, or more than one connecting rod.

A well-known example is the grindstone with its treadle; here one half of the revolution is 'dead'.

There is no trace of a crank of this sort in Antiquity, neither literary, archaeological, nor pictorial. A very fine intaglio showing Amor sharpening an arrow on a grindstone with a treadle, Fig. III.1, must be regarded as false, because the artist has placed the grindstone on a sort of wheelbarrow, and that is an unmistakable anachronism.

FIG. III.1

So we are left with the other sort of crank, the turning handle. About this Dr. Needham (p. 111) quotes Lynn White, Jr.[2]:

Continuous rotary motion [writes Lynn White in an admirable passage] is typical of inorganic matter, while reciprocating motion is the sole form of movement found in living things. The crank connects these two kinds of motion; therefore we who are organic find that crank motion does not come easily to us. . . . To use a crank, our muscles and tendons must relate themselves to the motion of galaxies and electrons. From this inhuman adventure our race long recoiled.

---

[2] Lynn White, Jr., *Medieval Technology and Social Change* (Oxford, 1962), p. 115.

To which is added a footnote *a*:

I am not sure that the statement is strictly true, as the rotifers may bear witness, but it is broadly so.

The rotifers, as will be shown, bear no witness against Lynn White, Jr.'s statement; but in some respects it needs correction all the same.

How are we to define 'rotation'? It is the motion of a wheel turning on its axle, it is characterized by the fact that a straight line in the plane of rotation will go through an angle of 360° for each rotation. A continuous rotatory motion is a repetition of the same rotation under the same conditions, Fig. III.2A.

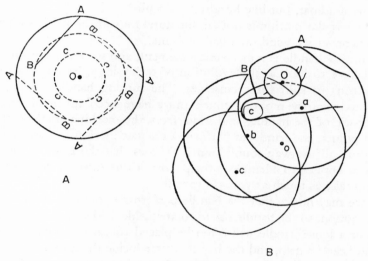

Fig. III.2

It is true, as has been said, that such a perpetual rotation does not exist in living organisms. This is because the rotating part of the organism must have possibilities for growth and maintenance, e.g. in the vertebrates be accessible to blood-vessels and nerves. But if the blood-vessels were to enter the rotating part from the side, they would be wound round it during the rotation; and if they entered from the end, they would become twisted. There is no anatomical possibility of a sliding connection.

But even if a continuous rotatory motion is excluded, a limited rotation is possible. We can rotate our head, and rotate our hand about 180°, but then it has to be turned back again. This short rotation then becomes in fact a reciprocating motion.

But there is nothing to keep the hand, which is connected with the body by three joints, at the shoulder, the elbow and the wrist, from performing a circular motion; it can do it in the shoulder-joint alone. But this circular motion is not a rotation as the one described above; it is characterized by the fact that each point of the moving body describes a circle, but the body itself keeps the same position in relation to the surroundings all the time. If you draw a circle with chalk upon a blackboard, the hand will perform a circular motion, but the back of the hand will be held in the same position throughout, Fig. III.2B.

As for the rotifers, their wheels consist of a circle of cilia, each of which performs a circular motion; when they work together, the circle itself seems to rotate, but it does not. It brings the animal through the water, not like the screw of a boat, but like hundreds of sculls.

But the circular motion is not an unnatural movement at all; we perform it e.g. whenever we wind up a ball of twine.

The turning handle is a mechanical element, and to understand its origin we shall have to look at the evolution of technology, leaving the galaxies and electrons to shift for themselves. The turning handle presupposes a continuous rotatory motion, which must have been invented first. It is characteristic of the handle, as distinct from the machine crank, that it has no dead points; the hand can perform a circular motion and so can act on the rotation all the way round. But it follows that the limit of the power that can be applied to a handle is the power of one man, while the machine crank can take as much as you can give it.

Also we may notice that the handle performs a rotation, but the hand a circular motion, so the handle has to rotate inside the hand. For this purpose a knob or a loose handle will often be placed round the stick, so that the hand can keep its grip, and the handle rotate inside the knob.

Evidence of technical matters from Antiquity is found in three ways: from finds, from pictures, and from the literature.

Since most of the crank handles were made of wood, we cannot expect to find many of them; but there is one place where they may have been present: in the rotatory handmills.

The oldest hand-mills consisted of a trough of stone and a round stone for crushing the grain.

The next hand-quern consisted of a flat stone and another flat stone with a funnel-shaped slit through it. By means of a wooden stick the upper stone was ground to and fro across the nether stone, which formed a sort of table.

Then, at about the same time, there came up the horse-mill with the hour-glass shaped upper stone, and the rotatory hand-mill, where a round upper stone was turned on a nether, stationary mill-stone shaped like a cone with a very obtuse angle. (Fig. III.3.)

The upper stone was turned by means of a wooden handle. In some stones a horizontal hole for the handle is found, and it is then possible to reconstruct the mill with a horizontal arm carrying a vertical handle; or you can deny the crank-handle and contend that the stone was turned to and fro by the horizontal handle alone, like the handquerns, and so was not provided with a crank. But in some of the stones the hole is vertical, and here it is certainly necessary to assume a vertical handle; but then it may be contended that the stone still was only turned to and fro by means of the vertical handle, and that here also there was no real crank.[3]

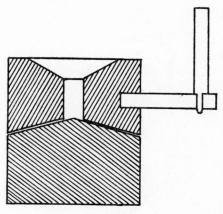

FIG. III.3

According to my definition the vertical handle on the horizontally-rotating stone is a crank, whether it has been used for rotating the stone half a turn or a whole turn, and since nobody has seen these mills in use, it is impossible to get an exact knowledge about it.

But there is one certain find, which is also exactly dated; there is a crank-handle found in one of the ships built by Caligula on the Nemi lake in + 40.

It is a round disk of solid wood, obviously a fly-wheel, with a square hole in the middle and another square hole near the rim; a long, round, wooden stick with a square foot fitted into this hole (Fig. III.4).

G. Ucelli, who reported on the Nemi ships,[4] has connected the apparatus with a bucket-chain meant for emptying out bilge-water; but Lynn White, Jr. points out[5] that the bucket-chain was found in the other ship, and that the stick is far too slender to work the bucket-chain. But even if this is as true as it is said, it does not allow the conclusion that no crank has existed before

[3] Op. cit., p. 108.

[4] Guido Ucelli, *Le navi di Nemi* (Rome, 1950), p. 181 and fig. 199.

[5] Lynn White, Jr., *Medieval Technology and Social Change* (Oxford, 1962), pp. 105–6.

the Middle Ages, for it is unmistakably a crank from +40, and *ab esse ad posse valet consequentia*.

We cannot tell what it was used for; but since it is found on a ship, we may guess that it had to do with rope-making.

So much for the finds; as for the iconographical material it simply does not exist. I do not know of a single picture of a crank-handle from Antiquity, be it in sculpture, mosaic, wall-painting, vase-painting or elsewhere. This indeed may be taken as a proof that it was not very much in use.

So we have to turn to the literature, and I shall begin with the technical works.

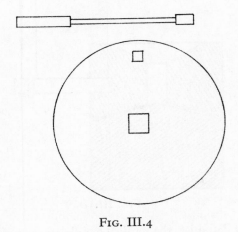

Fig. III.4

As I have said, it is only a very moderate power that can be used in a crank-handle; if more power is needed, you have to use handspikes. We have from Antiquity descriptions and pictures of wine-presses, where a horizontal drum is turned by means of long handspikes; we know of capstans, that is, vertical drums turned by horizontal handspikes, when one man or more walk round in a circle; also the catapults were drawn by hand-spikes.

For turning very small instruments, like drills, a bow with a string was used; this way was known already in Egypt.

What is left is instruments of middle size, and they are found in Oreibasios,[6] who about +362 wrote a large compendium of medicine; what interests us here are some engines for setting dislocated limbs and holding broken bones in place while a bandage was applied.

[6] Oribasii Collectionum medicarum reliquiae Vol. 4. Edidit Ionanes Raeder (Leipzig and Berlin, 1933) (Corpus medicorum graecorum VI, 2, 2).

The earliest engine is described by Hippocrates,[7] about −420; it consisted of a table on which the patient was placed; two ropes were tied to his body, two other ropes to the end of the fractured limb, and then the four ropes were pulled by means of drums and handspikes till the correct position of the bone was obtained. (Fig. III.5.) This was certainly not pleasant for the patient, but it was better than becoming a cripple for life. This engine, by the way, continued to exist as an instrument of torture. For the working of the engine two helpers were needed; so several surgeons tried to improve the construction, partly by leading all the ropes to a single drum, partly by introducing a gear and a pawl, so that the surgeon himself could work the engine and be independent of helpers.

They reached constructions that were not very different from the operation tables used by the surgeons of our times; only one thing they did not achieve: to make the patient fast to the table and only pull at the limb. There were always two ropes pulling the patient and two pulling the leg.

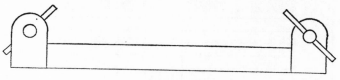

FIG III.5

Andreas was the name of a surgeon contemporary with Archimedes; he used the screw invented by Archimedes for his 'bench', as his instrument was called.[8] It contains a long spindle with two screw-threads, one right-handed, the other left-handed; when the spindle was turned, the two screw-threads moved two 'tortoises', that is, two wooden blocks running in grooves in the planks of the bench. When the spindle was turned one way, the tortoises were moved away from each other, when it was turned the other way, they came together. (Fig. III.6.) At Andreas' time it was not possible to cut a screw thread inside a hole in a wooden block; so a smooth hole ran through the block, and a wedge was driven in to engage the screw thread; later, when a method was found for cutting an inside screw thread, the tortoises became female screws.

The whole construction looks rather like the vice of a carpenter's bench; there is no reason to think that the screw was turned by means of a crank-handle, for the press of the carpenter's bench is still turned by means of a stick going through a hole in the head, a handspike, in fact. One reason for this is that the stick has to be out of the way, but another is that there may

[7] Ibid., 49:27.
[8] Ibid., 49:5.

be quite a lot of friction to overcome, and this has hardly been less in the twin screw in Andreas' bench. On the other hand the screw has the advantage that it does not need a pawl, it never recoils.

Nymphodoros was the name of another surgeon who invented a 'chest'.[9] A bench was a couch on which the patient could lie; a chest was a transportable instrument which the surgeon could take with him to the place of the accident, e.g., a stadium, where he would place the patient on a ladder

Nymphodoros is mentioned as a mechanic by Vitruvius (−25) 7 : introd. 14; and he must have lived after Archimedes, since he uses the endless screw; his name is found nowhere else.

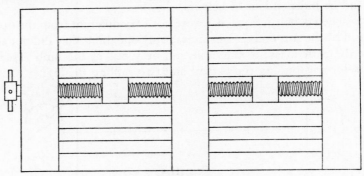

FIG. III.6

The instrument consisted of an axle with a toothed wheel which was turned by an endless screw. From this axle there came four ropes for resetting the dislocated limb. The axle of the screw came out through the side of the chest 'and it has a handle (*epitonion*) placed on its end for the turning.' Could this *epitonion* be a crank handle? (Fig. III.7.)

Oreibasios gives us a long list of terms relating to the resetting engines.[10] Here we find: '*periagogides* and *epitonia* and *skytalai*, all of them are tools for moving axles and screws. The *periagogides* come first, the *skytalai* next, the *epitonia* accomplish less than these; for most of them, the smallest, are turned by one hand.'

Skytale is a well-known word for a handspike, that is a bar or a rod or a stick inserted in a hole in a felloe or a drum for turning it.

According to Liddell and Scott *periagogis* is the same as *periagogeus*, which is found in Lucian[11] and means a windlass. This is, however, not possible, since *periagogis* in Oreibasios means 'an appliance for turning axles and screws'.

[9] Op. cit., 49:21.
[10] Ibid., 49:4: 26–7.
[11] Lucian, *Navigium* 5.

The word is also found in Soranos,[12] who describes a bench or stool for a woman in confinement. 'Some surgeons add to the lower part of the stool a visible axle provided with *periagogides* on either end and also with a peg.' The axle is used for drawing cords to extract an embryo, and the *periagogides* are handles, but for this purpose it is hardly necessary for them to be very strong.

Oreibasios uses the word once more.[13] The lentil-shaped screw, he writes, which turns the toothed wheel, is turned by a *periagogis* or an *epitonion*. Now the *periagogis* is stronger than the *skytale*, which again is stronger than the *epitonion*, but here, it seems, you cannot use the *skytale*, but either of the other two.

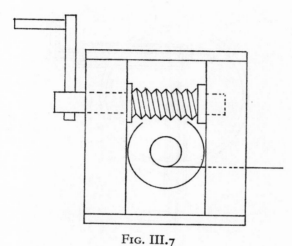

FIG. III.7

I suggest that the *periagogis* means four handles fixed to the drum or axle; they are more effective than the handspikes, because they are always in place, but they can be used in many sizes. But the *epitonion*, which is often turned by one hand, might well be the crank-handle.

Epitonion originally meant the peg round which the string of the lyre was laid to tune it. It was made like the screw of a violin, being no screw at all, but a conical peg in a conical hole, where it was held by friction. It had a finger-grip at its upper end.

Next the word is used for the plug of a cock, a conical piece of bronze in a conical hole.

[12] Soranos, Edidit Ioannes Ilberg (Leipzig and Berlin, 1927) (Corpus medicorum graecorum 4) Gyn. 2:3:3 (=68).

[13] Orobasii Collectionum medicarum reliquiae Vol. 4. Edidit Ioannes Raeder (Leipzig and Berlin, 1933) (Corpus medicorum graecorum VI, 2, 2) 49:5:7.

In Oreibasios *epitonion* means just a handle of some sort, the smallest of them made to be turned with one hand only. As for the *epitonion* of Nymphodoros' chest there is no reason to think that it should not be a crank-handle; but there is no direct evidence that it was.

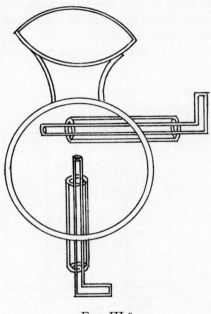

FIG. III.8

The word *epitonion* is found in Heron's Pneumatics and his Automatic Theatre.[14]

In the *Automatic Theatre* 13:5 and in the *Pneumatics* 1:31 in Pseudo-Heron's version it means a stopcock; in Heron's own text of 1:31 it is not found. In the *Pneumatics*[15] 2:17 it is the handles of two cocks, Fig. III.8; it is shown as a small peg at right angles to the tube that is to be turned. In 2:18 it is the handle of a syringe; here it is a cross handle, Fig. III.9. It is remarkable

[14] Heron, *Pneumatics*, Herons von Alexandria Druckwerke und Automatentheater. Griechisch und deutsch hrsg. von Wilh. Schmidt. (Leipzig, 1899) (Opera omnia Vol. 1).

[15] Heron, *Pneumatics*, MS. Marcianus 516.

that these two chapters are the only ones about which we can say with any certainty that they are taken from another book, to wit a medical treatise.

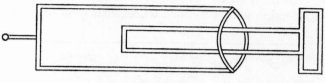

FIG. III.9

Heron's Barulkos is a queer chapter which is found as ch. 37 in his *Dioptra*[16] and ch. 1 in his *Mechanics*,[17] but it does not belong to either work. Heron describes how we can lift 1,000 talents by means of a power of 5 talents; it is done by a series of toothed wheels in the ratio of 1:5; the last axle is turned by handspikes. But in the version found in the *Dioptra* there is the remark that it is possible to fit a screw and turn it by means of a handle, *cheirolabis*. In two figures of the apparatus, one from Heron, Fig. III.10, the other from *Pappos*,[18, 19] Fig. III.11, it is shown as a cross-handle.

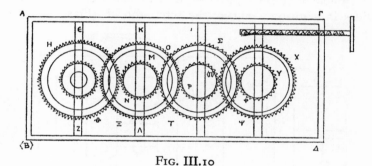

FIG. III.10

There remains now only Heron's *Mechanics*,[20, 17] which has come down to us only in Arabic. We have seen that this handle, to be turned by one hand only, seems to belong with screws. In the *Mechanics* there are several chapters

[16] Heron, *Dioptra*, MS. Codex Parisiacus inter supplementa Graeca no. 607 (Mynas Codex), Bibliothèque Nationale, Paris.
[17] Heron, *Mechanics*, Herons von Alexandria Mechanik und Katoptrik, Hrsg. von L. Nix und W. Schmidt (Leipzig, 1900). (Opera omnia Vol. 2, fasc. 1.)
[18] Pappos, MS. Bibliotheca Apostolica Vaticana, Greco 218.
[19] The same, text, Pappi Alexandrini collectionis quae supersunt edidit Frid. Hultsch. Vol. 3, Part 1 (Berlin, 1878), p. 1010.
[20] Heron, *Mechanics*, MS B. British Museum Add. 23390; MS C. Aia Sofia Nr. 2755; MS L. University Library, Leiden. Cod. or. 51.

on the use of the screw, which is the last of the 'five simple powers': the windlass, the pulley, the lever, the wedge and the screw.

The screw, Heron says, may be used in two ways, either together with a *tylos*,[21] or in connection with a toothed wheel, that is as an endless screw.[22]

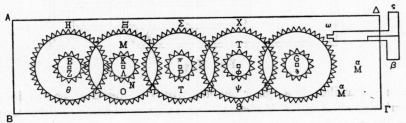

FIG. III.11

Tylos means a peg or plug; here it indicates the wedge that engages the screw-thread when the screw goes through a smooth hole; that is the method used by Andreas in his bench. In Heron it is just a theoretical example; the screw, like the other powers, has to lift a burden. So he places the screw-spindle vertically, provides it with handles, and lets the *tylos*, which slides in a groove, engage the screw-thread. (Figs. III.12, III.13.)

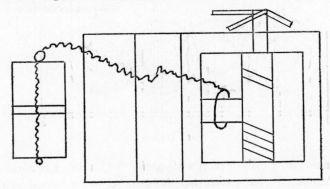

FIG. III.12

When the screw is turned, the *tylos* is lifted, and so is the burden tied to it. Instead of handspikes, he writes, we can put on the end of the screw a square block with handles in it, and turn the screw by means of this square block. The Arabic word for 'handle' here is a translation of *cheirolabis*, not of *skytale* or *epitonion*.

[21] Heron, *Mechanics*, Herons von Alexandria Mechanik und Katoptrik, Hrsg. von L. Nix und W. Schmidt (Leipzig, 1900). (Opera omnia Vol. 2, fasc. 1), 2:5.
[22] Ibid., 2:6.

The whole contraption is evidently theory; no one would lift burdens in this way. But the construction is found in the bench of Andreas, so it is not mere theory.

In the other method, the endless screw, the screw-threads engage a toothed wheel. For each turn of the screw the wheel is moved forward by one tooth; on its axle there is a drum round which the rope that carries the burden is wound.

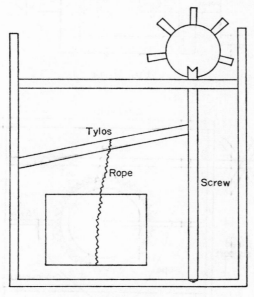

Fig. III.13

This machine element was invented by Archimedes and is still in use; it gives a very great gear ratio. Nowadays a crank-handle is generally used; Heron uses handspikes or the square block with the handles. I take it that this is his version of what Oreibasios later calls *periagogis*. (Figs. III.14, III.15.)

In a later chapter[23] Heron indicates how it is possible to combine four of the powers, to wit the lever, the tackle, the drum and the screw to form a single engine. This also is theory only; there is a lever, which is pulled down by a tackle whose rope goes round a drum which is turned by a screw engaging a toothed wheel.

The screw in the figure is marked L, and its upper end is provided with a handle marked M, according to the text. The Arabic word here also means *cheirolabis*. The figure from the MS L shows clearly an arm at right angles

[23] Ibid., 2:29.

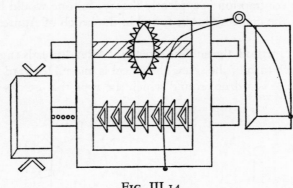

Fig. III.14

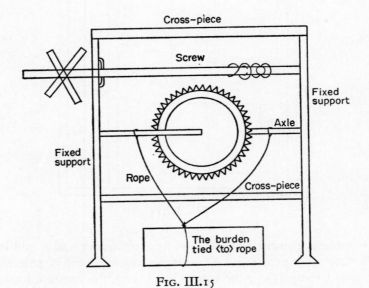

Fig. III.15

to the screw marked with an L; the bookbinder has cut off the end of the handle, so that we cannot tell if it carries an upright crank-handle. (Fig. III.16.)

In another MS, B, the figure is different; the upper end of the screw goes into a horizontal beam, and there is no arm jutting out. (Fig. III.17.) Now, which is right?

MS L is right, for in the text we find that there is a screw, L, and on it a handle, M. The L is clearly seen in the figure; if there was an M, the bookbinder has cut it out. But the handle is correct.

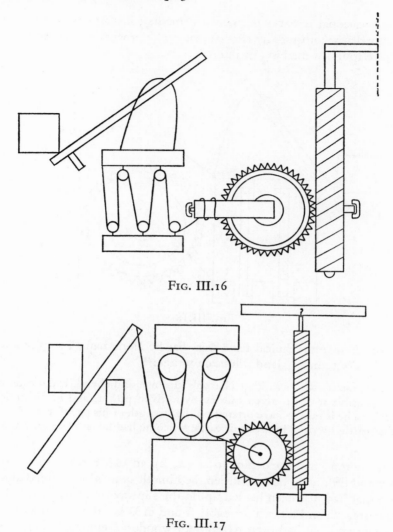

FIG. III.16

FIG. III.17

There is a third MS, C. The figure shows clearly an arm jutting out from the screw, and it does look as if there was also a crank-handle at right angles to the arm, parallel to the screw. (Fig. III.18.) But unfortunately it is no crank-handle, it is only the letter M, which we failed to find in the figure in the MS L. And now we are at an end. In none of the technical texts have we found decisive proof of the presence of a crank-handle.

There is still a field left for us to investigate, the non-technical literature. There is really an astonishing amount of technical information to be found through occasional remarks in poems, comedies, history and other works. But it is perhaps not quite by chance that it is in a work about Archimedes that I have found something of interest.

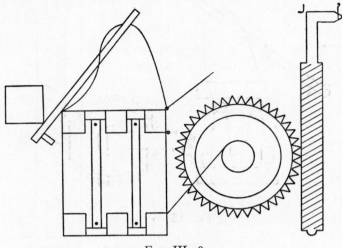

FIG. III.18

One of his many technical feats was that he alone hauled a fully loaded cargo ship along on dry land. Plutarch writes:[24]

> But Archimedes, who was king Hieron's kinsman and friend, had written that it was possible to lift a given burden by a given power, and in his youthful enthusiasm he is said to have uttered on the strength of his proof, that if he had another earth, he would move this earth, when he had taken place on the other one.

When Hieron challenged him to show, by an experiment, that a small power could lift a very heavy burden, he himself quite alone hauled along a big ship that had just been hauled up on the slipway by many men.

This story, which Plutarch probably found in Diodoros, whose books on Archimedes are lost, had been retold with wonderful embroidering.[25]

Athenaios[26] tells us that the very big ship presented by Hieron to King Ptolemy of Egypt was built under Archimedes' supervision, and when all the king's men failed to launch it, Archimedes hauled it into the water alone, using a screw he had invented himself. This is the endless screw; it consists of a drum provided with a toothed wheel; a screw engages this wheel, and

[24] Plutarch, *Vita Marcelli*, 14: 7–8.
[25] A. G. Drachmann, 'How Archimedes expected to move the earth', *Centaurus* 1958, 5, 278–82.
[26] Athenaios, *Deipnosophistae*, 5:206d, 207b.

it is moved one tooth for every full turn of the screw. If the drum has a circumference of one metre, and the wheel has 50 teeth, the burden will be moved one metre for every 50 turns of the screw. If the handle of the screw travels along a circle of one metre, the engine gives a gear ratio of 1:50.

Tzetzes[27] tells us that Archimedes hauled the ship with his left hand by means of a *trispastos*. This word generally means a triple pulley, which would seem rather inadequate for the job. But Oreibasios[28] tell us that Apellis and Archimedes invented the *trispastos* for hauling ships, but that it was later used in a resetting engine. The triple pulley was known to Aristotle,[29] before the time of Archimedes; but the *trispastos* described by Oreibasios is something else; it is a gear consisting of drums, wheels and ropes, and it is used in a crane described by Vitruvius, 10:2:5–7. The rope to pull the burden was wound round an axle, on this axle was placed a large drum, round which was laid another rope to be pulled by a capstan. Fig. III.19. If the diameter of the drum is five times the diameter of the axle, we get a gear ration of 1:5.

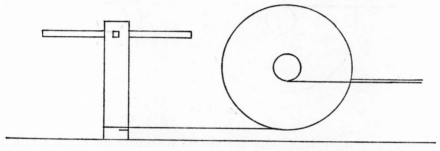

FIG. III.19

If we connect the endless screw with the *trispastos* we get a gear ration of 1:250 and the feat becomes quite possible.

About the boast of Archimedes that he would move the earth Simplicius tells us this:[30]

When Archimedes made the instrument for weighing called charistion by the proportion of that which is moving, that which is moved, and the way travelled, as the proportion went as far as it could go, he made the well-known boast: Somewhere to stand, and I shall move the earth.

---

[27] Tzetzes, *Chiliades*, 2, hist. 35, 107–8.

[28] Oribasii Collectionum medicarum reliquiae Vol. 4. Edidit Ioannes Raeder (Leipzig and Berlin, 1933) (Corpus medicorum graecorum VI, 2, 2), 49:23.

[29] Aristotle, *Mechanica*, 18.

[30] Simplicii in Aristotelis Physicorum libros quattuor posteriores commentaria, Edidit Hermannus Diels (Berlin, 1895) (ad 7:5), p. 1110.

Tzetzes mentions the moving of the earth twice; in one place he tells us that Archimedes moved the earth by means of a *trispastos*,[31] in the other he lets him say: 'Somewhere to stand, and I shall move the earth with a *charistion*.'[32]

Charistion means a steel-yard, and it would be quite unsuitable for moving anything big; but probably it is only that Tzetzes has misunderstood Simplicius; but just as his *trispastos* really had something to do with the launching of ships, so the *charistion* has something to do with the moving of the earth, since it gave Archimedes the idea of the Golden Rule of Mechanics.

If we try to get some sort of sense out of this tangle of partly contradictory, partly absurd statements, we may first remark that Archimedes did investigate

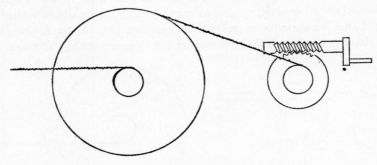

Fig. III.20

the theory of the balance; this led him to formulate the Golden Rule of Mechanics, which he expressed in this way: the centre of gravity of a magnitude consisting of two unequal magnitudes will be the point dividing the line connecting their centres of gravity in inverse proportion to their weight. On the strength of this proposition he made a practical invention, the *charistion*, or steel-yard, and he remarked in his book that it would be possible to lift a big burden of any given weight by a smaller burden of any given weight by regulating the arms of the lever. When Hieron remarked that surely there had to be a limit to the possibility, Archimedes answered: 'No! If I could have somewhere to stand, I could move the earth.'

Then it would seem that Hieron, to use a modern expression, tried to call his bluff, by suggesting that he should show by a practical proof that his theory really would work.

But Archimedes took up the challenge. He had already calculated the power of his differential gear, the *trispastos*, and of the endless screw, and

[31] Tzetzes, *Chiliades*, 3, hist, 66, 61.
[32] Ibid., 2, hist, 35, 130.

found that together they would give him a ratio of 1:250. So he arranged a test, hauling along single-handed a big ship on dry land, adding, for full measure, its load and its crew. Heron then said: 'In future we shall have to believe everything that Archimedes tells us.'

Fig. III.20 shows a diagram of the engine he used.

Here is Plutarch's description of the feat:[33]

> When Hieron wondered and wanted him to put the thing to proof in practice and show some big thing that was moved by a small power, he loaded one of the king's three-masted cargo ships, which had been hauled on land with great labour and many men, with a large crew and a full load, and then drew it along, smoothly and evenly, as if it was floating in water, not with great labour, but sitting down at a distance, gently swinging with his hand the end of a compound tackle.

The only way this could have been done is by means of a crank-handle driving an endless screw, and so I have put a crank-handle on my diagram. And there is nothing improbable in the thought that Archimedes invented the crank-handle, quite the contrary.

[33] Plutarch, *Vita Marcelli*, 14

IV

SIXTEENTH- AND SEVENTEENTH-CENTURY SCIENCE IN INDIA AND SOME PROBLEMS OF COMPARATIVE STUDIES

A. Rahman

Introduction

History of science and technology as we know today is by and large the history of development of science and technology in Europe. With the encyclopaedic contributions of Joseph Needham, on the development of science and technology in China, the myth that science and technology are essentially European has been exploded. There now remains the major task of studying the development of science and technology in other culture-areas, as systematically and intensively as has been done by Needham for China. Among the neglected culture-areas which require intensive studies India is foremost.

The availability of literature on science and technology in China, and to less extent from other culture-areas, has also brought to surface the need for comparative studies. It is only through such studies that we would be able to understand adequately the nature and character of science as a world-wide phenomenon. The study of growth of science, its patterns, and chief characteristics under different civilizations, would help us to understand the interaction of science with society, under different cultural influences and during different periods of history. Such studies are also relevant as science, for the first time in its history, is assuming a truly international character in so far as nearly all the countries of the world are beginning to contribute to discoveries, creating a world pool of knowledge and also drawing from it for their respective development.

The comparative studies, by necessity, would have to cover the development of specialized knowledge, as seen from the development of various branches of science and technology, concepts and theoretical framework as well as the organizational framework for acquiring knowledge and other related social and cultural features. The latter is particularly significant as it determines the capacity for further growth.

A brief description of the practice of technology towards the end of the sixteenth century and the general knowledge of biology in the early decades of the seventeenth, as given later in the paper, will reveal developments in India, which compare favourably with the contemporary developments in Europe. The comparison will also reveal that the failure of countries like India may not be in the degree of knowledge available or lack of capable persons in a particular period, but in not being able to develop the necessary potential for the growth of knowledge and an organizational framework to allow for the cumulative growth of science and technology.

For countries outside European culture-areas, like India, where despite significant contributions to knowledge and technological achievements, neither science nor technology could become a self-sustaining process, as they did in Europe, the study of organizational framework as well as other social and cultural features, assumes, therefore, the same significance as the study of the actual knowledge developed.

In this paper an attempt will, therefore, be made to give some idea of the development of science and technology in medieval India in the light of some of the recent studies and discuss some of the limitations of the scientific and technological tradition of the country, responsible for thwarting their further development.

Scientific and Technological Literature in Medieval India[1]

The medieval period in India begins with the eighth century and continues up to the eighteenth century. There is considerable scientific literature as will be evident from the number of books written in India and now available in Indian and European libraries.[2] Besides the manuscripts written on specific branches of science, there are large numbers of historical chronicles, memoirs of kings and other historical materials which are a rich source of information

[1] See for detailed discussion: 'Some problems of Source Material on History of Science in Medieval India', *Ithaca*, 26 VIII, 2 ix, 1962.

[2] A detailed, but by no means comprehensive bibliography of source material in science and technology, of books written in Sanskrit, Arabic and Persian languages has just been completed under the direction of the author, and is being processed for publication. It gives a fair idea of the extent and range of scientific and technological activity in India during the period.

on the development of science and in particular of technology. Very little work, if any, has been done to sift the scientific and technological material and work out a systematic account of their development.

The main languages of the manuscripts which have been catalogued are Sanskrit, Persian and Arabic. However, besides these there are other languages, numbering twelve, which developed during this period and in which a number of manuscripts on science and technology are available, but the work of preparing their bibliography is yet to be undertaken.

The literature available in Sanskrit, Persian and Arabic has been examined in each of the languages separately, in isolation from the material on the same subject and belonging to the same period in other languages. Sarton for instance in his pioneering work has given an extensive bibliography of Indian works on science and technology primarily in Sanskrit and Pali, but does not mention any significant work written in India in Persian and Arabic.[3] Consequently, we have the idea neither of the growth of knowledge in a particular branch of science in a defined period nor of the development of science as a whole in any given period. Such a treatment of the scientific knowledge has led to many unfortunate consequences, one of them being the belief that science and technology ceased to develop in India after the twelfth century, since hardly anything of significance was expressed in Sanskrit.[4]

The literature which is available is considerable, and suggests an intensive and continued activity during the medieval period. Much of this activity was, most probably, built around making summaries of books, writing commentaries or preparing annotated editions of works of well-known authors. The nature of writing in this manner should not lead us to presume lack of original contributions. Many of the scholars in annotating and commenting on the works of earlier savants seem to have made advances in specific branches of science. The source material has, therefore, to be critically studied in detail to understand fully the contribution to the knowledge of different branches of science and development of technologies.

A few studies recently published, however, give us some idea of the specific developments in India. A study of the late sixteenth century is now available which gives some idea of the technological development in the country.[5] Some idea of the development of biology till the early decades of the

[3] G. Sarton, *Introduction to History of Science.*

[4] This idea has been expressed by a number of Indian scholars who have worked on the development of science in India based on Sanskrit sources, see for example, Vishnu Mitre, 'Biological Concepts and Agriculture in India', *Indian Journal of the History of Science*, Vol. 5, No. 1, p. 160.

[5] M. A. Alvi and A. Rahman, *Fathullah Shirazi—A Sixteenth Century Indian Scientist* (National Institute of Sciences of India, 1968).

seventeenth century can be had from the personal diaries of emperor Jahangir (1605–1627). These have been compiled and published recently.[6] Basing ourselves on these studies, to give an idea of the extent of development of science and technology in India, we would like to raise a few problems of comparison and some questions, as referred to earlier.

Technological Contributions of Shirazi (1583–1588)

The main period of Shirazi's technological contributions begins with his joining the court of Akbar in 1583 and ends with his death in 1588. His major inventions were a wagon mill, a machine for cleaning gun barrels, a portable cannon, a seventeen-barrel cannon, and a travelling bath.[7]

The wagon mill, was probably developed for the army, to facilitate the grinding of the corn. It was a simple machine in which the movement of the wheels rotated the upper mill stone through the mechanism of pinion, cogwheel and the shaft.

More complex, but using essentially the same principle, was the machine for cleaning gun barrels (Figs. IV.1, IV.2). The idea behind the development was to use a bullock, thus releasing men for fighting or other work. The machine cleaned 16 barrels at the same time. It was worked by a bullock, whose movement, as in an oil press, rotated the axle, which in turn rotated the brush rod inside the barrels, through the movement of the cogwheel and pinions.

The portable cannon (Fig. IV.3) was developed to facilitate the movement of the heavy artillery to hill-tops. The entire cannon was fabricated in parts, each part was screwed to the other. It could be easily dismantled or assembled as required, and would have given greater mobility to the heavy artillery. There were a number of such cannons in the use of emperor Akbar.

The multi-barrelled cannon (Fig. IV.4) was developed by welding seventeen barrels in a row and could be fired in quick succession with one match-cord. It could be said to be the fore-runner of the machine-gun.

The travelling bath was made for the convenience of the king and his family. It was ingenious in being a huge construction containing the necessities for the emperor's bathing requirements: bath-room, dressing room, water containers, etc. and room for attendants!

These machines, in all probability, were constructed in the workshops (Karkhanas) of the emperor. These workshops were well-equipped, both with

[6] M. A. Alvi and A. Rahman, *Jahangir—The Naturalist* (National Institute of Sciences of India, 1968).

[7] For details, see above.

skilled workers, who were attracted from all over the empire, and other facilities.

These inventions, few as they were, are impressive particularly in the context of the time in which Shirazi lived. It will, therefore, be necessary to make a few remarks on the life of the inventor and the nature of his contributions.

Professionally, Shirazi started his life as a teacher of theology and ended as a highly respected scholar, courtier and statesman. In every sphere of life he had a significant contribution to make, whether it be currency reform,

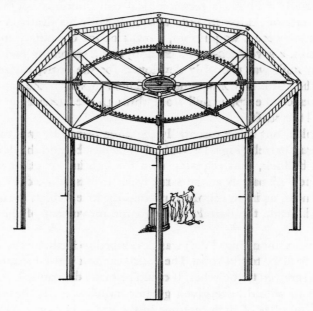

FIG. IV.1

revenue system, mechanical engineering, astronomy or any other field in which he was asked to take up the challenge.

There are three or four aspects which are interesting in the contributions of Shirazi. First, he seems to draw his knowledge from a wide range of traditions: Greek, Zoroastrian, Arab, Persian, and Indian. Secondly, being widely read and acquainted with different traditions, he seems to utilize the knowledge and insight gained for specific purposes of his own. For instance, he utilizes his knowledge of mathematics for the currency and revenue reforms in collaboration with Todar Mal at the court of Akbar, and knowledge of mathematics and of astronomy, in particular of the various calendars, for making a new calendar—*The Ilahi* Calendar.

But looking at his contributions one feature becomes apparent, that he limits his effort to utilizing the existing knowledge for making an improvement as in the case of the calendar or in applying it to a situation where it does not seem to have been applied before, as in the case of the gun-barrel cleaning machine, and in doing so he makes novel contributions. He does not seem to make any effort to collect new data either through observation or experimentation. This is clearly evident from his effort at reforming the calendar. In the reform of the calendar he does not carry out observations himself nor does he suggest the creation of observatories for making fresh

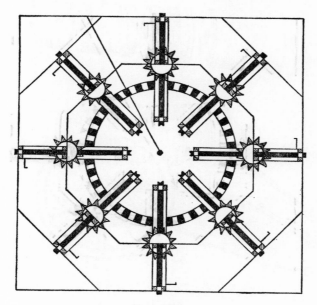

FIG. IV.2

observations. He uses the observations of others and uses his mathematical background and talents to evolve a more rational calendar which should be true to the observations made by the others and not subject to the same limitations as those of others. Perhaps the same method may have been followed in developing mechanical inventions.

It will be worth while to mention briefly some of the features of the period, as they emerge from the study of Shirazi.

Fathullah Shirazi's attainments suggest more than anything else that, given the necessary incentive by way of patronage, Indian genius could be directed to 'mechanical arts' and could contribute significantly to inventing mechanical gadgets even in the medieval environment. They also point out the existence and continuity of a tradition of craftsmanship and skill which could be directed and utilized in making complicated machines and constructions. Further, in view of the preferences of the times, the craftsmanship and skill also expressed itself forcefully in civil engineering works and architecture.

This brief and rather cursory description also poses a number of problems, and it would be worth while to mention them here briefly.

First, the sources of inspiration for the ideas developed by Shirazi; were

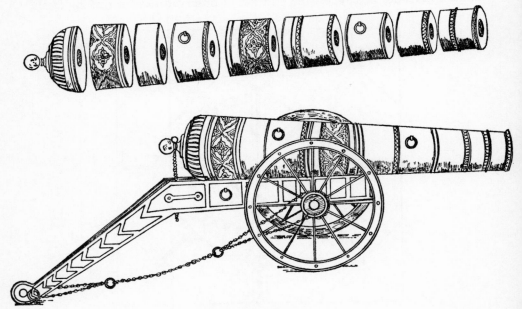

FIG. IV.3

they entirely novel ideas of his own, or had he borrowed ideas from other sources? If he had borrowed the ideas, what were the sources?

These inventions, few as they are, are remarkable for the range of ideas, detailed designing and skill in fabrication and are far ahead of contemporary developments anywhere else. Further, they suggest the existence of a well-established technological tradition, comprising detailed knowledge as well as the capacity to innovate. There does not appear to be any literature to suggest that the inventions had any predecessors, with the exception of the wagon mill. The wagon mill, according to Needham[8] was first developed in China

[8] *Science and Civilisation in China*, Vol. 1, 1954, p. 100; Vol. 4, Part II (Cambridge University Press, 1965), pp. 255-7.

in A.D. 340, and later in Europe around A.D. 1580. Would we be right in suggesting that the simultaneous appearance of the wagon-mill in Europe and India could possibly be due to a common source of inspiration? If that is so, what about the other inventions? In any case, an attempt at answering some of these questions would require a deep study of the nature of contacts between different countries and the nature and rate of transmission of ideas and techniques.

Secondly, the ideas may have been of Shirazi, but these could hardly be put to successful use had the knowledge of metallurgy not been developed, and if the necessary skills to apply it were not there. The existence of skill in foundry techniques, and the highly developed knowledge of metallurgy

FIG. IV.4

is evident from the iron beams of Konark temple, the iron pillar now at Delhi[9] and the casting of metal images. The question then necessarily arises, why this skill and knowledge was not applied in these or similar areas earlier? And what was the main motivation at the time of Akbar to make such contrivances? Was it scientific curiosity of an individual? Were they a result of the demands made by the patron and his needs? Were they a result of following up of an idea, like the extension of the principle of the Persian wheel to that of the machine for cleaning gun barrels and the wagon mill? It would be interesting to study the period more intensively to throw light on these problems and have an insight into the character of the scientific and technological activity of the period.

Thirdly, why was further development of these inventions not carried

[9] Recent developments in iron and steel making with special reference to Indian conditions, *Special Report 78* (Iron and Steel Institute, London, 1964), p. 28.

out? Or similar developments in other areas affected? The necessary skill and knowledge was there and a new set of ideas which were generated were successfully developed. If we are to believe Abul Fazal in his book, *Ain Akbari,* each of the inventions was manufactured and was a success. Then the question naturally arises why the ideas did not catch on? Does the answer lie in the lack of individuals, necessary institutions or the social and cultural climate?

Jahangir and the knowledge of biology

Considerable knowledge of biology was built around horticultural practices and animal husbandry. This would be particularly extensive in India in view of the range of climate, and varieties of plants and animals available in different regions. Further, in view of the interest of different kings, in different regions and periods, it may have resulted in pockets of specialization with regard to breeding of domestic animals, as well as those required for the purposes of war, and also in developing and cultivating special varieties of flowers, fruits and vegetables. Some idea of the knowledge can be had from the number of manuscripts on the subject. An inkling of the variety and depth of knowledge can be had from the personal diaries of Jahangir.

A cursory reading of his diaries would reveal his interest both in animals and plants. His description of the Indian crane, the gestation period of elephants, and physical features of birds and quadrupeds, are both detailed and meticulous and are a delight to read. They are supported by paintings, done at his instance, by the court painters. These paintings are accurate and of surpassing beauty and unmatched in the then contemporary world. He, it may be stated, had a huge establishment, containing exotic and wild animals as well as domesticated species. He observed the habits of animals directly as well as through gamekeepers. There is also an indication that he tried to cross-breed the animals.

Jahangir's observation on plants, in particular about fruits and flowers, are quite numerous, though not so extensive and detailed as on animals. His description is generally confined to mentioning any peculiarity he notices, colour and smell of flowers, size, weight and flavour of fruits. He gives a fairly good idea of the regional distribution of plants. He gives a list of plants of a region and mentions specifically the plants of one region which are not found in another. He also describes the fruits recently introduced to India. The pineapple was one such fruit brought by the Portuguese, and since he liked it he cultivated it in his *Gulafshan* garden at Agra, reaping as large a harvest as 1,000 fruits in one season.

His interest in picking or choosing an animal, plant, or natural pheno-menon for description, is guided by the novelty of the specimen or pheno-menon, a fact which he keeps emphasizing in different ways. Consequently, he mentions a number of tales commonly believed, or strange happenings, as and when they were reported to him. He, however, makes a clear dis-tinction, and states it unambiguously, as to whether he has been able to verify them or not.

We do not propose to discuss in detail the knowledge of animals and plants available from the diaries of Jahangir. We, however, do intend, by giving extracts from his writing, to indicate the degree of refinement of observation, detailed and specific description of observations made by him, the curiosity which impelled him to verify facts and the type of experiments he performed. This will, we hope, give an idea of the mental make up and calibre of a scientist of the period in India, if we assume that Jahangir shared the attributes with the scientists of the period.

Observation

One day on the hunting ground, the chief huntsman, Imam Wirdi brought before me a partridge that had a spur on one leg but none in the other. As the way to distinguish the female lies in the spur, by way of testing me, he asked whether it was a male or a female. 'A female', I promptly replied. Then they opened its belly. An immature egg appeared therein. Those present inquired incredulously by what sign I found that. I said that the end of the beak in the female is shorter than that of the male. This dexterity came from frequent observation and comparison.[10]

Describing in detail the habits of Indian cranes, he writes the following about the hatching of eggs:

Friday was the first day of the month of *Shahrivar* (August 22). From Sunday the third up to the night of Thursday (Mubarak Shambeh) it rained. It is strange that while on other days the pair of Sarus sat on eggs five or six times by turns, during these days and nights when there was constant rain and the air was somewhat cold, the male, in order to keep the eggs warm, sat without break from early in the morning until midday, and from that time until the next morning the female sat continuously, for fear that in rising and sitting again the cold air should affect them and the eggs became wet and be spoilt. . . . Stranger still that at first they kept their eggs close by each other underneath the breast, and after fourteen or fifteen days they made little gaps among them, lest their temperature should rise too high from contact and they get addled due to excessive heat.[11]

---

[10] M. A. Alvi and A. Rahman, *Jahangir, the Naturalist*, p. 55.
[11] Ibid., p. 71.

Detailed description

The western horned pheasant is described as follows:

> Among the animals observed in these mountains one is *Phul Paikar*. The Kashmirians call it *Sonlu*. It is one-eighth less than the pea hen. The back, tail and wings resemble those of the bustard and are blackish with white spots. The abdomen to the lower end of the breast is black with spots, some being red. The quills of the wings are fiery red, very lustrous and beautiful. From the end of the bill to the back of the neck it is brilliantly black. The two horns on the top of its head and its ears are of turquoise colour. The skin of the orbits and about the mouth (i.e. the skin of the cheek) is red. Below its throat there is skin (lappet) of a size enough to fill the palms of two hands. In the middle of this is a handful of skin, violet of colour with blue spots in the centre. Around it each streak is of turquoise colour, consisting of eight crucial (triangular) projections. Round these blue streaks it is red to two fingers, like peach flower. Again round its neck is a blue streak. The legs are red. Weighed alive the bird came to 152 *tolas* and 139 when killed and cleaned.[12]

The turkey was brought to India by the Portuguese, and Jahangir was quick to notice its features and habits, and describes them in detail:

> One of these animals (Muqarrab Khan brought from Goa) is one, larger than a pea-hen and smaller than a peacock. When it is in heat it spreads out its feathers like the peacock and dances about. Its beak and legs are like those of a cock. Its head, beak and parts under the throat (lappet) change to a different colour after every little while, quite red when in heat—one might say that it has adorned itself with red coral—now white like cotton, now of a turquoise colour. Like a chameleon it constantly changes its colour. On the head it has two pieces of flesh that look like the comb of a cock. A strange thing is this: when it is in heat, the aforesaid piece of flesh hangs down to the length of a span from the top of its head like an elephant's trunk. When again it raises it up it appears on its head like the horn of a rhino to the extent of two finger breadth. Round the eye it is always of turquoise colour, and does not change. Its feathers appear to be of different colours, differing from the colours of the peacock.[13]

Breeding of animals

Jahangir has an interest in breeding and he reports the efforts of his animal keepers as well as his own. He reports his own efforts, of pairing *markhor* goats with *Barbary* ewes, as follows:

> In Ahmedabad I had two male *markhor goats*. As I did not have a female in my establishment to pair with them, it occurred to me that if these could be paired with the Barbary ewes from Arabia, especially those from the port city of Darkhar (now Mirbat) young ones from their form and quality might be obtained. In short I paired them with seven Barbary ewes. After six months

---

[12] Ibid, pp. 60–1.
[13] Ibid, pp. 63–4.

each of the ewes had a young one at Fathpur, four females and three males, very pleasing in colour. Those that resemble the male (*bukeh markhor*) in colour have black linings on the base—like a dun—as well as red. They look more pleasing and beautiful than those having different colours and are more inclined to purity of breed.[14]

About the gestation period and the manner in which the baby elephant is born he has this to say:

A female elephant in the royal stud gave birth to a young one in my presence. I ordered them again to ascertain the period of their gestation. It turned out finally that a female young one is for one year and six months in the mother's womb and the male for nineteen months. As against the birth of human beings which, in most cases, is by head delivery, young elephants are usually born with their feet first.[15]

Verifying of facts

Jahangir's diaries are full of incidents and curious happenings reported to him by travellers and others. He records the unusual, mentioning clearly whether he has been able to verify them or otherwise. He, however, shows an urge to verify the reported facts, and wherever possible, he does so and records the result; for instance:

I have frequently heard from the hunters and those fond of the chase that at a certain time a worm develops in the horns of the wild ram which irritates him into fighting with his kind and that if he finds no rival, he sticks his head against a tree or a rock to allay the irritation. After investigation, the worm was found in the horns of the female sheep also, and since the female would not fight (on that account), the story does not appear to be based on truth.[16]

Another incident is rather interesting, since it involved the miraculous powers of a saint. Here also Jahangir's curiosity to know the facts, as against the common beliefs, and urge to verify them got the better of him, when he sent his personal scribe and surgeon to find out whether the body of Khwaja Yaqut, buried in a cave since the days of Mahmud Ghaznawi, had decayed or not, since the Saint was supposed to have miraculous powers and was said to be lying intact as he was buried. The report of the observation is faithfully recorded and the extent of decay is described.

Experiments

It would also be worth while to mention a few attempts at rudimentary experiments. Essentially the attempts were to confirm common beliefs, or to prove a point. The nature of the effort is, however, interesting.

[14] Ibid., p. 23.
[15] Ibid., p. 41.
[16] Ibid., p. 20.

In medieval Unani medicine bitumen (mumiya) was credited with considerable healing powers, particularly for healing broken bones. Jahangir reports the results of his efforts in the following words:

> I had heard much from the physicians, but when I tried it, no result was apparent. I do not know whether physicians have exaggerated its affects, or whether its efficacy had been lessened by its being stale. At any rate, I gave it to a fowl with a broken leg to drink in larger quantity than they said, and in the manner laid down by the physicians, and rubbed some on the place where it was broken, and kept it there for three days, though it was sufficient to keep it from morning till evening. But after I had examined it, no effect was produced, and the broken leg remained as it was.[17]

The climate of Ahmedabad was not agreeable to him and he proves its being bad in general in a rather ingenious way. According to him:

> Undoubtedly the water and air of Mahmudabad have no resemblance to those of Ahmedabad. By way of testing it, I ordered them to hang up a sheep on the bank of Kankariya tank after taking off its skin, and at the same time one at Mahmudabad, that the difference of air might be ascertained. It happened that after seven *gharis* (hours) had elapsed at that place (Ahmedabad) they hung up the sheep. When three *gharis* of the day remained, it became so changed and putrid that it was difficult to pass near it. They hung up the sheep at Mahmudabad in the morning, and it was altogether unchanged until the evening, and began to be putrid when one and a half watches of night had passed. Briefly, in the heighbourhood of Ahmedabad, it became putrid in eight hours and in Mahmudabad, in fourteen hours.[18]

Another example of his trying new things was the effort at using meteoric iron for the manufacture of a sword, dagger and a knife. His report is as follows:

> On the morning of Farwarden 30 (10th April 1621) a terrible noise was heard from the East in a certain village of the *parganah* (district) of Jullundur. . . . Amidst this noise a light fell from above on the ground. People thought that fire was raining down from heaven. A little while later when the noise (of the people) subsided and their hearts recovered from fright and bewilderment, they sent a fast runner to the Collector of that parganah, Mahmud Said, and informed him of the occurrence. Immediately, he rode and personally visited the spot. About ten or twelve yards of land in length and breadth was scorched up in such a fashion that no trace of grass or green was left. The effect of heat and burning could still be felt. He ordered them to dig soil, and the more they dug the greater the heat appeared to be, till they came upon a piece of heated iron. It was as hot as if it had been dug out from the spheres of fire. It took a

---

[17] *Tuzuk-e-Jehangiri, or Memoirs of Jehangir, from the 1st to 12th year of his reign*, translated by A. Rogers and edited by H. Bevridge (London, 1909), Vol. I, pp. 838–9.
[18] Ibid., Vol. II (London, 1914), p. 33.

long time to cool down. He brought it back with him to his residence, put it in a bag, sealed it and sent it to me. I ordered them to weigh in my presence. It came to 160 *tolas*. I ordered Ustad Daud to make of it a sword, a dagger and a knife and to show them to me. He represented that it did not stand the blow of the hammer and fell to pieces. I ordered him in that case to mix it with other iron and make use of it. Accordingly he mixed three parts of the meteoritic iron and one part of other iron, and having made two swords, one dagger and one knife brought them to me. The alloy had brought out its quality. Like the true swords of Yeman and the South, the sword could be bent and became straight again. I ordered them to try it in my presence. It cut very well.[19]

The preceding quotations from the personal diaries clearly reveal that Jahangir shared the qualities of acute observation, knowledge and wealth of detail, an irrepressible curiosity which urged him to confirm or reject the available knowledge, and some rudimentary desire to experiment, with the contemporary naturalists of the period, including those of Europe. His description of what he observes or hears is specific and meticulous, and he takes pains to indicate whether he has himself observed it or not, and whether he had been able to confirm what he had heard or not, like any meticulous scientist.

That with all the wealth of information he does not make any effort at systematization of the available knowledge, by way of classifying plants or animals, does not make any single observation the beginning of intensive studies in depth, or reveal any general philosophy of knowledge, could be attributed to his individual characteristics and his position. Nevertheless, even here he seems to share the attitude of the contemporary European naturalists, who show a similar level of understanding.

In Europe during this period the knowledge and method of acquiring knowledge was not dissimilar to those displayed by Jahangir. Information was being collected through voyages, undertaken by merchants and naturalists, classical knowledge was being revised, books on plants and animals were being written, covering information on known as well as new plants and animals of the far-off lands, herbaria, botanical gardens and museums were being organized by kings and nobles. However, the attempts at systematization of available knowledge through classification of plants and animals were rudimentary, and not yet scientific. There was no clear demarcation between facts and fancies, and both were fitted into the then existing framework of philosophy.

This being so we are then faced with the question: If the Indian mind was in no way inferior, in terms of acuteness of observation, curiosity, motivation to confirm the observation of others or commonly believed notions, experi-

[19] M. A. Alvi and A. Rahman, *Jahangir, the Naturalist*, pp. 139–40.

ment, and acquire knowledge or new plants or animals, then why could these characteristics not be further developed? Why were the new observations, new knowledge, and the wealth of available information, not utilized in generating more knowledge, and developing new methods of investigation and a philosophy, as happened in the West? Was it due to social factors or the general philosophy then prevalent? Or due to the character of the activity carried out by individual scientists which involved each effort beginning with an individual and ending with him, and which without an institutional base could not become cumulative?

Concluding Remarks

Having discussed the individual contributions of an engineer and a naturalist at the comparative level of development in India as compared to Europe and raised some questions engendered by the comparison, it would be worth while to turn to some broad conclusions.

The knowledge of science and technology in India and Europe of the late sixteenth and early seventeenth century compare favourably. The major differences may be not in the state of knowledge, but in the state of organization and activity. In Europe we begin to see the emergence of a different tradition, through the establishment of scientific societies, establishment of universities and academies, and printing, making possible the cumulative growth of knowledge and continuity of effort. It is these features which, with the passage of time, give vigour to the scientific tradition of Europe, while the absence of these in India continues to limit the scientific and technical activity to the medieval pattern.

In this context, the educational (and research) institutions developed for teaching and research, their continuity, the degree of their independence or subservience to other institutions of society, the number of scientists and the means available to them to interact with each other, within the same region as well as other regions, to disseminate their results and ideas and other facilities for their work, all assume as much significance as the examination of the end-product and require to be studied.

It would become clear from the preceding discussion that if the study were to be confined to the contribution of knowledge, as seen from the development of different branches of science, it may give a favourable picture of two culture-areas during a period, when a comparison is made purely in terms of knowledge, but this picture may not reveal the correct state of affairs in so far as in one culture-area it may represent an end-point while for the other it may be the take-off point. To understand the latter, science and technology requires to be studied as a sub-system of society in order to determine the former's capacity for further growth and development.

It would also be necessary not merely to limit the comparison to the 'achievements' of the ancients and the medieval men in terms of their comprehension of abstract problems and the technical solutions advanced to solve them, or the range of applications of solutions, but also to study the system of science and technology as a whole. Since it is the latter which would give an idea of the strength or otherwise of science and technology.

V

※

THE INNER ELIXIR (NEI TAN); CHINESE PHYSIOLOGICAL ALCHEMY*

Lu Gwei-Djen

The existence of two parallel traditions in Chinese alchemy has now been known or glimpsed in Europe for more than a century. In his pioneer paper of 1855 on Taoism Edkins was perhaps the first to mention it.

> The Taoists [he wrote] call the process of manipulating substances to obtain the elixir lien wai tan [1],[1] 'the obtaining by purification of the external elixir'. The corresponding process for rectifying the mind is denominated lien nei tan [2], 'to obtain by purification the inner elixir'. By the former the rank of earthly genii is attained, ti hsien [3]. But those who succeed in the latter become thien hsien [4] or celestial genii, and instead of enjoying their immortality in a grotto on some legendary mountain, they fly upward to Yü Ching [5], the abode of Yü Ti [6] [the Jade Emperor], or to Tzu Wei Kung [7], his lower residence.[2]

Edkins could not explain very clearly however what the second process consisted in. He knew indeed of a Taoist 'mode of self-training called lien-yang' [8] which had been founded by Ch'ih Sung Tzu [9] and Wei Po-Yang [10]—refining and nourishing—but conceived that it 'consisted of a hermit life and sitting cross-legged in a mountain cave' while repressing the passions. He realized however that 'making the breath return in a circle' had something

* The writer wishes to mark this occasion by placing on record her deep appreciation and enjoyment of the opportunities of collaboration in the study of the science, technology and medicine of her own civilization, during a period of over thirty years, with Dr. Joseph Needham, whose endeavours are so highly esteemed among the Chinese people. She also desires to express her most grateful thanks for many years of generous financial support from the Wellcome Trust of London.

[1] The romanization of Chinese terms and names in this paper follows the Wade-Giles system. Square-bracket references are to the table of Chinese characters on pp. 70–1. TT numbers refer to books in the Tao Tsang (Taoist Patrology) following Wieger's catalogue.

[2] It will be remembered that this is the astronomical name for the region of the circumpolar stars (cf. Needham, Vol. 3, pp. 259 ff. hereinafter referred to as SCC).

to do with it. He was aware, too, of another phrase, yang hsing [11], but associated it only with late ethical Taoism deeply influenced by Confucianism. The double pattern was also brought out clearly by Martin in his address to the American Oriental Society in 1868, often subsequently reprinted, and notable for the emphasis with which he supported Edkins' belief in a higher antiquity of alchemy in Chinese than in any other civilization. Although Martin could not give any more precise account of the nei tan tradition than Edkins, he provided a slightly different formulation of it:

> In the Chinese system [he wrote] there are two processes, the one inward and spiritual, the other outward and material. To obtain the greater elixir, involving the attainment of immortality, both must be combined; but the lesser elixir which answers to the philosopher's stone, or a magical control over the powers of nature, might be procured with less pains. Both processes were pursued in seclusion; commonly in the recesses of the mountains—the terms for adepts signifying 'mountain men' (hsien [12]).[3]

From what will appear in the following pages it is evident that Martin had been studying some nei tan texts, for he quotes (without precise reference) a sentence from Lü Tsu [13] (the late + 8th-century Patriarch Lü, Lü Yen [14], Lü Tung-Pin [15]): 'You must kindle the fire that springs from water, and evolve the Yin contained within the Yang.' Those who read this, such as Waite,[4] were duly baffled, and he wrote, quite understandably, 'We need to know much more than Dr Martin has told us about the spiritual processes in China which passed under the name of alchemy before we can take them into consideration on a quest of their (possible) correspondences with the groups of European texts.' Of the two Western-language books on Chinese alchemy in the early part of this century, Chikashige, as a plain blunt metallurgist, ignored the nei tan tradition altogether, but Johnson devoted a chapter to it which constituted a slight further advance in understanding. He translated nei tan as 'esoteric drug' and wai tan as 'exoteric drug', associating the former purely with the attainment of longevity and immortality, and the latter purely with the transmutation of metals. Nevertheless he knew that the nei tan procedures involved 'a comprehensive regimen of mental and physical discipline', gymnastic techniques, a regulated and selective diet including abstentions, e.g. from cereals, and also the consumption of unusual plant substances, and respiratory exercises including long holding of the breath. But he also thought that the 'esoteric drug' was a compound derived from minerals and metals.

But if the Western companion of metallurgical-chemical alchemy was psychological, its Chinese companion (nei tan) was essentially physiological.

[3] Vol. 1, p. 246.
[4] Pp. 57, 58, 61.

1 煉外丹	34 指歸集
2 煉內丹	35 心腎交會
3 地仙	36 精氣般運
4 天仙	37 存神閉氣
5 玉京	38 吐故納新
6 玉帝宮	39 或專房中之術精華
7 紫微宮	40 或採日月精華
8 煉養	41 或服餌草木妻
9 赤松子陽	42 或辟穀休妻
10 魏伯陽	43 修真秘訣
11 養性	44 類說
12 仙	45 木液
13 呂祖	46 金精
14 呂嵒	47 含和煉藏
15 呂洞賓	48 泥丸
16 金液	49 丹田宮
17 手太陰肺經	50 絳宮
18 精	51 百神
19 液	52 體殼歌
20 還丹	53 修真十書
21 返本還元	54 乾
22 返老還童	55 巽
23 修	56 還丹論
24 修補	57 鍾呂傳道集
25 復	58 鍾離權
26 逆流	59 呂洞賓
27 逆行	60 大還丹
28 相生	61 雲笈七籤
29 顛倒	62 元氣論
30 通幽訣	63 異類
31 氣能存生內丹也	64 形
32 藥能固形外丹也	65 無
33 吳悞	66 上清洞真品

FIG. V.a

67 神形
68 元一之氣
69 回風之道
70 關
71 邪
72 仙經
73 心
74 呼吸
75 導引
76 修福
77 修業
78 元氣之道
79 一陰一陽謂之道
80 三元二合謂之丹
81 蕭道存
82 修真太極混元圖經
83 靈寶真一經
84 黃庭
85 石泰篇
86 還原篇
87 真鉛
88 白虎脂
89 真汞
90 青龍髓
91 蕭廷芝
92 金丹大成
93 翕簫歌
94 鉛龍升兮汞虎降
95 鉛浮而銀沈也
96 還丹內象金鑰匙
97 參同契
98 悟真篇
99 張伯端

100 三黃道
101 四神道
102 黃道
103 赤水玄珠
104 孫一奎
105 本草綱目
106 李時珍
107 方外還丹
108 養生
109 天命
110 還丹秘要論
111 純全
112 一靈真性
113 交感精
114 呼吸氣
115 思慮神
116 天真
117 丹經
118 復
119 不可盡娈之天命
　　人定亦可以勝天也
120 全
121 化機
122 身心意
123 能奪天地造化
124 我命在我不在於天
125 養生延命錄
126 陶弘景
127 葛洪
128 抱朴子
129 離
130 坎

FIG. V.b

The Chinese adept of the 'inner elixir' did not seek psycho-analytic peace and integration directly, he believed that by doing things with one's own body a physiological medicine of longevity and even immortality (material immortality, for no other was conceivable) could be prepared within it. Thus there opens out before us the whole field of Taoist physiology, a proto-science not exactly the same as the physiology of the physicians down through the centuries, but not very far different from it. No greater mistake could be made than to analogize nei tan with the 'spiritual alchemy' of the West; it was physiological through and through, and though certainly not without parallelisms or even connections with Indian Yoga, it was generally more moderate, with more emphasis on hygiene, and always infused with characteristically Chinese empiricism and rationality.

One of the basic features of Chinese wai tan and nei tan alchemy was that a great many of their technical expressions were held and used in common. While it is possible, therefore, to categorize without hesitation certain particular texts as wai tan and others as nei tan, there are a good many where it is sometimes very difficult to be sure whether the writer is talking about laboratory operations or physiological techniques. Some texts indeed give the impression of having been designedly written ambiguously so that readers of either persuasion could take their choice. One has only to realize that chemical terms such as 'reaction-vessel', 'true mercury' or 'potable gold' were freely applied to physiological processes, as also to remember that the viscera and the metals were strictly associated together within the Five-Element symbolic correlation system[5] and the trigrams and hexagrams of the 'Book of Changes', to see that interpretation may not always be easy. And here it dawns upon the investigator that nothing short of a dual translation system will ever cope with the problems presented by nei tan alchemy in China. For reasons which will be given in detail elsewhere[6] my collaborator Dr. Joseph Needham and I have coined from Greek sources the word 'enchymoma' to designate the 'inner' or physiological elixir and distinguish it from the 'outer' or chemical ones prepared from metallic and mineral substances in the laboratory. Similarly, in wai tan contexts chin i [16], literally 'gold juice', has often been translated 'potable gold', but study reveals that in the nei tan context the two words must be Englished in an entirely different way, even involving the creation of a new, or the use of an unfamiliar word; so that here what we ought to say is 'metallous fluid', for it refers to the saliva, secreted at a point on the tract pertaining to the lungs,[7]

[5] See Table 12 in *SCC*, Vol. 2, p. 263.
[6] Cf. *SCC*, Vol. 5.
[7] The Cheirotelic pulmonic Thai-Yin tract (Shou Thai-Yin fei ching[17]). Cf. *SCC*, Vol. 6.

which belong to the element Metal. Thus we need special adjectives, in common use, other than those for the five elements, and we must be prepared to have 'aquescent', or some such coinage, to convey the idea of something under the sign of Water. As for the overlap of terms, it could almost be said that the nei tan experts took pleasure in punning usages which could put the uninstructed totally off the trail. Indeed they were bound not to transmit their knowledge except to disciples under an oath of secrecy.

Of course the nei tan texts can often be recognized because they give no clear instructions for manual chemical operations; it then becomes evident to the reader that they are using an abundance of chemical terms with purely physiological meanings. Here there is an interesting difference from Western writings. When a European alchemist speaks of 'true mercury' or 'our mercury' or 'philosophical mercury' we know that he is referring to some hypothetical entity or un-isolated constituent believed to exist invisibly behind the ordinary inorganic substances which he is handling in the laboratory. Just of this sort were the Tria Prima, the mercury, sulphur and salt, the 'three hypostatical principles' which Boyle combated in the 'Sceptical Chymist' as well as the four Aristotelian elements.[8] But when a Chinese writer speaks of 'true mercury' or 'true lead' he is sure to be speaking about the primary Yang and primary Yin essences within certain physiological secretions, organs or tissues. As we shall see in what follows, the most important of these were the ching [18] and the i [19], i.e. the semen and the saliva.

Take for example the basic idea of reversion, regeneration and return. For the proto-chemical alchemist the term huan tan [20] meant an elixir or part of an elixir prepared by cyclical transformation, such as may be brought about by repeated separation and sublimatory re-combination of mercury and sulphur, reducing cinnabar and re-forming mercuric sulphide. On the other hand, the phrase huan tan [20] was applied by the physiological alchemists to the Tao of regeneration of the primary vitalities. This meant techniques applied purposefully within the human body to bring about a reversion of the tissues (fan pên huan yuan [21]) from an ageing state to that with which the newborn child is endowed. These ideas go back far into Chinese antiquity; here one need only recall that pregnant phrase from the −4th century *Tao Tê Ching*: 'Returning to the state of infancy. . . .'[9] It was indeed one of the most ancient slogans of Taoism, and while the methods were more and more elaborated as the centuries went by, the fundamental idea probably changed but little, namely that there could be a reversion to youth, an attainment of longevity *because* of continued re-

[8] Cf. Leicester, pp. 97 ff., 110 ff.
[9] Cf. *SCC*, Vol. 2, p. 58, and Waley, p. 178.

juvenation—fan lao huan t'ung [22] in the proverbial phrase—worked for by means of hygienic and other physiological techniques.[10] There is no single key to physiological alchemy more important than the idea of retracing one's steps along the road of bodily decay. In order to signalize this in Western speech we have coined from the Greek the word 'anablastemic', convenient for qualifying the word enchymoma when that is used in Chinese in the special context of reversion (huan tan [20]). Huan and fan, regeneration and reversion, we have just met with, but there is also (and very prominently) hsiu [23], restoration, or hsiu pu [24], repair, as well as fu [25], replenishment, and several more. Further, this concept was connected with two others almost equal in importance, first a counter-current flow of some of the most important fluids of the body opposite to their normal directions, and secondly a thought-system which envisaged a frank reversal of the standard relationships of the five elements. The first idea, of flow in a direction opposite to the usual, is expressed by such terms as ni liu [26] or ni hsing [27], and was applicable, as we shall see, particularly to the products of the salivary and testicular glands. The second concerned the power which the physiological alchemists believed that their techniques could attain over the natural processes of mutual generation of the five elements (hsiang shêng [28]).[11] They dared to believe that by their efforts the normal course of events could be arrested and set moving backwards; this was called tien tao [29], 'turning nature upside down'. Thus 'to become as little children' was the nei tan ideal, and though one must not minimize the undertone of holy innocence which all true Taoists would have wished to recapture, the physiological alchemists of medieval China had, in our view, far more in common with those who attempt to halt the ageing of tissues and bodies today by biochemical means, endocrinological treatments and hygienic exercises than with those who think in terms of a purely psychological 'return to the womb'.

How did the physiological alchemists talk about the condition of vitality to which they wished to return? One just has to know the key, for the terms were ordinary words used as veils for a special meaning. In order to give a properly rounded idea of nei tan alchemy we can draw upon many interesting texts as examples, some strange, some surprising, some poetical and some of striking interest for the history of science. Let us begin with a short passage from a text which defines the nei tan and the wai tan. The *Tao Tsang* contains a short work of the Thang period entitled *T'ung Yu Chüeh* [30] (Lectures on the Understanding of the Obscurity of Nature),[12] which says:

---

[10] See again *SCC*, Vol. 2, p. 140.
[11] See *SCC*, Vol. 2, pp. 255 ff.
[12] TT/906, p. 18b. Ch'i may be compared with Gr. *pneuma*.

'The (primary) ch'i can preserve life, hence it is called the enchymoma (ch'i nêng ts'un shêng, nei tan yeh [31]). Chemical substances can strengthen the visible body, hence they are called the elixirs (yao nêng ku hsing, wai tan yeh [32])'. Some centuries later a Sung adept, Wu Wu [33], wrote another book of the same kind entitled *Chih Kuei Chih* [34] (Pointing the Way Home to Life Eternal; a Collection),[13] and also preserved in the *Tao Tsang*. The preface, written about +1165, says:

> The theory of the Nei Tan (enchymoma) is nothing more than the mutual conjunction of the heart (Metal) and the reins (Water), (hsin shen chiao hui [35]), the circulation of the ching (seminal essence) and the chhi (ching chhi p'an yün [36]), the preservation of the shen and the retention of the air (ts'un shen pi ch'i [37]), exhaling the old and breathing in the new (t'u ku na hsin [38]). Besides this, one may practise the special arts of the bedchamber (huo chuan fang chung chih shu [39]), or take the rays and emanations of the sun and moon (huo ts'ai jih yüeh ching hua [40]), or consume particular vegetable substances (huo fu erh ts'ao mu [41]), or again, it may be, abstain from cereal grains, or practise celibacy (huo pi ku hsiu ch'i [42]).

Here some of the additional alternatives are not strictly nei tan, but were included as facultative helps to the practice of it. The formulations in this passage occur over and over again in the nei tan texts, so that it becomes quite easy to recognize the procedures that the adepts are recommending. Another passage comes from a book called *Hsiu Chen Pi Chüeh* [43] (Esoteric Instruction on the Regeneration of the Primary Vitalities). The writer of this text is unknown, but it can be dated without fear before +1136, because parts of it were incorporated into the *Lei Shuo* [44] *florilegium* compiled in that year. It runs as follows[14]:

> The Inner and the Outer Macrobiogens.[15]
> Lao Chün (Lao Tzu) says that the changes in the atmospheric realm of the heavens and the earth are very difficult to fathom. There are two ch'i, one Yang, represented by the dragon and by the fluid of the element Wood (mu i [45]); the other Yin, represented by the tiger and by the essence of the element Metal (chin ching [46]).[16] When these two are brought into conjunction and made to react (with transformation)—then what results is called Elixir, the outer macrobiogen.

---

[13] TT/914.

[14] *Lei Shuo*, ch. 49, p. 5b, 6a (vol. 5, p. 3212). The same passage occurs in a slightly abridged form in the *Thi Kho Ko*[52] (Song of the Bodily Husk), collected in *Hsiu Chen Shih Shu*[53], TT/260, ch. 18, p. 7a.

[15] This is useful as a generic term for all elixirs and enchymomas.

[16] This refers to the association of elements and trigrams in the Wên Wang arrangement of the kua (cf. *SCC*, Vol. 4, pt. 1, p. 296). Metal corresponds in this to Chhien[54] and Wood to Sun[55]. For further understanding of the system of the trigrams (kua) of the *I Ching* (Book of Changes) see *SCC*, Vol. 2, Table 13.

But (the practices of) conserving and harmonising (the secretions), working alchemical transformations within the viscera (han ho lien tsang [47]), exhaling the old and breathing in the new, transmitting upwards to the brain (ni wan [48]), then showering downwards to the regions of vital heat (tan t'ien [49]), restoring and transforming in endless cycles, passing through the heart (chiang kung [50]) and there collecting the five ch'i (of the viscera) in order to nourish all the vitalities of the body (lit. the hundred archaei, pai shen [51]—this is called) Enchymoma, the inner macrobiogen.

For those who follow the Tao, the enchymoma can lengthen one's life, but the elixir can make one ascend to become an immortal. If the enchymoma succeeds, the elixir will necessarily be accomplished (lit. must respond to it), and this being so, the enchymoma will necessarily be strengthened. But neither of these two alone can succeed in effecting ascension.

In this passage the parallel character of the activities of elixir-making and enchymoma-making is well brought out, and there must have been a number of Chinese alchemists, perhaps a majority of them, throughout the Middle Ages, who pursued both objectives at the same time, believing them to be essential to each other.

It has now been possible to give an idea of some aspects of the complexities involved in the study of nei tan physiological alchemy, but the compass of an introductory paper will not permit more than a glimpse into the abundant literature that has come down to us. Passages are liable to appear contradictory, and terms confusing. For example, a discourse on the anablastemic enchymomas (Huan Tan Lun [56]) in the book *Chung Lü Chuan Tao Chi* [57] (Dialogue between Chungli Ch'üan [58] and Lü Tung-Pin [59] on the Transmission of the Tao), a work of the +8th or +9th century, lists more than a dozen names for various kinds of huan tan, all certainly physiological rather than chemical.[17] The differences between them depended largely on the starting-material used and the manner and duration of the carrying on of the practices. In the present account we are speaking mainly of the Ta Huan Tan [60] theory, which in general is very near the core of all anablastemic enchymoma thinking. Moreover there were undoubtedly several different ways of reading the same text; it could be interpreted as relating to the actual respiratory, gymnastic and sexual practices of the physiological alchemist; or again it could be understood purely metaphorically as describing the interactions of the Yang and Yin forces, their *conjunctio oppositorum*, or 'marriage of fire and water', within the organism of the physiological alchemist. The present paper has to confine itself mainly to what can be said about that reversion, regeneration and

---

[17] In *Hsiu Chen Chih Shu* (TT/260), ch. 16, pp. 2a ff. Some of the names are deceptively identical with wai tan elixir names.

return, in the fundamental process of rejuvenation, and hence perpetuation, which the adepts believed that they could implement.

It will be understood that all this was not without the backing of a great deal of classical philosophy, from which further developments, in cosmogony and microcosm-macrocosm doctrine for example, had proceeded. A treatise of this kind fundamental for the nei tan system occurs in the *Yün Chi Ch'i Ch'ien* [61] (collected in +1019); this is the *Yuan Ch'i Lun* [62] (Discourse on the Primary Vitality and the Cosmogonic Ch'i) by an unknown writer of the second half of the +8th century.[18] Maspero laid it under contribution for some interesting statements about the cosmic egg and the parallelism between the primary ch'i of man and the cosmogonic ch'i which formed the world,[19] so that here it will suffice to add a few further quotations. The style of the text is shown by the following passage:

> The primary (cosmogonic) ch'i (yuan ch'i [62]) had no appellation, but when change brought things to birth there arose names. The primary ch'i doubly embodied change and generation into the differentiated categories (i lei [63]) (of things). Of this double embodiment there was no sign, for the ch'i was unitary, and yet it may be considered the home of all original differences. When forms (hsing [64]) arose, then the myriad names were given, and their external characteristics were recognised; so that one can say that 'namelessness' was the Beginning of Heaven and Earth,[20] while 'naming' was the Mother of the Ten Thousand Things. He who is for ever without desires can penetrate with vision the Mystery, but he who harbours preconceived prejudices can see only superficialities. These are only the externals, but the Mystery lies within, and the Within is the foundation of all. The externals correspond to the beginnings; these can be called 'the Father' but the Mystery can be called 'the Mother'.[21] Such is the Tao.[22]

The writer presently describes how the natural endowment of yuan ch'i in every human being is dimmed and darkened by time and ageing.[23]

> The *Shang Ch'ing Tung Chen P'in* [66][24] says: 'Man at his birth embodies the primary ch'i of Heaven and Earth as his mind and body (shen hsing [67]), and he receives the ch'i of the primary unity (yuan i chih ch'i [68]) as his salivary and seminal essences (i ching [19], [18]). When the ch'i of Heaven wastes and

---

[18] *YCCC*, ch. 56.

[19] P. 207.

[20] Reminiscent of the 'nothingness' (wu[65]) of some Buddhist philosophical schools, which is full of all things in potentiality.

[21] Note the typical Taoist priority for the feminine.

[22] P. 3a.

[23] As we shall see in a moment (p. 82) this natural endowment was considered to be tripartite.

[24] There is nothing with exactly this title in the *Tao Tsang* now, though eight books have titles beginning with the first four characters.

decays the shen becomes dispersed. When the ch'i of Earth wanes and declines the hsing falls prey to diseases. When the primary ch'i ebbs and degenerates the life-span becomes exhausted. Thus the (wise) emperors used the technique of the returning wind (hui fêng chih tao [69]),[25] they opposed the natural directions of flow in the body; upwards they nourished the brain (ni-wan [48]), downwards they strengthened the primary ch'i. The brain being replete the shen was perfected, the shen being perfected the ch'i was at the full, the ch'i being at the full the hsing [64] was made an integral whole, and the hsing being made an integral whole the hundred gates (kuan [70]) were harmonised within and the eight malign influences (hsieh [71]) diminished outside. When the primary ch'i was fully present (in the body) then the marrow solidified to make the bones, and the intestines (supplied the means of) change for the muscles and nerves. Thus all was purified (and restored), the true ching, the primary shen and the primary ch'i were not lost from the mind and body. Therefore it was possible for (those wise emperors) to attain longevity (and immortality).[26]

Further on, the writer becomes more precise about some of the techniques involved.

The manuals of the immortals (hsien ching [72]) say: 'One's life-span depends upon oneself. If one can conserve the seminal essence (ching [18]) and obtain the ch'i, one may attain longevity without end.' And they also say: 'Maintain the form (hsing [64]) without (harmful) exertion, conserve the seminal essence without (harmful) agitation, restore the mind (hsin [73]) to ataraxy and peace. That is how longevity can be obtained.' The fundamental root of the life-force and life-span is set in this Tao. Although a man practises respiratory exercises (hu hsi [74]), gymnastic techniques (tao yin [75]), charitable acts (hsiu fu [76]) imitating or assisting works of public benefit (hsiu yeh [77]), and a thousand other techniques of experienced knowledge, and even though he manages to consume exalted medicines (elixirs), it will profit him nothing if he does not know the Tao of the primary unity (yuan ch'i chih tao [78]). He will be like a tree with fine branches and luxuriant foliage which yet has no proper roots, and so cannot endure. Is he not like a man who enjoys the pleasures of music and dancing-girls the whole night through, as well as all imaginable gastronomic joys? They will profit him nothing.[27]

This is evidently a criticism of those who practise many ancillary techniques while ignoring the principles of counter-current flow (tien tao), the enchymoma produced from secretions made to follow courses opposite to the normal. A little later we read more of this.

The manuals of the immortals say: 'One Yin and one Yang constitute the Tao (i Yin i Yang wei chih Tao [79]). The three primary (vitalities) and the union of the two components; that is the enchymoma (san yuan erh ho wei chih

---

[25] Another way of talking about huan tan.
[26] P. 8b.
[27] P. 11b.

tan [80])'. . . . And these manuals also say: 'The Tao of Yin and Yang is the prizing of the seminal essence and the saliva. If these are well and truly guarded, then longevity will be obtained'.[28]

Another illuminating passage is found in the book of Hsiao Tao-Tshun [81] entitled *Hsiu Chen T'ai Chi Hun Yuan T'u* [82] (Illustrated Treatise on the (Analogy of the) Regeneration of the Primary (Vitalities) (with the Cosmogony of) the Supreme Pole and Primitive Chaos). About + 1100 he wrote[29]—

> The *Ling-Pao Chen I Ching* [83] says: 'Heaven is like a covering basin. It is hard for the Yang to rise further, so it piles up and generates Yin. How can this be? It is because the Yang of the earth bears a true Yin hidden (within it); this is why (a Yin) can rise upwards. The earth is like a flat base of rock. It is hard for the Yin to descend into it, so it piles up and generates Yang. How can this be? It is because the Yin of the heavens hides and envelops a true Yang; this is why (a Yang) can come downwards. When Yin is at its maximum Yang is born, when Yang is at its maximum Yin is born—but Yin and Yang can be generated in a manner contrary to normal Nature; this is why the reversion of the Tao of heaven and earth can be brought about (i.e. the arrest and reversal of the ageing process). If a man understands the pattern-principle of the rise and descent of Yin and Yang, knowing that he can practise the Tao of reversion within (his own body's heaven and earth), then he can himself repair and recast (the ch'i). If the ch'i is recast, the primary ching can be formed; within the ching arises the ch'i, his own (primary) ch'i, and within the ch'i arises the shen, his own (primary) shen.'
>
> (Master) Liu explains this by saying that the heart corresponds to heaven and the reins to earth, the ch'i is like the Yang and the fluid (i [19]) is like the Yin. If the ch'i and the i do not come into conjunction there can be no union. When the ching (seminal essence) enters into the womb of a woman, then there occurs what is called the generation of a human being. But when the ching enters the Yellow Courts (huan t'ing [84], a region near the spleen)[30] of a man, there occurs what is called the generation of the (primary) shen. When this shen is collected, the (primary) ch'i brought together, and the embryonic ch'i released from its husk (the shell of the physical body), (a man can) ascend (to the heavens) as an immortal.

This is a good example of + 11th-century theorizing. It includes the basic ideas (a) that three primary vitalities have to be recreated within the body, (b) that a 'union' or reaction is necessary to form the enchymoma which does this, and (c) that the reactants are ch'i or juices in the body. It further shows (d) that in order to induce the geriatric reversal effect the secretions must be made to proceed in directions opposite to those in which they normally flow, and (e) that when they reach their extreme points both topographically

[28] Pp. 12b, 13a, 13b.
[29] TT/146, pp. 3b, 4a.
[30] Where the enchymoma is formed.

and quantitatively they undergo a change of sign, Yang into Yin and Yin into Yang.

In the light of what has now been said many statements which might otherwise seem incomprehensible or even contradictory become relatively easy to understand. For example, Shih T'ai [85] in the *Huan Yuan Phien* [86] about +1140, takes 'true, or vital, lead' (chen ch'ien [87]) to be a Yin thing (pai hu chih [88], the fat of the white tiger), and he calls 'true, or vital, mercury' (chen hung [89]) Yang (ch'ing lung sui [90], the marrow of the caerulean dragon).[31] This is obviously reasonable, fat and marrow representing in parable the Yin within the Yang and the Yang within the Yin. So also Hsiao T'ing-Chih in the *Chin Tan Ta Ch'êng* [92][32] says about +1250, in the 'Song of the Bellows and Tuyère' (T'o Yo Ko) [93] that this lead-dragon must go up and this mercury-tiger must come down (ch'ien lung shêng hsi, hung hu chiang [94]). And again: 'The lead rises and the (quick-)silver sinks' (ch'ien fou erh yin ch'en yeh [95]).[33] This refers to the counter-natural (tien tao [29]) process. An earlier statement can be found in the *Huan Tan Nei Hsiang Chin Yo Shih* [96], which goes back to +950, here already 'true' lead is associated with ching [18]. And as we shall note in the same place, various commentators on the *Ts'an Thung Chhi* [97] indicate that it is 'our lead' which has to be conveyed up the vertebral axis to the cephalic end of the body.

Once one has found the clue to the system of ideas of the physiological alchemists, everything falls into place and becomes understandable, even though various fluctuations and divergences remain (after all, the tradition was evolving through a millennium and a half). But there is nothing here concerning the minerals, metals and plants of the practical wai tan protochemists.

The same holds true of the *Wu Chen P'ien* [98], that remarkable poetical work written by Chang Po-Tuan [99] about +1075. This has often been taken as a discussion of laboratory alchemy. But we open the following quotations with a clarion call to abandon the chemical alchemy in favour of the physiological.

> st. 8   Desist from compounding and transmuting the Three Yellow Substances (san huang [100])[34] and the Four Wonderful Materials (ssu shen [101])![35] The common (medicines) of plant origin are even more different from the

---

[31] In TT/260, ch. 2, pp. 1b, 2a.
[32] In TT/260, ch. 9, p. 7a.
[33] Ibid., ch. 10, p. 7b.
[34] Sulphur, orpiment and realgar (according to Shang Yang Tzu, and TT/911, ch. 6, p. 13a).
[35] Cinnabar, mercury, lead and alum (according to Shang Yang Tzu). But TT/874 gives two lists, including variously malachite, magnetite, stalactitic calcium carbonate, and quartz, with orpiment and realgar.

true primary (vitalities). Yin and Yang, when of the same category, will respond to each other and come into conjunction. 'Two' and 'eight' (i.e. the Yin and the Yang meeting under appropriate conditions) will spontaneously unite in kinship and affection. Just when the Yin is strangely (and seemingly) destroyed, a red sun will appear at the bottom of the lake,[36] and the sprouts of the new medicine (the enchymoma) will appear like the white moon rising over the mountains. It is essential that people should recognise what is true lead and true mercury; they are nothing to do with common cinnabar and common mercury.

st. 18   First set up Ch'ien and K'un as the reaction-vessel and the apparatus, then heat together in it the crow (of the sun) and the rabbit (of the moon) as the chemical substances. When these two things are driven into the Yellow Way (huang tao [102]),[37] the metallous enchymoma will be formed and you need fear dissolution no more.

st. 22   To swallow saliva and inhale ch'i are well-known practices, but without the (right) reagents nothing truly vital can be brought into being. If the true seeds are not put into the reaction-vessel,[38] the operation will be as useless as having water and fire yet heating an empty kettle.

st. 32   Sun, in the Li [129] Kua, has femininity inside it; Moon (lit. the Toad Palace), in the K'an [130] Kua, has maleness within. Whoever does not understand the principle of inversion of the natural order (tien tao [29]) is like a man scanning (the broad heavens through a narrow) sighting-tube, and should cease to talk learnedly (about physiological alchemy).

The meaning of these passages was strikingly brought out in plain language by a medical writer of the late + 16th century. It was this passage which first enlightened us about the basic meaning of rebuilding, or reverting to, the primary vitalities. It occurs in a work called *Ch'ih Shui Hsüan Chu* [103] (The Mysterious Pearl discovered in the Red River), a system of medicine and iatro-chemistry by the eminent Ming physician Sun I-K'uei [104]. This was finished in + 1596, the same year that saw the publication of the greatest of all treatises on pharmaceutical natural history, the *Pên Ts'ao Kang Mu* [105] of Li Shih-Chen [106].[39] Towards the end of Sun I-K'uei's book, in chapter 10, he has an important section entitled Fang Wai Huan Tan [107] (Regenerative Enchymomas beyond all ordinary Prescriptions), and he tells us that he had been searching these out for the previous fifty years. The passage which we shall here give forms a prelude to a longish section on sexual

[36] Here of course the reference is to the bringing out of the Yang within the Yin (cf. p. 79).

[37] Not of course here the ecliptic, but the central region of the spleen.

[38] This refers to the Yang and Yin of the inner lines of the kua Khan and Li, *i.e.* the components of the primary unity (cf. p. 75).

[39] Cf. the biographical account by Lu Gwei-Djen.

practices and related iatro-chemical preparations. It is prefaced by a long
paragraph on the principles of redemptive hygiene (yang shêng [108]). Here
Sun contrasts the Buddhist acceptance of fate (t'ien ming [109]) and the idea
that chance and prayer alone determine whether death occurs a little earlier
or a little later, with the Taoist attitude, to which he himself inclines, that
people can do something actively and successfully about their life-span.
Only usually they do not start taking care soon enough. 'One cannot entirely
attribute events to fate' wrote Sun I-K'uei, 'on the contrary man can act in
such a way as to conquer Nature.'[40] Accordingly he counsels moderation in
everything, and gives detailed instructions on diet and regimen; only after
this are plant drugs any good at all, let alone the elixirs, even the most
precious. The passage is entitled Huan Tan Pi Yao Lun [110].[41]

A Discussion of the Mysterious Principle of the Anablastemic Enchymoma.
What can one say about the anablastemic enchymoma (huan tan [20])? It is
the Tao of reversion to the original state, the Tao of regeneration of the primary
vitality (fan pên huan yuan [21]).[42] All human life has an endowment coming
from the semen of the father and the blood of the mother.[43] The child at the time
of its birth possesses the primary ching, the primary ch'i and the primary shen
—all in a state of perfect purity (shun ch'üan [111]). But as it gradually grows up,
this numinous triune natural life-endowment (i ling chen hsing [112]) is attacked
and corrupted by the temptations of the four senses caused by colours, sounds,
perfumes and tastes, acting continually day by day and year by year. The
primary ching deteriorates into seminal essence of sexual intercourse (chiao
kan ching [113]); the primary ch'i changes into respiratory pneuma (hu hsi
ch'i [114]); and the primary shen is 'sicklied o'er by the pale cast of thought'
(ssu lü shen [115]). These three primary endowments being thus dribbled away,
it is exceedingly hard to regenerate the original innocence (t'ien chen [116]).[44]
    Therefore the teachers of old handed down their words in formulated doc-
trines, explaining in the various elixir and enchymoma manuals (tan ching [117])
the methods of repair (for this damage). Where the ching is deficient it must be
restored with (primary) ching, where the ch'i is deficient it must be restored
with (primary) ch'i, and where the shen is deficient it must be restored with
(primary) shen. This is applying the principle of 'reverting to the origin and
regenerating the primary vitality.'
    Such is the replenishment (fu [118]), but what really is replenishment? To
bring back the ching to perfection is like providing (a plant with) deep roots,
to bring back the ch'i to perfection is like giving it a firm stalk, and to bring
back the shen to perfection is like the bestowal of a marvellous harmony. To
be able to perfect (ch'üan [120]) (once again) these three (endowments), this is

---

[40] Pu kho chin wei chih t'ien ming; jen ting i k'o-i shêng t'ien yeh[119].
[41] Ch. 10, pp. 20b, 21a.
[42] Cf. *SCC*, Vol. 2, p. 76.
[43] An Aristotelian doctrine also; see Needham, *History of Embryology*.
[44] Cf. p. 74.

indeed (to use) the primary medicinal substances (i.e. the enchymoma) existing within the body itself. For example, many people have spoken of heaven and earth as 'furnace and reaction-vessel', of sun and moon as 'fire and water', of crow and rabbit as 'medicines and substances',[45] of Yin and Yang as 'the mechanisms of change' (hua chi [121]), of dragon and tiger as the 'mysterious application of techniques', of tzu and wu as 'the two solstices',[46] of mao and yu as 'the two equinoxes',[47]—all this is symbolism and parables, but in truth it does not go beyond the body, the heart and the mind (shen hsin i [122]). Of these three things the body is correlated with the ching, the heart with the ch'i and the mind with the shen.

Now what is this reversion? It is a renovation of these three things, contrary (ni hsing [27]) to the normal course (of ageing). What is regeneration? It is to bring about a replenishment (fu [118]) of the three primary endowments. To make these three vitalities perfect and primary (as they were at the beginning of life)—that is what is meant by the anablastemic enchymoma.

Thus what the physiological alchemists were talking about essentially was rejuvenation, and they believed that by their techniques they could 'make all things new'. However we may judge their physiological theories now, there is no reason for doubting that under appropriate conditions they could perform miracles of restoring physical and mental health.

Actually there was a great background to Sun I-K'uei's optimistic estimate of the power of man over Nature. That note had already been struck in the passage we gave from the + 8th-century *Yuan Ch'i Lun* (p. 78 above)—'one's life-span depends upon oneself'. As we go back through the centuries we see time after time how the Chinese physiological alchemists emphasized man's attainable mastery over his own fate. Take for example the *Hsiu Chen T'ai Chi Hun Yuan Thu* [82] already mentioned, composed by Hsiao Tao-Ts'un [81] about + 1100. The preface has a stirring phrase: the practice of nei tan, says Hsiao, 'can rob the power of the natural order of things (nêng to t'ien ti tsao hua [123])'.[48] This observation is not in accord with a conception of the Chinese mind which has been encouraged by many popular writers, namely that the Chinese always passively accepted Nature, reconciled himself with Nature, and adapted himself to Nature. It has often been said that it was not Nature over which the Chinese wished to acquire control, but rather himself; in calmness and resignation to fate. Yet this is contradicted by so many statements of the physiological alchemists, where we find repeated again and again: 'The length of one's life-span is not in the hands of Heaven, it is in one's own (wo ming tsai wo pu tsai yü t'ien [124]).' This is said in the *Yang*

[45] A reference to the legendary animals in the sun and moon respectively, hence to the Yang and the Yin, and the organs in the body corresponding to them.
[46] i.e. the two double-hours centring on midnight and midday.
[47] i.e. the two double-hours centring on 6 a.m. and 6 p.m.
[48] P. 2a.

Shêng Yen Ming Lu [125] by T'ao Hung-Ching [126] (ca. +490), quoting 'the manuals of the immortals'.[49] But the phrase is much older and must have been proverbial for centuries. Ko Hung [127] quotes it, ca. +300, in the *Pao P'u Tzu* [128] book.[50] Thus Chinese physiological alchemy, in spite of its medieval character and its use of ideas and practices so unfamiliar in the West today, was in a real sense akin to the optimistic and experimental outlook of modern science, especially biochemistry, endocrinology and geriatrics.

[49] In *Yün Chi Ch'i Ch'ien*.
[50] Ch. 16, p. 5b, from a Taoist book called *Kuei Chia Wên* (Divination Tortoise-shell Writings), which might or might not be the same as the *Kuei Wên Ching* cited in his bibliography, ch. 19, p. 3b. It is (or they are) otherwise unknown.

Bibliographical References

Chikashige, Masumi, *Alchemy and other Chemical Achievements of the Ancient Orient; the Civilisation of Japan and China in Early Times as seen from the Chemical and Metallurgical Point of View* (Rokakuho Uchida, Tokyo, 1936).

Edkins, J., 'Phases in the Development of Taoism', *Journ. Roy. Asiatic Soc. (North China Branch)*, 1855 (1st ser.), **5**, p. 83.

Johnson, Obed S., *A Study of Chinese Alchemy* (Commercial Press, Shanghai, 1928).

Leicester, H. M., *The Historical Background of Chemistry* (Wiley, New York, 1965).

Lu Gwei-Djen, *China's Greatest Naturalist; a Brief Biography of Li Shih-Chen*, *Physis*, 1966, **8**, p. 383.

Martin, W. A. P., *Hanlin Papers*, 2 vols.: Vol. 1 (Trübner, London, 1880; Harper, New York, 1880); Vol. 2 (Kelly and Walsh, Shanghai, 1894).

Martin, W. A. P., 'Alchemy in China', a paper read before the Amer. Or. Soc. 1868, Abstract in *Journ. Amer. Orient. Soc.*, 1871, **9**, xlvi. *China Review*, 1879, **7**, p. 242.

Maspero, H., 'Procédés de "nourrir le principe vital" dans la Religion Taoiste Ancienne', *Journ. Asiatique*, 1937, **229**, pp. 177 and 353.

Needham, Joseph (with the collaboration of Wang Ling, Tshao Thien-Chhin, K. Robinson, Lu Gwei-Djen, Ho Ping-Yü *et al.*), *Science and Civilisation in China*, 7 vols. in 11 parts (Cambridge University Press, 1954——. Six parts published, the eighth and ninth now passing through the press).

Needham, Joseph, *A History of Embryology* (Cambridge University Press, 1934) 2nd edition, revised with the assistance of A. Hughes (Cambridge, 1959; Abelard-Schuman, New York, 1959).

Waite, A. E., *The Secret Tradition in Alchemy; its Development and Records* (Kegan Paul, Trench & Trübner, London, 1926; Knopf, New York, 1926).

Waley, A., *The Way and its Power; a study of the 'Tao Tê Ching' and its Place in Chinese Thought* (Allen & Unwin, London, 1934).

VI

THE MEDICO-CHEMICAL WORLD OF
THE PARACELSIANS*

Allen G. Debus

At the death of Paracelsus in 1541 there was little to indicate that his work would become the focal point for debate among scholars for more than a century. True, he had been a controversial figure during his lifetime, but relatively few of his voluminous writings had been published while he had been alive. The flood of Paracelsian texts began to appear from the presses only later. In 1553 the important *Labyrinthus medicorum errantium* appeared, and in the following years a host of other manuscripts were sought out and published—often with extensive notes and commentaries. Before the end of the century the definitive ten volume edition of the *Opera* had appeared and this was to be reprinted twice in German and once in Latin translation by 1658.[1]

Along with the demand for books there was an ever-increasing need for physicians who practised the new medicine. In 1585 an English Parcelsian spoke of the large number of philosophers and physicians—including many who were formerly Galenists—who were currently following the doctrines

* The present study was completed with the aid of a research grant from the National Institutes of Health (LM 00046). In an earlier form the paper was read at a joint meeting of the History of Science Society and the American Historical Association in Washington, D.C. on 28 December 1969.

[1] The major editions are described by Walter Pagel in his *Paracelsus. An Introduction to Philosophical Medicine in the Era of the Renaissance* (Basle, S. Karger, 1958), pp. 31–5. Johannes Huser's edition of the *Bücher und Schrifften* of Paracelsus was printed at Basle by Conrad Waldkirch in ten quarto volumes in 1589–90. Lazarus Zetzner reprinted these works in three folio volumes at Strasbourg in 1603–05, and again in 1616–18. The Latin *Opera Omnia* was the result of the work of F. Bitiskius and was printed by De Tournes at Geneva in three volumes in 1658. In 1678 Richard Russell announced that he had already completed the translation of two of the three volumes of the works of Paracelsus. These were never printed. *The Works of Geber*, translated by Richard Russell (London, William Cooper, 1686) sig. A2 verso. From 'The Translator to the Reader' dated 3 May 1678.

and methods of the new 'chemical medicine'.[2] He listed these authorities by name in a fashion similar to that of his contemporary, George Baker, who informed his readers where they might find reputable apothecaries in London who would prepare the new chemical remedies.[3] On the continent other lists were compiled by George Bernard Penotus,[4] Oswald Crollius[5] and Olaus Borrichius.[6] For these authors it was paramount to point out that there was an alternative to the Galenism of the schools, that the texts and commentaries for this new system of medicine were available in print, and that there were physicians who practised the new medicine. For Crollius the future seemed so promising (1609) that he predicted that the Paracelsian dream of overturning the ancient doctrines of the schools would soon be a reality.[7]

The Paracelsian Universe

What did these men propose? Actually there is much in their work that is reminiscent of other Renaissance nature philosophers. Above all, they sought to overturn the traditional—and what they felt was the dominant—Aristotelianism of the Universities. For them Aristotle was a heathen author whose philosophy and system of nature was inconsistent with Christianity.[8] Like other nature philosophers they often seemed oblivious to the fact that many of their most favoured concepts were fundamentally Aristotelian. In spite of this they stated that his influence on medicine had been catastrophic since Galen had uncritically accepted his work and the Aristotelian-Galenic system had subsequently become the basis of medical training throughout

[2] R. B. Esquire (R. Bostocke), *The difference betwene the auncient Phisicke, first taught by the godly forefathers, consisting in vnitie peace and concord: and the latter Phisicke proceeding from Idolaters, Ethnickes, and Heathen: as Gallen, and such other consisting in dualitie, discorde, and contrarietie* . . . (London, Robert Walley, 1585), sigs. Ii verso–Iiii recto.

[3] See the 'Apologeticall Preface' to the *One Hundred and Fourteen Experiments and Cures of the Famous Physitian Theophrastus Paracelsus* in Leonard Phioravant, *Three Exact Pieces* (London, G. Dawson, 1652), sig. Ccc 4 recto–Ddd 4 recto.

[4] Conrad Gesner, *The New Jewell of Health* (London, 1576), sig. iv. From the prefatory 'George Baker to the Reader'.

[5] Oswald Crollius, *Discovering the Great and Deep Mysteries of Nature*, in *Philosophy Reformed and Improved*, trans. H. Pinnell (London, 1657), pp. 142–7.

[6] Olaus Borrichius, *De Ortu & Progressu Chemiae Dissertatio* in J. J. Manget, *Bibliotheca Chemica Curiosa* (2 vols., Geneva, G. De Tournes, Cramer, Perachon, Ritter and S. De Tournes, 1702), I, 1–37. Here sixteenth- and seventeenth-century authors are discussed on pp. 35 and 36.

[7] Crollius, op. cit., pp. 142–7.

[8] This theme is a common one in the literature of the period, but one might turn especially to Bostocke and Robert Fludd as examples of authors who speak at length against the 'heathnish philosophy' of the ancients.

Europe.[9] For them the Universities were hopelessly moribund and un-yielding in their adherence to antiquity. The Paracelsians hoped to replace all of this with a philosophy strongly influenced by the Christian Neo-Platonic and Hermetic texts. This was to be a philosophy which was to account for all natural phenomena. They argued that the true physician might find truth in the two-book concept of nature. He should turn first to the book of divine revelation—the Holy Scriptures—and then to the book of divine Creation—Nature.[10] Thus, the Paracelsians applied themselves on the one hand to a form of Biblical exegesis, and on the other to the call for a new philosophy of nature based on fresh observations and experiments. An excellent example of the latter approach may be found in the work of the important early systematizer of the Paracelsian corpus, Peter Severinus, who told his readers that they must sell their possessions, burn their books, and begin to travel so that they might collect observations on plants, animals and minerals. After their *Wanderjahren* they must 'purchase coal, build furnaces, watch and operate with the fire without wearying. In this way and no other, you will arrive at a knowledge of things and their properties'.[11]

One senses a strong reliance on observation and experiment in the work of these men even though their concept of what an experiment is and of its purpose might well vary from our own. At the same time one notes an underlying distrust of the use of mathematics on the study of nature. They might well—as Platonists—speak of the divine mathematical harmonies of the universe. Paracelsus, in addition, spoke firmly of true mathematics as the true natural magic.[12] But it was more customary, perhaps, for the Paracelsians to react with distaste against the logical, 'geometrical', method of argument employed by the Aristotelians and Galenists.[13] They condemned this 'mathematical method' along with the traditional scholastic emphasis

[9] Bostocke writes that 'The heathnish Phisicke of Galen doth depende uppon that heathnish Philosophie of Aristotle (for there the Philosopher endeth, there beginneth the Phisition), therefore is that Phisicke as false and injurious to thine honor and glory, as is the Philosophie'. Bostocke, op. cit., sig. Av verso, from 'The Authors obtestation to almightie God'.

[10] This approach is made evident in the *Labyrinthus medicorum errantium* of Paracelsus. See the *Opera Omnia* (3 vols., Geneva, J. A. and S. De Tournes, 1658), **1**, pp. 264–88 (275).

[11] Petrus Severinus, *Idea Medicinae Philosophicae* (3rd edition, Hagae Comitis: Adrian Clacq, 1660), p. 39. Although it would be a mistake to equate the Paracelsian 'experientia' with the modern concept of experiment, it may be noted that the terms 'experience', 'experiment', and 'experimental' were in common use in English Paracelsian texts in the sixteenth and the seventeenth centuries.

[12] From the *Astronomia Magna*. See the discussion in Debus, 'Mathematics and Nature in the Chemical Texts of the Renaissance', *Ambix*, 15 (1968), 1–28 (13–14).

[13] As an example see the discussion of the reaction of Severinus and van Helmont to mathematics in ibid., pp. 21–5.

on geometry and the study of local motion.[14] It was far better, they argued, for the scholar to recall that God had created 'all things in number, weight and measure'. Surely this was a more valid guide for the physician, the chemist, and the pharmacist—men who weighed and measured regularly in the course of their work.

If the Paracelsians rejected the 'logico-mathematical' method of the schools, they turned to chemistry with the conviction that this science might well be considered the basis for a new understanding of nature. This was an observational science—and in addition its scope was to be universal. These claims were to be found in the traditional chemical texts. For Paracelsus alchemy had offered an 'adequate explanation of all the four elements',[15] and this meant literally that alchemy and chemistry might be used as a key to the cosmos either through direct experiment or through analogy. In the *Philosophia ad Athenienses* (published first in 1564) the Creation itself was pictured as a chemical unfolding of nature.[16] The later Paracelsians agreed and amplified this theme. Gerard Dorn gave a detailed description of the first two chapters of *Genesis* in terms of the new chemical physics[17] while Thomas Tymme argued that the Creation had been nothing else but an 'Halchymicall Extraction, Separation, Sublimation, and Conjunction'.[18]

The chemical interpretation of *Genesis* helped to focus special attention on the problem of the elements as the required first fruit of the Creation. Although the Paracelsian *tria prima* (salt, sulphur and mercury) were a modification of both the earlier sulphur-mercury theory of the metals and other elemental triads, they have a special significance in the rise of modern science. The Aristotelian elements (earth, water, air and fire) served as the very basis of the accepted cosmological system. If they were used by the alchemists as a means of explaining the composition of matter, they were used by the physicians (through the humours) as a system for the interpreting

[14] J. B. van Helmont, *Oriatrike or Physick Refined*, trans. John Chandler (London, Lodowick Lloyd, 1662), pp. 176-7. This theme is further developed by the present author in his 'Motion in the Chemical Texts of the Renaissance', *Isis*.

[15] From the *Paragranum*; Paracelsus, *Sämtliche Werke*, edited by Karl Sudhoff and Wilhelm Matthiessen (15 vols., Munich and Berlin, 1922-3), **8**, pp. 55 f.

[16] The text of the *Philosophia ad Athenienses* will be found in Huser's *Bucher und Schrifften* (1616), **2**, 1-19; in the Latin *Opera Omnia* (1658), pp. 239-52; and in English translation in A. E. Waite's *The Hermetic and Alchemical Writings of Paracelsus* (2 vols., London, James Elliott and Co., 1894), pp. 249-81 and H. Pinnell's *Philosophy Reformed and Improved in Four Profound Tractates* (London, M.S. for Lodowick Lloyd, 1657).

[17] Gerard Dorn, 'Physica Genesi' from the *Liber de Natura luce Physica, ex Genesi desumpta* printed in the *Theatrum Chemicum* (6 vols., Strassburg, L. Zetzner, 1659-61), **1**, pp. 331-61.

[18] Thomas Tymme, 'Dedication to Sir Charles Blunt' in Joseph Duchesne, *The Practise of Chymicall, and Hermeticall Physicke*, trans. Thomas Tymme, Minister (London, Thomas Creed, 1605), sig. A3-A4.

of disease—and by the physicists as the basis for the proper understanding of natural motion. The introduction of a new elemental system thus ran the risk of calling into question the whole framework of ancient medicine and natural philosophy.

Although the new principles can properly be interpreted as part of an attack on scholastic philosophy, it is clear also that they led to considerable confusion. Paracelsus had not clearly defined these principles, and, indeed, they were of little value in the development of modern analytical chemistry since they were described as being qualitatively different in different material.[19] Nor had Paracelsus offered the principles specifically as a replacement for the elements. Rather, he had used both systems—and often in a seemingly contradictory fashion. By the fourth quarter of the century we find element theory in a state of flux with chemists choosing from observational evidence and Paracelsian texts as they saw fit.[20] Nevertheless, from the texts of this period we can see that the chemical physicians were turning in increasing numbers to the three principles as a means of explanation. Some were attracted by the trinitarian analogy of body, soul and spirit, while others turned to them in search for an alternative to the humours. For chemical theorists they represented philosophical substances which might never be isolated in reality,[21] while for the practical pharmacist they were nothing else but his distillation products.[22] It was not uncommon for a herb to yield a watery phlegm, an inflammable oil and a solid, and it was felt that these could be properly equated with mercury, sulphur and salt.

The concept of a chemical universe went beyond the chemical interpretation of the Creation and the problems of element theory. Those authors interested in meteorology explained thunder and lightning as a combination of an aerial sulphur and nitre which was analogous to the explosion of sulphur and saltpetre in gunpowder.[23] Similarly, Paracelsian authors were the first to offer a hypothesis meaningful for the development of agricultural chemistry. Seeking a cause for the beneficial effects of manuring in farming, they correctly postulated that the manure offered essential soluble salts to the soil.[24]

---

[19] From the *De mineralibus*; Paracelsus, *Sämtliche Werke*, **3**, 42 f.

[20] Allen G. Debus, *The English Paracelsians* (London, Oldbourne, 1965), pp. 26–9 and passim; note especially the indecision of Thomas Moffett (1584) described ibid., p. 73.

[21] An example of this may be seen in the discussion in Duchesne's *Practise*, sig. H3 verso.

[22] An example of this approach may be found in Penotus' 'Apologeticall Preface' in Phioravant's *Three Exact Pieces*, sig. Dd 1 recto.

[23] Discussed by Paracelsus in the *Von den Natürlichen Dingen* (in the *Opera Bücher und Schrifften* (1616), **1**, 1037).

[24] See Debus, 'Palissy, Plat and English Agricultural Chemistry in the 16th and 17th Centuries', *Archives Internationales d'Histoire des Sciences*, **21**, (1968), pp. 67–88.

Indeed, for the Paracelsians, the earth was seen as a vast chemical laboratory and this explained the origin of volcanoes, hot water springs, mountain streams and the growth of metals. The old concept of an internal fire was given as the explanation of volcanoes which were understood as the irruptions of molten matter through surface cracks.[25] But mountain streams were explained in an analogous fashion. Here they argued that subterranean water reservoirs were distilled by the heat of the central fire. As this vapour reached the surface, mountains acted as chemical alembics, and the results were the 'distilled' mountain streams.[2] Yet, some rejected the possibility of such a fire, arguing that the air requisite for such a conflagration did not exist within the earth. Henricus de Rochas suggested that the heat of mineral water springs derives from the reaction of sulphur and a nitrous salt in the earth,[27] while the English physician, Edward Jorden, offered a more comprehensive chemical alternative. A thorough vitalist, Jorden accepted the commonly held notion of the growth of metals which he tried to account for in a new way. In his investigation of the generation of the metals, he turned to the alchemical process of 'fermentation' which he explained as a heat-producing process which required no air and thus might easily give the internal heat requisite for this inorganic growth. With this new source of heat he was able to account also for volcanoes and mountain streams without dealing with the troublesome problem of the central fire.[28]

The Microcosm and Medical Theory

The Paracelsian chemical philosophy was considered to be a new observational approach to all nature, but from the beginning it carried a special appeal for physicians. Paracelsus had insisted that God rather than the constellations had created him a physician[29] while van Helmont repeated this and went on to add that because of its divine origin medicine stood above the other sciences.[30] Here they both reflected the priest-physician

[25] On seventeenth century geology and the central fire see Robert Lenoble, *La Géologie au Milieu du XVIIᵉ Siècle*. Les Conférences du Palais de la Découverte—Serie D, No. 27 (Paris, University of Paris, 1954).

[26] As an example see Michael Sendivogius, *A New Light of Alchymy* (London, A. Clark for Tho. Williams, 1674), pp. 94–5.

[27] As described by John French in his *Art of Distillation* (1650) (4th edition, London, E. Cotes for T. Williams, 1667), p. 177.

[28] Jorden's work is discussed in my paper 'Edward Jorden and the Fermentation of the Metals: An Iatrochemical Study of Terrestrial Phenomena', in *Toward a History of Geology*, edited by Cecil J. Schneer (Cambridge, Mass., M.I.T. Press, 1969), pp. 100–21.

[29] From the *Paragranum: Sämtliche Werke*, **8**, pp. 63–5.

[30] Helmont, *Oriatrike*, p. 4.

concept which was a fundamental part of Renaissance neo-Platonism, and it is likely that their ultimate source may be found in *Ecclesiasticus* 38:1, 'Honour the physician for the need thou hast of him: for the most high hath created him'. Indeed, for Paracelsus the rôle of the physician might be properly compared with that of the true natural magician. The *magus* transfers the powers of a celestial field into a small stone, and, in a similar fashion, the physician extracts the hidden virtues of herbs and prepares powerful remedies which he then uses to cure the sick and infirm, Thus, both the *magus* and the physician operate by means of God's Creation, nature.[31]

Fundamental for the orthodox Paracelsian was a belief in the macrocosm-microcosm analogy. Man is a small replica of the great world about him while within him are represented all parts of the universe.[32] At all times it was considered a fruitful field of inquiry to seek out correspondences between the greater and lesser worlds,[33] and the theory of sympathy and antipathy was employed to explain universal interaction. In contrast to Aristotelians who insisted on action through contact, the Paracelsians found no difficulty in explaining natural phenomena in terms of action at a distance.[34] Here, too, there was an important break with Aristotelian scholasticism, and it is easy to understand why Paracelsian-Hermeticists should have been among the first to defend the experimental research of William Gilbert on the magnet.[35] And applied to medicine, the involved seventeenth-century controversy over the weapon-salve may be seen as having overtones involving the broader implications of the possible validity of action at a distance.

For the Paracelsian the humoural theory of Galenic medicine was no longer considered to be adequate. The traditional explanation of disease as an

[31] See Walter Pagel's description of 'Magia Naturalis' in his *Paraclesus. An Introduction to Philosophical Medicine in the Era of the Renaissance* (Basel, S. Karger, 1958), pp. 62–5. Pagel's discussion is based largely on the *Philosophia Sagax*. See also D. P. Walker, *Spiritual and Demonic Magic from Ficino to Campanella* (London, Warburg Institute, 1958).

[32] Paracelsus' doctrine of the microcosm is discussed in the *Philosophia Sagax* [*Opera Omnia* (1658), **2**, 601] and is described by Pagel in Paracelsus, pp. 65–8. In his attack on Paracelsus Thomas Erastus placed special emphasis on the doctrine of the microcosm (1572), ibid., pp. 323–4.

[33] In practice this meant most commonly the doctrine of signatures. There were many works devoted to this subject. Two Paracelsian works of this genre might be offered as examples: Joseph Duchesne (Quercetanus), *De Signaturis Rerum Internis seu Specificis* in *Liber De Priscorum Philosophorum verae medicinae materia, praeparationibus modo, atque in curandis morbis, praestantia* (n. 1, Thomas Schürer and Barthol, Voigt, 1613), pp. 104–53; Oswaldus Crollius, *Tractatus, De Signaturis Internis Rerum, seu de Vera et Viva Anatomia majoris & minoris mundi* (with separate pagination) in *Basilica Chymica* (Frankfurt, Godefrid Tampach, n.d.).

[34] Robert Fludd turns specifically to the magnet as an example of sympathy and antipathy in his *Mosaicall Philosophy* (London, Humphrey Moseley, 1659), p. 200.

[35] See Debus, 'Robert Fludd and the Use of Gilbert's *De Magnete* in the weapon-salve Controversy', *Journal of the History of Medicine and Allied Sciences*, **19** (1964), pp. 389–417.

internal imbalance of the humours was rejected by Paracelsus, who preferred to emphasize local malfunctions within the body which were dependent on the three principles.[36] Some Paracelsians, most notably Joseph Duchesne, proposed a reformed system of three humours which were to be closely connected with the three principles.[37] Others discussed traditional urinalysis and argued that if the analysis of urine had any value at all, it should not be treated in humoural terms, but rather from the standpoint of chemical distillation.

The Paracelsians sought the cause of disease less in internal imbalances of fluids than in external factors which entered the body and became localized in specific organs.[39] Here an analogy could be drawn between macrocosm and the microcosm. In the same fashion that metallic 'seeds' in the earth resulted in the growth of the veins of metals, so, too, 'seeds' of disease grew within the body while they combated the local life force—or *archeus*—of a specific organ. This *archeus* acted much as an alchemist separating the pure from the impure.[40]

The relationship of the macrocosm to man had further chemical implications. Duchesne reflected both the Aristotelian position on catarrh and the persistent search for chemical analogies among the Paracelsians when he spoke of respiratory diseases in terms of the same distillation analogy utilized by other iatrochemists when explaining the origin of mountain streams.[41] And special significance was attached to the air which was recognized as essential for the maintenance of both fire and life. If, on the one hand, an aerial sulphur and nitre might combine to cause thunder and lightning in the sky or hot springs in the earth, on the other hand, they might react within the body when inhaled to generate diseases characterized by hot and burning qualities.[42] By the early years of the seventeenth century the aerial nitre had become associated with a life force requisite for man.[43] Indeed, this life force was on occasion identified with the *spiritus mundi*.[44] It was postulated that after having been separated from gross air, this

[36] Pagel, *Paracelsus*, pp. 126–33, 189–96.

[37] Duchesne, *Practise*, sig. L1 recto.

[38] Pagel, *Paracelsus*, pp. 189–96; Debus, *English Paracelsians*, pp. 157–8.

[39] Pagel, *Paracelsus*, pp. 130–1 (*Paramirum*); pp. 140–1 (*Labyrinthus mediscorum* and elsewhere).

[40] Pagel, *Paracelsus*, pp. 105–11.

[41] Joseph Duchesne, *Traicté de la Matiere, Preparation et excellente vertu de la Medecine balsamique des Anciens Philosophes* (Paris, C. Morel, 1626), p. 183. First Latin edition, 1603.

[42] Paracelsus, *Bertheonae* in the *Chirurgische Bücher* (Vol. 3 of the *Opera*) (Strasbourg, Zetzner, 1618), p. 354. Some fistulas and ulcers are due exclusively to Sal nitrum. See the *Grossen Wundarznei* in ibid., p. 85.

[43] See Debus, 'The Paracelsian Aerial Niter', *Isis*, 55 (1964), pp. 43–61.

[44] See especially the work of Nicasius Le Febure, Johann Rudolph Glauber and William Simpson described in ibid., pp. 58–9.

substance was formed into arterial blood.[45] Maintaining this concept—or modifications of it—it is little wonder that we find seventeenth-century Paracelsians and Helmontians rejecting the common practice of blood-letting.[46] This operation, they argued, would only diminish the essential life force of the patient. At the same time, the rejection of blood-letting served to reflect the Helmontian opposition to traditional humoural pathology.

If the Paracelsian chemical philosophy of nature provided a conceptual framework for the iatrochemist to work within, it also provided a basis for his practical work. Because of the importance of heat and of fire analyses both the new chemical analysis of urine and the new chemical doctrine of signatures were to be characterized by distillation procedures. Similarly, in a search for the ingredients of medicinal water spas the Paracelsians did much to further the development of analytical chemistry. A long medieval tradition in this field had resulted in the development not only of isolated tests, but real analytical procedures—and it is understandable that the Paracelsians quickly adopted this tradition and added to it. By 1571 Leonard Thurneisser was using quantitative methods, solubility tests, crystallographic evidence and flame tests, while early in the next century Edward Jorden was advocating the red-blue colour change of 'scarlet cloth' as a regular test for those liquids which we would classify as acids and bases. The work of these men provided the basic information necessary for Robert Boyle's analytic research later in the century.[47]

The results of the new chemical analyses were put to practical use. Chemists could now give directions for the preparation of artificial mineral waters for those who could not travel to the health-giving spas,[48] and at the same time this analytical information added an argument in favour of the use of chemically prepared medicines. The Paracelsians argued passionately that theirs was a new and violent age—one that had spawned ravaging diseases unknown to the ancients (they were particularly appalled by the venereal diseases).[49] As a result they needed new and more potent medicines than the traditional Galenicals. Their meaning was clear, these medicines

[45] Robert Fludd, *Anatomiae amphitheatrum effigie more et conditione varia* (Frankfurt, 1623), p. 266.

[46] The examples of Noah Biggs (1651) and George Starkey (1657) are discussed by Debus in 'Paracelsian Doctrine in English Medicine' included in F. N. L. Poynter (ed.), *Chemistry in the Service of Medicine* (London, Pitman, 1963), pp. 1–26 (17).

[47] These developments are described in Debus, 'Solution Analyses Prior to Robert Boyle', *Chymia*, 8 (1962), pp. 41–61.

[48] John French describes the preparation of artificial Tunbridge and Epsom waters in his *Art of Distillation*, pp. 182–4.

[49] See the preface to Phillip Herman's *An excellent treatise* trans. by John Hester (London, 1590), p. 31.

were their chemically prepared metals and minerals. It is important to point out that the Paracelsians were not innovators in this because there was an important medieval distillation tradition in medicine—a tradition that was still strong in the Renaissance.[50] And—regarding metals—there is evidence of the widespread use of antimony pills, solid metals, and even elemental mercury in the non-Paracelsian literature of the period.[51] However, as Bostocke pointed out in 1585, the true Paracelsian could be distinguished from others through his careful attention to dosage and his use of the chemical art to extract only the valuable essence of dangerous minerals.[52] Furthermore, in his defence of these medicines (1603), Joseph Duchesne thought it important to refer to the spa water analyses which clearly indicated that minerals had beneficial medicinal effects.[53]

The defenders of the traditional *materia medica* were far from appeased by those who wrote apologies for the new chemical medicines, and, in truth, it must be admitted that the Galenic fear of the new drugs was not groundless. Paracelsus had broken with the Galenic dictum that 'contraries cure' and turned instead to Germanic folk-medicine which insisted that 'like cures like'.[54] Now, rather than seeking out bland vegetable concoctions, the physician was told to investigate poisons. That poison which causes a disease should now—in proper form—become its cure. And although the chemists sought to remove the toxic qualities, the medical establishment felt little reassured by this claim. For them many of the proponents of the new drugs were uneducated charlatans. In a Galenic text the very name 'Paracelsian' had an unsavoury connotation. Thomas Erastus accused Paracelsus of advocating the internal use of lethal poisons[55] while John Donne—in his comparison of the innovations of Copernicus and Paracelsus—admitted only the latter to the inner sanctum of Satan's lair as the governor of the 'Legion of homicide Physicians'.[56] In answer there were surely some chemists who

[50] Robert P. Multhauf, 'John of Rupescissa and the Origin of Medical Chemistry', *Isis*, **45** (1954), pp. 359–67; 'The Significance of Distillation in Renaissance Medical Chemistry', *Bulletin of the History of Medicine*, **30** (1956), pp. 329–46; 'Medical Chemistry and the Paracelsians', ibid., **28** (1954), pp. 101–26.

[51] John Evans published a tract on the virtues of drinking from his 'antimoniall cup' in 1634 and Bostocke cited the use of raw metals among those who were not Paracelsians, op. cit., sig. Eiv verso.

[52] Ibid., sig. Ev verso.

[53] Duchesne, *Practise*, sig. Q3 verso.

[54] Discussed by Paracelsus in the *Paragranum, Sämtliche Werke*, **8,** 107. See also Pagel, *Paracelsus*, pp. 146–8.

[55] Thomas Erastus, *Disputationes de Medicina Nova Paracelsi* (1572), Part IV, as cited by Pagel, *Paracelsus*, p. 313.

[56] In John Donne's *Ignatius His Conclave*, 1610, the innovations of Paracelsus are judged to merit more reward from Satan than those of Copernicus. John Donne, *Complete Poetry and Selected Prose* (Bloomsbury, 1929), pp. 362–9.

spoke ever more forcefully in defence of their medicines and methods of cure. No less an authority than van Helmont suggested that several hundred sick poor people should be taken from the hospitals and the military camps. These should then be split in half and lots cast to decide which group would be treated by the Galenists and which by van Helmont. The number of funerals would determine whether the chemical or the traditional medicine had triumphed.[57] This was a challenge which was to be repeated often by the chemical physicians of the seventeenth century.

The new drugs became a subject of intense debate at the university level in the early years of the seventeenth century. And yet this debate is characterized also by an ever-increasing recognition of the need for moderation. In 1603 Joseph Duchesne, then lecturing on iatrochemistry in Paris, published a book in which he defended not only the new remedies, but also the whole chemical approach to medicine.[58] This was anathema to the conservative Galenists who controlled the medical faculty, and he was answered immediately by the elder Riolan.[59] Within three years this subject had stimulated enough authors to set their views to paper for Andreas Libavius to be able to prepare a seventy-page folio discussion of the books and pamphlets that had appeared to date.[60] A much longer volume on the same subject appeared from his pen a year later. In the course of his own defence of the new medicines Libavius suggested that the wise physician need neither choose the narrow path of the Galenist nor the mystical road of the Paracelsian theorist. Rather, he should take a middle way which included the best of the theory and the practice of both groups.[61] It is this spirit of moderation

---

[57] Helmont, *Oriatrike*, p. 526.

[58] This is Duchesne's *De Priscorum* . . . cited above (footnote 33) in the 1613 edition.

[59] J. Riolan ('the elder', *Apologia pro Hippocratis et Galeni medicina adversus Quercetani* . . . (Paris, Hadrianus Perier, Off. Plantiniana, 1603).

[60] Andreas Libavius, 'Proemium Commentarii Alchymiae Ipsiusque Artis Apologeticum, in quo Examinatur Censura Scholae Parisiensis per Ioannem Riolanum de Alchymia, annis 1603, 1604 edita', in *D.O.M.A. Commentariorum Alchymiae Andreae Libavii Med. D. Pars Prima* (Frankfurt ad Moenum, Excudebat Ioannes Saurius impensis Petri Kopffij, 1606), pp. 1–70. This theme is discussed at greater length by the present author in the following two articles: 'The Paracelsians and the Chemists: The Chemical Dilemma in Renaissance Medicine' in the Symposium volume of the Meeting of the International Academy of the History of Medicine, London (August, 1970), ed. F. N. L. Poynter (in press) and 'Guintherius-Libavius-Sennert: The Chemical Compromise in Early Modern Medicine', *Science, Medicine and Society in the Renaissance*, ed. Allen G. Debus (New York: Neale Watson Academic Press/London: Heinemann, in press).

[61] From Libavius' *Alchymia triumphans* (1607) as quoted by W. P. D. Wightman, *Science and the Renaissance* (2 vols., Edinburgh/London/New York, Oliver and Boyd/Hafner, 1962), 1, p. 260.

that is seen again in the first London Pharmacopoeia of 1618.[62] At Paris Duchesne had been defended first and most fervently by his colleague, Theodore Turquet de Mayerne. After the assassination of his patron, Henry IV, Mayerne had moved to London as the chief physician to James I. Here he became a member of the London College of Physicians which by this time had a large number of Fellows who were favourable to the new remedies of the chemists. Mayerne was highly influential in reactivating a long dormant plan of the College to publish its own pharmacopoeia. When this appeared in 1618, it was found to include sections on chemical remedies of all sorts as well as the more traditional medicines of the Galenists. The London Pharmacopoeia thus represents not only the compromise position of the English physicians who had outlined such a volume some thirty years earlier, but also that of Andreas Libavius in the Parisian dispute. As the first national pharmacopoeia, and also as an official publication of the widely-respected College of Physicians of London, it was to carry great weight in making respectable the preparations of the iatrochemists throughout Europe.

We may then speak properly of an increasing polarization between the Hermetic physicians and the Galenists. The reaction of the Paracelsian authors to the criticism of Thomas Erastus—or the heated debate at Paris— may serve as examples. Yet, at the same time, the position of the London College of Physicians shows a tendency towards compromise on the difficult question of the internal use of the new medicines. Among the chemical physicians themselves there was an ever-increasing number who sought to maintain chemistry as the basis of a new philosophy of nature, but rid it of its most mystical and least experimental aspects. Influential iatrochemists such as Daniel Sennert and Andreas Libavius agreed with Paracelsus that chemistry was a proper basis of medicine and that this was the chief of all sciences. But for them the works of Aristotle, Galen and Hippocrates were not to be discarded and burned in the market place. Rather than resorting to polemics, the true physician should examine both the old and the new medicines and accept the best of both. For many seventeenth-century iatrochemists the chemical philosophy could be safely followed because this provided a fundamentally observational basis for the new science. These men were disturbed no less than the Galenists—or the mechanical philosophers—by the mystical alchemical cosmology of some of their fellows. Thus,

---

[62] The standard works on the 1618 London Pharmacopoeia are the publications of George Urdang: 'How Chemicals Entered the Official Pharmacopoeias', *Archives Internationales d'Histoire des Sciences*, 7 (1954), pp. 303–14 and 'The Mystery About the First English (London) Pharmacopoeia (1618)', *Bulletin of the History of Medicine*, 12 (1942), pp. 304–13. To this should be added Urdang's lengthy introduction to his reprint of the *Pharmacopoeia* (Madison, 1944). I have discussed the compromise nature of the work in *The English Paracelsians*, pp. 145–56.

although the contemporaries Robert Fludd and Jean Baptiste van Helmont may properly be called 'chemical philosophers', there were grave differences between them. The work of the former glorifies the occult knowledge of the mystical alchemists while that of van Helmont is characterized by its rejection of much of what he considered to be the unsound theory of Paracelsus and other alchemists. Indeed, van Helmont and chemists like him found themselves more in agreement with Father Mersenne and Pierre Gassendi than the alchemical cosmologists of the early seventeenth century.[63]

Conclusion

The middle decades of the seventeenth century represent the formative years of modern science. It was then that Harvey became recognized as the founder of a new approach to physiology. He would no longer be interpreted as a lesser disciple of Robert Fludd as he had been earlier. It was then that the impact of Bacon and Descartes began to be felt, while the elements of the mechanical philosophy were forged by men who distrusted both the mysticism of Renaissance natural magic and the scholastic addition to antiquity even though they borrowed from both. Already the stage was being set for the replacement of the vast system of chemical analogies by a clockwork universe. And yet, it would be incorrect to imply that the work of the Paracelsians and the iatrochemists had played anything less than a major rôle in setting this stage—or, for that matter, that their influence suddenly ceased in the third quarter of the seventeenth century. The *Principia philosophiae* of Descartes appeared in print the same year as the first major collection from the pen of van Helmont, the *Opuscula medica inaudita* (1644). The former document stands as a landmark in the development of the mechanical philosophy and the latter was to serve as a preamble to a revitalized iatrochemistry. Indeed, the chemical medicine of the late seventeenth century was to be predominantly Helmontian in tone and its practitioners were men who found ties with the earlier 'reformed' chemical medicine of Libavius and Sennert. They rejected the macrocosm-microcosm analogy of traditional Paracelsianism and in so doing they broke part of the chain binding man to the greater world. If they still insisted on a search for a 'chemical' understanding of man, they had lost much of their interest in the 'chemical' interpretation of the macrocosm.

There is no time to discuss this later iatrochemistry here. At best we can

[63] Note, for instance, van Helmont's lengthy letter to Mersenne (19 December 1630) criticizing Robert Fludd. *Correspondance du P. Marin Mersenne*, edited by Cornelis De Waard and René Pintard (Paris, P.U.F., 1945), **2**, pp. 582–93.

summarize a few major themes seen in the century between the death of Paracelsus and the impact of the Helmontian texts. What had the pre-Helmontian iatrochemists accomplished? How had they influenced medicine and science in this period?—Above all, Paracelsian medicine represents a reaction against the traditional veneration for antiquity. The early Paracelsians spoke harshly of Aristotle and Galen (if not Hippocrates) and they turned instead to the recently translated Hermetic Alchemical and Neo-Platonic texts. A vitalistic universe founded on the truths of the macrocosm-microcosm analogy and the divine office of the physician seemed to be the basis for a new Christian understanding of nature as a whole. In their drive for reform the Paracelsians proceeded to strike at the very foundations of the older system. Both the Aristotelian elements—upon which the old cosmology was founded—and their attendant humours—upon which Galenic medicine depended—were questioned. Chemists now turned to the three principles as an explanatory device, while Paracelsian physicians spoke in terms of local seats of disease governed by internal *archei* rather than the imbalance of fluids.

The Paracelsian answer to antiquity was best expressed in the emphasis on observation and experience as a new basis for the study of nature. Surely the Paracelsians were not alone in this plea, but their special interest in chemistry as a guide for the study of man and the universe distinguishes them from other Renaissance philosophers of nature. Their extensive use of chemical equipment in distillation experiments and their constant reference to chemical analogies as a means of understanding all natural phenomena clearly indicate their connection with their Hermetic-alchemical heritage.

The medicine of the Paracelsians was strongly tinged with chemistry, but not with mathematics. While they might still pay lip service to the certainty of mathematical proof, in fact their concept of quantification was closest either to neo-Pythagorean mystical relationships or to practical measurements by weight. Mathematical abstractions of natural phenomena and geometrical proofs savoured strongly of scholasticism and this was plainly to be avoided. Logic itself was suspect as a form of the 'mathematical' science and medicine of antiquity. The new medico-science of the Paracelsians thus tended to be a less rather than a more mathematicized approach to nature than that of the past.

The opinions of these chemical physicians were set forth with conviction, but often with little tact. They decried the current overreliance on antiquity. They called for a new medicine and a new science based on chemically oriented observations and experiments. And they demanded educational reforms so that their 'Christian' concept of nature might be taught at the universities. On these points they came into direct conflict with tradition.

Yet, they argued no less vehemently amongst themselves. Here they debated questions we would call essentially geological, the place of mathematics in the formation of a new philosophy, the truth of the elements, the reality of the macrocosm-microcosm analogy, and the meaning of astral emanations. We can, of course, credit the Paracelsians with specific advances—their concept of disease or their recognition of the importance of chemistry for medicine (both as a basis for the understanding of physiological processes and as a new source for medicinal preparations) might serve as excellent examples. And surely there is little question that many of the 'modern' concepts of the late seventeenth century have their roots in the 'non-modern' speculations of the iatrochemists of the preceding century. Nevertheless, the significance of the Paracelsians is to be found in the provocative nature of the issues they raised as well as in specific concepts to be found in their texts. By proposing and by defining their vision of a new science based on medicine and interpreted through chemistry they found themselves engaged in a debate which resulted in the definition of significant aspects of modern science and medicine.

VII

THE SPECTRE OF VAN HELMONT AND THE IDEA OF CONTINUITY IN THE HISTORY OF CHEMISTRY

Walter Pagel

The history of chemistry may be said to be concerned in the first place with the lives of great chemists, with their theories and with the identification of the substances which they have described and processed. On the other hand historians should not overlook the general ideas which form the background to chemical research in the sixteenth and seventeenth centuries. This period may not be taken as all too significant in the development of modern chemistry. And yet it encompasses Robert Boyle who may well be called the father of modern chemistry. In fact it is the time when chemistry may be said to have stood at the crossroads. It was the very time in which it became a science and no longer a complex of what seemed to be fortuitous observations. With this we have raised the fundamental question of continuity in the early history of chemistry and thereby in the history of science and medicine at large. Some continuity must exist, but all too often it is taken for granted without adequate substantiation. Looking back from the present, we are apt to construct a historical sequence by selecting those observations and results which appear to be relevant stepping stones in a continuous line that leads from the technical experience of prehistoric and ancient civilizations up to the present day.

When examined in their own environment, however, such data may assume a quite different complexion. When they were acquired they may have had a different meaning altogether. What appears to be relevant in them today, may not have been intended to be conveyed by the ancient author. Or else it may have been accidental to what he intended to convey or it may have formed an inessential part of it. Or, finally, it may not have been there at all and it is only ourselves who infer that it was there.

For example: It is generally accepted that *gas* was discovered by Jean-Baptiste van Helmont (1579–1644). This is true but it is not the whole truth. Van Helmont observed and described volatile substances which he judiciously distinguished from air and water vapour. Historians of chemistry often refer to carbon dioxide and sulphur dioxide as the most notable examples. To him, however, gas had a much more complex meaning than a number of chemicals produced in the laboratory. The main point for him was the demonstration and comprehension of *specificity* in natural bodies at large. Air and vapour meant to him media that were common to all objects of nature. Gas was *specific* to the individual object and its species. It was in fact the spiritual enmattered spark that makes the object tick and reach its destined end. Hence it could be captured *in vitro* when its coarse cover was removed by heating—*per ignem*—van Helmont called himself the philosopher *per ignem*. When coal is heated in a closed vessel, van Helmont observed it to be converted entirely into a smoke. This smoke he called a new entity deserving a new term, namely that of *gas*. The substance heated had lost nothing except its shape and coarseness. In fact it was present in its purest form possible. It revealed its innermost nature. What remained is still matter but not matter pure and simple (which was water, according to Helmont) but matter that was disposed to fulfil a certain function and to assume certain shapes. It was matter spiritualized as much as it was spirit materialized. Van Helmont realized the significance of his observations in the study of the nature of matter—in other words in chemistry. What was more important to him, however, was the solution of the perennial philosophical problem of the interaction of matter and spirit. Here he came to a *monist* solution: there was no separate soul that acted *on* matter but only a spiritual impulse *in* matter and inseparably interlocked with it. It was a chemical version of the time-honoured Aristotelian *entelechia*—the entelechia made visible and given to the laboratory worker *in vitro*.

Van Helmont voiced opposition to Aristotle in abusive terms. The same is true of Paracelsus and many members of the Paracelsians to whom he belonged. However, if his position (and for that matter the position of Paracelsus) was in some way related to Aristotelian vitalism the search for this relationship in other Paracelsians should be rewarding. The example of Duchesne (Quercetanus, ?1544–1609) reveals not only its existence, but also its awareness and expression in clear terms in a Paracelsian natural philosopher. He takes Galen to task for the emphasis which he laid on the material (elemental) qualities of components as determining the essential 'form' of a natural object. Following the erroneous view of Empedocles, Galen had thus represented Nature as the product of mixing and the change of mixtures (*naturam nihil aliud esse quam mixtionem et alterationem mixtorum*). Aristotle had controverted this opinion. *Crases* of this kind are neither efficient nor formal causes, but merely instrumental—they are no more than *temperamenta et mixtiones qualitatum elementarium* that acquire definiteness only by the *substantial forms* which also give the end for the sake of which an object is formed. This is the *physis* or nature. It constitutes the individual object embracing its matter that is inseparable from this unit (*substantia*). This *logos* or principal cause of *naturalia* is not an element or something composed of elements, or anything that is 'developed', but a divine principle. What does 'develop' is the material part which together with and inseparable from the principle constitutes the individual object (*substantia*). The latter as a whole is endowed with a power (*dynamis*) to act. The elemental mixture of a herb such as rhubarb is merely the instrument by means of which its specific power—the purging of bile—is exercised. Vision is due to the 'perfection' (*entelecheia*) of the eye. Its function as an organ cannot be the product of the summation, juxtaposition or mixture of its material (elemental) components, but only from its action as a whole. The principle is not visible, but recognizable by its effects—this in itself proving that not matter, but something acting *in* matter and inseparably bound up with it, accounts for specificity, for quality and property. This makes the stomach produce chyle, senna purge black bile and pepper make the tongue hot, although neither its surface nor any of its particles are hot to touch. Quercetanus notably refers to Jacob Schegk (1511–1587)—'the second Aristotle' who had been a 'father and teacher' to him (*Ad veritatem hermeticae medicinae ex Hippocratis decretis*, cap. XI, Paris 1604, p. 128 seq).

Quercetanus was undoubtedly an important source for van Helmont. Perhaps it was reminiscent of this self-confessed Aristotelian Paracelsist when van Helmont, too, spoke of the *Ens primum* of any object in matters chemical (*in chymicis*) in unmistakable Aristotelian terms. This is found for example in his answer to a question addressed to him by Marinus Mersennus —as to whether the 'world-soul' is the *primum ens* of each object, whether sulphur is the *primum ens* or salt. Van Helmont's answer is in the negative. The *ens primum* of anything *in chymicis* 'saves' (*servat*) the determination of its compound, wherefrom it is (*determinationem sui concreti, unde est*). Hence it (i.e. the *primum ens*) is multiple corresponding to the number of individual objects and is mostly in liquid form containing the *seminal essence* and *crasis or entelechia* of the compound in highest purity, simplicity and subtility of matter. It is also *Ens primum* in nature at large (*in physicis*) what Paracelsus calls *iliastes*. Van Helmont does not deny a force dispersed throughout the world, analogous to the Platonic *anima mundi*. He calls it *magnale* (elsewhere: *Blas*) and upon it respiration and pulse depend. It is the occult factor that gives air its vital quality. It seems historically related to the 'MM'—the

mysterium magnum—that Paracelsus and the Paracelsians (notably the author
of the *Philosophia ad Athenienses*) located in the air. All this occurs in the
same letter of 6 February to Mersenne (*Correspondence du Père Marin Mersenne*,
ed. Mme. P. Tannery and C. de Waard (eds.), Vol. III, 1631–1633, Paris 1969
(2nd edition), pp. 81–3). However, van Helmont's answer to Mersenne's
question emerges from an Aristotelian position. This is evident from his
assertion of individual *seminal logoi* and *entelechiae*. These are responsible for
a new body to be created when two other bodies are mixed. The result of
this *mixis* or *crasis* is that the new body does not consist of the same elementary particles as all other bodies, but of a homogeneous *substance* that is specific
for the body in question. It cannot be reproduced by mere re-constellation
of particles of matter in general, i.e. of atoms. It is true that van Helmont
uses the term atoms for finest particles (see the *loci* collected from his works
in *Mersenne, Correspond.*, loc. cit., 1969, p. 45, note 1), but this does not
provide evidence for any atomistic principle that van Helmont may be
thought to have used as basis for his natural philosophy. In this the Aristotelian substantial forms and their chemical interpretation remain paramount.
They are also of greater significance than the welter of animistic and 'magic'
ideas which can be found in van Helmont's collected works. The subject
requires further study and no more than a preliminary interpretation of the
passage quoted from van Helmont's letter can here be given. For the
antagonistic position of Atomism versus Aristotelianism in seventeenth-
century chemistry we refer to the detail (notably on the Aristotelian *mixis*)
given in H. Kangro, *Joachim Jungius' Experimente und Gedanken zur Begründung
der Chemie als Wissenschaft*, Wiesbaden, 1968 and to this the present author in
Ambix, 1969, XVI, 100–8.

It would be difficult to overlook the strong religious overtones in van
Helmont's chemical discovery. Gas was first and foremost the divine gift
which every object had received when it was created, the power by which it
lived, the power of unfolding a sequence in form and function that was
unique, in other words the specific power which informed its life. Here we
see the intimate connection of a *vitalist* idea with divine creation. What is
today a chemical text-book item, then, was to its discoverer something quite
different. We say that the discovery of gas and pneumatic chemistry was due
to the observations of van Helmont. We forget, however, that this discovery
was a rather incidental part of a religious and vitalistic system of natural
philosophy.

Examining the results of van Helmont's chemical work in terms of
progress achieved over preceding knowledge of volatile substances we forge
a chain of continuity that links up with those who came after and superseded him. This continuity, however, is the product of abstraction and

construction dictated by modern standards. It is not a true historical continuity. For the chemical results in van Helmont's case are the by-products of his religious and vitalistic philosophy and can only be opened to our understanding when so viewed.

Another example concerns the position of *Alchemy*. This is often said to be a forerunner of chemistry. If its main tenet was transmutation it would seem to have contributed to one of the earliest systems of chemistry. This is the *Alchemia* of Andreas Libavius of 1597. Libavius was an Aristotelian and his belief in transmutation was indeed well in line with the principles laid down in Aristotle's *Meteorologica*.

Again it would be difficult to uphold the idea of continuity in this example. Before Libavius a system of chemistry had been attempted. This was the *Archidoxis* of Paracelsus. The latter, however, was no alchemist. He was no believer in transmutation. 'Alchemy' to him was the search for the active essential in natural substances which would serve in medicine more effectively than the unrefined Galenic 'soups'.

We return to van Helmont as the most successful chemist among the Paracelsians. His attitude towards alchemy is puzzling. At one time he confessed his belief in transmutation and to have carried out a successful projection personally by means of a $\frac{1}{4}$ grain of the Philosopher's stone given to him by a vagrant chemist. And yet in his early work on the Waters of Spa (1624) he had provided *in vitro* a conclusive proof against transmutation. This concerned the precipitation of copper from a vitriol solution upon addition of iron. Libavius and Sennert as late as 1629 believed that this was a case of transmutation, namely of iron into copper. This had been opposed prior to van Helmont by Guibert and Angelus Sala. However, the interpretation of the process in terms of an exchange of iron particles by copper particles was van Helmont's. He compared this to the recovery of silver—invisible in nitric acid solution—by adding copper. These ideas were brought to a conclusion by Joachim Jungius (1587–1657).

Again, the example of alchemy has brought us face to face with *discontinuity* rather than continuity in the early history of chemistry.

We have, of course, simplified matters through identifying alchemy with transmutation and ignoring its multifarious and protean connotations which have little to do with chemical history. A final point: When did chemistry become a science? And how can this be seen in the perspective of continuity? It may be said that chemistry became a science through the revival of the corpuscular hypothesis, i.e. of *atomism*. This took place largely due to the work of Boyle. But it had been foreshadowed in some respects by Joachim Jungius. Atomism was opposed to Aristotle. The essential point was: what happens when two substances are mixed and a third substance seems to have emerged? To Aristotle the decisive change was not one of the material components but the introduction of a new form that directed them. By contrast the atomist did not believe that a new substance had been produced and that a change in the quality of components had taken place. To him there had been merely a change in the position of the components. These—the atoms—are unchangeable in themselves. What had changed was the way in which they moved and touched each other—their juxtaposition. This prompted *syn-diacrisis*—a procedure of primitive analysis and synthesis: dissolving substances and recovering them from solution. We mentioned the precipitation of copper from a vitriol solution—it was the crowning witness for the corpuscular hypothesis. It may be regarded as the decisive step of chemistry into scienceness.

Yet again the course of continuity is not as smooth as it may seem. It is again van Helmont who disturbs it. His basic position was vitalist and hence Aristotelian, however much he inveighed against the 'heathen' Philosopher. At the same time and to this extent van Helmont found himself opposed to the corpuscularians, and through this to the scientific progress of chemistry. It was this very attitude, however, which made him lay the foundation of *Pneumatic Chemistry*. In this respect Boyle referred to him as a worthy predecessor in many places. Indeed it cannot be denied that one of the roots of scientific chemistry was thus through *van Helmont, the vitalist and the Aristotelian.*

I fear that I have presented a number of hurdles in what seems to the present-day observer a clear highway of unimpeded progress. It has been said that all history is contemporary history. In other words we are unable today to overcome the influence of present day standards in describing the past. Our account of the past will thus be of necessity incomplete and the product of selection. This is of course true—up to a point. But it is the task of the historian to reduce this to the possible minimum and to revive the past in all aspects. This can be achieved by the effort of making oneself contemporary to the savants studied. In other words what the scientist looking back into the history of his subject has cut out of context the historian has to replace. He must do so unafraid of complicating something which appeared to be simple.

The Diagrams

In these an attempt is made at integrating van Helmont's results in science and medicine with his philosophical and religious-mystical ideas.

Fig. VII.1. His basic penchant towards *mysticism* was fed by his resentment of the ruthless Spanish occupation of his country with the consequent domination of her cultural life by the Jesuits and their wordy and formal-logical argument (*'artificiose altercari'*). His disgust with the pseudo-truth sold by the ruling powers in state and university expressed itself in *scepticism* towards the School-tradition as a whole. Assuming the attitude of the Cusanian *idiota* he replaced the scholar's false security by unlimited belief—listening to an inner voice, inviting visions as the fruit of mystic resignation (*Gelassenheit*) and joining all this with modest 'knocking' at the door of Nature, i.e. unprejudiced observation and testing of divine creatures (*Empiricism*).

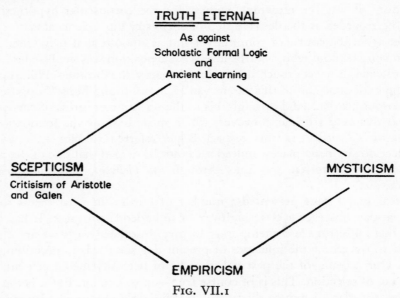

TRUTH ETERNAL

As against
Scholastic Formal Logic
and
Ancient Learning

SCEPTICISM

Critisism of Aristotle
and Galen

MYSTICISM

EMPIRICISM

Fɪɢ. VII.1

Fig. VII.2. The way to real, that is divine, truth aims at uncovering the divine spark in objects. This is their *scientia*, their *archeus,* their *gas.* To visualize it *in vitro* is a stepping-stone on the way towards the union of the soul of the human observer with the divine in the object and thus with the Creator. This way may be followed in the opposite direction and thus lead the adept to dominate nature, to curing disease or upgrading base matter and metal.

Fig. VII.3 and Fig. VII.4. From the attitude outlined there follows the revolutionizing break-away from time-honoured antitheses and generalizations as implied in the old ideas of elements, humours and qualities. Instead he concentrates on each individual object and its divine message as expressed in specificity. He believes in sympathy up to a point, but will have nothing of antipathy (*natura contrariorum nescia*).

MYSTICISM

Aims at UNIO with 'Deus Sive Natura'

THE WAY UP	THE WAY DOWN
From Creature to Creator Natural History and Science	From Creator to Creature Union with Creator
= Search for the DIVINE IN NATURE (The 'Seeds'). Revealed by Combustion of Coarse Material Cover, the essential of Object remaining as GAS	(i) Provides SOPHIC KNOWLEDGE of Hidden Structure and (Immanent) Function of Creatures (ii) Enables Adept to

IMPROVE NATURE

Curing Disease   (Al-) Chemical Operations

Fig. VII.2

SCEPTICISM	MYSTICAL EMPIRICISM
General Media and Forces	Specific Individual and Generic Forces of Seeds
Abolition of:–	Introduction of:–
Sympathy – Antipathy	Each object in its own right
Makrokosm – Mikrokosm; elements	Inert matter [Water]
Humours 'Soul' as Entity Super-added to Body	Organised matter, immanent (= not added) impulses = function (Archei, Fermenta)
Air and Water Vapour as Biological Agents	Gas as Vector of Specificity (Individual, generic)
Humoral Pathology (Catarrh)	Local seats = Morbid Anatomy Exogenous Agents = Aetiology
Prognostication [Astrology, critical days]	Diagnosis (Ontological pathology)

Fig. VII.3

The attention given to the individual object and specificity makes itself palpable in all branches of science and medicine which he studied and notably in his concept of disease. The latter, directed towards exogenous agents (*spina infixa*) and local changes accessible to morbid anatomy, pathological chemistry (e.g. the rôle of acid in pus formation) and specific therapy, ushers in our present-day concept.

RESULTS

In terms of Modern Science and Medicine

(a) <u>CHEMISTRY</u> _____

(b) <u>PHYSIOLOGY</u>

Gastric Acid
Alkaline Digestion (Duodenum)
Importance of Bile
Rhythmic movement of Pylorus
Respiration – for the disposal as volatile salts of a <u>Residue in Venous Blood</u> by combination with a ferment from the air
Vegetative Nervous System (Tonus, endocrine impulses)

(c) <u>PATHOLOGY</u>

Acidity in Pus Formation.
Local seat and Exogenous cause of Disease
Occupational Diseases
Asthma
Specific gravity of Urine for diagnosis of Kidney Disease

(d) <u>GENERAL</u>

Biological time. Pendulum for time measuring, thermometer

FIG. VII.4

Again the development to modern concepts in medicine was not straightforward. Van Helmont held an *ontological* view of disease. Rejecting the ancient theory of humoral imbalance (*dyscrasia*) he regarded each disease as an object in its own right, an *Ens morbi*. This is begotten by the vital principle (*Archeus*) when irritated by an outside agent, but the *Ens* is not simply the *Archeus* and its individually varying reactivity. The morbid *Ens* is in the first place an *image* or *semen* embodying an ideal *plan of action*. As such it is conceived by the *Archeus* in a state of morbid imagination. The idea 'becoming flesh' is externalized and as it were from outside invades and occupies the vital principle. It now behaves like a *parasite* making the *Archeus* subservient to its own particularist life-schedule. The parasitic *Ens morbi* is not simply identical with the outside agent that touched off the whole process—

such as a poison, the saliva of a rabid dog or the virus of plague. It is, however, specifically conditioned by the agent. The latter decides which organ is attacked, the nature of the local anatomical and bio-chemical changes and the outcome of the disease. The latter varies with the spirituality and penetrability of the poison and hence the ability of the vital principle to exercise its natural healing power. On this ontological and parasitistic basis Van Helmont developed indeed important and quite modern ideas and observations on the seats and causes of diseases.

VIII

WILLIAM DELL
AND THE IDEA OF UNIVERSITY

Charles Webster

'The throne of the beast in these Nations; are the universities, as the fountaine of the Ministry.' *Tryal of Spirits* (1653) p. 43.

The Elizabethan and Laudian statutes established a pattern of education at Oxford and Cambridge which was to endure for centuries. After making adjustments for the requirements of modern education, these institutions have shown a remarkable capacity to influence the course of higher education. The modern university system has thus absorbed many of the traditions established by the reformation humanists. Although these trends have not received universal approbation, there have been few sustained attempts to completely redirect the course of higher education. Undoubtedly, the most dramatic interruption in the smooth course of university education occurred during the civil and political turmoils of the Puritan Revolution. In the radical atmosphere of this period, reformers were prompted to look beyond traditional institutions, to elaborate models of university education suitable for the utopian communities which they sought to establish. William Dell (1606/8–1669), distant predecessor of Joseph Needham as Master of Caius College, Cambridge, was one of the few important academics to sympathize with these goals. This remarkable but enigmatic figure is thus an ideal subject to illuminate the crisis facing the universities during the Puritan Revolution. Furthermore, his writings give an important insight into the motivations of the advocates of the new science, who made such a spectacular impact during this period.

Pressure for university reform emerged from various directions during the early phases of the Civil War, but the exponents of reform had diverse goals in mind. The universities under Laud had become an increasing irritant to the puritan party. Parliament therefore recognized that the universities must be 'cleansed' before they could once more become efficient seedbeds for the puritan ministry and lectureships. With this in view a parliamentary committee was established in 1641 to examine the state of the universities. There were even signs that Parliament would take a more general educational initiative. It was also clear that the provincial middle classes were anxious to obtain wider opportunities for higher education, petitions for university rights being submitted from many regional centres. At a lower social level, the sectaries identified the universities as the agents of Antichrist, with disabilities intensified by intimate association with the established church.

The Civil War quickly extinguished hopes of immediate practical reforms, but there was no diminution in reforming enthusiasm. Indeed the military successes of Parliament prompted idealists to make plans for the complete reconstruction of secular and religious institutions. An important focus for this endeavour was the parliamentary army, which recruited an increasing number of chaplains with antinomian affiliations. This chaplaincy came to assume great importance in the army, its influence extending from the rank and file to the generals, much to the embarrassment of the Presbyterian orthodoxy.[1] William Dell moved quickly from the obscurity of a Bedfordshire living to a prominent rôle in the chaplaincy, his abilities as a 'learned and pious' minister attracting the interest of Cromwell and Fairfax.[2] At Marston near Oxford he was selected to conduct at the marriage of Cromwell's sister to Ireton. Such connections would normally portend a successful career in the Independent ministry, but from the outset Dell's sermons were highly critical of orthodox puritan ministry. He regretted that ministers generally lacked the power of the Holy Ghost, the only necessary qualification for preaching. Their sermons hence lacked 'fire' whereas the spiritual converts 'were as men made all of fire, running through the world and burning it up'.[3] His assertion that the human learning of the clergy was irrelevant to salvation was undoubtedly comforting to his military audience and potentially important for his later confrontation with the universities. The above sentiments also indicate the urgency of his spiritual message. The saints were destined:

[1] Leo F. Solt, *Saints in Arms: Puritanism and Democracy in Cromwell's Army* (London, 1959).

[2] For the first accurate and detailed biography of Dell, see the recent admirable book, Eric C. Walker, *William Dell Master Puritan* (Cambridge, 1970). Dell was a graduate of Emmanuel College, Cambridge, the centre of puritan influence. In 1641 he was granted the living of Yelden, Bedfordshire. See also H. R. Trevor-Roper, 'William Dell', *English Historical Review*, 1947, **62**, pp. 377–9.

[3] William Dell, *Power from on High, Or the Power of the Holy Ghost dispersed through the whole body of Christ* (London, 1645), p. 18. This was Dell's first published work, dedicated to his patroness, Elizabeth, Countess of Bollingbroke.

to carry the light of heaven up and down this darke world; among the people that sit in darkness and shaddow of death, to show then the way to life and salvation: you are to turne the world upside downe; to change the manners and customes of the people; . . . you are to reduce the earth unto conformity with heaven, and set up Gods Kingdome here in this present world.[4]

Thus the army was encouraged to believe in the sanctity of its mission, all exertions preparing for the final victory over Antichrist and the establishment of the millennial kingdom on earth.

Dell's message must have seemed particularly appropriate to the army besieging Oxford in May and June 1646. This graphically illustrated the strong relationship between the universities, the church and alien political forces. During the siege the chaplains no doubt elaborated their views on the deficiencies of the universities. A typical product of this period must have been Dell's sermon to Fairfax and the Army at Marston, preached shortly before the surrender of Oxford. He reported that the army had witnessed an unparalleled increase in 'assemblies of saints', once freed from the influence of the established church, whose 'old professors' had become the greatest obstacle to godliness.[5] This was regarded as the fulfilment of Isaiah's prophecy that the Lord would teach directly during the final days of the church. In these circumstances 'connatural learning', wisdom and ceremonies were worthless, the only valid ministerial function being the preaching of the word in 'clear and evident' terms, as a guide to inward understanding. This attention to the scriptures was necessary to prevent subjective 'revelations and sparklings' from misleading converts. Although Dell was well aware of the intrinsic dangers of enthusiasm, his reservations were insufficient to satisfy some of his listeners. His opponents laid charges against him, which resulted in a House of Lords enquiry in July 1646. Dell was excused and in an uncontrite mood he published the offending sermon, inserting a preface containing an inflammatory attack on the orthodox Presbyterian and Independent parties, while enumerating the spiritual virtues of the army.[6]

The sermons of Dell and his colleagues reinforced the aggressive religious elements in the army, which lost no time in challenging religious authority when Oxford capitulated at the end of June 1646. Anthony Wood, one of the numerous unfavourable witnesses, gave a graphic impression of the startling impact of the invading army and its notorious chaplains.

[4] Ibid., p. 5. Original italics are not given in the following quotations, unless necessary for the meaning.

[5] Dell, *The Building and Glory of the truely Christian and Spiritual Church* (London, 1646), pp. 1–2. This sermon was preached on 7 June.

[6] Ibid., p. 14.

When the forces belonging to the parliament were enter'd, who were all presbyterians, independents or worse, were among them their chaplains of the same persuasion, who forthwith, upon all occasions, thrust themselves into pulpits, purposely by their rascally doctrine to obtain either proselytes, or to draw off from their loyal principles and orthodox religion the scholars and inhabitants. Among them were Hugh Peters that diabolical villain and pulpit-buffoon, Will Dell chap. to sir Tho. Fairfax, John Saltmarsh, Will. Erbury, &c. and what they did there besides, during their stay, is too large a story now to tell you.[7]

Convinced of the merit of their spiritual experience, the soldiers were sufficiently confident to apply Dell's maxims, taking to the pulpit and attacking the educated clergy. Wood's account indicated the receptivity of the army to the points of emphasis in Dell's sermons:

The soldiery did declare their impudence so much, that they forebore not to preach in some of the pulpits, and to thrust themselves into the public schools, and there, in the places of lectures, speak to the scholars against human learning, and challenge the most learned of them to prove their calling from Christ.[8]

This behaviour, which was thoroughly consistent with Dell's views, scandalized the Anglican Wood. Other reports confirm Wood's impressions about the influence of the chaplains. At the celebrated debate between the chaplain William Erbury and the orthodox Francis Cheynell, Erbury's popular following insisted that the meeting should be transferred from the public schools to the university church. Thus occupation of Oxford by the Parliamentarian army not only represented a challenge to conventional religious attitudes, but was also a popular movement threatening the central functions of the colleges and university. This radical outburst offended not only the defeated Anglicans, but also the Presbyterians, whose own preachers had little influence in the army. Their primary weapon was a series of fierce pamphlets directed against the chaplains.

Parliament was faced with conflicting advice on university reform. On the one hand the chaplains and their allies were deeply suspicious of scholastic education and the clerical associations of the universities. They were opposed by the Presbyterians and Independents who upheld traditional education and the learned ministry. Although Parliament determined on a course of action which led to wholesale expulsion of its political opponents, it was cautious to replace them with academics sympathetic to the moderate puritan view of education. Consequently the great political changes of the interregnum made no impact on the course of scholastic education at Oxford and

[7] Anthony à Wood, *Athenae Oxonienses*, ed. P. Bliss, 4 vols. (London, 1813–20), 'Fasti Oxonienses', Vol. 4, p. 100, 24 June 1646.
[8] Ibid., p. 101.

Cambridge. The radical expectations of the army in 1646 were completely disappointed on this as many other issues.

The chaplains remained temporarily in the army, eventually returning to their civil congregations. Hugh Peter became an important public figure as chaplain to the Council of State, while John Webster preached obscurely in London and Lancashire. In the case of Samuel Hartlib, Parliament actually established a committee to 'consider some present maintenance for Mr. Hartlib; who hath done very good service to the Parliament; and also of some Place at Oxford for him for his future Support'.[9] However, no place at Oxford was created for him, or for John Dury his associate, although both were granted state pensions. At about the same time Dell was awarded £50 when he delivered the articles of surrender from Oxford on behalf of Fairfax.[10] Thus Parliament was anxious to reward its loyal servants, but it was less willing to place them in positions of public authority in the universities. For their part the reformers were not enthusiastic about such preferment. The preachers generally preferred to return to their congregations; Hartlib regarded London as the best base for his philanthropic activities. There was an increasing interest in proposals for entirely new educational institutions contrasting with the universities which were settled in their customary traditions.

Dell was probably the only unorthodox puritan to gain an important university post and retain it for the entire period of Parliamentarian dominance. His career ideally illustrates the difficulties arising from involvement in an uncongenial institutional situation. He became Master of Caius College Cambridge in 1649. It is somewhat surprising that he was offered or induced to accept this appointment. Evidently his unorthodox opinions had not yet caused friction with his influential patrons. Acceptance of a college position would not be expected from the sermon which he preached to Parliament on 25 November 1646.[11] This reinforced the views which he had expressed at Oxford on the dangers of education for those seeking an inward spiritual faith. He argued that church reformation would not be achieved through human institutions or traditional learning in the final days. His audience was reminded that Christ had operated through:

poor illiterate, mechanick men [who had] turned the world upside downe; they changed the manners, customes, religion, worship, lives, and natures of men.[12]

---

[9] *Commons Journals*, 25 June 1646.
[10] Ibid, 22 June 1646.
[11] Dell, *Right Reformation: Or the Reformation of the Church of the New Testament, Represented in Gospel-Light* (London, 1646).
[12] Ibid., p. 14.

This was exactly the point of emphasis of the 'mechanick' preachers in the army, who attracted the support of such educated chaplains as Dell against the established clergy. Dell's sermon ended with advice to the new rulers of the nation. This was primarily concerned with the defence of liberty of conscience, but also contained his first pronouncements on secular matters. He appealed to Parliament to have:

> regard [for] the oppression of the poore, and the sighing of the needy: Never was there more injustice and oppression in the Nation then now; I have seen many oppressed and crushed, and none to help them.[13]

While Dell's priority was for spiritual regeneration, he was fully aware that good men must involve themselves in ameliorating the severe social problems of the Civil War. It was not sufficient for the saints to ignore their social responsibilities. Dell impetuously published his sermon without licence, earning further rebukes from the Presbyterians, as well as official Parliamentary censure. However no obstacles were placed in the way of his appointment to Caius College.

Dell's response to his new situation was quite unlike that of other intruded heads of colleges. His familiarity with the university produced no mellowing of his earlier convictions. Accordingly he never moved beyond the periphery of university life. As the first married head of his college, he opted to leave his family at Yelden, his own participation in college affairs being routine and limited. His greatest efforts were reserved for public diatribes on the deficiencies of the universities. The academic life appears to have had little appeal, his intellectual outlook remaining that of a Parliamentary chaplain whose sermons appealed primarily to congregations outside the university. From this position he examined more fully the spiritual consequences of current academic practices.

This critical examination of university education reached a climax in 1653, the year which witnessed numerous expressions of educational idealism, aimed at influencing the policies of the Saints' Parliament. This was the first and only Parliament of the interregnum to display a real educational initiative, establishing a committee in July 1653 charged with responsibility for examining all petitions appertaining to the advancement of learning. The constitution of the parliament and its committee provoked a mood of optimism among the reformers, who responded with practical proposals and general advice. With the collapse of this parliament in December 1653 this reform momentum was dissipated and not again recaptured during the remaining years of the interregnum.

Two forceful sermons, *The Stumbling Stone* and *The Tryal of Spirits*, were

---

[13] Ibid., pp. 27–8.

Dell's contribution to the reform initiative of 1653. Characteristically, each sermon was in two parts, the first preached to a university congregation, the second to townspeople. Neither audience was spared his extreme views on the dangers of learning. These sermons expanded the 'Advice' on the reform of church and state which had formed the concluding section of *Right Reformation*, preached to the Long Parliament in 1646. The universities, by virtue of their ecclesiastical and secular associations, were relevant to both areas of this advice. While the general themes of these sermons derive from his earliest writings, their expression is intensified by millennial convictions. Consequently he pressed for the immediate reduction of the church and state to conformity with the highest spiritual standards:

> For I Judge it most necessary for the true Church to be acquainted herewith, especially after so many of the *Seals* have been opened, and also seeing the time of the *Restitution of all things* makes haste upon us, and we hope is even at the doors.[14]

Dell's experiences convinced him that the universities lay behind many of the malignancies of the times. In *The Stumbling Stone* he declared at the outset that he was against '*All* Universities' as well as those townspeople 'baptised into the University Spirit'.[15] Similar views are known to have commanded support in Parliament, but this must not be taken to indicate general pressure for abolition of universities. This was probably a position adopted by only a small fringe of the sectarian movement. However, Dell's critics would have regarded his opposition to the traditional ecclesiastical rôle of the universities as tantamount to abolitionism.

The Stumbling Stone expressed the central theme in Dell's theology, that 'Wit, Wisdom, and Parts, and Learning' were a positive barrier in the religious aspect of life. Reference to the New Testament church indicated successful religious witness by men with none of the academic attributes thought essential for ministry within the dominant factions of the later church. Jesus himself had been a carpenter, 'a Mechanick man', or 'as any ordinary Tradesman of this Town'.[16] His disciples were of an equally mean social origin, their success being due solely to the power of the Holy Spirit. Dell therefore conjectured that the spiritually most receptive element in the community was not 'the Great and Honourable and Wise, and Learned, but mean, plain, and simple people'.[17] Society consisted of two divisions,

---

[14] Dell, *The Stumbling-Stone, or, A Discourse touching that offence which the World and Wordly Church do take against 1. Christ himself. 2. His true Word. 3. His true Worship. 4. His true Church. 5. His true Government. 6. His true Ministry. Wherein the University is reproved by the Word of God* (London, 1653), Sig. A2v.

[15] Ibid., Sig. A2r–v.

[16] Ibid., pp. 12, 22–3.

[17] Ibid, p. 21, see also pp. 6–8.

believers and unbelievers, irrespective of social rank. However, scholastic education placed higher classes at a serious disadvantage, and the clergy at the severest disadvantage. The religion of 'the dull and drousie Divinity of Synods and Schools' buttressed by authority, degrees, and ceremonies, was an insuperable barrier to spiritual faith. Dell singled out the teaching of pagan Greek philosophy as a particularly inimical aspect of scholastic education. These defects could only be overcome if the universities renounced all ecclesiastical connections:

> If the Universities will stand upon an Humane and Civil account, as Schools of Good Learning for the instructing and educating Youth in the knowledge of the Tongues, and of the liberal Arts and Sciences, thereby to make them useful and serviceable to the Commonwealth, if they will stand up on this account, which is the surest and safest Account they can stand on, and will be content to shake hands with their Ecclesiastical and Antichristian interest, then let them stand, during the good Pleasure of God.[18]

Dell's last contribution to the 1653 debate and indeed his final significant work was *The Tryal of Spirits*.[19] This exceeded his previous sermons in extremity of tone and directness of message, identified as fanaticism by his critics. His passion was undoubtedly provoked by the unsympathetic attitude of his colleagues and the inactivity of Parliament. Unlike earlier writings, the *Tryal* was dedicated not to influential patrons, but to the 'truely Faithful' who would 'undertake impossible works to flesh and blood'. If this was confidence, it was confidence born of desperation at a crucial stage in the millennial sequence. He regretted that his former allies in the army had been diverted from their revolutionary aims 'to comply with the world and worldly Church and the teachers thereof, and can persuade themselves that there is enough done for their time'.[20] In the search for allies against the clergy and universities, he expressed confidence in 'poor plain Husbandmen and Tradesmen' rather than his educated colleagues. He rose to Miltonic eloquence in denouncing the current state of the church, linking scholastic learning with authoritarianism, corruption and monopolism:

> They [the clergy] especially desire to Preach to rich men, and great men, and men in place and authority, that from them they may have protection, favour,

---

[18] Ibid., p. 29.

[19] *The Tryal of Spirits Both in Teachers & Hearers. Wherein is held forth the clear Discovery, and certain Downfal of the Carnal and Antichristian Clergie of these Nations. Testified from the Word of God is added a . . . Confutation of . . . Mr. Sydrach Simpson. . . . And lastly, The right Reformation of Learning, Schools and Universities, according to the state of the Gospel, and the light that shines therein* (London, 1653).

[20] Ibid., Sig. aa2r.

preferment, and a quiet life, and care not so much to Preach to the poor, plain mean people by whom they can expect no worldy advantage. . . . To this end also, they speak in the words which mans wisdom teacheth, and so mingle Philosophy and Divinity, and think to credit the Gospel with Termes of Art, and do sprinkle their Sermons with Hebrew, Greek, Latine, as with perfume acceptable to the Nostrills of the world.[21]

Thus Dell saw scholastic education, terminating with ordination, as a source of the great spiritual defects of the ministry and higher social classes. This education produced alienation from the spiritual faith:

By University Education (as things have hitherto been managed) youth is made more of the world than they were by Nature, through the high improvement of their corruptions, by their daily converse with the Heathens, their vain Philosophers, and filthy and obscene Poets.[22]

Having established the religious dangers of traditional education, it was a short step to asserting the general worthlessness of 'fleshy wisdom, Rhetorical Eloquence, and Philosophical Learning'.[23]

Up to this point, Dell's educational pronouncements were mainly negative. An important aspect of his religious outlook was complete scepticism about prevalent intellectual and social standards. This could lead to apathy and introspection, but with Dell and many puritan radicals it resulted in strong commitment to social reform. With liberation from traditional standards and diminished respect for established social institutions, it was possible to evolve models of social organization consistent with their millennial dreams.

Dell's views on the education appropriate to the reformed community were given in an appendix to the *Tryal by Spirits*.[24] He expressed the characteristic puritan conviction that ease and idleness were conducive to rapid intellectual decline. Education was designed to counteract this tendency. It was therefore recognized as the basis for any programme to reform the commonwealth. This priority was recognized by many other writers ranging from Hartlib to Winstanley; education was therefore one of the main themes of the flood of reform proposals which were announced during this period.

Although Dell recognized that education was necessary for all social classes, he made only passing reference to pre-university education. His views on the earlier stages of education were in line with other puritan

[21] Ibid., pp. 30–1; see also p. 54.
[22] Ibid., p. 55.
[23] Ibid., p. 66.
[24] The section had a separate heading, 'The Right Reformation of Learning'. It is included in the second paginated part.

authors. He would almost certainly have approved of the ideas of Dury and Comenius on the treatment of children. Formal elementary education was regarded as necessary for all in order to provide the foundations for the unaided interpretation of the scriptures. The children would be under women at the earliest stages of education, but under men at the later critical phases. While most education would be conducted at village or parish level, Dell suggested that specialized schools were needed in larger centres to provide a classical or vocational education. Although he recognized that Latin, Greek and Hebrew were needed for advanced biblical studies, he made his customary strictures about the dangers of heathen authors.[25]

The Right Reformation of Learning gave the most expansive treatment to higher education. This is not surprising in view of Dell's direct experiences; furthermore it was this level which was taken by so many writers to epitomize the character of scholastic education. Therefore Dell, like Webster, Hall, Peter and Dury, regarded the modification of Oxford and Cambridge as an essential condition for the establishment of an enlightened religious and social order. Dell's proposals were in the general spirit of the other reformers, but with many interesting and original features. He supported the moves to extend higher education to provincial centres, regarding the centralization in Oxford and Cambridge as a monopoly prejudicial to the welfare of the commonwealth. Like other writers he opposed monopolistic tendencies in all spheres of social and religious organization. In the case of higher education Dell argued that the separation of students from their families caused unnecessary financial strains and exposed the young to subversive moral pressures. He therefore advocated the establishment of universities or colleges in every major town, London, York, Bristol, Exeter and Norwich being mentioned as suitable locations. Thus the whole educational process would take place within the natural community:

> And this the *State* may the better do (by *provision* out of every County, or *otherwise*, as shall be judged best) seeing there will be no need of *indowment* of *Scholarships*, inasmuch as the *people* having *Colledges* in their own *cities*, neer their own *houses*, may maintain their children at *home*, whilst they *learn* in the *Schools*; which would be *indeed*, the greatest *advantage* to *learning* that can be *thought* of.[26]

Many writers agreed with these sentiments, but there was great difference of opinion about the kind of education to be offered in the new colleges. Many authors agreed with William Sprigge, favouring the relaxation of the scholastic curriculum to include middle-class diversions of the kind familiar from the French academies.[27] This was certainly not the opinion of

---

[25] 'Right Reformation', pp. 26–7.
[26] Ibid., p. 28. Original italicization is given in the passages from 'Right Reformation'.
[27] William Sprigge, *A Moderate Plea for a Free and Equal Commonwealth* (London, 1659).

such earnest puritans as Winstanley, Hartlib and Dell. Radical changes were made inevitable by the latters' insistence that the faculty of divinity should be abolished and all activities relating to the training of the clergy be abandoned by universities. On the other hand the faculties of medicine and law were to be retained in view of their social utility, but it was recognized that both were in need of drastic reform.

In Dell's mind the scholastic curriculum was firmly linked with the education of the clergy. With the abandonment of the latter, the value of the traditional arts curriculum was open to question. As indicated above, many aspects of scholastic education were a positive danger to the religious spirit, while the remainder had little positive value for an active public life. The traditional intellectual accomplishments of the gentleman and divine were dismissed as idleness:

> Fo *commonly* it falls out, that *youth* lose as much by *idleness*, as they gain by *study*. And they being *only* brought up to *read books*, and *such books* as onely continue *wrangling, jangling, foolish,* and *unprofitable Philosophy*; when they have continued any *long time* in the *University* in these *unwarranted* courses by *God*, they are commonly in the *end*, fit for no *worthy imployment* either in the *world*, or among the *faithful*.[28]

Since members of society had an immediate obligation to seek their 'lawful calling' it was necessary for the university to avoid superfluities of the kind encouraged by the arts course. A much better foundation for education was:

> *Arithmetick, Geometry, Geography*, and the like, which as they carry no *wickedness* in them, so are they *besides* very *useful* to *humane Society*, and the *affaires* of this *present life*.[29]

As the basis for many vocational studies the sciences had a positive value, while their freedom from traditional error caused no impediment to the spiritual faith.

As a further incentive to civil involvement, Dell proposed that all students should devote part of the day to study and the rest to vocational tasks.[30] Thus the universities would be fully integrated into the community. Like Petty, Dell realized that familiarity with practical skills was a great asset to the pursuit of the sciences. The division between the philosopher and the craftsman was widely recognized as a barrier to scientific and technological progress. The return to unified knowledge which the education advocated by Dell would involve, was analogous to his proposals to reintegrate the

[28] 'Right Reformation', p. 29.
[29] Ibid, p. 27.
[30] Ibid., p. 29.

ministry with the community. Preachers would be restored to the apostolic condition only when active participation in 'an honest calling and employment' was attained. Thus it was only possible to attain spiritual and intellectual enlightenment when fully involved in the life of the community. This edification was not possible when intellectuals or ministers were a separate 'sect' alienated from society in the unreformed universities.

Dell's *Right Reformation of Learning* develops a theory of education based on his unchanging theological convictions. It was the only aspect of social policy which aroused his detailed comment; its inclusion in his final and most important work is an indication of his recognition of the importance of education in the progress towards the millennium. The search for a pattern of education consistent with his theology reinforced his antagonism to the universities and other established social institutions. Having worked out the educational clause in his theology he relapsed into silence, leaving many others to restate this position, from Webster in 1654 to Milton in 1659.[31] The cost to Dell was probably isolation from the academic community and the loss of patronage from his former Parliamentarian associates. His writings were attacked by a wide spectrum of divines, from Independents to Anglicans, who temporarily allied to defend traditional education and the learned ministry. Dell and others accepting his position formed no coherent group. Most retreated into obscurity even before the Restoration. Dell himself retired from active writing, quietly resigning his office at the Restoration of Charles II.

Dell was not a direct participant in the scientific movement of the Puritan Revolution; he neither experimented nor contributed to theoretical natural philosophy. As with the majority of his contemporaries his priorities and preoccupations were religious. The quest for salvation and preparation for the millennial kingdom dictated attitudes to all aspects of social behaviour, from economic life to philosophy. While Dell's particular theology represented only one segment of puritan opinion, these priorities would not have been questioned by his many critics. It was therefore inevitable that the scientific movement among the puritans should have strong religious overtones, although the particular scientific motivation and outlook depended on the precise theological viewpoint adopted. It is important to remember that 'puritanism' is a convenient descriptive term embracing diverse theological viewpoints. On a range of issues puritans were united, but on many fundamental questions they exhibit strong internal divisions. Therefore generalizations about the entire movement's reaction to an issue

[31] John Webster, *Academiarum Examen, or the Examination of Academies* (London, 1654). John Milton, *Considerations touching the likeliest Means to Remove Hirelings from the Church* (London, 1659).

are hazardous. Any correlation must be proved independently for each theological position. Dell and associated advocates of free grace theology provide interesting insights into the theological determination of viewpoints on many secular issues. In particular they show a uniform enthusiasm for experimental science, recognizing its value for educational and social reconstruction. While they are not known to have made significant scientific innovations and have therefore not earned the attention of historians of science, these writers are important indicators of the appearance of scientific awareness in an important section of the community. From the previous description of Dell's career, it is apparent that his theology had a strong popular appeal to a group embracing fighting soldiers, mechanick preachers, as well as the educated chaplains and army commanders. With minor differences of emphasis, Dell's position was expressed in the influential writings of John Webster, Hugh Peter and Gerrard Winstanley. All displayed a positive enthusiasm for radical educational reform and Baconian science. At a slightly less radical standpoint, this social programme was adopted by the extremely active group associated with Samuel Hartlib. While individualism was the keynote of this movement, it was one of the most original and aggressive aspects of puritanism, making a substantial contribution to the tract literature during the Commonwealth and Protectorate.

The above account of Dell's sermons leaves no doubt about his priorities. The overwhelming emphasis of his sermons was for men to liberate their minds from traditional constraints in order to receive the holy spirit. This would increase the body of saints, who would triumph over Antichrist to establish the millennial kingdom on earth. It was felt that the Civil War was the resolution of this conflict into military terms, thus indicating the imminence of God's final dispensation. Much of the passion and urgency of Dell's sermons is to be explained by such beliefs. This was not a time for dispassionate estimates or involvement in trivialities.

Dell saw religious authority, supported by a professional ministry, as the great barrier to spiritual freedom. Only with the abolition of religious uniformity would the laymen attain spiritual responsibility. It was recognized that such a change would involve complete social reorientation. Each community would be freed from the financial liability of maintaining the church and would have greater responsbiilities in ordering its economic fortunes and religious life.

A completely new view of education was necessary if a radically different pattern of social life was to evolve. Each member of the community had to attain sufficient literacy to exercise spiritual responsibility and maintain an active social life. The scholastic educational system of the seventeenth century was totally inadequate for these purposes. A great extension of educational opportunity was required and the scholastic curriculum needed radical revision.

As indicated above, Dell's objections to the scholastic curriculum were strongly expressed in his sermons. He regarded most classical studies as positively harmful, while other subjects were superfluous. This education assisted neither the religious or vocational life. The functions which it served were totally out of place in the society which Dell advocated. Like other idealists, he was obliged to seek new foundations for education in keeping with his religious and social views. His adoption of the sciences as an alternative basis for education was not altogether surprising considering the publicity which the 'new science' had received in the philosophical and educational writings of Bacon and Comenius, two writers who attained great popularity during the Puritan Revolution. Experimental science was presented as the study ideally suited to the religious spirit of the age. It was free from the errors of traditional learning; it was not necessarily related to classical texts of a spiritually harmful nature; it had a close reciprocal relationship with practical life. Empirical craft procedures formed the natural basis for the methodology of the experimental sciences, which in turn would yield practical benefits. The experimenters exhibited precisely the qualities which Dell extolled and which he found so absent among the clergy. In social life experimental science was therefore the analogue of the 'experimental' faith which he advocated.

Having established main principles, Dell was content to leave most of his social and educational programme in outline. Other reformers, arguing from similar premises, gave more expansive treatment to these topics, providing an insight into detailed projects which would have earned Dell's approval. Not surprisingly the universities and their scholastic curriculum were subject to considerable critical comment. The universities were presented as standing apart from society in 'Monkish isolation', as vestiges of pre-Reformation society. Following the arguments used by Dell, scientific and technical studies were advocated as a means to the social rehabilitation of the universities. The most detailed and notorious expression of this view came from John Webster, whose writings reinforce the impression from Dell that new thinking on the university curriculum had a Baconian inspiration.

Webster's writings are widely taken to indicate the enthusiasm for experimental and utilitarian science among the radical puritans. However it is important not to interpret this as a general movement away from a religious world-view. If an increasingly rational approach to nature was involved, the motivation remained religious. *Academiarum Examen* was the climax of a series of writings defending spiritual religion against religious uniformity and the learned ministry, written shortly after Dell's *Right Reformation of Learning*. Thus *Academiarum Examen* and the *Right Reformation of Learning* may be regarded as an educational application of principles articulated in writings defending free grace theology. Webster and Dell adopt utilitarian science not for its own sake, but as a necessary function of the spiritual community which they seek to construct.

This religious context enables us to detect the motives underlying the fierce censures which the writings of Dell and Webster provoked. As defenders of free grace they had critics, but as opponents of the learned ministry and scholastic education, using science as a Trojan horse to overthrow the existing order, they emerged as a deadly threat to a wide section of the puritan intelligentsia. Thus the widespread criticisms of the universities which appeared in 1653 evoked numerous replies. The defence of the scholastic curriculum against science became the focal point of the wider attempt to maintain an existing social and religious order against radical social change. While the critics were almost entirely outside the universities, the defenders included influential academics, whose dual rôle as clerics and teachers was undermined by the proposed reforms. Most of the traditionalists were content with well-worn themes in their defences of the scholastic curriculum and learned ministry. A noticeable exception was the able and interesting apologia from an unexpected source, John Wilkins and Seth Ward, two of the most able academics appointed to Oxford by the Parliamentarians.[32] As promoters of the new science they were in a strong position to criticize Webster, whose enthusiasm outstripped his abilities when framing a detailed but ill-digested scientific curriculum for the universities. *Vindiciae acedemiarum* skilfully concentrated on a point-by-point refutation of Webster, in which the authors were able to display their superior command of natural philosophy, giving the impression that science was being actively pursued at the reformed universities without interference with traditional education. By concentrating on specific points raised by Webster they aimed to throw into disrepute the entire party of critics, without becoming directly involved in the defence of scholastic education and the learned ministry. Since Dell had concentrated on those latter points, making only general statements about the scientific curriculum, Wilkins and Ward produced a perfunctory and ill-composed critique of the *Right Reformation of Learning*, the issues of central importance for Dell being mentioned only in passing. Nevertheless, in spite of their pronounced interest in science,

[32] John Wilkins and Seth Ward, *Vindiciae Academiarum, Containing some briefe animadversions upon Mr. Websters Book, stiled The Examination of Academies. Together with an Appendix concerning what M. Hobbs, and M. Dell have published on this Argument* (Oxford, 1654).

Wilkins and Ward were orthodox on the main educational issues. They opposed the reconstruction of the scholastic curriculum, the extension of university education to the provinces, non-residential education and reference to social function. On the learned ministry they accepted the arguments of their scholastic allies.

Modern estimates of the 1653-4 debate over the universities have concentrated on the Webster-Ward exchange and the immediate scientific issues which they debated. On the interpretation of the new science, sympathies have tended to lie with Ward, rather than the 'frenzied and uncouth . . . puritan treatises'.[33] On the other hand, Debus has recently given a more sympathetic view of Webster's science.[34] But more important than the immediate empirical issues were the underlying principles separating the disputants, which emerge only through examination of the whole debate. Here Dell's rôle is crucial and representative of a widely held theological point of view. Theological motivation does much to explain the vehemence of the dispute between parties with obvious strong affinities on basic scientific questions. However, such was the underlying tension that the disputants were forced to exaggerate their disagreements over the new science.

This debate underlines the hazards in generalizing about the intellectual attitudes of the puritan movement. It indicates deep theological divisions within puritanism, which entailed friction on the whole range of spiritual and secular issues. Wilkins and his allies were guardians of the central puritan tradition, maintaining the rôle of the ordained clergy and the universities as the training ground for the church. The universities had the additional function within the secular order of providing for the 'vertuous education' of the gentry. On the other hand Dell, Webster and Winstanley looked forward to a new millennial social order which had no place for the ordained ministry or scholastic education. Universities were therefore to be completely transformed to assist in the preparation of a wider cross-section of the community for effective public service. This would have represented a complete break with tradition and completely changed the relationship between universities and the state.

In view of these radical aims, it is not surprising that widespread contemporary propaganda represented Dell and his associates as advocates of anti-intellectualism, favouring the abolition of universities and a reaction against civilized standards. As indicated above, anti-intellectual sentiments

[33] For a representative opinion, see R. F. Jones, *Ancients and Moderns* (St. Louis, 1961), p. 113.
[34] A. G. Debus, *Science and Education in the Seventeenth-Century* (London, 1970); Introduction to reprints of the *Examen* and *Vindiciae*.

were indeed prevalent, but primarily in relationship to spiritual matters, approximating to the widely-expressed fideist position held by such cultivated figures as Montaigne. Paradoxically it was this outlook which prompted the highest aspirations for the development and application of science. It also engendered pressure for the complete reorganization of education in accordance with new social and intellectual ideals. Thus there was an immediate sympathy for the philosophies of Bacon and Comenius, the advocates of social improvement and intellectual regeneration. This educational consciousness of the spiritual reformers made the universities an inevitable focus of attention in the general religious and social debates among the puritans. The reforms demanded by Dell and the popular movement within the army were not undertaken. Indeed the Parliament confirmed the traditional rôle of the universities, to the approval of Wilkins and Ward. It was therefore appropriate that Clarendon at the Restoration expressed a debt of gratitude to the educational record of the Parliamentarians. Wilkins and Ward had earned the bishoprics granted by Charles II, while Webster, Hartlib and Dell, having failed to induce a major educational reorientation, returned to obscurity.

IX

REASONS AND EVALUATION IN THE HISTORY OF SCIENCE

Mary Hesse

. . . our proper conclusion seems to me to be that the conceptual framework of Chinese associative or coordinative thinking was essentially something different from that of European causal and 'legal' or nomothetic thinking. That it did not give rise to 17th-century theoretical science is no justification for calling it primitive. Joseph Needham, *Science and Civilisation in China*, Vol. 2, p. 286.

I

The historiography of science, more than the history of other aspects of human thought, is peculiarly subject to philosophic fashion. This shows itself in two ways: first in the way historical studies reflect views of the nature of science current in contemporary philosophy of science, and second in the philosophy of history presupposed. An analysis of the first kind of influence was carried out by Joseph Agassi[1] in showing how inductivism and conventionalism generate distinctive kinds of history of science. The second kind of influence is exemplified by the application of Butterfield's category of 'Whiggish history'[2] to the kind of history of science that sees science as essentially cumulative and progressive, or by Collingwood's use of history of science in *The Idea of Nature*[3] as a paradigm case of doing history according to the Hegelian prescription of 'thinking men's thoughts after them'.

Historians of science have been on the whole less self-conscious about such issues than their colleagues in general history, and philosophers of history have devoted little attention to the special historiographical problems

[1] J. Agassi, 'Towards an historiography of science', *History and Theory*, **2** (1963), p. 1.
[2] H. Butterfield, *The Whig Interpretation of History* (London, 1931).
[3] R. G. Collingwood, *The Idea of Nature* (Oxford, 1945).

of science. In the last few decades, however, history of science has come of
age as a sub-discipline of general history, and consequently some of these
problems have begun to be discussed, not least in the sensitive comparisons
of methodology to be found in Joseph Needham's work from which I
have quoted above. My aim in this paper is to bring out into the open one
such problem, which can best be described as the question of whether
evaluations of the truth and rationality of past science are proper parts of the
historian's task.[4]

That this is only now becoming a problem in the history of science may
well come as a surprise to general historians, who have long wrestled with
the problem of evaluation in other fields, and arrived at some implicit
understanding that a kind of evaluation of the past is a necessary condition
of good history. As W. H. Dray puts it:

> Historians . . . who concentrate largely on showing us how things appeared
> to the participants, it might be noted, are seldom regarded by their fellows as
> attaining the highest rank. They seem to be regarded, indeed, in much the same
> way that theoretical scientists regard those in *their* field who do not go beyond
> the level of classification and empirical generalization. In both cases, although
> in different ways, the idea of the inquiry requires a further interpretation of
> the materials thus provided.[5]

That the problem of evaluation is on the other hand a problem at all may
come as a surprise to philosophers of science who instinctively regard any
historical enquiry into science as incomplete which does not pose and answer
the questions 'Was it reasonable?', 'Was it true?', and who in their more
reconstructive moods are sometimes justly accused of preferring these
questions to a more pedestrian investigation of 'Did it happen?'.

There are, I think, three reasons why the best contemporary historians
of science are not in a position to assimilate with simple piety the convergent
wisdom of the historians and philosophers that evaluation of the past is
both possible and necessary. The first is a form of backlash against the
naïveties, both historical and philosophical, of the type of history of science
which Agassi has called *inductivism*. This presupposed a philosophy of
science according to which nature, when investigated with a properly
open mind, reveals an ever-increasing accumulation of hard facts, which can
be progressively better understood by means of cautious and tentative
generalizations and modest theories. The process of scientific theorizing

---

[4] For historical illustration I have drawn in this paper on an example developed at
greater length in my 'Hermeticism and Historiography', *Minnesota Studies in the Philosophy
of Science*, ed. R. Steuwer, Vol. 4, 1971. I am grateful to the Minnesota University Press for
permission to reproduce some paragraphs of that paper in the present one, principally
in section III.

[5] W. H. Dray, *Philosophy of History* (Englewood Cliffs, New Jersey, 1964), p. 38.

is in this view dangerous and has usually proved mistaken, therefore what is interesting in past science is the sum of its facts which still form the basis of modern science. The present is in this crude sense the standard of truth and rationality for the past, and gives the inductivist historian grounds for the reconstruction of past arguments according to an acceptably inductive structure, and for judging past theories as simply false and often ridiculous. Such inductive history is of course, among its other defects, self-defeating, because if all theories are dangerous and likely to be superseded, so are the present theories in terms of which the inductivist judges the past.

Replacement of inductivism by more sophisticated history and philosophy of science does not however entail that such history must avoid all judgements of truth or rationality. Inductivist judgements of the truth of past science as seen from the present may be anachronistic, but it has still been thought possible to evaluate the scientific character and reasonableness of past scientific inference by taking account of its more limited access to facts and different general presuppositions. Most of the best work in recent so-called 'internal' history of science has been conducted according to this recipe, which, again after Agassi, may be called *conventionalist*. Duhem rediscovered the continuity of astronomy and mechanics with the rational philosophy of Greece, Koyré and Burtt reconstructed its internal coherence in the period before Newton, Lovejoy described the 'archaic' background of biological science before the nineteenth century in terms of the 'great chain of being'. The tacit assumption of all this work has been that certain canons of rationality can be recognized even in long out-moded conceptual systems, and that the injunction to 'think men's thoughts after them' can be followed as easily in history of science as in history of philosophy where it seems most at home, for natural science is just the arena of man's rational commerce with the world. It has also been presupposed in practice that this kind of internal history of science is relatively autonomous, that is to say that it can be carried on for the most part independently of the social and political environment of science and of the biographies of scientific researchers. But even if the autonomy of internal intellectual factors could be sustained, the tendency of internal history would still be towards relativizing canons of rationality along with the scientific ideas themselves, for an investigation such as Evans-Pritchard's into the metaphysics of witchcraft among the Azandi[6] is not very different methodologically from, say, a study of Stoic physics.

However the autonomy of internal history with respect to environmental factors has not gone unchallenged. The second type of influence making for relativism in regard to history of science is the increasing application to it

[6] E. E. Evans-Pritchard, *Witchcraft, Oracles and Magic among the Azandi* (Oxford, 1937).

of categories of social history and psychological analysis, and resulting emphasis upon so-called 'external' or non-rational factors in the understanding of scientific development. I shall suggest later that the distinction between 'external' and 'internal' factors is by no means clear cut, since it has to accommodate a whole spectrum of influences from such social factors as the standards of education of a society, through unconscious psychological motivations, to the metaphysical commitments of an age and its accepted forms of scientific inference and logic. But there is no doubt that excitement over the comparatively new task of investigating the complex interactions between all these types of factor has challenged the independence of older types of internal history, and has tended to obscure the question of rational evaluation of scientific ideas and arguments.

A third development tending to relativize the historian's criteria of a rational science has been the failure of current philosophy of science to provide a generally acceptable account of scientific rationality which could serve to delimit the subject matter of internal history. If we consider the *deductivist* analysis of science which has been almost universally accepted by philosophers of science until recently, little help can be found for the internal historian seeking criteria of autonomy. The deductivist view has been characterized by a radical distinction between the sociology and psychology of science on the one hand, and its logic on the other, or as it is sometimes expressed, between the contexts of *discovery* and of *justification. How* a hypothesis is arrived at is not a question for philosophy of science, it is a matter for the individual or group psychology of scientists, or for historical investigation of external pressures upon science as a social phenomenon. The question for philosophy or logic is solely the question whether the hypotheses thus 'non-rationally' thrown up are viable in the light of facts, that is, whether they satisfy the formal conditions of confirmation and falsifiability adumbrated by deductivist philosophers, conditions which are themselves at present in a considerable state of disarray. Although this view places a heavy straitjacket on the philosophy of science, it appears to exert little restraint upon its history, and in particular it does not help historians to recognize timeless normative criteria of scientific argument. It allows historians to take seriously as scientific whatever theories were contemplated in the past, arrived at by whatever external or internal influences, and however apparently bizarre, just so long as these theories were, at least in intention, logically coherent and empirically testable. The use of the terms 'logical' and even 'rational' in the deductivist analysis is indeed far narrower than in the intellectual historian's 'internal logic of science', or in his view of the history of science as the history of man's rational thought about nature. For deductivism characterizes all influences leading to discovery as non-logical or even non-rational, and leaves the whole context of discovery to the efforts of the historian without offering him any criteria of distinction between kinds of influence on discovery. And since, for given evidence, a theory satisfying the deductive criteria is never unique, the particular kinds of concepts adopted are always dependent in this view on non-logical influences. Hence within deductivism as a view of science it is even impossible to make the distinction between intellectual and social influences on discovery, and *a fortiori* no general claim to internal autonomy of the history of scientific ideas can be sustained.

What has in fact happened is that, far from philosophy providing criteria for history, all forms of historical investigation, internal as well as external, have led to radical questioning of all received philosophical views of science. For they have revealed the impossibility of drawing any sharp line between the contexts of justification and discovery, between the 'rational' arguments as defined by deductivism and the psychological and cultural processes which determine what kinds of theory are contemplated, and even between the 'hard facts' which must be respected as tests of theory and the way these facts were interpreted in a given cultural environment. It is no accident that the current attacks upon these entrenched dichotomies of modern philosophy of science come either from historians of science (explicitly and recently from Kuhn, but also implicitly from Duhem), or from philosophers who are deeply immersed in history of science and conduct their discussions by means of detailed case histories (for example, Popper, Feyerabend, Hanson and Toulmin). Some of these writers hold a conventionalist position in philosophy of science which parallels what is described above as conventionalism in historiography, with stress on the rôle of intellectual factors in scientific development, but without any implication that external causes of change are excluded, or that there is any intrinsic distinction between the approaches of internal and external history.

II

How then are we to understand the evaluative task of the historian with regard to the rationality of the science of the past? The practice of historians is usually wiser than their theoretical self-reflections, and all these pressures towards scepticism about criteria of scientific rationality do not prevent the occurrence of evaluation and interpretation in the best historiography of science. But the pressures are not unimportant, because they sometimes tend to impoverish the self-understanding of historians of science, and sometimes to bias their practice and conclusions. In justification of these claims I shall consider in some detail a particular development in the historiography of seventeenth-century science. This is the effect of renewed interest in the hermetic and natural magic traditions, stimulated by Frances Yates' pioneering studies, particularly her book *Giordano Bruno and the Hermetic Tradition*, and by Walter Pagel's detailed work on Paracelsus and his period.[7]

The hermetic writings were a group of gnostic texts actually dating from the second and third centuries A.D., but believed in the sixteenth century to be contemporary with or prior to Moses, and originating in Egypt. They consequently carried all the ancient authority so much revered in the Renaissance; they were quite non-Aristotelian in spirit and hence reinforced the anti-scholastic tendencies of Renaissance thought; and since they were in fact written in the Christian era, they contained some elements of Judaism and Christianity which were regarded as prophetic and so enhanced still further their authority. P. M. Rattansi epitomizes well the main tenets of hermeticism in contrast with the careful distinctions maintained in medieval scholasticism between the natural and the marvellous, the magical, and the miraculous—

> For Hermeticism, by contrast, man was a *magus* or operator who, by reaching back to a secret tradition of knowledge which gave a truer insight into the basic forces in the universe than the qualitative physics of Aristotle, could command these forces for human ends. Nature was linked by correspondencies, by secret ties of sympathy and antipathy, and by stellar influences; the pervasive nature of the Neo-Platonic World-Soul made everything, including matter, alive and sentient. Knowledge of these links laid the basis for a 'natural magical' control of nature. The techniques of manipulation were understood mainly in magical terms (incantations, amulets and images, music, numerologies).

Of this tradition Rattansi comments: 'It was not completely vanquished by the rise of the mechanical philosophy. Without taking full account of that tradition, it is impossible . . . to attain a full picture of the "new science".'[8]

In *Giordano Bruno* Miss Yates herself had specifically disclaimed the intention of contributing to the history of science proper: 'with the history of genuine science leading up to Galileo's mechanics this book has nothing whatever to do. That story belongs to the history of science proper . . . The phenomenon of Galileo derives from the continuous development in Middle Ages and Renaissance of the rational traditions of Greek science.'

[7] F. A. Yates, *Giordano Bruno and the Hermetic Tradition* (London, 1964); W. Pagel, *Paracelsus* (Basle, 1958).

[8] P. M. Rattansi, 'The intellectual origins of the Royal Society', *Notes and Records of the R.S.*, **23** (1968), pp. 131–2.

More recently, however, she has made bolder claims for the relevance of the hermetic tradition—

> I would thus urge that the history of science in this period, instead of being read solely forwards for its premonitions of what was to come, should also be read backwards, seeking its connections with what had gone before. A history of science may emerge from such efforts which will be exaggerated and partly wrong. But then the history of science from the solely forward-looking point of view has also been exaggerated and partly wrong, misinterpreting the old thinkers by picking out from the context of their thought as a whole only what seems to point in the direction of modern developments. Only in the perhaps fairly distant future will a proper balance be established in which the two types of inquiry, both of which are essential, will each contribute their quota to a new assessment.[9]

The measured statements of Miss Yates and Rattansi only hint at the potential difficulties. Lying not far below their surface, and explicit in some younger writers, is the suggestion that the enterprise of internal history of science, as pursued for instance in the history of the seventeenth-century mathematico-mechanical tradition, is a mistake. It is a mistake because various kinds of non-rational factors are so closely bound up with rational argument that a history which tries to concentrate on the latter is necessarily distorted. It is a mistake also because, so it is implied, it reads and evaluates seventeenth-century science as part of a tradition which is our tradition, instead of understanding it in its own terms.

In order to investigate these issues we need to look more closely at the notions of 'rational science' and its 'internal history'. The way in which these terms are commonly intended by historians may be indicated in the hermetic example by contrast with some of the historiographical elements which are claimed to challenge them. These are (1) the social and political affiliations of certain religious sects, and the schools of Paracelsian and Helmontian doctors and chemists, (2) the full-scale hermetic and natural magic tradition as a way of thought and life in such writers as Paracelsus himself, Bruno, and Fludd, and (3) the doctrines of extended spirits and powers of matter which persisted in later seventeenth-century science, including Newton's work, in opposition to corpuscular mechanism.

Of the first of these factors it does not seem to be anywhere claimed that they provided more than the occasion and the motivation for certain developments connected with the new science. Such sectarian figures as Hartlib, Dury and Comenius helped to encourage Baconian allegiances in the early Royal Society, and the anti-establishment circles in which they moved

[9] F. A. Yates, 'The hermetic tradition in Renaissance science', *Art, Science, and History in the Renaissance*, ed. C. S. Singleton (Baltimore, 1968), p. 270.

go some way to explain the suspicion with which the Society was viewed by
Royalists and Churchmen in the Restoration period. But none of this seems
to impinge essentially on the internal tradition of the history of the mechanical philosophy as found in the writings of such historians as Duhem, Koyré,
Burtt and Dijksterhuis. Indeed in the debates which have followed the
related theses of Robert Merton and Christopher Hill regarding the influence
of puritanism on seventeenth-century science, several commentators have
remarked that the argument suffered from too little conceptual clarity about
what was to count as 'science' (and indeed as 'puritanism'). Far from
suggesting a restructuring of the internal tradition, these debates presupposed
its existence, and the disputants were counselled to look at what had been
achieved in internal history in order to acquire some internal specification
of what 'science' is.[10]

The case with the second and third elements of the hermetic complex is
different, because here it is not a question of interacting social factors, but
of intellectual factors which might be held to be necessary ingredients of
the history of science seen as the history of thought. Their close relation to
the social factors just mentioned has obscured the fact that the real challenge
to the received internal tradition comes not so much from social and political
factors, as from within history seen as 'thinking men's thoughts after them'.
In the passage quoted above Miss Yates excludes some *ideas* from the
'rational tradition', and yet these ideas are undoubtedly in men's heads and
presupposed in much of their literature. The first question is, then, can the
notion of 'internal history' be more closely defined in terms of some understanding of what is to be 'rational science', in contrast to complexes of ideas
such as hermeticism? Only when this is answered can we judge whether the
pursuit of relatively autonomous internal history is a viable proposition. If
this question is taken in a philosophical sense it is a request for some
perennial logical criterion which can be used to delimit the rationality of
science at different periods, and it carries also the suggestion that a distinction of historical method between internal and external history is appropriate, since science pursued according to certain internal norms of rationality
is likely to have a structure of development different from that of the
contingent clash of events, actions and thoughts which are the stuff of
general history.

In his book *Foundations of Historical Knowledge*, Morton White has given a
useful and relevant analysis of the idea of intellectual history. He makes a

---

[10] R. K. Merton, 'Science in seventeenth century England', *Osiris*, 4 (1938), p. 360;
C. Hill, *Intellectual Origins of the English Revolution* (Oxford, 1965); M. Purver, *The Royal
Society: Concept and Creation* (Oxford, 1967); and articles by C. Hill and H. F. Kearney in
Past and Present, 1964, 1965.

three-fold distinction between causal explanation, which may be rational or non-rational, and what he calls 'non-causal rational' explanation. 'Causal non-rational' explanations refer to the particular occasions upon which some belief comes to be held, for example the presence of certain Continental social reformers in England which encouraged some members of the earlier Royal Society to adopt Bacon's philosophy of science. On the other hand when a historian asks 'Why did Descartes believe in the existence of God?', he may give a 'non-causal rational' explanation in terms of 'the reasons stated by the thinkers or half-stated by them, or the reasons they would have stated if they had been asked certain questions'.[11] But, White goes on, it may be the case that 'whereas he *said* he believed p because it followed from certain other propositions which he thought were true, the *real* cause of his believing p was the fact that he wanted to believe it or was scared into believing it or was in the grip of some neurosis'. This kind of explanation White calls 'causal rational' explanation. Like all causal explanation in his view causal rational explanation falls under the regularity or covering-law model of historical explanation, whereas non-causal rational explanation does not. White appears to believe, further, that there can be no interaction between these different kinds of explanation, but at most a temperamental disagreement between different sorts of historians about what they want to call the *real* explanation. If this analysis were acceptable for history of science we could make the distinction between internal and external history coextensive with White's distinction between non-causal and causal explanation, and in terms of this definition internal history would be logically independent of all types of external factors.

Before discussing the adequacy of the analysis for history of science, it should be remarked in parenthesis that White's distinction is not the same as that commonly made in terms of explanation of human thought and action in terms of *causes* and in terms of *reasons*. Indeed his view of historical explanation is directly opposed to the view that history deals in reasons and not primarily in causes, that is that understanding of historical action (including belief) must involve a sort of sympathetic rehearsal of the intentions of the historical characters on the part of the historian, and that this method is logically distinct from the type of causal explanation appropriate in science. White's category of *causal* rational explanation on the other hand is intended to assimilate some beliefs that are commonly taken to be reasons (whether valid or mistaken) to causes as understood within the regularity view of explanation. It is only when reasons are appealed to as norms that they count as *non-causal* explanations; moreover appeal to norms must be by the *historian*, not only by the historical character, for the character's appeal

[11] M. White, *Foundations of Historical Knowledge* (New York and London, 1965), p. 196.

might simply be reported by the historian as a fact to be inserted in a possible *causal* explanation. Thus it is only in connection with the history of intellectual pursuits, in which the existence of such atemporal norms is plausible, that White's category of non-causal rational explanation is applicable. Mathematics might be a fairly uncontentious candidate for such explanation, but philosophy and science are much more doubtfully so, as I shall now try to show.

If we attempt to apply White's distinction to the demarcation of internal history of science, we are immediately in difficulties. He appears to take for granted what constitute 'reasons' in the case of Descartes' argument for the existence of God. Are these, then, self-evident and timeless norms to which all philosophers *qua* philosophers are bound to conform? This can hardly be so, because even in cases where a classic philosopher can, as it were, be argued with as with a colleague, it is not self-evident that his own understanding and use of, for example, the ontological argument was the same as that of a modern analytic philosopher, and indeed some historians have argued that Descartes' understanding of this argument was not. And when it comes to the history of science, we have seen that neither science nor philosophy of science provide us with timeless norms of rationality ready-made.

Does White perhaps mean to assert that non-causal rational explanation must be given in terms of what were seen as reasons at the time? This is the kind of 'self-transcendence' or 'self-emptying' suggested, for example, by Lovejoy and Butterfield (both primarily historians, not philosophers), and by Collingwood. But White clearly regards this type of history as providing causal explanation, not reasons. A few pages before his three-fold distinction he says:

> The savage's belief that a person is a witch may be causally explained by reference to another belief of the savage which, as *we* might say, does not logically support it. There are explanatory deductive arguments that connect one mad, false, or superstitious belief with others. And the point is that the historian is engaged in causal explanation *in the same sense* whether he is explaining the beliefs of a sane or insane man, a civilized man or a primitive man, a genius or a fool.[12]

The explanation is given in the same sense, not because all alleged logics are equally logic, but because the regularity model of explanation is the same for all types of cause, whether the causes are beliefs or not, and whether the beliefs are true or false.

It is clear that any proposal to try to make a logical distinction between atemporal rational explanation, and hence internal history, and causal

[12] Ibid., pp. 188–9.

explanation in terms of the rationality of a particular period will not stand up to a moment's investigation. In the sixteenth or seventeenth centuries there was no agreed 'scientific rationality of the time', and even the intellectual practice of individuals was a species of sleepwalking. What for example, in terms of this proposal, could we make of Copernicus' inference that the sun is at the centre of the planetary system because it is analogous to the king at the centre of his court, or to God the still centre of the universe? Indeed, as the historian has found when he has tried to specify seventeenth-century scientific method in its own terms, a very great variety of modes of acceptable argument emerge, some of them almost as remote from our views of rationality as the tenets of hermeticism itself. Moreover, metaphysics was inseparable from method: what was understood as 'science' was to some extent constituted by the mathematical, mechanical, non-animist, and non-teleological approach to nature. It is useless even to appeal to a general 'concern for facts' among those who count as 'scientists', for many of them are explicit that the mathematical and mechanical framework determines what is to count as a natural phenomenon, all else is excluded from science as supernatural or miraculous. Part of the historians' task is precisely to *discover* how far various kinds of inferences were acceptable at the time and why, and to investigate the *changing* conceptions of rationality which themselves partly constituted the scientific revolution.

The only remaining possibility of distinguishing internal history in terms of a logical demarcation of rational explanations is that reasons are just what appear to us to be reasons, whether or not we can explicitly formulate these, and whether or not there is any agreement about their timelessness or normative character. This might be called the *rational reconstruction* view of intellectual history, and it does indeed seem to be the standpoint adopted by White with respect to his non-causal explanations. It is, however, in conflict both with the judgements of historians quoted earlier, to the effect that autonomous internal history is necessarily distorted, and also with the actual tradition of internal history of, for example, seventeenth-century mechanical philosophy, where interpretations in terms of now-unacceptable forms of inference are taken for granted. Rational reconstructions, on the other hand, can be seen as a species of latter-day inductivism in the history of science, in which, although past theories and alleged facts are not seen wholly in the light of their correspondence with present theories and facts, modes of scientific inference are so seen.

A case might be made for rational reconstruction along the following lines: there are sometimes in the development of science 'deep reasons' why one sort of theory is intrinsically preferable to another, or likely to be more progressive than another, and these may not be reasons that anyone was or could have been aware of at the time. By 'deep reasons' I mean logical or normative relations which emerge only if models of logically possible structures of science are investigated independently of their historical incarnations, and also mathematical relations which turn out to have interesting consequences which could not rationally have been foreseen at the time. As examples from normative philosophy of science we might refer to Popper's demonstration that a certain kind of power of theories is related logically to their simplicity, or to the proof within a probability theory of induction that inference by analogy from instance to instance gives high probability of predictions in some cases where theories understood merely as deductive systems do not. The first example 'explains' (in the sense of White's non-causal rational explanation) why scientists generally prefer simple theories, the second explains in the same sense why they usually proceed by analogical models rather than by formal deductive systems. The mathematical type of deep reason is perhaps applicable only to history of mathematics and the mathematically-oriented sciences. An example would be Maxwell's introduction of the displacement current, for which he gave various relatively unconvincing reasons at the time, including an apparently superficial analogy with magnetism and also an argument depending on a dubious mechanical model of the aether. But though he could not have explicitly known it, the most significant feature of the displacement current was, and remains, that it renders Maxwell's equations Lorentz-invariant. It may not always be improper to 'explain' a scientist's apparently baseless intuition by such mathematical truths as this, particularly if our later knowledge of them enables us to interpret hints in Maxwell that could not have been intelligible to his contemporaries.

Passmore[13] has remarked that history of science differs from history of philosophy in that many histories written by philosophers are polemical, and are intended as first-order contributions to philosophical debate, but there are few polemical histories of science. Perhaps there ought to be more. Rational reconstructions may be a contribution to contemporary science, but their function in this respect should be distinguished from their contribution *to history*, where they should surely be seen only as ancillary to approaches which are more sensitive to different conceptions of rationality in the past.[14] Indeed no history of science can avoid mixing these ingredients in some proportion; the best history from Aristotle through Whewell and Duhem to Koyré has been such a judicious mixture, and has also been

---

[13] J. Passmore, 'The idea of a history of philosophy', *History and Theory*, 5 (1965), p. 8, n. 18.

[14] The distinction is made very sharply in J. Dorling, 'Scientist's history of science and historian's history of science', forthcoming.

explicit about the normative philosophy of science adopted. To return to the question of the autonomy of internal history, it must be concluded that if 'internal' history is interpreted in any other way than as pure rational reconstruction, then the question whether it can be pursued independently of external and non-rational factors is a historical, not a philosophical, question, and the answer to it will vary from case to case. A little further consideration of the hermetic debate may illustrate this.

III

When all has been said about the absence of *a priori* criteria for 'rational science', our intuition remains that however varied may be the explicit and implicit methodologies of seventeenth-century science, they are still worlds away from hermeticism. This intuition is confirmed by several examples in the period of intellectual dispute between adherents of the two traditions in which we find moral, political, and theological arguments deployed along with philosophical ones in the course of vigorous repudiation of the hermetic cults.

In an exchange of polemics with the English Rosicrucian doctor Robert Fludd, Kepler dissociates himself from the interpretation of mathematics found in the hermetic writers.[15] Kepler does indeed himself believe in a mathematical harmony of the cosmos as the image or analogue of God and the soul, but his geometry is Euclidean, his conclusions require proof, and they must correspond with facts (that is, the kind of facts Kepler inherited in Brahe's planetary tables). According to Fludd, on the other hand, Kepler merely 'excogitates the exterior movement . . . I contemplate the internal and essential impulses'. Fludd *complains* that geometry is dominated by Euclid, while arithmetic is full of 'definitions, principles and discussions of theoretical operations, addition, subtraction, multiplication, division, golden numbers, fractions, square roots and the extraction of cubes'. There is, he goes on, no 'arcane arithmetic', no understanding of the significance of the number 4, deriving from the sacred name of God.

In less measured tones than Kepler, Mersenne devotes himself to combating the arrogance and impiety of the terrible magicians.[16] Their arbitrary

[15] The documents have been presented by W. Pauli, 'The influence of archetypal ideas on the scientific theories of Kepler', in C. G. Jung and W. Pauli, *The Interpretation of Nature and the Psyche* (English trans., London, 1955), p. 151. See also A. G. Debus, 'Renaissance chemistry and the work of Robert Fludd', *Ambix*, 14 (1967), p. 42, and 'Mathematics and nature in the chemical texts of the Renaissance', ibid., 15 (1968), p. 1.

[16] R. Lenoble, *Mersenne ou la naissance du mechanisme* (Paris, 1943), p. 7; F. A. Yates, *Giordano Bruno*, ch. 22.

numerologies do not even agree among themselves; they do not understand that words are mere *flatus voces*, merely conventional signs or sounds, not images or causes. The proportion of the planetary distances may exhibit harmony, but whether it does or not is a matter of fact, not of cosmic analogies. Moreover, astrology, magic and the cabala are not just harmless games, they reduce human freedom to cosmic determinism and hence are morally reprehensible. Although some alleged examples of sorcery may be facts, use of sorcery is morally detestable; the magicians are guilty of arrogance and impiety in their claim that the human intellect is divinely inspired and is the measure of things. When Fludd replies to this onslaught with equal violence, Mersenne requests Gassendi to take up the cause, and he, slightly reluctantly but for friendship's sake, drops what he is doing in order to study Fludd's writings.[17] That is a measure of the externality of the hermetics at this period to the new philosophy.

Another such polemical exchange is Seth Ward's *Vindiciae Academiarum*, written in reply to an attack upon the academic activities of the University of Oxford by John Webster.[18] Webster berates Oxford for its neglect of the new science, citing indifferently as representatives of that science Bacon, Copernicus, Galileo, Paracelsus, Boehme, Fludd, and the Rosy Cross. Ward replies with careful distinctions between the true natural language or universal character 'where every word were a definition and contained the nature of the thing', and 'that which the *Cabalists* and *Rosycrucians* have vainly sought for in the Hebrew'. Hieroglyphics and cryptography were invented for *concealment,* grammar and language for *explication.* Magic is a 'cheat and imposture . . . with the pretence of specificall vertues, and occult celestiall signatures and taking [credulous men] off from observation and experiment. . . . The discoveries of the symphonies of nature, and the rules of applying agent and materiall causes to produce effects, is the true naturall magic'. Both Mersenne and Ward take Aristotle for an ally against the magicians: it is not Aristotle, rational though wrongheaded, who is the enemy of the new philosophy, but 'the windy impostures of magic and astrology, of signatures and physiognomy'.[19]

Rattansi characterizes the situation accurately when he contrasts 'the emotionally-charged and mystical flavour of Hermeticism, its rejection of corrupted reason and praise of "experience" (which meant mystical illumination as well as manual operations), and its search for knowledge in arbitrary,

---

[17] G. Sortais, *La philosophie moderne depuis Bacon jusqu'a Leibniz,* Vol. 2 (Paris 1922), p. 43.

[18] S. Ward, *Vindiciae Academiarum,* Oxford, 1654; J. Webster, *The Examination of Academies* (London, 1654).

[19] Ward, op. cit., pp. 22, 34, 36; Lenoble, op. cit., p. 146.

scriptural interpretation', with 'a sober and disenchanted system of natural knowledge, harmonized with traditional religion', and goes on

> To move from one to the other was to change one conceptual scheme for ordering natural knowledge to another, with an accompanying shift in the choice of problems, methods, and explanatory models.[20]

The change of sensibility is also a contemporary view. For example Glanvill: 'among the Egyptians and Arabians, the Paracelsians, and some other moderns, chemistry was very phantastic, unintelligible, and delusive, . . . the Royal Society have refined it from its dross, and made it honest, sober, and intelligible. . . .'[21] And Sprat's plea for a 'close, naked, natural way of speaking' is directed as much at the 'Egyptians' as at the Aristotelians.[22]

In view of all this, the suggestion of a *confluence* of hermeticism and mechanism into the melting pot of the new science is a mistake. In all that constituted its essence as a way of thought and life, hermeticism was not only vanquished by the mid-century, but had provided the occasion for the new philosophy to mark out its own relative independence of all such traditions. The style of argument required in the polemics is itself significantly different from that adopted in domestic scientific disputes. It involves rhetoric and ridicule, and appeals to theological and moral principle, and sometimes political and pragmatic test. A new form of rationality can be seen to be distinguishing itself from traditional modes of thought, this can be perceived and described by the historian, and in some cases its application can be seen historically to have delimited a relatively autonomous area of study. Thus the question about the autonomy of internal history is not a question of imposing external norms, but of investigating actual historical influences.

IV

In the light of the debate about hermeticism let us now look at three types of scepticism regarding rationality that have been drawn from the history of science. All three raise issues already familiar in the general philosophy of history, and I shall conclude by examining the question whether there are any peculiar features of natural science which make some assumptions

[20] Rattansi, op. cit., p. 139.
[21] J. Glanvill, *Plus Ultra* (London, 1668), p. 12.
[22] T. Sprat, *The History of the Royal Society of London* (London, 1667), sec. 20; see also sec. 3.

of the general debate inapplicable to this case. The three sceptical conclusions are

1. Since no generally acceptable normative criteria of rationality are forthcoming either from philosophical or historical analysis of science, notions of what constitutes scientific rationality are historically relative.

2. Since we find in the history of science no guarantee of autonomous internal and external sectors but a complex of interactions between many different types of factor, to select some of these and omit others is necessarily to distort the picture.

3. Value-judgements about the rationality and truth of past systems of scientific thought should be avoided by the historian.

1. I have already accepted that there are no *a priori* normative criteria for science suitable for providing a logical demarcation of internal and external history. It does not follow from this, however, that the relative autonomy of internal history, or of the history of a certain kind of scientific tradition, may not be established on historical grounds in particular cases. It is possible that it can be established in the hermetic example, and I would indeed judge that none of the literature on the hermetic tradition has yet shown the contrary. The possibility of finding within the seventeenth-century debates themselves claims to partial independence of one tradition from others; the difference in character of the argument within the mechanical philosophy and the polemics between it and other traditions; and already established results regarding the internal development of the physical sciences seen in terms of a relatively independent rational tradition implicitly defined by themselves, all suggest that relative autonomy may be justified on historical grounds. A proper historical perspective neither involves uncritical accumulation of every minor writing of forgotten figures, nor is it necessarily vitiated by the imposition of our standards of rationality on an alien age. We cannot, to be sure, merely adopt Butterfield's advice to use the judgements of importance of the period rather than our judgments,[23] for how in that case could we judge the relative weight in the history of science to be given to Kepler's arguments as against Fludd's? To reply that the seventeenth century itself clearly accepted Kepler's and Mersenne's view rather than Fludd's is not sufficient, for this would be like relying on the popular verdict of Athens upon Socratic philosophy to dictate our judgements of intrinsic importance. So long as we select science as our subject-matter, we are bound to write forward-looking history in the limited sense that we regard as important what we recognize as our own rationality, having some historical continuity with our own science. This is likely to mean that given

---

[23] Butterfield, op. cit., p. 24.

a choice of forms of explanation between what we regard as rational methods on the one hand, and social pressures or psychology on the other, we shall regard the former as more significant. But this does not imply that we impose our own theories or even our own views of method on the science of the past. And if it seems in danger of becoming a circular definition of internal history as that which is continuous with our science according to our internal history, the only cure is to look more closely at the historical record to see whether the relative autonomy of internal history can be maintained in spite of possible disturbing factors. We may, indeed, sometimes be shown wrong in our imputation of our sorts of reasons even when the conclusions were in our eyes correct, for we may find strong evidence that ideological or psychological motives were in fact more influential.

2. The problem of the distorting effects of selection is of course a perennial one for all types of historian, and might seem to be by now sufficiently well understood to make it unnecessary to give it special attention in relation to the history of science. But there have been so many unguarded comments on the hermetic tradition, and in other assessments of the relation of social history to science, that historians of science may almost be said to be unconscious of the problems. It often seems to be assumed that by adding to the picture all influences that fed into it, of all conceivable degrees of relevance, we get nearer to some form of complete description or complete understanding of the 'whole picture'. But the view of history as complete description, or 'telling it like it was', is an error analogous to the error of inductivism in science. It presupposes that history is a search for hard facts, which are relatively independent of each other, and that the full picture is attained by accumulating as many of these as possible. As a philosophy of history, this is the temptation of the conventionalist rather than the inductivist, for the inductivist has already his principle of selection, whereas the conventionalist is likely to think that the total internal coherence of a period will emerge more clearly as more factors are taken into account. But even the suggestion that it is possible to get nearer the true picture by accumulating factors should be treated with caution. Throwing more light on a picture may distort what has already been seen. The immediate enthusiasm for searching for the names of 'Hermes', 'Orpheus', and other pseudo-priscine authors in seventeenth-century scientific writings may be more distorting than the received internal tradition itself, unless careful distinctions are made between several different types of use of these names. They may be used as pious archaisms; or as familiar labels for ideas about non-mechanical forces which are in fact central to the new scientific tradition; or as thinly disguised Christian piety, owing little to gnosticism and magic, as in the

case of Kepler; or on the other hand as really indicating full-blooded adherence to a hermetic cult. Once all these distinctions have been made, the received internal tradition recovers something of its autonomy and is seen not to have been seriously distorted by their previous neglect.

3. Thirdly there is the question how far evaluations of the truth and rationality of past observations, theories, and forms of scientific inference are desirable or even unavoidable in the historiography of science. In the philosophy of the social sciences there are familiar arguments purporting to show that the selection of areas of interest to study, and even the occurrence of value-loaded concepts in these areas, does not imply that the investigator himself makes the value-judgements. 'Hysterical' may be a value-loaded term in the history of medicine, but in this view the historian of medicine can report the circumstances in which the term was used and with what implications, without himself pronouncing upon the desirability or otherwise of hysteria. This view should, I think, be kept distinct from the view that often accompanies it, namely that fact and *theory* can be kept distinct. It may be argued that the circumstances of application of 'hysteria' can be described without either making value-judgements or using other value-loaded words (which would make the elimination of value impossible), but this does not entail that *theory*-loaded terms can be explained factually without appeal to other theory-loaded terms. Therefore rejection of the distinction of fact and theory (which I should myself want to reject) does not entail rejection of the distinction of fact and value.

Without pursuing that particular question further, let us assume that the consensus of analytic opinion is right, that in general the description by the historian of the use of value-judgements by historical characters does not commit the historian to making these judgements, and in particular that various kinds of scientific inference and truth-judgement can be described by the historian of science without any commitment on his part to the rationality of the inference or the truth of the judgement. It may, however, still be the case that there are peculiar features about the history of *science* which relate judgements of rationality more intimately to the subject matter than in the case of other types of intellectual history. I think there are such features, and I shall conclude by mentioning some of them to present a case for a limited inductivism in the history of science.

First it may be argued that the relativist interpretation of scientific methodology undermines the very assumptions of historical method itself. In the covering-law or regularity view of historical explanation, the deductive analysis of scientific explanation is explicitly taken as the model, and if the latter is abandoned there seems little case for retaining the former. The opposing view, associated with the ideas of irreducible human action, intention, and reason, on the other hand, has generally been defined in *contrast* to the notion of scientific explanation in terms of causes. As Nagel puts it 'even extreme exponents of the sociology of knowledge admit that most conclusions asserted in mathematics and natural science are neutral to differences in social perspective of those asserting them, so that the genesis of these propositions is irrelevant to their validity'.[24] And even extreme exponents of the doctrine of *verstehen* as the mode of historical explanation require to use the results of so-called 'scientific history'—the provenance of documents, the dating of archaeological remains, the recovery of ancient languages, the changing ecology of environments, and so on. The modern practice of history would be impossible if it were not sometimes assumed to be meaningful to ask the question 'Did it happen?' in the expectation that generally speaking various kinds of historical investigation will converge upon the same answer. At least this expectation is similar to that generally made in science when an experiment is taken to yield an answer to the question 'What happens?' The two expectations have similar logical structures, even though it may not be correct to describe either of them in terms of a naïve inductivism. It is therefore viciously circular to use historical findings to undermine the timeless validity of these currently accepted forms of scientific practice.

In the particular case of the history of science this argument is even stronger. For if we are to explain, say, the fact that Priestley believed in the phlogiston theory, we have to consider what his data were, and what inferences he claimed to draw from them. But we know what his data were partly because he described them, and described them in terms such as 'I could not doubt but that the calx was actually imbibing something from the air; . . . it could be no other than that to which chemists had unanimously given the name of *phlogiston*'.[25] From this we have to recover *what* it was Priestley was observing, and this can only be done in the light of the best scientific information *we* have about what happens when lead oxide is heated in an atmosphere of hydrogen. Or take Oersted's twenty-year-long attempts to demonstrate what any schoolboy can now do with a battery, a piece of wire and a compass needle.[26] In order to understand Oersted's difficulties, we have not only to know that he believed in a Newtonian theory of central attractive and repulsive forces and applied it to produce the wrong expectations in this case; we have also to reconstruct what his equipment must

[24] E. Nagel, *The Structure of Science* (New York, 1961), p. 500.

[25] Quoted in S. Toulmin, 'Crucial experiments: Priestley and Lavoisier', *Journ. Hist. Ideas*, 18 (1957), p. 209.

[26] See R. C. Stauffer, 'Speculation and experiment in the background of Oersted's discovery of electromagentism', *Isis*, 48 (1957), p. 33.

actually have been like in order not to have immediately revealed what is obvious to us about the direction of rotation of the compass needle. This is not an undesirable inductivism—it is a requirement of the programme of taking the facts and ideas of a period seriously. And it is incidentally one of the comparative tests of our science that it can not only explain what was differently explained in the past, but explain why other things were not explained or even observed when they should have been. It may of course be replied that this comparison is reciprocal, because *we* are doubtless also neglecting things we should be seeing, and a different rationality might detect and explain this neglect. But with regard to *our history* the point remains valid, for no one has yet suggested that *we* should write Aristotelian history of twentieth-century science, even if such a project were conceivable, and we certainly cannot write it from the point of view of a rationality of the future.

Examples from the history of science remind us that science has to do with happenings as well as ideas, and if the interpretation of past happenings in terms of modern theories is to be more than an arbitrary imposition of our own standards, it must be presupposed that there is something absolutely preferable in our own science to that of the past. That in other words there is a sense in which the development of science stands outside historical relativity and is absolutely progressive. Can this sense be perceived without losing sight of much that is acceptable in the relativist analysis?

Though the progressive character of science is not straightforwardly a question of having more knowledge about more facts, it is perhaps less misleading to put it that way than to refer to the progression of better and better theories or increasing adequacy of methods. What has been illuminating in the relativists' account is their demonstration that conceptual theories undergo deep revolutions and do not converge continuously towards some fundamental truth. For the same reason the progress of science is not a simple expansion of the number of true observation statements accounted for in successive theories, because the language of the observation statements is itself permeated by the theory, and there is no simple comparison between the way two theories describe the same events. Neither of course is it simply the case that we know more facts now than in, say, 1600. New facts may be learned by cataloguing, botanizing, collecting moon samples, or devotion to high-speed data processing systems. Some of these may be used by science and some may depend on science, but they do not in themselves *constitute* the progressive character of science. Moreover post-1600 science has deliberately discarded and forgotten many facts that were then known—numerical facts about the Pyramids or Old Testament chronology, and facts about alchemical and magical operations for example. If science is in some sense progressive with respect to facts, the facts must be specified more closely as those that are significant for science.

There is however a sense in which we do know more facts, and that is the sense most closely related to technological control, although it is not simply identical with it. We know more facts that are accounted for and interpreted by very general and systematic theories, and we know them in the sense that we can use these theories to provide us with verifiable expectations about what will happen next. This is exploited in technology, but is not identical with it, because there could be rule-of-thumb technology without it, and on the other hand it need not be so exploited. Because this sense of the progress of science is about controlled happenings, it is independent of the way facts are described relative to different theories. It is a positive and absolute development of the last few centuries, and it constitutes the core of truth in inductivism, whether that is taken as a philosophy or a historiography of science. It provides an absolute criterion of distinction of our rationality, although it does not of course imply any *moral* evaluation of that rationality.

In summary, then, I have suggested three sources of scepticism in regard to the historian's treatment of scientific rationality: the collapse of inductivism, the failure of philosophical analysis to present a normative model for science which is generally acceptable and clearly relevant to the history of science, and the historical investigation of relations between history of science and social and ideological history. I have argued that the notion of 'internal history' with which intellectual historians of science have worked is not definable in terms of external normative criteria of rationality, nor by simply taking the accepted criteria of the time as normative. Relative autonomy of internal history may, however, be justified in some cases, as with the mechanical versus the hermetic tradition in the seventeenth century, by historical investigation itself. The historian of science then inevitably adopts as his principle of significance that tradition which has some historical continuity with our own, although this does not mean that he imposes either our accepted theories or our accepted methods on the past in the fashion of the old inductivism or the newer rational reconstructions. Finally, I have suggested that our understanding of scientific rationality must have a special place in historical study, and especially where questions about what was actually observed have to be answered in the context of the history of scientific theories. In this sense, history of science is inevitably evaluative and forward-looking. Moreover this does not indicate a relativity to our science merely, because there is a relevant sense of 'progress' in which science is absolutely progressive in its understanding of facts. The sceptical conclusions lately drawn are not justified.

X

SOME EVALUATIONS OF REASON IN SIXTEENTH- AND SEVENTEENTH-CENTURY NATURAL PHILOSOPHY

P. M. Rattansi

I

The association between nature-mysticism and science . . . is to be found embedded in the very foundations of modern (post-Renaissance) scientific thought. Joseph Needham, *Science and Civilisation in China*, Vol. 2, p. 95.

It is some decades since Walter Pagel published a 'Vindication of Rubbish'[1] and Otto Neugebauer a defence of 'The Study of Wretched Subjects'.[2] Dr. Mary Hesse's paper in this volume indicates that an apology has now become necessary for the study of what has come to called 'Renaissance Hermeticism'.

Since modern science is regarded as the antithesis of magic, it has seemed self-evident to many historians of science that the study of such 'rubbish' is not their concern. Hermeticism offered anxious men in a turbulent age the delusive hope of controlling nature through the power of the stars. It represented a pathological resurgence of irrationalism, succeeding the decay of scholastic-Aristotelian rationalism. The basis of such superstition was destroyed by the 'scientific revolution' of the seventeenth century, which established rationality and objectivity of a radically different sort. It is perverse, then, to grant Hermeticism any significance for the history of science. It was as alien to the scholastic-Aristotelian rationalism as to the new scientific rationalism which replaced it.

Dr. Hesse's closely-argued rejection of the significance of Renaissance

[1] *Middlesex Hospital Journal*, 1945.
[2] *Isis*, 1951, 42.

Hermeticism for the study of the history of science makes some other anxieties aroused by recent studies clearer. Those who engage in such studies are seen as part of a convergent movement (in the history as well as the philosophy of science) which has relativized judgements of the truth and rationality of past science. To grant Hermeticism any prominence in the history of sixteenth- and seventeenth-century science is tantamount, apparently, to challenging the rationality of science. The number of times, and the variety of contexts, in which the word 'rational' recurs in Dr. Hesse's paper is quite striking. Besides affirmations of the 'rationality' of science, a modified 'rational reconstruction' is urged as a historical method, and even the familiar debate on 'internal' and 'external' factors in the history of science is restated in terms of 'rational' and 'non-rational' factors. Whenever there is a choice between 'rational methods' and 'social pressures or psychology' in explaining past scientific developments, we are counselled to choose the former. Such iteration, when accompanied by the suggestion that those who study Hermeticism are thoroughgoing relativists, gives the argument a powerful emotional resonance: the issues assume the form of a defence of 'scientific rationality' against a host of detractors and enemies—a defence all the more vital in a current situation in which the values of science are, indeed, under attack from sections of the rebellious young.

There is no reason why those who have thought it important to assess the impact of Hermeticism on natural philosophy in the sixteenth and seventeenth centuries should accept such a characterization of their motives or 'objective rôle'. Some of the most exciting work in the history of early modern science has come for a generation from intellectual historians like E. A. Burtt, Ernst Cassirer, and Alexander Koyré, who showed how the study of nature is related to larger metaphysical assumptions and is involved in complex ways with other areas of intellectual culture. Their work has seemed to require extension in at least two directions. First, by a closer study of Renaissance Neo-Platonism and Hermeticism. Both Burtt and Cassirer attached great importance to Renaissance Neo-Platonism, and its assimilation and development needs greater study in relation to the various religious and philosophical currents of the sixteenth century. The other, and related, concern, is the social context of the study of nature in the sixteenth and seventeenth centuries.

Neither concern indicates such a radical break with the prevalent style of the historiography of science as Dr. Hesse implies. The recent explorations of Renaissance Neo-Platonism and Hermeticism and of particular thinkers in relation to those currents have, indeed, been guided by questions about the sort of science *we* consider significant and continuous with our own. In that sense, 'Whig' criteria will generally be decisive in determining what seems most interesting and important to us in the study of the natural philosophy of the past—a situation which is no different in other branches of the history of ideas. But 'Whig' criteria in the *selection* of historical subject-matter do not sanction a 'Whig' *treatment* of the subject-matter.

When carried out in any depth, the historical study of a particular natural philosopher must resolve the sharp dichotomy which has been suggested between those historians who are guided by 'Whig' canons of significance, and those who urge reliance on the judgements of contemporaries. To acquire some feeling for the contemporary resonance of an individual's ideas and concepts, the historian must trace the course of his intellectual formation and development. That involves a knowledge of authors who loomed large at that time, however much their works may constitute 'the minor writings of forgotten authors' for us. A host of lesser writers have to be studied, because they enjoyed crucial importance within a contemporary environment to which those on whom we focus reacted in complicated ways, and which their own work was to transform.

There are good reasons for dissatisfaction with a prevalent 'Great Books' approach which precludes the development of greater sensitivity for the fine texture of the history of scientific thought. Walter Pagel's subtle and meticulous studies of Paracelsus, Helmont, Harbey, and other thinkers of this period have best demonstrated the way in which our knowledge of their work, and of their intellectual environment, is transformed when there is a total engagement with their thought, rather than a crude selection of what now seems valuable, while the rest is dismissed as the fustian of an unenlightened age. That is by no means incompatible with assessing the importance of their contributions as judged by our own standards of value and worth, or of their modes of inference by our canons of scientific inference. Paracelsus, Helmont, and Harvey demand consideration on even the narrowest of 'Whig' criteria of significance; but in studying their thought the historian's task cannot be that of isolating 'rational' and 'irrational' components, but of regarding it as a unity and locating points of conflict and tension only on the basis of an exploration in considerable depth.

The study of the thought of the natural philosophers of the past as a complex unity rather than as a Manichean search for grains of light enveloped by an overwhelming darkness is an ideal which, of course, guided the work of Cassirer and Koyré. Perhaps the twin threats to intellectual culture in their own time—from a new brutal irrationalism as well as modes of analysis asserting the primacy of the non-logical—do something to explain their neglect of aspects of sixteenth- and seventeenth-century natural philosophy which are attracting renewed attention. In examining the sixteenth century for sources of the change from one world-picture to another by the mid-seventeenth century, the pervasive influence of the Neo-Platonism of the Florentine thinkers appeared an important candidate for consideration. Burtt and Cassirer attached major importance to it in re-defining man's relation to the universe, with momentous consequences for the conception of nature and of the ways in which this was to be studied and explained.[3]

It is now clear that the Neo-Platonism of Ficino and Pico was deeply intertwined with the magical doctrines of the *Corpus Hermeticum* and the later Neo-Platonists. Pico's attack on astrology and his 'Oration on the Dignity of Man' have often been cited as an illustration of the new view of man in nature. But he condemned astrology because its determinism ignored the influence which man could exercise on the stars and, through them, on the universe. The dignity of man consisted for him precisely in his status as a *magus*, at the centre of a universe interlinked by cosmic sympathies and harmonies.[4] As in the later Neo-Platonists, the Neo-Platonic cosmology and ontology was believed to provide philosophical support for theurgic techniques which, among other things, gave man control over nature.

The significance attached to the magical component of Florentine doctrines varied interestingly from one European culture-area to another as these doctrines were assimilated and came to constitute an important basis for the 'Northern Renaissance'.[5] Florentine Neo-Platonism reconciled the ideals of Italian humanism with a spiritual world-picture in a synthesis which had a special appeal for the religious sensibility of the trans-Alpine lands, and their influence on art, literature, and ideas of life-styles and conduct has been much explored.[6] Its image of an inexhaustibly creative God and his harmoniously interlinked universe in which man occupied an exalted place; of Eros as the bridge between man and the divine; of the contemplation of the beauty of nature as a vehicle to divine beauty; of artistic creativity as the imitation of divine creativity—all these themes blended the humanist thirst for experience and the enjoyment of the sensible world with the emotional religiosity of the 'Northern Renaissance'. They were rooted in

[3] E. A. Burtt, *The Metaphysical Foundations of Modern Science*, 2nd edition (London-New York, 1932), pp. 42–4; E. Cassirer, *Individuum u. Kosmos in der Philosophie des Renaissance* (Leipzig, 1927).

[4] 'Oration on the Dignity of Man', English trans., in E. Cassirer, P. O. Kristeller, and J. H. Randall, Jr., *The Renaissance Philosophy of Man* (Chicago, 1948), pp. 223–54.

[5] D. P. Walker, 'The *Prisca Theologia* in France', *J. Warburg & Courtauld Insts.*, 1954, **17**, pp. 204–59.

[6] Discussion and bibliography in P. O. Kristeller, *Renaissance Thought* (New York, 1961 edition), pp. 48–69, 148–51.

a Neo-Platonic view of the world, and the importance of the Florentines in insinuating a Neo-Platonic ontology and cosmology in place of the basically Aristotelian one cannot be overrated. The cosmos was a graded hierarchy, where the hypostases at each level emulated divine creativity by generating subordinate entities which were like a pale reflection of themselves in the ontological scale. The relation between each level was thus one of cause and effect rather than the final causality of the Aristotelian cosmos. All power and activity was infused into the lower world from the higher ones. All things in the natural world were dim shadows of eternal realities. The forms of objects in the sublunary world came from the soul of the celestial world which, contemplating the perfection of the angelic or intelligible world, tried to transmit something of their perfection to the world of matter. In a famous comparison, Pico described how things reflected and were dependent on the levels above them: what existed as the element fire in the corporeal world was the sun as the heat-diffusing power in the celestial world, and the seraphic intellect in the highest realm: 'But behold how they differ. The elemental burns, the celestial fire gives life, the super-celestial loves.'[7]

The study of sixteenth-century natural philosophy has to take account both of the Neo-Platonic cosmology and of the Hermeticism based on it. But it is by no means necessary to regard every thinker influenced by the former as an enthusiastic advocate of the latter. Of the three thinkers who were important in propagating the Florentine doctrines beyond the Alps, John Colet reflected the cosmology in his ideas on church reform; Lefèvre d'Étaples showed extreme caution about Hermetic magic; and only Johannes Reuchlin professed open enthusiasm for the Cabbalistic number-magic of Pico. Sir Thomas More translated the life and a few letters of Pico into English but it is unlikely that the deep influence of Florentine teachings on his circle included their magical doctrines.[8] Similar considerations apply when assessing the significance of Florentine Neo-Platonism for natural philosophy.

The Neo-Platonic cosmology is indispensable, for example, in understanding the argument from Copernicus in favour of his heliostatic cosmology cited by Dr. Hesse: the sun must be at the centre of the planetary system because it was analogous to the king at the centre of his court, or God at the still centre of the universe. The lower levels of the cosmic hierarchy must imitate and reflect the higher levels, and the macrocosm of the universe must serve as the visible symbol of the divine pattern. In *De Revolutionibus*, Copernicus appealed to some characteristic Neo-Platonic tenets: the study of astronomy as the vehicle for drawing the mind to the contemplation of

[7] Pico, 'Heptaplus', in *Opera* (Basle, 1572 edition), Bk. 1, Ch. 1, p. 7.
[8] 'Life of John Picus, Earl of Mirandula', in *Works*, ed. W. Rastell (London, 1557), pp. 1–34.

the highest good; lack of harmony as the strongest reason for rejecting the Ptolemaic cosmology; the sun as the symbol of divine power. Kepler much later explained his initial acceptance of Copernicanism by his realization that if God was not, as for Aristotle, a Prime Mover and final cause of all things, but (as in Neo-Platonism) an ever-active and ever-generative God diffusing his power into all things, then the only appropriate symbol for these relationships in the visible world would be a heliocentric system.[9] Kepler's attempts to explain the number of the planets and the size of their orbits in terms of the relation between the planetary spheres and the five regular solids is well known. His so-called Third Law (together with a great many other 'laws') emerged from his attempt to retrieve the failure of the earlier enterprise to match observation by searching for the 'music of the spheres'.[10] Acceptance of the Neo-Platonic schema sanctioned a method based on tracing analogies between the intelligible or archetypal world, the macrocosm of the universe, and the microcosm of the world and of the human body.

Paracelsus, too, based his programme for the reformation of medicine on tracing analogies between the microcosm and the macrocosm. Through his study of astronomy, the physician was to study the disposition and relation of the parts of the universe; through cosmography that of herbs and minerals on the earth. Such a knowledge would equip him to understand the structure and function of corresponding parts of the human body and the occurrence of diseases within it. As Pagel has pointed out, Paracelsus' ontological conception of disease and his 'seat of disease' view, both sharply contrasting with ancient opinion, emerged from this analogy between the universe and the human body.[11] Kepler regarded Paracelsus as one of the greatest ornaments of the German nation,[12] and shared with him not only the basic conception of the study of nature as tracing correspondencies, but also the Neo-Platonic view in which they were rooted. Kepler cannot be turned into a 'modern' by citing his strong criticism of Robert Fludd; he was steeped in the Renaissance Neo-Platonism of which Hermeticism was an extension, and the Kepler-Fludd debate is better understood as a debate between various lines of development within sixteenth-century Neo-Platonism than as one between a 'rational scientist' and an irrational Hermetic magician. It has seemed regrettable to some historians that valuable and often funda-

[9] A. O. Lovejoy, *The Great Chain of Being* (1936) (New York, 1960), pp. 103–16.

[10] J. L. E. Dreyer, *A History of Astronomy from Thales to Kepler*, 2nd edition (New York, 1953), pp. 405–12.

[11] W. Pagel, *Paracelsus, An Introduction to Philosophical Medicine in the Era of the Renaissance* (Basle, New York, 1958), pp. 65–71; W. Pagel and P. Rattansi, 'Vesalius and Paracelsus', *Medical History*, 8 (1964), pp. 309–28.

[12] Discussed by W. P. D. Wightman, *Science and the Renaissance* (Edinburgh, London, 1962), pp. 284–302.

mental scientific discoveries should have resulted within such a disreputable structure of ideas, but historical comprehension demands some deeper effort at understanding its framework rather than its curt dismissal as 'a species of sleep-walking'.[13] The argument by analogy survived in what we regard usually as quite a respectable form in Boyle's, Newton's,[14] and especially Locke's, defence of corpuscularianism by reference to the Great Chain of Being.[15]

Grappling with the sixteenth and early seventeenth centuries as an interregnum, between the decline of the Aristotelian-scholastic philosophy (accepted by the church and taught at the schools) and the mechanical philosophy, has provided the most important motive for the study of Renaissance Neo-Platonism and Hermeticism. It has seemed reasonable to assume that a deeper study of that period will cast greater light on much that still remains obscure in explaining the importance assumed by the study of nature in the seventeenth century and the triumph of the mechanical view.

The remainder of this paper is devoted to a discussion of some aspects of the work of Sir Isaac Newton to show that a knowledge of Renaissance Neo-Platonism and Hermeticism, and its vicissitudes through the sixteenth and seventeenth centuries, is essential in resolving some major problems and puzzles which have resisted the narrowly 'Whig' approach for a long time.

II

Newton's undergraduate notebooks at Cambridge in the early 1660s show him absorbing the Aristotelian learning (prescribed by the Statutes) mainly through the German neo-scholastic texts and commentaries of the sixteenth century. But he was, at the same time, rapidly assimilating the mechanical philosophy which had transformed the intellectual landscape, through the works of the great continental masters Descartes, Galileo, and Gassendi, and also through such English versions as those of Sir Kenelm Digby,

[13] The phrase is borrowed from Arthur Koestler's *The Sleepwalkers, A History of Man's Changing Vision of the Universe* (London, 1959), where no such contrast between the styles of scientific inference as between the sixteenth- and seventeenth-century natural philosophers is intended.

[14] M. Mandelbaum, 'Philosophy', *Science and Sense-Perception* (Baltimore, 1964), pp. 61–117; J. E. McGuire, 'Atoms and the "Analogy of Nature" ', *Studies in History and Philosophy of Science*, 1 (1970), pp. 1–58.

[15] M. Hesse, *Forces and Fields* (London, 1961), pp. 122–5.

Walter Charleton, Robert Boyle, and of the Cambridge Platonists.[16] The last influence was of basic importance for his fundamental assumptions about the nature and scope of the mechanical philosophy. The Cambridge Platonists attempted to restate classical Neo-Platonist philosophy to safeguard a religious vision of the world in their own time, as the Florentine thinkers had in theirs. Although the Hermetic component of late-antique and Florentine Neo-Platonism initially occupied a very attentuated place in the version of the Cambridge thinkers, it became more prominent later on. These features of Cambridge Platonism are essential for explaining one of Newton's major—and most puzzling—activities: his interest in—or obsession with—the literature of alchemy. It is now well-known that a great deal of Newton's energies during his most creative period were devoted to his alchemical studies and to Biblical chronology and prophecy. Historians of science have felt able to exclude the latter from their purview; but the alchemical studies, it has seemed obvious, must have some connection with his scientific work.

Newton's alchemical studies have been explained in terms of two apparently incompatible explanations. The first is that he was compelled to resort to alchemical writings, rather like Robert Boyle, for the large body of chemical information which the alchemists had amassed in the course of their deluded quest. The other, that Newton was in some sense an alchemist, fully sharing its ideas and its vision of the world.[17] Neither position is free of difficulties. The large body of Newton's alchemical notes, which has survived in manuscript form, does not support the conclusion that his interest was centred upon metallurgical information. But to call Newton an alchemist is to split him into 'rational scientific' and 'irrational Hermetic' selves which have nothing in common. To resolve these paradoxes a closer examination of Cambridge Platonism and of its influence on Newton is a necessary preliminary.

The shaping of Cambridge Neo-Platonism and shifts in emphasis within its doctrinal structure must be seen against the background of the crisis in social, religious, and political affairs signalled by the civil wars and the overthrow of monarchy and the established church. The Cambridge thinkers believed that religion was in dire peril from two different directions. It was menaced by a growth of doubts of the existence and providence of God, and the immortality of the soul; as in much other polemical literature

---

[16] A. R. Hall, 'Newton's Note-Book, 1661–5', *Camb. Hist. J.*, 9 (1948), pp. 239–50; R. S. Westfall, 'Newton's Philosophy of Nature', *Brit. J. Hist. Sc.*, 1 (1962), pp. 171–82.
[17] E.g. F. S. Taylor, 'An Alchemical Work of Newton', *Ambix*, 5 (1956), pp. 59–84; A. R. and M. B. Hall, 'Newton's Chemical Experiments', *Arch. inter. d'Hist. des Sc.*, 11 1958), pp. 113–52.

published during the 1640s, the source is alleged to be the Aristotelian-Averroism of Italian thinkers like Pomponazzi and Cesalpino. It was equally menaced by enthusiastic sectaries who grounded their heresies in the claims of supernatural illumination and, among other opinions, propagated the idea of a Hermetic science of nature.[18] A Neo-Platonism reconciled with the mechanical philosophy of Descartes would counter both dangers. 'Averroism' was undercut when its Aristotelian basis was replaced by a mechanical view of nature which was, however, free from the atheistic implications of ancient atomism since a 'rightly-understood' mechanical philosophy placed the emphasis not on the 'blind and stupid' matter, but on the power and activity which characterized the visible universe and which only could flow into it from a hierarchy of *incorporeal* principles. The sectaries were refuted by demonstrating the possibility of a purely rational defence of religion and by shifting the emphasis heavily from extra-logical means of cognition to sense-experience and reason; the claims of religious knowledge were defended by a distinction between the irrational and the suprarational.[19] That the Cartesian system was Platonic was thought to be proved by its reliance on innate ideas, and, more generally, by the contrast it made between the dead and inert world of matter and the incorporeal world of power and activity. Indeed, Cartesianism completed or complemented the Platonic system. The deficiencies of Platonism as regards positive knowledge had resulted in Aristotle's work becoming the basis of intellectual culture in the 'three religions' of Judaism, Islam, and Christianity at a critical point in their development. Borrowing a characteristic tactic of the Florentine Platonists, the Cambridge Platonists proposed that Descartes' 'rediscovery' of a complete Platonic system of natural knowledge showed that the philosophy of Plato and the Cartesian natural philosophy had once formed part of a complete Mosaic system of knowledge which had been lost after the Fall. While its philosophical doctrines survived in Plato, its scientific doctrines were perverted by the Greek atomists and their successors Lucretius and Epicurus.

But even during his fervent early enthusiasm for the Cartesian system, Henry More was aware that it required important modifications before it could fully be integrated with Neo-Platonism. Above all, Cartesian dualism must be replaced by the Neo-Platonic cosmic hierarchy. The publication of Thomas Hobbes' works, expounding a materialistic mechanism, completed

---

[18] Henry More published *An Antidote Against Atheism* (1652) and *The Immortality of the Human Soul* (1659), as well as two works against Thomas Vaughan (1651 and 1652) and *Enthusiasmus Triumphatus* (1656); on the mortalist debate in France, H. Busson, *La pensée religieuse française de Charron à Pascal* (Paris, 1933) and R. Lenoble, *Mersenne ou la naissance du mécanisme* (Paris, 1943).

[19] E.g. More in *Enthusiasmus Triumphatus*, secs. li–lv.

More's disenchantment. The Cartesian account remained the only satis-factory account of natural phenomena. But its dualism and its acceptance of the possibility of a perfectly consistent and purely mechanical explanation of nature meant that it could serve as the basis of a threat to religion which was perhaps even greater than that posed by Aristotelian 'naturalism'. The greatest challenge to religion and society came now not from the Averroists or the Hermetic sectaries, but from 'perverted' mechanism. There was a significant change of emphasis within More's Neo-Platonism by the 1650s. Although he kept clear of Hermetic magic, he was led to give greater attention to such topics as the Cabbala, the prophetical books of the Bible, the 'Teutonic philosophy' of Jacob Boehme, and concern with spiritualist phenomena as empirical proof of the existence of an incorporeal substance capable of influencing matter.[20]

More stressed the very limited character of mechanical explanations, which guided the mind to the realization that 'the primordials of the world are not mechanical, but *spermatical* or *vital*'.[21] The Cartesian mathematical analysis of nature ultimately rested on the law of the conservation of the quantity of motion, which was linked, in turn, with the sharp division between thinking and extended substance. For More, by contrast, all motions came from spirits in a descending hierarchy of increasing materiality, and an ontological difference between agent and patient was essential to true causality.[22] Nature could not be explained in terms of a self-sufficient or conservative mechanical system. When the chain of mechanical causes is traced far enough, 'im-mechanical principles', transcending 'the nature and power of matter',[23] must be invoked. When the rectilinear motion of a bullet fired from the surface of the earth, in accordance with the Cartesian laws of motion, became a curvilinear trajectory, some other power was involved: the 'hylarchic' or 'plastic' principle which was a subordinate agency of God.[24] In such criticisms, More was pointing at concealed dynamic assumptions within the Cartesian system which had attempted to do without the embarrassing animism of the notion of 'force'; but it is important to notice that for him 'the *vis agitans* that pervades the whole body that is moved'[25] became a basic spiritual reality, a continuous manifestation of the divine in the sensible world.

The distinctive character of the Cambridge Platonist doctrines about the

[20] E. Cassirer, *The Platonic Renaissance in England* (English trans., London, 1953), pp. 129–56.

[21] *Divine Dialogues* (London, 1668), p. 122.

[22] Cassirer (note 20), p. 146.

[23] *An Antidote Against Atheism*, Ch. II.

[24] *Remarks Upon Two Late Ingenious Discourses* (London, 1676).

[25] *Divine Dialogues*, p. 98.

relation between natural philosophy and metaphysics will perhaps be clearer if we compare them with the ideas of Descartes and Leibniz. Both the latter thinkers were aware of the danger that a mechanistic account of nature could lead to atheism, as in ancient atomism. In his correspondence with More in the 1640s Descartes has shown that he was fully conscious of the perils. His solution lay in distinguishing between two levels of discourse: at one level, all change in nature could be explained in purely mechanical-mathematical terms; but at the deeper, metaphysical, level, it involved the continuous creative action of God. Descartes had chosen not to expound this complex notion in his writings for fear that he might 'seem to be supporting the opinion of those who regard God as *anima mundi*, united to the matter of the world'.[26]

As emerged very clearly in the Leibniz-Clarke correspondence, Leibniz made a similar distinction of levels in his very complex system. The sphere of efficient and final causality, in his terms, must be kept distinct. At the efficient level, all events in the material universe were completely explicable in terms of matter, motion, and force; in terms of final causality, only by the free activity of an infinite number of individually differentiated monads.[27]

Such divisions appeared untenable as well as dangerous to the Cambridge Platonists. If a purely mechanical explanation of nature was granted autonomy, even when founded in a theistic metaphysics, it could drift away from such moorings, as the teachings of Hobbes had demonstrated. More's solution confounded the two levels which Descartes and Leibniz had sought to maintain analytically distinct.

The influence of More's view of the mechanical philosophy is pervasive in the early work of Newton which has been published as 'De gravitatione'.[28] This influence continued to be reflected in the ontological presuppositions guiding his natural-philosophical programme and some of its basic conceptual and explanatory features. Like the Cambridge thinkers, he remained convinced that the relation between scientific and theological knowledge could not be too close. A purely mechanistic account of nature on Cartesian lines reduced everything to blind necessity, but it was 'unphilosophical to pretend that it (the universe) might arise out of Chaos by the mere laws of nature'.[29] The business of the mechanical philosophy was to proceed from

---

[26] *Oeuvres*, Adam and Tannery edition, 5, pp. 403–4; discussed by N. Kemp Smith, *New Studies in the Philosophy of Descartes* (London, 1963), pp. 199–209, 313–14.

[27] *The Leibniz-Clarke Correspondence*, ed. H. G. Alexander (Manchester, 1956), pp. 41, 42–3, 85–8, 91–7; Cassirer, pp. 150–55.

[28] A. R. and M. B. Hall, eds., *Unpublished Scientific Papers of Isaac Newton* (Cambridge, 1962), pp. 89–156.

[29] *Opticks*, Query 31 of 1730 London edition (New York, 1952), p. 402.

effects to causes until it reached the first cause which was 'manifestly immechanical'.[30]

Cartesianism, however, left no possibility of ascending from effects to causes in an ontologically ascending series up to God.[31] It had banished from nature all but mechanical causes, whereas the principles which infused all power and activity in the universe were certain 'active principles' which could not be reduced to such causes.[32] In the post-*Principia* period, Newton began to make a distinction between the 'passive' laws of motion of bodies, arising from their *vis inertiae*, with a 'very potent active principle' which generated 'new motion'.[33]

Newton's ontology is anchored, thus, in the scientific-metaphysical-religious synthesis which the Cambridge Platonists had sought to achieve. The Cambridge Platonists had eagerly seized upon phenomena that were difficult to explain satisfactorily in mechanical terms, especially cohesion, capillarity, gravitation, and magnetic and electrical actions. A rightly-understood mechanical philosophy would regard such difficulties not as a cause for embarrassment, but for exultation: they served to demonstrate the necessity of a spiritual factor. The famous analysis of the nature of space by Henry More served the same purpose. Space, which Descartes had identified with matter, was really intermediate between the divine and the material, and bridged the gap between the two sorts of substance since both were extended. Newton accepted More's idea of space and employed it in his anti-Cartesian arguments in his 'De gravitatione'.[34] The notion of 'forces', far from carrying the embarrassing animistic associations it had for the first generation of the mechanical philosophers, served for Newton as a manifestation of the divine in the sensible world. Concentration on the dynamic rather than structural aspects of phenomena and, at the microscopic level, on the motion of the vibrating medium and of particles rather than the configuration and geometrical properties of the particles in explaining changes; the introduction of unexplained attractions and repulsions in hypotheses; an orientation towards non-conservative modes of action in the universe—all these features could derive support from a Neo-Platonic ontology. At the same time, Newton's 'mathematical way' offered an avenue out of the total confusion which Descartes had apprehended unless the spheres of the mental

---

[30] Ibid., Query 28, p. 369.
[31] 'De gravitatione', pp. 142–5; *Opticks*, Query 28, p. 370.
[32] *Opticks*, Query 31, pp. 399–404.
[33] J. E. McGuire, 'Force, Active Principles, and Newton's Invisible Realm', *Ambix*, 15 (1968), p. 172.
[34] 'De gravitatione', pp. 131–48; Cassirer, *Platonic Renaissance*, pp. 146–50; A. Koyré, *From the Closed World to the Infinite Universe* (New York, 1958 edition), pp. 110–89.

and the corporeal were rigidly separated and non-material entities totally excluded from the realm of mathematical-mechanical analysis.

Acceptance of the Neo-Platonic ontology seems to have been important in guiding Newton to some of the Hermetic doctrines with which it had been historically associated. Matter was dead and inert and, indeed, there was far less of it, compared with the void in the universe, than was usually imagined.[35] The laws associated with it were merely the 'passive laws of nature'. The study of higher levels of causal efficacy would give far deeper insight into the greatest secrets of nature. Gravitation was one of the fundamental 'active principles', but it was concerned with the passive and inertial characteristics of bodies. The laws of motion of the mechanical philosophy merely described the mechanical transactions between bodies at the gross or microscopic levels. Far more important were the real changes or transmutations wrought in the small parts of bodies by such active principles as those of fermentation, and of 'life and voluntary motion'.[36] How was their influence communicated to matter?

Newton differed from the Cambridge Platonists in one important respect: so closely did he link God with his creation that he did not favour intermediaries like the 'world soul' or the 'plastic principle'.[37] The problems of theodicy which the Cambridge Platonists had attempted to solve by an intermediary agency were met for Newton by a conception of God which stressed His will rather than His reason, and met the demand for cosmic justice by a stress on the hidden Providential design of God.[38] But even if the relation of the 'active principles' to God was different from that of the 'hylarchic principle', Newton found it necessary to postulate an intermediary between those principles and inert matter. Like the Cambridge Platonists, he drew upon the idea of the *spiritus* which had been greatly revived during the sixteenth and seventeenth centuries, and which combined Platonic, Peripatetic, and Stoic, as well as Judaic and Christian conceptions.[39]

The *spiritus* was a central component in Renaissance Neo-Platonism as well as in the Hermeticism which received metaphysical support from it. When Gassendi attacked Hermeticism (as personified for him by Robert

---

[35] A. Thackray, ' "Matter in a Nut-Shell": Newton's *Opticks* & Eighteenth-Century Chemistry', *Ambix*, **15**, pp. 29–63.

[36] Cambridge University Library, Add. 3970 f.252^v.

[37] 'De gravitatione', p. 142.

[38] More, *Immortality*, Chs. xii–xiv; draft 'Scholium Generale', Halls (note 28) p. 357; Roger Cotes, Preface to the second edition of *Principia,* 1713, esp. pp. xxvii, xxxi-xxxii, in Motte-Cajori trans. (University of California Press, 1966).

[39] See e.g. G. Verbeke, *L'Evolution de la doctrine du pneuma* (Paris-Louvain, 1945); H. Siebeck, 'Neue beiträge zur Entwicklungsgeschichte des Geist-Begriffs', *Arch. f. Gesch. Philos.*, **20** (1914), pp. 1–16; F. Sherwood Taylor, 'The Idea of the Quintessence', in E. A. Underwood edition, *Science, Medicine, and History* (1953), I, pp. 247–65.

Fludd) in 1629, he found one of its most characteristic tenets in the belief in a certain *spiritus* which traversed the universe and penetrated all things and, being itself intermediary between matter and soul, carried the powers of the superior world to the inferior one.[40] Ficino had revived the idea of the *spiritus* as a basis for magical operations. The *spiritus* (as the animal-spirits) mediated in man, as in the universe, between soul and matter, and man could use his *spiritus* to work on the cosmic *spiritus* and attract beneficial celestial influences.[41]

The widespread use of the idea of intermediary spirits in physiological, neuro-physiological, and more general natural-philosophical discussions during the period tended in two different directions. There was a tendency, on the one hand, to conceive the physiological spirits as material substances: a tendency which was carried to its culmination in the mechanistic physiology of Descartes.[42] The apparent success of a late-medieval distillation tradition in extracting the cosmic spirit from herbs and minerals[43] and rendering it familiar in the form of extracts was probably important for this development. In a Neo-Platonic context, on the other hand, the concept of the *spiritus* was employed in the service of a religious vision of the universe which detected a residual materialism even in Aristotle's doctrine. Cornelius Agrippa summarized an influential position (reflected, among many others, in Jean Fernel[44]) when he argued that the Aristotelian 'elementary' analysis failed to reveal the most important powers and virtues of a thing—the occult qualities—because these did not depend so much on its elementary composition as on a particular planet, whose influence was mediated through the *spiritus*.[45] For Paracelsus, that was proof of the total inadequacy of the logical-dialectical method in the study of nature, and it buttressed his natural-historical programme, well-adapted to the search for new therapeutic agents: only by an empirical search through nature and an 'overhearing' deeper than the operations of reason could the hidden virtues of things be brought to light.[46] Helmont much later transformed the *spiritus* idea by denying importance to the planetary influences, while refusing to substitute for them an explanation of 'occult' qualities based on the rearrangement of

[40] 'Examen Philosophiae Roberti Fluddi Medici' (1629), in *Opera Omnia* (Leyden, 1653), III, pp. 219–59; B. Rochot, *Les travaux de Gassendi sur Epicure et sur l'Atomisme, 1619–1658* (Paris, 1944), pp. 55–7.

[41] D. P. Walker, *Spiritual and Demonic Magic from Ficino to Campanella* (London, 1958).

[42] E. Gilson, *Etudes sur le rôle de la pensée médiévale dans la formation du système Cartésien* (Paris, 1930), pp. 51–100; Walker (note 44).

[43] Pagel, *Paracelsus*, pp. 263–4; Taylor (note 39).

[44] D. P. Walker, 'The Astral Body in Renaissance Medicine', *J. Warb. & Court. Insts.*, 21 (1958), pp. 119–31.

[45] *De occulta philosophia* (1533) esp. Lib. 1, Caps. 1–14.

[46] Pagel, *Paracelsus*, pp. 53–71.

material atoms.[47] Matter was transformed into a specific substance when a 'ferment' acted on the seeds scattered in matter and transformed it into a 'gas' or an 'acting body' to which the dualistic separation of body or soul could not be applied.[48]

The *spiritus* notion played a key rôle, again, in the conception of Hermetic science as part of a much greater social and religious reform which was envisaged by various thinkers in the Germanic lands, connected with the more radical wing of the Reformation, in the latter sixteenth and early seventeenth centuries.[49] Determined to base natural philosophy on a 'Christian' basis, they drew its fundamental principles from an interpretation of the text of Genesis which made the *spiritus* the counterpart, on the lower ontological level, of the Second Person of the Trinity. It was the spirit that had hovered upon the waters and had been God's instrument at Creation; it had inspired the Holy Writ, and its descent opened a channel through which man could receive suprarational knowledge. The pedagogic, social, religious, and intellectual reform of these thinkers was based on eschatological premises. They assumed that the millennium was imminent and would be marked by the greatest progress and flourishing in the arts and sciences.[50] The knowledge over nature which Adam had lost after the Fall would be recovered, partly by the reversal of confusion of tongues and regaining the 'Adamic' language.[51] The workings of the Holy Spirit had been given the greatest importance by apocalyptic writings in these developments.[52]

An early letter of Robert Boyle to a thinker who had brought the ideas of these German reformers to England well illustrates the religious-utilitarianism and the extensive use of explanations in terms of the *spiritus*. In a letter almost certainly written in the 1640s, Boyle wrote to Samuel Hartlib that the considerable labours bestowed upon exact observations in astronomy could not be justified unless 'some end, benefit, use or advantage' resulted;

[47] Pagel, 'van Helmont's Science and Medicine' Suppl. 2, *Bull. Hist. Med.*, 1944, pp. 16–22; *Paracelsus*, pp. 104–5; Newton's notes on Van Helmont's 'gas' are to be found in notes from the *Ortus medicinae*, Keynes Ms. 16 at King's College, Cambridge.

[48] Pagel, *Harvey's Biological Ideas*, pp. 167–8.

[49] F. A. Yates, 'The Hermetic Tradition in Renaissance Science', in C. E. Singleton, ed., *Art, Science, and History in the Renaissance* (Baltimore, 1968), pp. 155–274; A. G. Debus, 'The Chemical Dream of the Renaissance', *Churchill College Lecture 3* (Cambridge, 1968).

[50] E. L. Tuveson, *Millenium and Utopia* (New York, 1964); Michael Fixler, *Milton and the Kingdoms of God* (London, 1964), esp. Chs. 1–4; C. Webster (ed.), introd. to *Samuel Hartlib and the Advancement of Learning* (Cambridge, 1970), pp. 1–72, esp. pp. 33–7.

[51] H. Aarsleff, 'Leibniz and Locke on Language', *Am. Phil. Q.*, I, 1964, pp. 165–88; H. Ahrbeck, 'Einige Bemerkungen über "Mosaische Philosophen" des 17. Jahrhunderts', *Wiss. Z. Univ. Halle, Ges. Sprachwiss.* VII/5, 1958, pp. 1047–50.

[52] G. F. Nuttall, *The Holy Spirit in Puritan Faith and Experience*, Oxford, 1946.

otherwise 'we know them only to know them'.[53] However, it seemed that the planets transmitted virtues with their light, and were the 'real property' of its light:

> Not only the air, by reason of its thinness and subtility, is capable of being thus penetrated, moved and altered by these planetary virtues and light; but foreasmuch also as our own spirits, and the spirits likewise of all mixed bodies, are really of an aerious, ethereal, luminous production and composition, these spirits therefore of ours, and the spirits of all other bodies, must necessarily, no less suffer an impression from the same lights . . . but rather as our spirits are more near and more analogous to the nature of light than the air, so they must be more pure and easy to be impressed than it. And if our spirits, and the spirits of all mixed bodies may be altered, changed, moved and impressed by these superior bodies, and their properties; than these spirits being the only principles of energy, power, force and life, in all bodies wherein they are, and the immediate causes through which all alteration comes to the bodies themselves. It is impossible therefore spirits should be changed, and yet no alteration made in the bodies themselves. . . .[54]

Henry More, in his adaptation of Cartesianism, had identified the First and Second Elements of Descartes with the *spiritus*: they constituted 'that true Heavenly or Aetherial Matter which is every where, as Ficinus some-where saith Heaven is; and is that *Fire* which *Trismegist* affirms is the most inward vehicle of the Mind, and the instrument that God used in the forming of the World, which the Soul of the World, wherever she acts, does most certainly still use.'[55]

Newton's own ideas on an aetherial medium which would explain gravitation, cohesion, capillarity, chemical reactions, combustion, electrical attraction, and even body-mind interaction, and the beating of the human heart,[56] have usually been examined in terms of variations on the Cartesian ether. But they conflate the Neo-Platonic ideas of the *spiritus* which have been outlined above. Newton's deep study of alchemical literature, too, makes much greater sense in that context, because it constituted the largest body of literature which purported to describe the operations of that spirit and ways of capturing it through practical laboratory operations.[57] His acceptance of the vast and mysterious literature of alchemy as embodying the deepest wisdom depended on an assumption equally reflected in his

[53] 'Of celestial influences, or effluviums in the air', Title xiii, in *The General History of the Air* (1692) in T. Birch, ed., *Works* (London, 1744), Vol. V, pp. 124–8.
[54] Ibid., p. 126; cf. Boyle's later contrasting views in e.g. *Usefulness of Experimental Natural Philosophy*, Vol. III, pp. 452–3, and *Experimental Essays*, I, p. 310.
[55] *Immortality*, Bk. II, ch. 8, prop. 6.
[56] H. W. Turnbull ed., *The Correspondence of Isaac Newton* (Cambridge, 1959) I, pp. 362–89; (1960) II, pp. 288–96.
[57] Noticed by F. S. Taylor (note 17).

studies of the prophetical books of the Bible and of 'pristine' natural philosophy among the ancients.[58] Truth had been given by God in the beginning but had been fragmented and corrupted in the course of time; its traces survived in enigmatic form in these different sorts of literature, but had to be recovered by a sort of dialectic between hard, disciplined inquiry and the ancient sources; such recovery took place in accordance with the Providential design in history. In alchemy, that involved practical laboratory work guided by the basic text of the art, the 'Smaragdine Tablet' which was simultaneously an account of Creation and of the alchemical work, guided by the writings of the succession of alchemical masters.[59]

In the alchemical collection which Newton knew intimately, Elias Ashmole had described the operations of the *spiritus*: '. . . the Power and Vertue is not in Plants, stones, Minerals &c. (though we may sensibly perceive the Effects from them) but 'tis that universal and All-piercing Spirit, that One operative vertue and immortal Seede of worldly things, that God in the beginning infused into the Chaos, which is everywhere Active and still flowes through the world in all kinds of things by universall extension. . . . Which Spirit a true Artist knows how-so to handle . . . as to take it from Corporeity, free it from Captivity, and let it loose that it may freely worke as it does in Æthereal Bodies.'[60] Newton himself at one stage believed that the *spiritus* was light. The alchemists had most often compared the spirit to light; the identification of spirit and light would fit the Creation account in Genesis; Newton seems to have believed that it would also explain the fermental activity which was ascribed to the Philosopher's Stone.[61] Even in the *Opticks* he referred much of the activity of bodies to the particles of light embedded in them, pointed to the interconvertibility of matter and light, and justified the inclusion of chemical topics in the Third Book as hints about 'light and its effects upon the frame of nature. . . .'[62] Light probably satisfied the state intermediate between matter and soul demanded of the Neo-Platonic *spiritus* by being, for Newton, composed of the finest sort of materiality while at the same time being associated with some of the strongest forces in nature,[63] and forces were ultimately the

---

[58] J. E. McGuire and P. M. Rattansi, 'Newton and the "Pipes of Pan"', *Notes & Records Roy. Soc. Lond.*, **21** (1966), pp. 108–43.

[59] Lists of 'best authors', each headed by the *Tabula Smaragdina*, in Keynes Mss. 13 and 49, King's College, Cambridge.

[60] E. Ashmole, *Theatrum Chemicum Britannicum* (London, 1652), pp. 446–7.

[61] Ms. in the Burndy Library, headed 'Of natures obvious laws & processes in vegetation'; on the Stone as a ferment, see 'Pretoisa Margarita Novella' and remarks in Pagel, *Paracelsus*, p. 261, and H. Mètzger, 'L'Evolution du règne métallique d'après les Alchimistes du xviiᵉ siècle', *Isis*, **4** (1921–2), pp. 466–82.

[62] *Opticks*, p. 405; Query 30, pp. 374–5.

[63] Ibid., Query 21, pp. 351–2; McGuire (note 33), pp. 159–61.

manifestation of the divine presence throughout the universe. Newton conjectured that light was the grand 'vegetative spirit' in the universe involved in all processes that could not be ascribed to the effects of merely mechanical rearrangements of the small parts of bodies.[64]

III

There are a host of problems connected with Newton's ideas on the ether and the active principles, and changes in them over time, which cannot be discussed here. The purpose of this discussion has been to point out that the context of ideas which is relevant for understanding some of Newton's fundamental scientific ideas is much broader than is usually realized and certainly includes the Neo-Platonism and Hermeticism of the Renaissance. Nor can that context be excluded from the province of the history of science by suggesting that it merely explains the ways in which relatively independent 'scientific inferences' were subsequently reconciled with religious dogma. The interplay between science and religion was far more subtle than that. By insisting on banishing causal hypotheses from 'experimental philosophy' and conceiving its task as that of formulating quantitative laws by the rigorous analysis of phenomena, Newton may seem to have insulated his natural philosophy far more effectively from metaphysical and theological considerations than Descartes or Leibniz. If that was the whole story then we would have to invent another Newton, a mystical and crankily fundamentalist one, to supplement the 'hard-headed' and phenomenologically-inclined Newton, in order to explain the theological and alchemical studies which absorbed so much of his creative life. But the link between these two aspects of Newton is much greater if we recognize one of the formative influences which set Newton free from some of the restrictions of the classical mechanical philosophy of the mid-seventeenth century: a justification for the mathematical study of forces which rested on the Cambridge Platonist rejection of a purely mechanical explanation of nature, and reached back to the Florentine Platonist vision of a continually active and creative God whose love for His creation was imperfectly mirrored in human love as it was, at a lower level, in the cosmic sympathy, immanent in a *spiritus*, which bound the universe together.

Like some of the other great successors of the first generation of mechanical philosophers, even the cold and austere Newton seems to have combined an intoxication with the possibilities of the mechanical approach to nature

---

[64] Burndy Ms. (note 61), 'This spt perhaps is ye body of light because both have a prodigious active principle, both are perpetual workers'.

with a chilling sense of its inadequacies as a world-view to live by. The alternatives he sought, the materials he drew upon to supply what he saw as its deficiencies, and his final achievement in this respect may seem to us crude and ramshackle when compared, say, to the system of Leibniz. But they are part of an interlocking structure of ideas in which some of Newton's greatest achievements are rooted. The recognition of our own sort of 'rationality' and of a great pioneering rôle in the creation of our kind of science has justified the great concentration of historical study on the thought of Isaac Newton. But the historian's task cannot be that of demonstrating the timeless rationality of Newton's scientific inferences when they are lifted out of the time- and place-bound 'irrelevancies' in which they were embedded. His interest must focus much more on those 'historical' structures: on the complex and changing interplay between Newton's scientific concerns and a whole variety of other concerns, and between them and the society and intellectual culture of his own time.

It is striking how demands that the historian not merely explain but judge the past and direct his attention to that segment of the past which seems most 'relevant' from our own point of view—demands which are currently radical battle-cries in the field of general history—serve a conservative function in the history of science. They are used to sanctify one particular conception of the boundaries of the subject, to 'close the context' in one preferred direction.[65] Even when full weight is given to its distinctive characteristics, the history of science is not immune to the changes of perspective which other historical disciplines have long learned to accept. These changes can either be regarded as deplorable changes of 'fashion', or as an enrichment of our understanding by bringing different aspects into view and developing skills and conceptual tools for dealing with them. Philosophers of science can play an important rôle in evaluating such changes; but only if they keep their minds open to the possibility that such changes may be opportunities rather than disasters. They are likely to cut themselves off from some exciting new developments if they detect epistemological relativism in every attempt to study the social context of scientific knowledge, or a plea for a 'mystical' approach to nature in every exploration of Renaissance Neo-Platonism and Hermeticism, or a crypto-revivalism in the study of the impact of specific religious beliefs on scientific styles and presuppositions.

---

[65] Cf. John Dunn, 'The Identity of the history of Ideas', *Philosophy*, **33** (1968); Quentin Skinner, 'Meaning and Understanding in the History of Ideas', *History and Theory,* **8** (1969), pp. 1–53.

XI

EXPLANATION AND GRAVITY

Gerd Buchdahl

I

It has been recognized for a long time that the range of factors determining the acceptability of scientific explanations and explanatory concepts extends beyond what is provided by mere inductive confirmation, or by the ability of an explanation to account for the observational data it is invoked to explain. Already Aristotle stressed the existence of such additional factors when he claimed that scientific explanation involves premises that must be, not only true, but necessarily true, and even seen to be necessarily true; and at the start of the modern period, people as diverse as Kepler and Galileo, Descartes and Locke, are found to echo such a view. In more recent time, philosophers like Mach, Meyerson and Burtt, and latterly historians like T. S. Kuhn and logicians such as W. C. Salmon, to mention just a few, have similarly stressed the importance of these extra-inductive factors, but their most celebrated exponent was really Kant, whose *Metaphysical Foundations of Natural Science* provided the first systematic attempt to elucidate such additional components of scientific theories.

Of course, as a list containing such a diversity of writers implies, we cannot expect unanimity about the range and nature, let alone logical status, of these extra-inductive determinants. What seems a mere pragmatic device to one, or an expression of psychological instinct to another, will to a third be a display of rational insight, and at the extreme end of the spectrum may even be claimed to involve theological backing. What is most wanting among such a bewildering diversity of approaches is a methodological classification, to serve as a guide towards a clearer grasp of the significance of the different ways in which a belief in the existence of extra-inductive criteria can express itself.

The provision of such a classification is perhaps not simply a descriptive exercise; to classify, is to take a point of view; to subsume a species under a classificatory genus is to engage in interpretation and re-evaluation. To do this in detail, would require a volume. In what follows, I must limit myself to a bare outline; furthermore, I shall support and explain my classification by exploring just one important example, viz. the historical and logical vicissitudes that led eventually to the acceptance of the Newtonian conception of gravity; hoping thereby to test the usefulness of my schema for illuminating certain well-known ambiguities in the classical writings on this problem, especially those of Newton. Our case-study will be shown strongly to exemplify the methodological structure I am proposing, whilst at the same time being a reminder that the components of that structure have a dynamic of their own in a given particular historico-critical context, as is indicated by the violently controversial nature of the history of our concept.

But let me first elucidate the bare outlines of this methodological structure. In addition to the procedures that define and lead to inductive[1] confirmation (or corroboration), and thus provide the physical content, I want to distinguish—following some of the writers previously mentioned—a metaphysical dimension. But I want to suggest (and show) furthermore that this metaphysical dimension itself has a finer structure, consideration of which leads us to distinguish (in addition to the inductive component) two further components that determine acceptance of a scientific theory. One of these may best be described as 'conceptual explication'; the other (the third of our group of three) I shall label 'regulative determination'. Furthermore, whilst corroboration or confirmation yield a degree of 'empirical strength', or 'empirical content', I shall say that conceptual explication determines the 'intelligibility' of a concept, whilst regulative determination articulates its 'rationale'. For brevity I shall also sometimes refer to these three components as the conceptual, the inductive, and the architectonic components.[2]

The whole classification is reminiscent of certain structural layers implicit in the Kantian system, where my notion of 'conceptual explication' appears in the context of the foundations of Newtonian mechanics, as a conceptual analysis of its basic terms (matter and motion), linked in a complex way to the transcendental concepts and principles which (in Kant's view) govern *all* experience, whereby what he calls objective, or real, or constructive (as

---

[1] 'Inductive' is here used so as to include 'hypothetico-deductive' patterns.

[2] I use the term 'architectonic' as a reminder that Leibniz introduced some of the regulative principles here involved under the label of 'architectonic'; but more relevantly, because in Kant, architectonic denotes 'the art of constructing systems' under the guidance of regulative principles of 'systematic unity' that 'first raise ordinary knowledge to the rank of a science' (K653), belonging to the province of theoretical reason.

distinguished from mere logical) possibility is guaranteed.[3] Similarly, the label 'regulative determination' reminds us of the Kantian 'regulative or hypothetical employment of reason',[4] whilst the empirical instances that fall under an explanatory hypothesis determine what Kant calls its inductive 'probability'.[5]

The architectonic component is for Kant a function of the employment of certain regulative maxims (e.g. simplicity, economy, continuity, etc.),[6] as well as of the presence of theoretical or ideal concepts (e.g. air, fire, water; chemical principles; gravity), built into a consilient network of empirical laws, whose systematic interconnectedness or 'unity' and mutual affinity is guaranteed by the theory of the subject.[7] This 'unity' may also express itself as the demand that the plurality of phenomena should be reducible to an underlying identical 'substance'; a reduction which determines the rationale —here literally the 'rationality' of the procedure.[8] Moreover, according to Kant the systematic unity of such an arrangement may also be represented as teleological; a point that will become important in the sequel. Not only teleology, but also analogy will here play an important part. Thus the fact that electrical and magnetic attraction display similarities to gravitational attraction is adduced by Newton in the latter's favour.[9] Similarly, Kant uses mathematical and optical analogies in his attempt to cement the gravitational conception.[10]

[3] Cf. CP.20, and M., *passim*, especially ch. III; for 'real possibility', see K.27n., 110, 503; M.140. The most easily grasped instance of one of Kant's foundational 'demonstrations' of a Newtonian law is his proof of the law of inertia as a 'special case' of the law of causation. Thus: Every change has a cause. But in mechanics, the only changes relevant are changes in the velocity of material particles; and the only causes relevant are 'external', and called 'forces'. Hence all change of velocity is due to the action of an external cause: which is equivalent to the law of inertia. (Cf. M.222-23.)

[4] K.546, 535.

[5] K.535, 625-7; also K.613, L. 75-6, where a distinction is drawn between the probability of an hypothesis based on the verifiable consequences that can be drawn from it, and the *possibility* of the supposition itself.

[6] Cf. K.538, 543-4; J.18.

[7] Cf. K.542, 545; P.83, the case of Newton's theory of universal gravitation, where this consilience is so impressive 'that no other law of attraction than that of the inverse square of the distances can be properly devised [schicklich erdacht] for a cosmic system'.

[8] E. Meyerson already remarked upon this connection between the rationality of an explanation and the reduction to an underlying identity.

[9] O.376; Newton expresses this by saying that 'Nature is very consonant and conformable to herself'; a fact to which at P.398 he refers as 'the analogy of Nature'.

[10] Cf. M.194-6; cf. Pr. 82-3. Lacking our trichotomous schema, this (like much else in Kant's methodology) has been misunderstood, and Kant has been accused of wanting to deduce *a priori* the inverse square law from geometrical considerations, whereas the aim of the analogy was no more than to make the empirical result plausible, something which one 'may well conceive *a priori*, but which one may not pretend to assume as real', since the law of attraction—as Kant expressly notes—must be 'inferred from the data of experience' (M.201, 212).

I do not want to make too much of these Kantian parallels; in any case, I am using Kant in this chapter only as a kind of fixed point or frame of reference, against which to measure the much less well-defined intellectual developments that I am about to chart and which only eventually cumulate in the Kantian position. But as usual it will be helpful to assume that this contrast with what was to be is a considerable help towards any understanding of the earlier episodes, as though these formed part of the ancestral history of the later and more sophisticated scheme.

In any case, the interpretation of the Kantian parallel is itself highly controversial, and the whole Kantian schema of such complexity that its proper understanding has usually eluded its commentators who have therefore been unable to appreciate its relevance for our understanding of modern philosophy of science.[11] Furthermore, even if the schema provides a rough guide, the whole classification emerges only as the result of centuries of groping, during which there is by no means sufficient clarity about the status of its individual members. Besides, it is often difficult to draw a definitive dividing line between the different methodological components, and certainly their denotation is not invariably clear-cut. For this reason, I have chosen the somewhat less rigid terminology of conceptual, inductive and architectonic components, determining respectively, intelligibility, empirical strength, and rationale, with the intention of using them as a rough guide in the attempt to interpret the history of the subject.

My contention will be that these three components act like vectorial weights that determine the growth and acceptance or rejection of some given explanatory concept or theory. Each of them may affect the other; individually, they will often resist one another's influence; and this all the more, since the true significance of the different factors is not initially grasped very clearly. It may therefore be useful here to anticipate in summary fashion our eventual findings for the case of the concept (or hypothesis) of gravitational attraction ('attraction at a distance'). Initially, the intelligibility (for which here read: empirical meaning) of the notion of bodily action was so conceived as to militate against the empirical employment of the concept of action at a distance. This was for instance the position of Huygens and Leibniz. In the development of Newton's own thought we find that whilst the limits of intelligibility, or rather a conviction of the total lack of intelligibility, of gravitational attraction were never in doubt, the weight of considerations affecting the estimate of its 'rationale' (i.e. its regulative determination) counterbalances the effect of the conceptual deficiency, and bestows enough confidence on the gravitational concept for it to count as real 'explanation', given its independent inductive articulation. For Newton, this rationale is

[11] See my *Metaphysics and Philosophy of Science*, Chapter 8.

based not so much on explaining attraction by means of the action of an interphenomenal ether (although this will play its part), but rather on the belief that *actio in distans* may be regarded as the expression of a supra-phenomenal—here, teleological—state of affairs. Finally, in Kant, we shall find a clear recognition of the possibility of a real alternative to such an architectonic foundation, amounting to a novel definition or 'analysis' of the concept of matter, whereby the property of *actio in distans* is so to speak 'packed into' that concept. This is achieved by using what I have called the conceptual component, i.e. by an exposition of the concept of matter under the guidance of the categories; in the present case, of the category of quality (reality, negation, limitation).[12]

In respect of this category, matter is 'explicated' as that which 'fills space', a notion in turn analysed as 'resistance to anything movable'. But, Kant goes on, matter does not resist just in virtue of its 'mere existence, but of a particular moving force'.[13] In the present case, the moving force involved is shown to be 'repulsive force'.[14] Moreover, so the argument continues, the possibility of matter also requires a universal attractive force, to counteract repulsion.[15] The important point here is that both repulsion and attraction are now conceived as being *logically prior* to the notion of physical contact between individually bounded volumes of material bodies. Dynamic action (whether by way of repulsion or attraction) does not *presuppose* touching or contact; on the contrary, the latter presupposes the former.[16] It follows that one does not have to invoke impact action, in order to make distance-action intelligible and it will not longer be necessary to frame any corresponding 'hypotheses' using impact assumptions.[17]

Evidently, force is here written into the very concept of matter, a concept which—it should be noted—is expressly declared to be 'empirical'.[18] And in reality it is obvious that Kant is looking over his shoulder at the relevant concepts of Newtonian science. This fact does not however diminish the importance of Kant's step which consists in his modification of the concept of matter itself, thus paying express attention to the possibility of alternative

---

[12] Cf. M.199.

[13] M.169–70. Cf. Kant's pre-critical tract on the *Clarity of Principles*, and particularly the 'example', CP.18–21; and the even earlier *Monadologia Physica* (Cf. below, p. 200).

[14] M.172.

[15] M.182–3.

[16] M.187; cf. CP.20.

[17] Cf. M.189–90.

[18] M.140. The *a priori* ingredient here concerns only the kinematical and causal aspects of the situation, as well as the contention that what is real in our perception of matter has *some* degree of intensive magnitude which can vary from zero to any positive number (Cf. K.201 ff., and especially K.206–7).

conceptual explications rather than employing only the purely architectonic aspect in the manner of Newton, as we shall see in what follows.[19]

Kant is indeed perfectly aware of the step he has taken, which (in the *Clarity of Principles*) he describes as a need for awareness to 'take note of every changed application of a concept'.[20] Thus, taking the problem of the 'correct' analysis of the concept of bodily action as his example, he argues that before we can assert dogmatically that all bodily action must take place through contact, by touching, we should first enter upon a careful analysis of the term 'touch'.[21] But when we do this, we realize that the concept of 'touch' does not primarily involve reference to spatial 'contact', but rather to 'resistance of impenetrability', and with this, to force. This shows, Kant argues, that force is here the basic aspect; and if experience and induction involve the concept of attraction without contact, we cannot argue from ill-understood 'surreptitious definitions' that the notion of *actio in distans* is otiose. Earlier philosophers were at fault by employing an insufficiently refined explication of the concept.[22]

In the *Metaphysical Foundations* Kant comments on the approach of the physicists of his time (Lambert and others) who regard 'solidity' as though it entailed already analytically a kind of 'resistance', as a property which belongs to physical substance 'by its very conception'. And again he complains that this notion has been linked insufficiently to the results of considerations from empirical science, i.e. the inductive component, which demand that resistance should be viewed as a 'repulsive force', subject no doubt to the usual mechanical laws. That is to say, do not imagine that one body resists another in virtue of a concept, grasped perhaps only opaquely, but rather in virtue of an understanding which involves reference to the inductive component![23]

Kant was not, of course, the only philosopher to pursue the ramifications of the conceptual argument concerning the proper analysis of 'matter', although by being closely integrated with the principles of his vast philosophical system of transcendentalism, it comes to occupy a more significant place as an important key for our understanding of that system. In England itself we find a school of thinkers who early recognize the implications of the notion of force for our conception of matter. The most prominent figure was Priestley, in whose *Disquisitions relating to Matter and Spirit* (1777) we find some arguments which though no doubt arrived at quite independently

---

[19] Cf. particularly the remarks of M.200, which explicitly register this extension of possibilities.

[20] CP.22.

[21] 'I now ask: what do I mean by touching?' (CP.20).

[22] CP.20.

[23] M.170–1.

from those of Kant bear a remarkable resemblance to the latter's views as expressed in *Clarity of Principles* of 1763, not to mention the later writings.

Like Kant, Priestley argues that solidity and impenetrability, when properly considered, relate to nothing other than resistance, and with that, to the power of repulsion; and in support Priestley appeals to Newton's first two Rules of Philosophizing which enjoin us to employ always the smallest number of 'causes' in our explanation of the phenomena, and always to assign to the same effects the same causes. Similarly he argues—though he give no clear grounds for his contention—

that without a power of attraction . . . there cannot be any such thing as matter; consequently, that this 'foreign property' as it has been called, is in reality absolutely *essential to its very nature and being*.[24]

The appeal to Newton's Rules of Philosophizing is a clear indication here that we are dealing with conceptual explication; Priestley is concerned to show that we can devise a concept of matter in which attraction and repulsion are not extraneously introduced 'foreign powers' but form an 'essential' part of the notion itself. Like Kant, he has come to realize that it is both possible and often necessary to enlarge a concept under the weight of inductive considerations.[25]

II

Now I began this essay with an account of the Kantian standpoint in order to present the more graphically what is missing for instance in Newton, testifying to the fact that although scientific meanings—as has recently been argued[26]—are not invariant under change from one theory to another, we meet here an interesting instance where there is considerable resistance to change. Change is not automatic, however strong may be the pressure from the inductive component, but presupposes—at least in the case of basic concepts—severe philosophical upheavals, as our subsequent story will show. Thus in Newton's early tract *De Gravitatione et Aequipondio Fluidorum* (*ca.* 1664 to 1668), body is distinguished from empty space by its quality of 'being

[24] Op. cit., pp. 103, 105. Cf. Heimann and McGuire, Sect. V.

[25] The injection of empirical factors into the meaning, particularly of scientific terms, has become a prominent subject for discussion in modern philosophy of science.

[26] Cf. Feyerabend, and for a critical assessment of this, Shapere.

impervious' (impenetrable) to other bodies.[27] Here our concept is again so delineated as to exclude any consideration of the notion of force and its laws in general, which would have opened the way to an argument like Kant's.

The need to distinguish between three different components is then quite evident, for such a distinction clarifies the nature of the logical controversies that arise during periods when new scientific conceptions seek entrenchment. Equally clearly, it is often difficult to draw sharp boundary-lines between our components; a difficulty which is heightened by the incidence of considerable cross-traffic between them. Indeed, as our case-study will show, there is a kind of dialectical interplay, because the status and definition of each component is never altogether clear. Hence perplexity surrounding each of the components may cause us to seek support from the others; deficiencies of the conceptual interpretation, for instance, calling forth supplementation from architectonic considerations. Nothing exemplifies this better than the history of Newtonian gravitational theory.

Let us consider then in greater detail, through our Newtonian case-study, first the process which involves attempts at conceptual explication (under whatever name it may parade), determinative of the limits of intelligibility. This may best be introduced by contrasting Newton's conception of matter and related notions with its contemporary alternatives.[28] For our purposes, we need to single out only a few of these. For the Cartesians as well as for Newton and his followers, matter is 'inert', and a material body will be set in motion by another body only through impact action, such action taking place in accordance with certain mechanical laws. But here their views part. For Descartes, there is only one *essential* property of matter, viz. extension, as a consequence of which his is a *plenum* theory. In addition, all matter is characterized by the *universal* property of motion, defined as volume times speed, the sum total of all motion in the universe being constant through time. It is also a cosmological assumption of most Cartesians that the universe could have evolved unaidedly from any primeval chaotic state into its present form, in accordance with the properties and laws of matter and motion.

By contrast, for Newton, in his mature view,[29] matter is conceived

[27] Halls, p. 139.

[28] It is not my intention here to be historically exhaustive, but to select what is philosophically illuminating. I have discussed some of the issues that follow in an earlier essay, 'Gravity and Intelligibility: Newton to Kant', but without attending there to any of the historical *minutiae* so important for an understanding of Newton's thought as a historical and philosophical phenomenon.

[29] Newton's views mirror those of a number of seventeenth-century natural philosophers, but for our purposes it is sufficient here to indulge in the historical fiction that these are the ideas of a single philosopher.

atomistically, involving the postulation of empty space. Moreover, in addition to extension, atoms have further essential properties: in the *Principia* he mentions hardness, impenetrability and inertia.[30] Finally, the motions of and impacts between these atoms are again subject to the laws of mechanics; here, Newton's Laws of Motion. He sometimes refers to these as 'passive laws of motion', suggesting a contrast with 'active principles' that will occupy us later.[31] Matter thus does not strictly 'act', even on impact. All genuine sources of motion must be immaterial.[32] Moreover, all such motion must be communicated by material means, since any 'gaps' would imply that matter *can* 'act' independently of a non-material source.

Now this, as already anticipated, for Newton exhausts the concept of matter and determines the framework of what is regarded by him as 'intelligible'. This manner of proceeding is an instance of 'conceptual explication'. No doubt the 'meaning' of the concept of matter that emerges is determined also by pre-scientific influences, e.g. the fact that we cannot normally act on another body without the use of our hands and instruments. On the other hand, inertia is quite a sophisticated notion, expressed clearly only *via* its corresponding laws (the three laws of motion). Meaning, we here note, is a function of pre-scientific as well as nomothetic and systemic considerations, implying that interaction takes place between the inductive and the conceptual components.

This shows incidentally that Kant had characterized the difference between the Newtonian and his own approach too sharply. It is not the case that the Newtonian concept of matter and action took *no* note of inductive considerations, e.g. the laws of motion; but rather that, being unconscious of the fact that determination of meaning involves the inductive component, it had cast its net insufficiently wide, and indeed put up—as we shall see—considerable resistance to its enlargement. For as our story shows, although there is in fact considerable cross-traffic between the three components, each frequently

---

[30] P.399.

[31] O.401. In fact, 'impact action' has rather an uneasy logical status in these systems of mechanics. Both Cartesians and Leibnizians are plenists. For the former, matter is ultimately incompressible; for the latter, dynamic interaction is at best only something 'phenomenal', since ultimately all takes place in accordance with a pre-established harmony. Locke will still say (and rightly so on his principles) that interaction is 'incomprehensible', however much it is an observable phenomenon. However, for practical purposes, these logical difficulties were ignored until the time of Kant, and here we shall do so likewise.

[32] Cf. *Opticks*: 'The *vis inertiae* is a passive principle. . . . By this principle alone there never could have been any motion in the world. Some other principle was necessary for putting bodies into motion; and now they are in motion, some other principle is necessary for conserving the motion.' (O.397; cf. below, pp. 179, 193 f.)

seeks to retain (at least for a considerable time) a separate identity, and to preserve immunity from change despite considerable pressure, for instance, from the inductive and regulative (architectonic) components; indeed, intellectual conservatism often prevents these thinkers from realizing that meanings are capable of change at all.

Thus in the present case, the properties in Newton's original list—and these alone—soon come to be labelled 'essential properties', expressive of the 'nature of matter'. As a result, despite the fact that Newton believes and affirms gravitational action to be a *universal* characteristic of matter (as implied by the inductive component), it is excluded from his list of 'innate' or 'essential properties',[33] although there are of course additional reasons, as we shall see, for his traditionalism.

Certainly Newton declares expressly that any notion of one body acting on a second without the interposition of a physical medium or other body is something 'unintelligible', in the sense here used. This is what he affirms in the well-known passage of his third letter to Bentley, where he says that action of 'brute matter . . . without mediation of something else, *which is not material*', that is to say, on 'other matter without mutual contact', is 'inconceivable'.[34] The idea that gravity is 'innate', he adds, would mean that one body could 'act upon another at a distance through a vacuum, without the mediation of anything else'; and this is for him an 'absurdity'.[35]

But no sooner is this said, than the pressure from the inductive component makes itself felt. As Kant's discussion implied, the notion of causal action is not so sharply defined as Newton makes it appear. And Newton certainly does not wish to deprive himself of *the use of the phrase* 'matter attracting other matter at a distance', except that his inability to make sense of it leads him to add at frequent intervals the proviso that such a phrase does not imply any claim on his part to an understanding of 'the cause' of this attraction,[36] or of its 'physical reason'[37]; all locutions which lead the

[33] P.399–400; Thayer, p. 53; cf. Koyré, pp. 156–63.

[34] Thayer, p. 54; my italics. In their edition of Newton's unpublished scientific papers, the Halls, following Koyré, seem to interpret the italicized words as though they were equivalent to 'which is immaterial', suggesting that Newton means here to imply that bodies might after all be able to act by means of immaterial or spiritual forces. (Op. cit., p. 194.) Newton's emphasis *here* is however certainly on a body's inability to act without material means, the only way in which, strictly speaking, it can act at all. The fact that Newton will subsequently look for a *rationale* of gravity must not be confused with his unchanging conviction regarding the unintelligibility of bodily action without contact; an interpretation, that will at once clarify a number of apparent contradictions in Newton's approach.

[35] Ibid.

[36] P.546–7; O.376.

[37] P.6.

reader to expect from Newton further *physical explanations*, when in fact—as we shall see—he is just as often hinting at a teleological or theological supplementation through the use of the architectonic component, balancing the deficiency of the conceptual component.[38]

III

But let us look further, for any deeper motives lying behind the conceptual explication which led Newton and his circle to consider the action of matter *in distans* unintelligible. In part, it was a tacit acceptance of some of the ideas of the Cartesians. The rejection of any hylozoic account, endowing matter with a quasi-animistic power to act, a kind of internal life of its own, was too precious an achievement easily to be jettisoned again. Basically, matter is conceived of as inert and passive, and what the seventeenth-century thinkers vaguely call its 'essence', determined by its essential properties and mechanical laws of motion.[39] The very term 'essence' warns us of the vagueness of the logical situation. In what sense are the laws of motion 'essential'? Newton will of course *say* that they are 'deduced from phenomena and made general by induction',[40] but one can never be sure of the influence of the fact that these laws, and especially the law of inertia, were in part the outcome of conceptual explications; witness the method by which Descartes first formulated the law.[41]

At any rate, the limits of intelligibility more or less coincide with this notion of essence, and its very fluidity lends itself well to a dialectical interplay between it and the other two factors, determining rationale and empirical content. Thus Locke will say that gravitational attraction 'cannot be conceived to be the natural consequence of that Essence', and that it cannot be explained or made 'conceivable by the bare Essence . . . of matter

---

[38] The famous reference in the *General Scholium* of the *Principia* (P.547) to the need to keep out of 'experimental philosophy' '*hypotheses*' that are 'metaphysical or physical', and 'of occult qualities or mechanical', is indeed a conflation which joins the requirement to provide a *physical explanation* with the quite different enquiry into a *philosophical* theory that can provide an *interpretation of, and thus a rationale for, the concept* of gravity. Cf. below, p. 187.

[39] Cf. the passage from Locke, quoted immediately below, Koyré, p. 155. The same conception is echoed in Kant's early writings; cf. OG.100, where he says that 'the laws of motion of matter are absolutely necessary', deriving as they do from 'the possibility of matter as such and of its essence [Wesen]'.

[40] Letter to Cotes, Thayer, p. 6.

[41] Cf. my 'The Relevance of Descartes' Philosophy for Modern Philosophy of Science'. For Newton's concept of inertia in relation to Descartes, see I. B. Cohen (1964).

in general, *without something added* to that Essence *which we cannot conceive*.[42] The only 'action' which Locke can *'conceive'* is action 'by impulse'.[43]

This does not exhaust all the considerations that fix the limits of intelligibility of the action of matter. Both Locke and Clarke at times hold action at a distance to be *logically* self-contradictory, since they interpret it as the action of a body 'where it is not'; which is—as Locke says in the *Essay*—impossible to conceive.[44] But again the basis of the argument is not clear. Locke, in the first three editions of the *Essay* explains in more detail that it is 'impossible to conceive that body should operate on what it does not touch'.[45] But he does not explain what has determined the meaning of 'operating' or 'touching' so as to lead to the contradiction.[46]

Newton's own insistence on a narrow definition of matter and the list of its essential or 'innate' properties is partly motivated by the consideration that as a consequence of such a position, the missing property has to be anchored in a different dimension, unless the kinematical aspect of gravitation (mutual acceleration)[47] be accountable for in terms of an intervening material mechanism—a possibility about which the later Newton was often in considerable doubt.[48] The 'dimension' just referred to, and as we shall see again presently, belongs to the architectonic component, governing the 'rationale' of gravity, and concerns the mataphysically-interpreted notion of teleology

[42] Koyré, p. 155; my italics. The nature of this 'addition' will occupy us presently.

[43] Ibid. This must be distinguished from Locke's general position which is that *no* contingent relations, including those of action by impact, can be rationally intuited, being in this sense all equally inconceivable. Which is not to say that these two cases of 'conceivability' are not very intimately related, and that the general case may not have helped to lend plausibility to the special conceptual worry. Locke himself employs the generalized, or degenerate, notion of inconceivability as a model for empirical contingency; furthermore, God's will is invoked as a model for describing the case of those contingent matters of fact that are universal and lawlike. Kant later generalized the Newtonian position to that of Locke, by again questioning the intelligibility not only of action at a distance but also of impact action (action and reaction); it was the task of the third chapter of the *Foundations* to save the intelligibility of impact action by explicating it in terms of the Third Analogy of Experience.

[44] Op. cit., ii.8.11; Clarke-Leibniz, Fourth Reply, 45; Alexander, p. 53. Clarke says explicitly that it 'involves a contradiction'.

[45] Op. cit., ed. A. C. Fraser (Oxford, 1894), I, 171, note 1.

[46] Cf. Kant's critique of this narrow definition of 'touch', above, p. 172.

[47] In what follows the ambiguity in the use of the verb 'gravitate' should always be kept in mind. It may denote either kinematical aspects of gravitation, or dynamical ones; a fact which makes it all the easier for Newton to insist with firmness that 'gravity does really exist' (P.547), whilst at the same time denying the intelligibility of the concept.

[48] We may distinguish three periods. In the 1670s Newton was quite prepared to formulate 'hypotheses' of an ethereal mechanism. During the 1690s and up to 1706 he became increasingly sceptical about this, but after 1715, under the influence of new researches into electrical and magnetic phenomena, the ether began to play again a more prominent part. (Cf. H. Guerlac, 'Newton's Optical Aether'.)

as divine providence, the 'thinking and willing' of a designing Deity. Such a view was not particularly strange, given the intellectual atmosphere of Newton's period, and as expressed by some of his mentors, among them especially the Cambridge Neo-platonists. Thus, in his *True Intellectual System of the Universe* (1678), Cudworth in a very similar manner argues that because matter includes in its concept nothing but figure, size and motion, it requires supplementation by a second range of ideas such as understanding, sensation, soul, life,[49] ideas which he links moreover to the influence of the Deity.

Now God, for Newton, is always a concept used to indicate the existence of what we might call an 'ontological anchor', or ontological basis; and the 'will' of God supplies what without it would be a deficiency in the element in question, here, of gravity. In a similar manner, in *De Gravitatione*, the early Newton—still half in the throes of Cartesianism—holds that because what 'exists eternally' is only space and time (albeit as 'an emanent effect of God'),[50] it requires the 'divine will' to cause any portion of space to be 'impervious', and thus to produce (as has already been noted)[51] existing material bodies— a theological way of saying that matter is not equivalent to extension. However, subsequently impenetrability becomes for Newton part of the concept of matter—theologically expressed by him when he says that—

> God *in the beginning* formed matter in solid, massy, hard, impenetrable movable particles, of such sizes and figures and with such other properties, and in such proportion to space, as most conduced to the end for which he form'd them.[52]

This implies that matter as such (and with its 'essential properties' noted above), once created, is left to itself. But once again Newton, just as before in the case of space and time, regards such a state of affairs as an invitation to atheism, since it would not seem to leave anything for the Deity to do, short of the initial act of creation. As so often in such cases, a virtue is made of the deficiency, and God is given the task of supplementing it. The argument operates however at two different levels, a physical (empirical and theoretical) and a metaphysical level. It is physical when Newton argues that the growing irregularities in astronomical motions, as well as the continual 'loss of motion',[53] require a power that will 'reform' the parts of the universe

---

[49] Op. cit., e.g. p. 862. Cf. also below, p. 194, for another passage from the same work.

[50] Halls, p. 132; though Newton adds that since 'we have an absolute idea of space without any relationship to God', it is easy to overlook this relation, and slide down the slippery path towards atheism (p. 143).

[51] Cf. above, p. 174.

[52] O.400; my italics. For the continuation of this passage, see below, p. 193.

[53] The argument in the last Query of *Opticks* suggests that this may be a confused reference to loss of energy, loss of momentum, and increase of entropy (loss of negentropy); cf. O.397, 402–3; above, p. 175, and below, p. 193 f.

at suitable moments in time; a lack of perfection in God's original design that Leibniz will later use as the starting point of his criticism in his correspondence with Clarke.[54] To this Newton further joins an argument from the display of adaptation in the biological universe (plant and animal kingdom) to a designer.[55] On the other hand, the 'metaphysical' argument is quite different, though Newton conflates the two reasonings. He contends that the existence of gravitational phenomena becomes rational (and thus real) only on the supposition that they are an expression of divine providence, often described as an 'active principle', which however—unlike the cause of impenetrability—operates *continually*, a fact which Clarke (Newton's spokesman) expresses by saying that God's conserving power is 'an actual operation and government, in preserving and continuing the beings, powers, orders, dispositions and motions of all things'.[56]

To the details of the notion of 'active principle' I shall return.[57] But it should be noted here that divine power operates rather differently in the two cases of impenetrability and gravity; the former belonging to the conceptual explication, the latter, to the regulative determination of gravity (i.e. to the conceptual and the architectonic components, respectively). And so, although both impenetrability and gravity may ultimately depend on God, they do so in rather different ways. If this were not so, the difference between an 'innate' and a merely 'universal' property of matter would be lost and with that the methodological thought of Newton gravely misrepresented.[58]

IV

Newton's architectonic approach will be developed further below.[59] I mention it here only in order to point to an additional motive for his narrow

[54] Clarke seeks to avoid the difficulty by arguing that the 'correction' required is not with regard to God, but to us only' (Second Reply, 8. Alexander, p. 22). But the concession weakens the semantic force of Newton's general argument, the employment of the concept of design as a factor intended to provide a rationale for the ontological aspect of gravitation, since it is the empirical cases that give a meaning to the whole idea.

[55] O.403, 369–70.

[56] Second Reply, 11; Alexander, p. 23. Metaphorically, both Clarke and Newton express this by denominating God 'a governor' (ibid.; cf. P.544–5).

[57] Cf. below, p. 193.

[58] I cannot therefore agree with the Halls, who overlook the difference by labelling both kinds of properties 'spiritual', thus implicitly over-physicalizing gravitational action. This false identification is aided and abetted by their view that 'matter in its merely passive properties of impenetrability and inertia exists only by the *continued exertion* of the divine will', interpreting 'dependence' of matter on God, as 'continuous', whereas in fact only 'active principles' such as gravity are. (Cf. op. cit., p. 197.)

[59] Cf. below, p. 190.

definition of the conceptual boundaries of matter. The situation which emerges is somewhat quaint: the physico-teleological approach, whilst eventually supplementing a foundation which is too weak owing to over-narrow conceptualization, in its turn itself comes to determine and fix rigidly the outlines of that conceptualization. This shows again that despite the choice of an authoritarian term like 'the essence of matter', the considerations which lead to a delineation of this essence are fluid and contentious in the extreme.

There is however a further complication: once the 'essence' has become entrenched, it may react on beliefs held also as regards the status of the third component, dealing with the inductive basic. This is well shown by the position that Leibniz takes on the question of gravity.

In his controversy with Leibniz, Clarke had contended that attraction, 'considered as a general law of nature'—a locution which echoes Query 31 of Newton's *Opticks*[60]—acts 'regularly and constantly', and hence 'may well be called natural'.[61]

Now to call something 'natural' (given the fluidity of this term) may well imply a reference to the conceptual limits of a term. If so, this would mean that Clarke holds (what in fact he does not hold) that attraction is after all an essential characteristic of matter. This Leibniz of course denies, as much as Newton and Clarke. However, for Leibniz the pressure from the limits of the concept extend further, affecting more deeply the status of the inductive component, and throwing doubt even upon the phenomenal (inductive) situation itself. His comment on Clarke makes this quite clear—

> But it is regular (says the author), it is constant and consequently natural. I answer; it *cannot be regular*, without being reasonable; nor natural, unless it can be explained by the natures of creatures.[62]

The expression 'nature of created things', or 'nature of things', occurs also in certain other letters of Leibniz, e.g. to Conti and Hartsoeker,[63] and the context makes it clear that it has reference to the essential properties and mechanical laws of matter. Thus, in the letter to Hartsoeker he says that a planet moving in a non-linear orbit round a centre must 'by the nature of things' move from its orbit along a tangent, since it will be subject to centrifugal forces; a phenomenon which in his *Tentamen de mutuum coelestium causis*,[64] is also described as being in accordance with 'the laws of nature'.

---

[60] O.401.
[61] *Correspondence*, Fourth Reply, 45, Alexander, p. 53; Clarke refers to the Newton passage in his fifth reply, cf. 110 ff., Alexander, p. 115.
[62] Fifth Paper, 121; Alexander, p. 94; my italics.
[63] Cf. Koyré, Appendix B, pp. 144, 141.
[64] *Mathematische Schriften* (ed. C. J. Gerhardt), VI, 149; cf. Koyré, p. 128.

Elsewhere in the *Tentamen* Leibniz refers to these laws as 'corporeal laws'; what he means to imply by this phrase is his assumption that these are laws that belong, so to speak, to the body by itself, in isolation; or if not that, at least they manifest themselves only during impact between bodies. The possibility of any law of attraction is thereby excluded; representing the same kind of conservatism that prevented Newton from including gravity among the list of essential properties. And once again, the conceptual aspect of the matter becomes manifest in Leibniz's phraseology when he says that only action in accordance with corporeal laws, involving 'corporeal impressions' is *'understandable'*.[65]

At the inductive level, Leibniz here follows essentially the kind of theories found in Descartes and Huygens, involving the conception of an aetherial vortex, a revolving fluid in which the planet is embedded. Centrifugal forces result in the production of centripetal forces, since the matter that is pushed outward in turn causes other matter to be driven inward, thus producing the 'phenomenon' of gravitation, not to say gravitational attraction. It is not 'really' an attraction, but at best, what Leibniz calls a 'solicitation of gravity'.[66]

This whole theory is described by Leibniz as one that 'conforms to reason'.[67] Presumably this is so mainly because it links up with his general scheme, according to which nothing happens without a sufficient reason, a principle invoked, among others, in connection with his contention that God cannot have left any space empty; just as He has placed in all things such 'powers, orders, dispositions, and motions' as conform to the general harmony and perfection of the whole.[68] This aspect of teleology (belonging, as it does, to the architectonic component) will occupy us presently. For the moment, we note that the various powers are permitted here to work only *via* what Leibniz calls 'material causes'; 'immaterial' ones, he tells us, are not to be used within the explanations of physics.[69] God's design must express itself, so to speak, only immanently; all activity (including all forces) is an expression of the life of each individual substantial thing, acting in accordance with the principle of perfection.

This very brief sketch is meant to do no more than sketch the context of Leibniz's reply as to what is 'reasonable' and 'natural'.[70] Evidently we meet once again the spectacle of a rather conservative attitude towards certain kinds of physical theory supported by the vast background of Leibnizian metaphysics, together forming the ground on which the concept of matter

[65] Koyré, p. 127.
[66] Koyré, p. 131.
[67] Koyré, p. 134.
[68] Third Paper, 16; Alexander, p. 29.
[69] Koyré, p. 134.
[70] Cf. above, p. 181, quotation from the fifth paper, 121.

is supported. The important thing however is that, unlike in Newton, the framework determining the 'intelligibility' of the concepts is here so powerful as to inhibit the acceptance of gravitation even at the phenomenal level, as something 'acting regularly and constantly', as a phenomenon belonging to nature. For as the passage from the Fifth Paper showed, Leibniz now goes so far as to claim that since gravity is nothing 'reasonable or natural' it 'cannot be regular and constant'; at least, not unless it be made intelligible first. In other words, the constraints imposed by considerations from the side of the conceptual component are here so powerful as to militate against any inductive developments; and they certainly militate against an inductive theory which would not explicitly employ the mechanism of an ether. Evidently, given the same conceptual foundation, Newton's only course will be to adjust the architectonic component in a different way; Newton's teleology and general metaphysics will have to differ sufficiently to loosen the grip.[71]

However, for the moment we need to enter a proviso. We must be careful not to misunderstand the range of Leibniz's denial, to get the proper contrast with Newton who after all often gives vent to similar sceptical remarks on the aspect of the physical reality of gravity. For there is a sense in which Leibniz is *not* denying a regularity—what he called 'solicitation to gravity'; indeed in the passage in which this expression occurs, he even refers to it as a 'special law of attraction'.[72]

At times, therefore, Leibniz does not seem to differ so much from Newton, when for instance he describes the planets as 'gravitating' towards the sun—like Newton availing himself of the kinematical sense of this term; meaning that the planets have a component of acceleration directed towards the sun. Here however the similarity stops. For the sense of the unreality (due to the unintelligibility) of gravity is for Leibniz so strong that it affects the construction of his physical theory, the inductive component. Newton was later to protest that however much the 'acceleration of gravity' might need further 'explanation', either inductively by way of hypotheses concerning ethereal action, or architectonically, by way of consideration of the provident design of the physical world, this need not affect the form of the inductive enterprise, of what he calls 'experimental philosophy'.[73] In the *Principia* the 'action of

---

[71] Of course, it will be said that Newton took great pains to equal Leibniz in his denial of the 'physical reality' of gravity. But although Newton did issue such denials quâ natural philosopher, unlike Leibniz he was prepared to speak of the attraction of gravity as an 'explanation of the phenomena' (P.546); and *that* is the essential point, however much psychologically he was torn between two conflicting positions.

[72] Koyré, p. 131.

[73] P.547. It is only 'natural philosophy' that needs to concern itself with the architectonic question; cf. P.546.

gravity' stands on its own feet. Not so for Leibniz whose *physical account* is so constructed as to necessitate the reference to the ambient ethereal medium, of whose action the gravitational phenomenon are only the result.

It is to ram home this difference, that Leibniz emphasizes that gravitation should not be regarded as a physical reality, as a cause of gravitating. To state his difference with Newton in this way was however misleading, since they both agreed on the question of the lack of physical reality. What distinguishes them really is a difference concerning the consequences of this denial for the significance of the inductive component; first, whether any inductive structure could be cast in the form of the *Principia*: a formal difference; secondly, whether Newton was entitled to think of the law of gravity as a genuine law; lack of intelligibility in Leibniz's eyes evidently implying a lack of inductive foundation for the assertion of lawlikeness—a metaphysical difference. No wonder the parties, and subsequent commentators, have become muddled about the true nature of the intellectual gap that divided them.

V

At this point it is important to remind ourselves—however briefly—of the logical structure of Newton's exposition in the *Principia*, leading to the law of gravitation. Here there is, as we noted, no mention of any 'ether', but not unlike Leibniz (though the details differ widely) Newton, from the laws of planetary motion and using his three laws of motion, deduces a centripetal force acting on the planets, proportional in magnitude to the inverse square of the distance. This explains perhaps Newton's claim that not only the laws of motion, but also the law of gravity itself had been 'inferred from the phenomena';[74] a claim which is less puzzling if we remember that in the *Principia* the planetary laws are labelled 'phenomena'.[75] Having thus derived the formula for the centripetal force by a purely deductive process, Newton goes on to show that the magnitude of this force as it acts on the moon would, at the earth's surface, yield an acceleration equal to that found experimentally for all terrestrial objects.

Newton's next step is to introduce a principle of simplicity: assume the minimum number of forces responsible for identical accelerations; furthermore, given the 'assumption' that to identical effects there corresponds a unique cause, Newton concludes that the centripetal force computed from the motion of the planets is identical with the 'force of gravity'.[76]

[74] P.547.
[75] Though they had been labelled 'hypotheses' in the first edition.
[76] P.408–9.

The two assumptions just mentioned are of course the first two of the four 'Rules of Reasoning in Philosophy', laid down at the beginning of Book III of the *Principia*, and to which references has already been made in connection with Priestley. Broadly speaking, they deal with what Newton conceives to be the major presuppositions of inductive and analogical reasoning. Apart from the principles of simplicity and causality (Rules I and II), the third rule sanctions the inferential step from the qualities *observed* to belong to all bodies within our experience to *all* bodies, including those transcending observation in virtue of their smallness, e.g. atoms. The fourth rule enjoins us to accept inductive inferences of the kind just discussed 'notwithstanding any *contrary hypotheses*', unless and until requiring modification or rejection in the light of new 'phenomena'.[77] With the use of these additional rules (III and IV), Newton then expands his results to yield the generalization which makes the force of gravity responsible not only for the planetary accelerations towards the sun, but indeed holds it to act universally between all bodies.[78]

VI

So far, except for technical differences between their respective mathematical treatments, Leibniz might have assented. But as we have seen, Newton does not insert *into the technical body of the Principia* any mention of what it is that causes the planets and all other bodies to gravitate. For instance, he does not invoke—as Leibniz and Huygens had done—any circumambient ether. On the contrary, he short-circuits any such questions by introducing into his technical comments no more than the purely logical notion of a cause of acceleration (such causes being labelled 'forces'), leaving open the question of the identity of the causal agent. Thus, a 'centripetal force' is defined at the outset as 'that by which bodies are drawn or impelled, or any way tend, towards a point as a centre'; and any 'impressed force' in general is defined as 'an action exerted upon a body, in order to change its state, either of rest, or of uniform motion in a right line'.[79] The concept of force is hence for Newton so far no more than that of 'something' (whatever that something might be) productive of a stated kinematical effect.

Now it is true that the *physical* significance of such a force might be unpacked, for instance, as the action of a medium. But the impressive formal apparatus, and the care which Newton expends on stating the *inductive*

[77] P.398–400.
[78] P.410, 413.
[79] P.2, Definitions V and IV.

foundations of his reasoning, as summarized in Section V above, have at least the strong *psychological* tendency of suggesting that even without the introduction of hypotheses of any mechanistic or etherial medium the conception of the force of action of gravity may be taken seriously, regarding it *as though* it were a physical agent in its own right; thus obscuring the difficulty about its unintelligibility, however much Newton when reminded of this might rebel against such an interpretation. Certainly in speaking of gravity as a 'cause', Newton already goes further than Leibniz would have countenanced, since for the latter 'gravity' was—as we have seen—undoubtedly no more than a *façon de parler*. Nor are the Newtonians always consistent in their use. Sometimes, as above, Newton speaks as though gravity *were* a cause, in the sense of being a causal agent; and in the General Scholium he says that in the *Principia*, 'the phenomena of the heavens and of our sea' have been *'explained . . . by the power of gravity'*.[80] At other times however he thinks of gravity as an effect, though it is not always clear whether he means that the *force* of gravity itself is an effect, or only the kinematical phenomenon of mutual acceleration in accordance with the gravitational formula. Similarly Clarke, when pressed by Leibniz, speaks of the 'phenomenon' of 'gravitating bodies' as the 'effect' of an unknown cause; but he immediately slides into saying that attraction is a tendency, whatever its cause, of bodies 'towards each other, *with a force* which is in direct proportion to their masses. . . .'[81] Sometimes Newton as much as tells us that the term 'attraction' does not mean what it seems to say, as when in Query 31 of the *Opticks* he says that he has been using this word 'to signify only in general any force by which bodies tend towards one another, whatsoever be the cause' ('force' thus not standing for 'cause'), and that what he has *called* attraction 'may be performed by impulse, or by some other means unknown to me'.[82]

Sometimes Newton is even more hesitant or phenomenalist, as when in the famous last paragraph of the explanation to Definition VIII of *Principia* he explains that terms like 'attraction, impulse, or propensity of any sort towards a centre', where they occur in the *Principia*, must be understood to stand only for quantitative measures of a theoretical entity which though we call it by the name 'force', need not and should not be given what we now call a

[80] P.546, my italics.
[81] Fifth Reply, 118; Alexander, p. 118.
[82] O.376. Once again we note (what is important for our understanding of Newton's treatment of the architectonic component) that the phrase 'cause or force' may refer us to a physical, e.g. mechanical, explanation of the kinematical phenomena, or, alternatively, it may designate the 'nature' of that causal agent referred to by the general causal term. This ambiguity allowed Newton to move either towards specifying some mechanical causal explanation, or instead towards entertaining a more mathematical, Platonic, teleological point of view.

physical interpretation. However, this last and rather instrumentalist account squares badly with the idea that the 'power of gravity' provides a physical or causal explanation, and we shall ignore it here.

At any rate, if we put all these various formulations together, we may perhaps say that they imply a view according to which gravity, or the attraction of gravity, is an expression that stands for some causal agency or other (called a force) responsible for the mutual lawlike acceleration (or attraction) of bodies, and that furthermore, the theory of the *Principia* does not involve any further specification of the nature of that which is represented by this expression. This is not to say that Newton thought such a specification unnecessary; what he maintained was that this did not need to be included in the inductive part of his theoretical treatise.

VII

From all this emerges that Newton's insistence either on an instrumentalist account, or at least on the careful inductivist presentation of the *Principia*, was not so much an exercise in empiricism, or an anticipation of a phenomenalist or positivist philosophy—though his approach might quite easily lend itself to such an interpretation in other contexts—as a deliberate attempt to insulate the inductive component from the pressure brought to bear on his thinking by the lack of intelligibility of the conception of attraction; concealing some of the difficulties by operating with an unspecific and rather uncritical use of the notion of 'cause' or 'force'.[83] Quite obviously, this is also the main significance of the celebrated cry that in experimental philosophy we are to 'feign no hypotheses'—

> for whatever is not deduced from the phenomena is to be called an hypothesis; and hypotheses, whether metaphysical or physical, whether of occult qualities or mechanical, have no place in experimental philosophy.[84]

Commentators have had a field-day with this passage; and there is no need here to go once more into the question of the many and varied meanings which Newton attached to the term 'hypothesis'. All we need to remember is that this passage—as so often with Newton—conflates a number of conflicting considerations, and that the context makes fairly clear the relevant

---

[83] If the cause of action at a distance could not lie in the body (for reasons we have explained), it was easy to physicalize the concept of a 'causal *relation*' holding between two bodies, the kind of notion Hume was later to attack implicitly. Newton's pseudo-positivist manoeuvres should not be thought of as anticipating such a Humean approach.

[84] General Scholium. P.547; cf. Note 38.

uses of the term hypothesis that are involved. And it seems to me that there are just three major strands in Newton's programme.

1. In part, we have simply a comment on the procedure of the *Principia*, which 'demonstrates' the law of gravitation as an induction from the phenomena, in the way described above.

2. A second strand, often confused with the first, is Newton's insistence that the treatment of the *Principia* can be insulated from any question as to whether or not an explanation of the action of one body on another in terms of an intervening ethereal medium can or has to be given. Such an explanation had been attempted by Newton in the 1670s, and it appears again in the third edition of his *Opticks* (Queries 17–22) where the ethereal medium is given certain explanatory functions in connection with optical phenomena (refraction and reflection), and where Newton's confidence concerning the use of an ether had re-established itself again following certain researches into magnetism and electricity.[85] At the time of the second edition of the *Principia*, where the General Scholium first appears, Newton's belief in the ether had reached its nadir. Particularly with this sceptical attitude in mind, Newton may also be interpreted as wanting to recommend the adoption of a phenomenological as against an interphenomenal type of theory. This consideration, involving judgements of preference in regard to theory-types, leads naturally to a third strand in Newton's thinking.

3. To this end, let us refer back to the teaching of the Fourth Rule, according to which an inductive foundation stands unimpugned, 'notwithstanding any contrary hypothesis that may be imagined'.[86] Here is involved a sense of hypothesis which we frequently find in the philosophical literature of Newton's contemporaries. Writers like Cudworth for instance often speak of the 'Atomic Hypothesis', or the 'Atheistic Hypothesis', or again, of the 'Pythagorean Hypothesis'. They are then addressing themselves to certain general philosophical systems, whether they happen to approve of such systems or not. Now this is the third sense intended by Newton's declaration against hypotheses. He is in fact conflating two quite different points. On the one hand he is again repeating the contention which I discussed under (2), but this time arguing that the inductive component does not and ought not to require the support of *philosophical* superstructures. Or rather, he is actually only claiming—what is much less—that the inductive account does not stand in need of certain *particular* philosophical ideas, viz. those four types of approaches referred to in the General Scholium as 'hypotheses'. Thus his claim is really that certain philosophical or methodological notions are either unnecessary or unacceptable as ways of shoring up the inductive component.

[85] Cf. Note 48.
[86] P.400.

He opposes both the notion that gravity is a physical reality, and the notion that it is an Aristotelian or alchemical occult quality; nor will he have any truck with the mechanistic scheme of a Descartes, itself based on a metaphysical foundation.

However, it does not follow from this that Newton was necessarily opposing *all* attempts to supplement the inductive account as such; indeed, if my general scheme is correct, we should expect him to opt for some alternative supplementation at the architectonic level; and Newton's vehement opposition to the four-fold group of 'hypotheses' would then be a left-handed way of inviting the natural philosopher to accept his own doctrine, his own supplementation, with greater readiness.

This suggestion differs from some recent views on the subject. I. B. Cohen, and following him N. R. Hanson, have drawn attention to the 'philosophical' or 'metaphysical' use of the term 'hypothesis' in Newton.[87] But they have been too ready to assume that because Newton's particular examples of 'hypotheses' concern philosophical views which he opposes, thus suggesting a pejorative use of the term even with this new meaning, he must have been opposed quite generally to philosophical supplementation. One may perhaps agree with Hanson on the mistake of 'people like Mach and Kneale and Russell' who tried either to praise or berate Newton for his proscription of hypotheses, when they fail to realize that this was a sense of hypothesis entirely different from what they supposed; and that Newton's 'hypotheses non fingo' is hence quite neutral to the question whether he had used or should or should not have used hypotheses in our modern sense of the term. But Hanson infers that, having disposed of these misleading uses, we must suppose that Newton really meant to keep philosophy out of the subject. Whereas the situation is rather more complex, since it is clear that Newton's *veto*—with its use of the pejorative and philosophical ascription of 'hypothesis'—was directed only against certain *particular* philosophical views.

To summarize: although Newton employs this sense of hypothesis to characterize philosophical doctrines inimical to his own, and although he holds that the *inductive* presentation *can* be insulated from *all* philosophical doctrines, it does not at all follow that Newton did not think *some* philosophical views to be relevant to the question of the rationale of conceptions involved in such an inductive presentation, e.g. the conception of gravity.

[87] Cohen, 'Hypotheses in Newton's Philosophy', pp. 173–9; Hanson, 'Hypotheses Fingo', pp. 31 ff.

VIII

Now, as I have already suggested, Newton's opposition to these hypotheses in fact links up with his own attempt to supply a rationale of gravity, particularly during the years covering the second editions of the *Opticks* (1706) and the *Principia* (1713), when his belief in an interphenomenal explanation of gravity was at its lowest ebb, and when he plays with the view that—

> it's necessary to empty the heavens of all matter, except perhaps some very thin vapours, steams or effluvia, arising from the atmospheres of the earth, planets and comets.[88]

Newton thereupon refers to the authority of the Greek atomists, involving the scheme of the vacuum and the void, adding that these theories had been 'tacitly attributing gravity to some other cause than dense matter'.[89] This reference to 'the ancients' is a favourite traditional device of those times to support a writer's own opinions, the views attributed to 'authority' not necessarily having any historical basis. At any rate, it prepares us for Newton's own attempt to provide a rationale that is meant to overcome the deficiency of the conceptual basis, as already described. Moreover, it is an account that involves philosophical positions (e.g. Neo-Pythagorean views) which he sees as competing with those philosophical tendencies referred to as the four kinds of 'hypotheses' that are to be avoided.

The composition of the whole passage is curious. After his reference to the doctrine of the atomists, he mentions the contrary view of the Cartesians, who—he says—have instead been 'feigning hypotheses for explaining all things mechanically, and referring other causes to metaphysics'. He then goes on to contrast this with his own position, first defining his approach at the inductive level—

> The main business of natural philosophy is to argue from phenomena without feigning hypotheses, and to deduce causes from effects, till we come to the

---

[88] Query, 28, O.368. This is the text as it stands in the 2nd edition. In the third, a clause is added to this sentence: '. . . and from such an exceedingly rare aethereal medium as we described above', a reference to the new hypothesis of an elastic ether that had been newly added to this edition. This addition makes of course nonsense of the opening sentence of the next paragraph, 'And for rejecting such a Medium, we have the authority of the . . . philosophers of Greece and Phoenicia, who made a *vacuum*, and atoms, and the gravity of atoms, the first principles of their philosophy. . . .' (O.369). It shows that Newton either remained undecided or maintained contradictory opinions to the end. At any rate, the paragraph beginning with the passage last-quoted was written with the view in mind that *no ether-explanation* was satisfactory, so that here the need for a rationale would be most pressing.

[89] O.369.

very first cause, which certainly is not mechanical; and not only to unfold the mechanisms of the world, but chiefly resolve these and such like questions.

The last five words are ambiguous; does Newton have in mind the technical problems involved in the inductive treatment, or other questions? (The sentence concludes with a full-stop and not a colon!) I think we may however safely assume that he is referring to the major questions of '*Natural*', and not just 'Experimental Philosophy', questions which he indeed immediately proceeds to pose, though the transition is peculiarly abrupt—

What is there in places almost empty of matter, and whence is it that the sun and planets gravitate towards one another, without dense matter between them?

And then, suddenly, even more abruptly—

Whence is it that nature does nothing in vain; and whence arises all that order and beauty which we see in the world?[90]

After mentioning his old difficulties about the *raison d'être* of comets, and the fact that the planets, unlike the former, all move in the same way in concentric orbits, he then draws attention to evidence of design in the animal kingdom, seemingly (as suggested earlier)[91] in the belief that this evidence *from* observable design implies that we may also argue *to* the existence of design to account both for the astronomical problems and for those more general difficulties surrounding the concept of gravity.[92]

Here evidently we have the promised rationale of gravity formulated in terms of the adaptedness and fitness which Newton claims to be a general characteristic. So we see: the conceptual account having been limited by rejecting gravitational attraction as a physical reality (as an 'innate property' of matter), and also the corresponding theory-type which this would have implied, i.e. phenomenological, Newton offers us instead a teleological rationale, as part of the architectonic component. The postulate of design, and indeed, of a designer, is not of course for Newton a hypothesis, as is the construct of the ether, but it is clearly an alternative to the rejected philosophical 'hypotheses'—a fact which Newton represents by saying that 'it appears from the phaenomena that there is a Being incorporeal, living, intelligent, omnipresent. . . .'[93]

It is this notion of God 'appearing from the phenomena' which Kant will

[90] Ibid.
[91] Cf. above, p. 180.
[92] For the distinction between the arguments *from* and *to* design in Newton, cf. R. H. Hurlbutt III, *Hume, Newton and the Design Argument*, Chapter 1.
[93] P.370. And he adds in the *Principia* that 'to discourse [of God] from the appearances of things, does certainly belong to Natural Philosophy' (P.546).

later reinterpret through his regulative methodology. But already in Newton we find an interesting distribution of 'logical stress' between our three components. The conceptual deficiency, coupled with the stronger weight attached to the inductive component, builds up sufficient pressure to demand a rationale which is supplied by the physico-theological account, itself here not called into question; despite the fact that on occasion Newton's intentions seem also to go the other way, as when he tells Bentley, that 'when I wrote my treatise about our system, I had an eye upon such principles as might work with considering men for the belief of a Deity'.[94]

The clearest summary of the whole position can be found in Cotes' Preface to the second edition of the *Principia*. Both the school of Descartes and that of Leibniz are denounced for having imposed on the inductive enterprise of mechanics the implications (apparent or real) of their metaphysical doctrines. Descartes' 'presumption' lies in the claim that he can discover 'the true principles of physics and the laws of natural things by the force alone of his own mind and the internal light of reason'. Leibniz is equally presumptuous in his claim that man can *deduce* the *minutiae* of the will of God, and the laws involved therein, by reference to the principle of fitness, again without reference to the phenomena. On the contrary, Cotes insists, 'all sound and true philosophy is founded on the appearances of things'. It is true that the laws of nature flow from 'the perfectly free will of God directing and presiding over all', and that they display 'the most wise contrivance', and 'clearly manifest to us the most excellent counsel and supreme dominion of the All-wise and Almighty Being'.[95] However, so Cotes clearly wants to say, we must not *argue from* design to the phenomena, but rather note the design *displayed by* the phenomena, themselves first 'discovered' inductively. The principle of fitness *must not anticipate* the extent of the phenomenal powers; it can do no more than provide their 'rationale'; the phenomena so to speak being made intellectually respectable by being regarded as displaying providential design.

Newton's teleological approach is thus deeply entrenched in the philosophical thought of the period, and manifests itself under a number of guises. We have seen that he conceives of matter uncompromisingly as endowed only with a 'passive' nature. Any conception of live or living force, of the kind which informs Leibniz's teleological approach to the concept of physical substance, is foreign to him. Yet, in the end he does use a corresponding notion, though introduced more extraneously, and linked as

---

[94] Thayer, p. 46. Typically, he does not think of his own suggestion as an alternative 'hypothesis' (in the philosophical sense)!

[95] P. xxxi–ii.

already noted, to God's creative will.[96] Query 31 yields the clearest statement of this position.

> God in the Beginning formed Matter in solid, massy, hard, impenetrable, movable Particles. . . . These Particles have not only a *Vis inertiae*, accompanied with such passive laws of motion as naturally result from that force but also . . . are moved by certain active principles, such as that of gravity, and that which causes fermentation, and the cohesion of bodies.[97]

Why speak of '*active* principles'? Firstly, no doubt, this is suggested as a contrast with 'passive'. Secondly 'passive laws' (the three laws of motion) *assume* that there should *be* motion in the universe, but they are silent on the *source* of this motion. But Newton adds yet a further consideration: we need a source, he says, to replenish the motion that is constantly lost to the world, not to mention certain 'irregularities' in the celestial motions which require occasional redress.[98]

Let us note once more the special aspect of Newton's philosophical reasoning here: we start with perfectly general cases (gravity, inertial motion), where the need for a 'source of activity' is, so to speak, metaphysical. (After all, there was the 'atheistic' alternative—for Newton *horribile dictu*—that motion and gravity might be eternal and uncreated or unmaintained.) But these metaphysical cases are mixed up with others of quite a different logical order, involving technical difficulties peculiar to Newton's particular system. Newton's teleological approach involves a similar shift from the point of view which 'sees' the systemic structure of dynamics as an expression of teleology, to the argument which seeks an aid from considerations of technical adaptations in the realm of animal and plant biology.[99]

But there is finally yet a further, and very special connection between the conception of 'active principles' and 'final causes' (teleology). McGuire and Rattansi have recently drawn attention again to the many similarities between

---

[96] In the third letter to Bentley, Newton says that 'gravity must be caused by an agent acting constantly according to certain laws', but leaves it open 'whether this agent be material or immaterial' (Thayer, p. 54). In his commentary on Rohault's *System*, Clarke is more outspoken: since matter cannot act at a distance we need to postulate 'the action of some immaterial cause which perpetually moves and governs matter by certain laws'. (Op. cit., 1723, Vol. I, p. 58n.)

[97] Op. cit., pp. 400–1. (For this passage, see also above, p. 179.)

[98] Cf. O.396–7; cf. above, p. 179, and Note 53.

[99] Cf. above, pp. 180, 192. Once again it is interesting to note how an echo of this dual aspect is caught in the system of Kantian philosophy. For here also we have the interpretation of 'systematic unity' as 'purposive unity', being an expression of the regulative approach in *general*, linked with the employment of a rather more *special* regulative principle of teleology in the biological field; each supplementing the deficiencies of the other in its own peculiar way.

Newton's philosophy and that of Cudworth;[100] and some of the central notions of Cudworth's philosophy are—as already noted[101]—of considerable help in any attempt to understand the significance of Newton's reasoning. Like Newton, Cudworth insists that in addition to the 'passive' principles of matter, we require an 'active principle' or 'active power', which he describes as the course of life as well as of reason.[102] The central point about these vital principles is that they manifest themselves by way of effects to be understood by us in terms of final causes. This position Cudworth expresses clearly when discussing 'gravity';

> There are many phaenomena in nature, which being partly above the force of these mechanical powers, and partly contrary to the same, can therefore never be salved by them, *nor without final causes, and some vital principle*. As for example, that of gravity . . . the motion of the diaphragma in respiration, the systole and diastole of the heart. . . .[103]

This is an instance of that large body of seventeenth-century Neo-Platonist doctrine which influenced Newton, and which lies behind the brief passage in the General Scholium, when he writes that 'we know [God] only by his most wise and excellent contrivances of things, and final causes'.[104]

In other words, we are able to save the rationale of a concept if we can link it with, or if we can view it as, the expression of divine activity working towards wise ends—with the emphasis on ends, rather than God's will, growing as we proceed down the century, *via* Leibniz, and towards the Kantian finale. Newton's thinking here is suffused with this notion of apt disposition, harmony, analogy and nature's consonance with itself, a phrase that recurs continually throughout his writings. It has its hold already on the inductive foundations which Newton formulates in his Rules, as when in Rule III which sanctions analogical and inductive inference, he writes of 'the analogy of Nature, which is wont to be simple, and always consonant to itself'.[105] Again, 'Nature is very consonant and conformable to herself' he writes in Query 31, when he notes the analogies between the various laws of attraction, and predicts the existence of many 'attractive powers' of this kind.[106] In this way the phrase gets its more generalized justificative function, when Newton concludes, in the same Query—

---

[100] Cf. 'Newton and the Pipes of Pan', especially pp. 132, 134.

[101] Above, p. 12.

[102] Cf. *Intellectual System*, p. 27.

[103] Op. cit., p. 684; my italics. Note again the running together of logically different cases.

[104] O.397.

[105] P.398–9.

[106] O.376.

And thus Nature will be very conformable to herself and very simple, performing all the great Motions of the Heavenly Bodies by Attraction of Gravity which intercedes those Bodies, and almost all the small ones of their Particles by some other attractive and repelling Powers which intercede the Particles.[107]

IX

In his unpublished Scholia to Propositions IV to IX of Book III of the Principia, Newton was even more outspoken in linking gravitation to the general harmony of nature.[108] Thus he refers with approval to the Pythagoreans' view of divinity, 'inspiring this world with harmonic ratios like a musical instrument'; and this is linked by him with the ratios through which the force of gravity expresses itself.

Subsequently, MacLaurin in his account of Newton's philosophy again quotes with approval this supposedly Pythagorean doctrine (it is only a cryptical gloss) of the analogy between the ratios of stretched strings, musical chords and the 'gravities of the planets'.[109]

The whole intention of these teleological and Neo-Platonic pleadings is to supply sufficient weight for the rationale of the conception of gravity, given its conceptual deficiencies. But the process of weighting down the notion of gravity through the architectonic component, and thus fixing its rationale, is itself full of complexities, owing to the fact that Newton wishes to steer a middle path between giving gravity a rationale, and avoiding the accompanying suggestion of its physicalization; the whole exercise teaching us something about the complexities surrounding the architectonic component itself.

The ambiguities surrounding the term 'principle' here give Newton plenty of scope for confusion. In Query 31, the active principle is said to be 'the *cause of* gravity'. [110] Two pages later he speaks of 'certain active principles, such as is *that of* gravity'. Does the notion of 'principle' here stand for 'cause' or for 'law'? It would seem the latter, since Newton goes on to say that he considers these principles 'as general laws of nature', and that they correspond to 'manifest qualities', their 'truth appearing to us by phaenomena'. But

---

[107] O.397.
[108] Cf. McGuire and Rattansi, op. cit., where it is shown that Newton sought support for this interpretation in the writings of antique and early medieval philosophy—much in the style of Cudworth and others during this period.
[109] MacLaurin, p. 34; cf. McGuire and Rattansi, p. 117.
[110] O.399; my italics. 'Active principle' then supplies what we would now call 'an inductive foundation' for what otherwise might be a merely 'accidental regularity'; cf. the like move on Leibniz's part, above, p. 184, end of Section IV.

since the other meaning can hardly have been expelled altogether, we must regard this ambiguity itself as a way of bestowing additional weight (or rationale) on the conception or law of gravity. Newton comes here as close as is for him possible to making its status basic, and *almost* innate. But in the very act of pronouncing it a phaenomenon (or 'derived from phanomena') he draws back again from what he considered an abyss. To make clear that when he calls gravity a property he has no wish to regard it as a physical reality, i.e. 'innate', he continues to insist that he does not yet know its 'cause', despite the fact—remember—that he has called the 'active principle' itself a cause. Is this not a contradiction? Newton here avoids it, by interpreting the request for the 'cause' as being for the 'how', the 'mechanism' of attraction. Thus, after having affirmed that attraction is an expression of nature's 'consonance', he adds:

> How these attractions may be perform'd, I do not here consider. What I call attraction may be perform'd by impulse, or by some other means unknown to me.[111]

In other words, the teleological move is not quite strong enough, and he still requires additional help to safeguard the rationality of his concept.

This may be put down simply to intellectual indecision. Such an interpretation is supported by a remark of one of the members of Newton's circle, Fatio. Writing in the context of an ether hypothesis to which he claims Newton at one time subscribed (it differs from the one given in the third edition of *Opticks*, Query 21), Fatio adds—

> [Newton at one time] did not scruple to say that there is but one possible mechanical cause of gravity, to wit that which I had found out: Tho he would often seem to incline to think that gravity had its foundation only in the arbitrary will of God.[112]

Newton's 'inclination' reminds us of course of Locke's approach to gravity to which we have alluded; lack of conceivability or rational transparency being described positively as forcing us to conclude that the connection is due to the fact that God has 'put into bodies' this power of acting.[113]

An alternative account of this apparent inconsistency in Newton's thought has been given by Koyré, who says that Newton in fact took fright from these attempts to let God put the power of gravity into bodies, since this seemed to make it too much like a primary quality, approaching the otiose

---

[111] O.376.
[112] Quoted in Halls, p. 206.
[113] Koyré, p. 155, cf. Note 43.

position of it being 'innate'.[114] This would explain the tenor of Advertisement II to the 3rd edition of the *Opticks,* which warns the reader against such an interpretation, by handing him yet another ether hypothesis.

However, instead of saddling Newton either with indecision or psychologically self-contradictory attitudes, it may also be the case that he saw nothing incompatible between a teleological approach tied to an interphenomenal ether explanation. It was after all not unreasonable to assume God's actions to take place in a manner as closely analogous to the general 'analogy of nature' as possible. And we can point in fact to an interesting explicit attempt at such a reconciliation on the part of one of Newton's disciples, viz. MacLaurin, already mentioned. Like Newton, MacLaurin emphasizes the importance of 'the laws of nature' that are 'constant and regular'. But immediately following we get the move from 'law' to 'power', when MacLaurin says that possibly 'all of [these *laws*] may be resolved into one general and extensive *power*' a power, moreover, that 'derives its properties and efficacy, not from mechanism, but in great measure, from the immediate influences of the first mover'.[115]

Two pages later, however, we find that we must also assume these 'powers' to be circumscribed and regulated in their operations by

> mechanical principles; and that they are not to be considered as mere immediate volitions of His (as they are often represented) but rather as instruments made by Him, to perform the purposes for which He intended them. If, for example, the most noble phaenomena in nature be produced by a rare elastic *aetherial medium,* as Sir Isaac Newton conjectured, the whole efficacy of this medium must be resolved into His power and will, who is the supreme cause. . . . Nor is there any thing extraordinary in what is here presented concerning the manner in which the Supreme Cause acts in the universe, by employing subordinate instruments and agents, which are allowed to have their proper force and efficacy; [and] subject to the like laws as other elastic fluids.[116]

Though Malebranche is not named, the doctrine here opposed, 'immediate volitions of the supreme cause', is no doubt his. But the opposition to Malebranche reminds us also that MacLaurin was not only a pupil of Newton's but also a follower of George Berkeley,[117] where the notion of divine

---

[114] Koyré, p. 156, in connection with Newton's objections to Cheyne's phraseology which is similar to Locke's. Newton resisted strongly the suggestion that even God could put into bodies the power to attract. This would have placed attraction either at the conceptual or the inductive level; whereas the only solution acceptable to him was that which affects the rationale of the concept.

[115] MacLaurin, p. 406; my italics.

[116] Pp. 408–9.

[117] For Berkeley's objections to Malebranche, cf. for example *Principles of Human Knowledge,* 53.

providence similarly forms the way out of the awkward possibility—suggested by Berkeley's phenomenalist metaphysics—of a purely phenomenal universe, lacking any sort of 'hidden mechanism', as contrasted with one in which—as suggested by science—such a mechanism (corpuscularian in kind) has a rightful place. According to Berkeley, whilst God *might* have created a universe containing no more than is displayed by its manifest phenomena, the rationale of a hidden mechanism is established, first, because this makes it possible to give a rational *explanation* of the phenomena, secondly, because the production of the phenomena in accordance with the 'rules of mechanism' is an expression of the 'wise ends established' by the deity. In general, Berkeley argues, 'if we follow the light of reason' we may be sure that we can 'from the constant, uniform method of our sensations, collect the goodness and wisdom of the spirit that excites them.'[118] Once more, teleology forms the kingpin of the structure.

There were moreover suggestions also in the Neo-Platonist writers of a possible reconciliation between the idea of divine providence and the supplementary means of realizing its objective; means that were 'internal' to God's creatures. Thus Cudworth postulates a 'plastic nature' which acts as an internal 'energy', through which God as it were indirectly produces ends that are desirable and fitting. In the sequel, this plastic nature becomes an ethereal medium which is diffused throughout the universe; except that it is a medium which, from being 'plastic', ends up as a corpuscularian structure.

In the end—as is already apparent in MacLaurin—once the teleological ingredient had reconciled the scientists to the rationality of distance-forces, one of the prime tasks that had led to the original introduction of the ether is lost, and its particles can now be subject to distance-forces. In the words of MacLaurin, although the general mechanical principles (the laws of motion) operate universally, they require a medium which to some extent 'circumscribes and regulates the powers' involved in the operation of the structural medium; but these powers themselves 'surpass mechanism'.[119]

Whether it was Newton's ultimate view that the ether involves itself action at a distance will never be clear.[120] Ostensibly, the third edition of the *Opticks* introduces—as we have seen—the ether only as a therapeutic device, to remind us that gravity is not an innate property. Beyond that, it is left open to us to suppose that the *modus operandi* of the ether itself was 'rationalized' by Newton's teleological approach.

[118] *Principles*, 62, 72.
[119] MacLaurin, p. 408.
[120] The third edition (Query 21 of the last edition) speaks of the 'supposition' that the ether 'may contain particles which endeavour to recede from one another' (O.352). Although attraction is here ultimately founded on repulsion, this itself involves action at a distance.

X

We see then how in general a very delicate balance is maintained between the conflicting stresses arising from the complex interpretation of the three components of the hypothetical or explanatory notions of science. Let us test this against the immediate sequel to Newton's speculations. If we allow a position of relative independence to our three components, what would be the result of dropping the teleological approach, whilst retaining the traditional concept of matter?

If we study just one of the prominent histories of science of the eighteenth century, that of Savérien (1775), which takes the story down to about 1752, we find a general attempt to revert back to some form of ether explanation, with hints that the concept of *vis viva* may somehow bear on the question of the nature of gravity.[121]

What of the philosophers? Apart from Berkeley, Hume is here the most interesting. The latter has replaced Locke's confused notion of the 'inconceivability' of contingent relations as well as Newton's vague conception of cause by a theory of constant conjunction coupled with that of natural belief. But this argument concerns solely the inductive level; and Hume is still thoroughly traditional in his conception of matter, in which he appears to follow Newton, being opposed to action at a distance. In consequence, if my analysis has been correct, he ought to espouse an hypothesis of the ether —which is precisely what he does, referring with approval to Newton, for his 'modest hypothesis' of an ether which accounts for gravitational attraction.[122] That, and that alone, makes the latter rational—a consideration quite independent of its being a well-established inductive generalization. Once again, without my distinction, the argument makes no sense.

The second test of my account lies in predicting that once the concept of matter itself is attacked and modified, the ether ought—at least temporarily —to drop out of the picture. This development centres primarily on Leibniz and his successors, e.g. Wolff and Hamberger,[123] and independently, on Boscovich. From Leibniz dates the conception that substance has to be understood in terms of monads, whose physical mode is described in terms of 'forces', in the present instance, the so-called 'passive forces' (primitive and derivative). In Leibniz's pupils, this notion of the monad is altogether physicalized. It is not until we come to Kant (and less importantly, Priestley)

---

[121] Savérien, pp. 42–59. See Heimann and McGuire for a long list of similar eighteenth-century views.

[122] *Enquiry*, E.73n.

[123] For Hamberger, cf. Adickes, *Kant als Naturforscher*, vol. i, pp. 89 f., pp. 128 f., p. 146n.1.

that we find a clear distinction between the inductive level of enquiry, and the stage which precedes it and which lays what Kant calls the 'metaphysical foundations', which (as was shown at the start) consists in an explication of the concept of physical substance involving essentially repulsive and attractive forces. On this, let me add a final comment.

Kant's point of view dates back in fact to an early period of his development, and is succinctly made in his *Monadologia Physica*, where what I have called 'conceptual explication' is most graphically exemplified, as yet quite divorced from the encumbering difficulties of transcendentalism. We must distinguish, he tells us in the Preface, between the discovery of the laws of nature by way of the testimony of experience, and 'the original ground and cause of these laws', to trace which is the task of metaphysics. Bodies consist of parts; and metaphysics has to investigate how this combination into a whole can be made intelligible, whether in virtue of 'mere coexistence of original parts' or of 'mutual interaction of forces which fill all space'. But if one analyses this situation, one realizes that no physical substance can be present to another except by virtue of the forces which it exerts on the latter, and vice versa.[124]

Evidently, the essential considerations here are conceptual and epistemological. What is it for one body to coexist with another? How can one physical system be aware of another? Clearly, the answers which such questions require from the start bear on the concept of force, which thus becomes a part of the conceptual boundaries of the concept of matter.

One final point. In their article, McGuire and Rattansi note that there is only an apparent contradiction between Newton's inductivism and his Neo-Platonic philosophy. But although they seem to agree that Newton realized that his forces 'required a different category of existence for their *explanation*' (i.e. reference to philosophical considerations), they describe this as a 'restriction' on the part of Newton upon our 'pretensions to *knowledge* of the natural world'.[125] But it is important to realize that it is one of the glories of Newton's intellectual history that he (perhaps only obscurely) realized that our understanding of attractive forces occurs at three different levels, only one of which (the inductive) can be characterized by terms like 'explanation' and 'knowledge'. The correct account is that the architectonic and conceptual components are as such not concerned with explanation, but rather, with explications at the conceptual and architectonic levels. Confusion of these different levels is however easy, since the non-inductive components are themselves not immune from considerations of the results of scientific theorizing.

124 Op. cit., i, pp. 475–6. For Kant, cf. above, pp. 5–7.
125 Op. cit., pp. 124–5; my italics.

It will be noted that this discussion has proceeded by enquiring into, and bringing to light, the components which (so I claim) determine the acceptance of hypotheses. The outcome of the discussion shows that the satisfaction of the usual inductive criteria is by no means a sufficient condition, though it is of course a necessary one, for acceptance. But I must stress that I have here taken a 'subjective' point of view, indulging in 'descriptive methodology'. I have assumed that the various components are *in fact* criteria which people consider before deciding to accept an hypothesis; and I have moreover indicated—though only sketchily—that these criteria involve reasoned argument and beliefs about a number of general scientific and epistemological questions. But I have not asked whether the satisfaction of these criteria forms a ground which would *justify* us to assert 'the truth' of the hypotheses so accepted. I am not actually certain that this distinction is a tenable one, but in any case, I have not thought profitable to discuss it in the context of this essay.

Bibliographical References

Adickes, E., *Kant als Naturforscher*, 2 vols. (Berlin, 1924–25).

Alexander, H. G. (ed.), *The Leibniz-Clarke Correspondence* (Manchester, 1956).

Berkeley, G., *A Treatise Concerning the Principles of Human Knowledge* (1710).

Buchdahl, G., *Metaphysics and Philosophy of Science. The Classical Origins: Descartes to Kant* (Oxford, 1969 and Cambridge, Mass., 1969).

'The Relevance of Descartes' Philosophy for Modern Philosophy of Science', *Brit. J. Hist. Sc.*, 1963, **1**, pp. 227–49.

'Gravity and Intelligibility: Newton to Kant', in R. E. Butts and J. W. Davis (eds.), *The Methodological Heritage of Newton* (Toronto and Oxford, 1970), pp. 74–102.

H'istory of Science and Criteria of Choice', in R. Stuewer (ed.), *Minnesota Studies in the Philosophy of Science*, Vol. V (Minneapolis, 1970), pp. 204–30.

The present essay is an expanded and more historically orientated version of the Minnesota chapter, whilst the latter considers later sequences of our case-study.

Burtt, E. A., *The Metaphysical Foundations of Modern Physical Science* (London, 1932).

Cohen, I. B., ' "Quantum in se est": Newton's concept of inertia in relation to Descartes and Lucretius', *Notes and Records of the Royal Society of London*, Vol. 19, 1964, pp. 131–55.

Cohen, I. B., 'Hypotheses in Newton's Philosophy', *Physis*, 1966, **8**, pp. 163–84.

Cudworth, R., *The True Intellectual System of the Universe*, 1678 (repr. Stuttgart-Bad Cannstatt, 1964).

Feyerabend, P. K., 'Explanation, Reduction and Empiricism' in H. Feigl and G. Maxwell (eds.), *Minnesota Studies in the Philosophy of Science* (Minneapolis, 1962).

Guerlac, H., 'Newton's optical aether: his draft of a proposed addition to his *Optiks*', *Notes and Records of the Roy. Soc. London*, 1967, **22**, pp. 45–57.

Hall, A. R. and M. B. (eds.), *Unpublished Scientific Papers of Isaac Newton* (Cambridge, 1962) [Referred to as Halls].

Hanson, N. R., 'Hypotheses Fingo', in R. E. Butts (ed.), *The Methodological Heritage of Newton,* op. cit., pp. 14–33.

Heimann, P. H. and McGuire, J. E., 'Newtonian Forces and Lockean Powers: Concepts of Matter in Eighteenth-century Thought', *Historical Studies in the Physical Sciences*, 1971, **3**, pp. 233–306.

Hume, D., *Enquiries concerning the Human Understanding and concerning the Principles of Morals*, L. A. Selby-Bigge (ed.) (Oxford, 1902) [Referred to as E].

Hurlbutt, R. H. III, *Hume, Newton, and the Design Argument* (Lincoln, 1965).

Kant, I., *Disputation on the Monadologia Physica*, 1756; *Works*, Akad. ed., 1910, i, pp. 473 ff.

   The only possible Ground for a Demonstration of the Existence of God, 1763; *Works* (Akad. ed., 1912) ii, pp. 63 ff. [Referred to as OG].

   Examination concerning the Clarity of the Principles of Natural Theology and Ethics, 1763; in *Kant, Selected Pre-Critical Writings*; trans. G. B. Kerferd and D. E. Walford (Manchester, 1968) [Referred to as CP].

   Critique of Pure Reason, 1787; trans. N. Kemp Smith (London, 1933) [Referred to as K].

   Prolegomena to any Future Metaphysics, 1783; trans. P. G. Lucas (Manchester, 1953) [Referred to as Pr].

   The Metaphysical Foundations of Natural Science, 1786; trans. E. B. Bax (London, 1883) [Referred to as M].

   Critique of Judgment, 1790; trans. J. H. Bernard (New York, 1951) [Referred to as J].

   Introduction to Logic, 1800; trans. T. K. Abbott (New York, 1963) [Referred to as L].

Koyré, A., *Newtonian Studies* (London, 1965) [Referred to as Koyré].

Kuhn, T. S., *The Structure of Scientific Revolutions* (Chicago, 1962).

Leibniz, G. W., *Tentamen de motuum coelestium causis*, 1689; in C. J. Gerhardt (ed.), *Leibnizens mathematische Schriften* (Halle, 1860), vi, pp. 144 ff., 161 ff.

Locke, J., *An Essay concerning Human Understanding*, 1690; A. C. Fraser (ed.) (Oxford, 1894).

Mach, E., *The Science of Mechanics*, 1883; 6th edition (Illinois, 1960).

MacLaurin, C., *An Account of Sir Isaac Newton's Philosophical Discoveries*, 1748; (3rd edition, London, 1775).

McGuire, J. E. and Rattansi, P. M., 'Newton and the "Pipes of Pan" ', *Notes and Records of the Roy. Soc. London*, 1966, **21**, pp. 108–43.

Meyerson, E., *Identity and Reality* (London, 1930).

Newton, I., *Mathematical Principles of Natural Philosophy* (*Principia*), 1687; trans. F. Cajori, 2 vols. (Berkeley, 1962) [Referred to as P].

   Opticks, 1704 (Dover edition, 1952. Based on the 4th edition, London, 1730) [Referred to as O].

Priestley, J., *Disquisitions relating to Matter and Spirit*, 1777, in J. A. Passmore (ed.), *Priestley's Writings on Philosophy, Science and Politics* (New York, and London, 1965).

Rohault, J., *System of Natural Philosophy*, 1671; trans. J. Clarke (with Samuel Clarke's Notes), 2 vols. (London, 1723).

Salmon, W. C., *The Foundations of Scientific Inference* (Pittsburgh, 1967), ch. 7.

Savérien, A., *Histoire des Progrès de l'Esprit Humain dans les Sciences Naturelles* (Paris, 1775).

Shapere, D., 'Meaning and Scientific Change', in R. G. Colodny, *Mind and Cosmos* (Pittsburgh, 1966).

Thayer, H. S., *Newton's Philosophy of Nature* (New York, 1953) [Referred to as Thayer].

XII

TRAGEDY IN THE HISTORY OF SCIENCE

Jerome R. Ravetz

We can take our problems from an affirmation of the late George Sarton:

Definition: Science is systematized positive knowledge, or what has been taken as such at different ages and in different places.

Theorem: The acquisition and systematization of positive knowledge are the only human activities which are truly cumulative and progressive.

Corollary: The history of science is the only history which can illustrate the progress of mankind. In fact, progress has no definite and unquestionable meaning in other fields than the field of science.[1]

This credo, charming in its naïveté, redolent with echoes of both Comte and Condorcet, stands as a monument of a particular ideology of science, in whose terms the genuine history of science was impossible. It matters little that Sarton himself was deeply appreciative of other precious aspects of human life, or that the qualifying clause in the definition destroys the whole thesis. The faith was simple, and blindingly clear. Our present task is to begin to achieve an understanding of science in whose terms we can comprehend its fascinating past, and also help to cope with its difficult future.

It is possible to have studies of past events in which human intention plays a minor or negligible rôle, as when the traces of past activity are so fragmentary that only the most broad and general inferences can be made from them, or when statistical evidence is used for mass phenomena that can be considered as simple reactions to possibilities and threats in an environment. Both types of study come into the history of science, although at its margins: the one in the case of lost civilizations and the other in the analysis of works by masses of men whose individual names are too numerous and too insignificant to be recorded. But the history of the science of our civilization

---

[1] G. Sarton, *The Study of the History of Science*, 1936 (New York, Dover, 1957), p. 5 of second pagination.

and of those other literate cultures with which it has interacted, will necessarily focus on the men recognized as great by their personal achievements. To write their stories as a series of undiluted triumphs for 'cumulative and progressive' knowledge (even allowing for temporary adversity in some cases), requires the faith of Sarton, that science is truly unique among human activities; and fortunate is the blessed band that can do such great good at such little risk.

It would be dangerous to under-estimate the pervasive influence of this conception of the history of science, on our own approach to the study, and on the very materials that we rely on for our research and teaching. Until very recently indeed, all the influences which conditioned the writings of the history of science acted in concert to produce accounts that confirmed and reinforced this conception. The sources of biographical materials, in elegies and hagiographical writings; the technical histories, constructing folk-histories for particular fields; the assumption that the true discoveries of the past are reproduced in the textbooks of the present; and the long-standing identification of Science with the noble struggle for Reason against Dogma and Superstition;[2] all these combined to produce a conception and practice of the history of science of which Sarton's is but the most revealing statement. The whole process of the creation of historical materials, from the choice and interpretation of primary documents, through the establishment of accepted facts, to the most general syntheses, has been conditioned by these influences; and my generation of historians will not succeed in escaping from them.

Misfortunes and Errors in Science

Once we cease to identify Science with the Good and the True, we are faced with a painful task of reconstruction and salvage. As a small contribution to this, I shall explore the theme of 'tragedy' as it relates to the scientific endeavour. For this, I will use a conceptual apparatus I have developed elsewhere, seeing the work of science as the investigation of problems, proceeding through complex cycles of coming-to-be and passing-away, and only exceptionally pursued in an environment insulated from human concerns.[3]

[2] See Karl Pearson, *The Grammar of Science*, 1892, Preface to First Edition; he explains the 'obscurity which envelopes the *principia* of science' by the difficult warfare of science with 'metaphysics and dogma', in which it 'conceived it best to hide its deficient organisation'.

[3] J. R. Ravetz, *Scientific Knowledge and its Social Problems* (Oxford, The Clarendon Press, 1971).

To begin with the easiest cases, we may review the common circumstances in which misfortune can occur in science. For example, a scientist may invest much time and resources in a problem which turns out to be incapable of an adequate solution. All scientific research, except the most routine and banal, is a gamble, and not every gamble wins. Such lost ventures rarely find a trace in the public record of scientific achievement; and they are of historical importance only in the context of a close reconstruction of a particular investigation. A related misfortune is when a scientist finds himself 'anticipated' by another; his research, on the way to success or perhaps even completed, is nullified as intellectual property when someone else publishes the same or a closely related result. We should note that such a misfortune can exist only in the framework of a certain conception of knowledge, and of personal benefit. The system of 'priority' is a reminder that in spite of the undoubted idealism of the creators and leaders of science over past centuries, the social system of science has needed to assume the existence of discrete bits of original knowledge, belonging by right to their respective discoverers. It may be that the activity of science on a large scale requires such a metaphysics and social ethics; if so, this characteristic only strengthens the identification of modern science with European capitalist culture.

Philosophers of science have recently come to realize that the understanding of the scientific endeavour requires an appreciation of larger 'elements' than the single assertion or even the single problem. We may speak of 'paradigms', or 'research programmes' or 'strategies', to describe a cluster of problems and results, whose inner coherence is defined as much by expectations of the future as by information about the past and present. Even to rise to this level of complexity, raises troubling difficulties for the inherited ideology of science. For such larger units are incapable *in principle* of the ready assessment of quality, to which individual results are (ideally) amenable. In these circumstances, there is, in Lakatos' words, an end to 'instant rationality', and with it the assumptions which gave the positivist philosophy of science whatever plausibility it ever had.[4]

With this enriched conceptual framework, historians of science can begin to come to grips with the problems of *error* in science. To be sure, historians of science have always been aware of mistaken judgements, sometimes made by distinguished men; for in every great controversy, one side can be seen to have lost. But such errors were always an embarrassment, and needed to be explained away. Hence the history of a controversy would be carefully shaped to present the (retrospectively) victorious side as the advocates of obvious, if bold, theories; while the losers could be smeared as 'reactionaries',

[4] These new ideas derive from T. Kuhn and the Popper school; see I. Lakatos and A. Musgrave (eds.), *Criticism and the Growth of Knowledge*, Cambridge University Press, 1970).

prevented by personal failings from opening their eyes to the truth.[5] The burying of cases of honest error was made easier by the common assumption that in scientific education, 'ontogeny recapitulates phylogeny'; hence the task of the history of science was to recount the victories leading to the establishment of standard textbook knowledge. Even there, anomalies occur, as in the case of Newton's corpuscular theory of light; but any skilled scientist knows how to handle anomalies in such a way as to reduce their significance.

An honest error about a research programme, involving a commitment of one's time, talents and resources to a large piece of work which eventually proves fruitless, involves elements of tragedy. To show how such errors can occur, I must use some technical terms developed more fully elsewhere. We know that scientific research does not consist of the collection of facts, but rather of the posing and solution of problems. Since there is no formally valid pattern of argument leading from particular data to a general conclusion every solution to a problem is governed and assessed in terms of criteria of adequacy. These relate to the rigour of control of data, and to the nature of the inference-links whereby the data are converted into information. This is in turn used as evidence in an argument, while the argument itself has partly deductive elements, but also inductive, probabilistic and analogical elements as well. The criteria of adequacy applied to the results of a particular problem-investigation depend very strongly on the field in which the work is embedded; the need for a competent referee to be someone engaged on the same sort of research, is an indication of their particularity and subtlety. If we ask, whence come these criteria of adequacy, and what confirmation do they have, we find that they are a social possession, developed in a largely tacit fashion from successful experience, and incapable *in principle* of being demonstrated in a 'scientific' way.

Any deeply novel piece of research will necessarily involve its author in a recasting of criteria of adequacy; new ones will have to be developed for the control of unprecedented aspects of the work, and old ones will have to be reduced in significance lest the new work be smothered in its inevitable anomalies. Hence the debate over a challenge to existing orthodoxy will involve not only the evidence, but also methods and methodology, inherently incapable of straightforward resolution. In the decision on investment of resources in one programme rather than another, another set of criteria are necessarily involved: those of value. These have the same 'non-scientific' characteristics as the criteria of adequacy, to a more intense degree; and hence

[5] A challenging survey of the techniques used by earlier philosophers of science to resolve the problem of error, is J. Agassi, *Towards an Historiography of Science* ('s Gravenhage, Mouton, 1963), *History and Theory*, Beiheft 2.

debates on value are prone to be even more subjective, partisan, and subject to error.

Because of the influences I have mentioned above, the history of science as practised hitherto has provided few properly studied examples of errors both reasonable and important, in decisions on research programmes. There are of course inherent difficulties in any such study; in addition to being capable of seeing a situation in terms of its own obsolete concepts and tools, and that from all sides, the historians would also need to be able to play a counter-factual game, to see what sort of fruitful research could have been done at that point in the development of the science. All the historian can do at this stage is to observe those cases where the arguments of the 'reactionaries' had more cogency than propagandist historians have allowed, and where the 'revolutionary' features of a new programme were not so crucial for shaping an ongoing reform, as protagonists and disciples claimed. One obvious case in point here is Lavoisier's grand theory of explaining combustion by identifying it with the traditional notion of acidification, and importing this theory into an existing programme for reform of chemical nomenclature. How long it would have taken other chemical philosophers to replace 'phlogiston' by, say, 'pyrogène', and also to accept a multiplicity of distinct chemical species, I cannot say. But the vitality of chemical research in Britain as compared to France, and the absence of important work in 'articulating the paradigm' of Lavoisier's theory, indicates that the 'conservative' school were not hopeless reactionaries.

Of course, when a brilliant and influential man launches a research programme, and attracts talented and zealous disciples, something is bound to result. The assessment of 'success' or 'failure' for an endeavour extending over decades or generations is difficult in the extreme, and generally to be attempted only with great caution. Yet here too it is possible to find cases where the achievement fell so far short of promise, that one can speak of at least a significant lack of success, and the corresponding element of tragedy in the lives devoted to the programme. One clue for the historian is the absence of traces of the programme in the standard, general-purpose materials used in later times. For although it is misleading and unfair to re-create and judge the remote past simply on the basis of the materials which are considered useful at the present, the survival of results is an important indicator of their quality. I have argued this at length elsewhere; for the present let it suffice to say that if a particular result can survive through all the transformations of its theoretical content, and through all changes of problems, including the demise of the field in which it originated, then this is evidence, and the best we have, that the scientific result does have real contact with the external world.

Conversely, the vanishing of results from the research tradition of a few decades later is evidence for their superficiality. Such 'negative evidence', like the 'argument from silence' of general historiography, is very dangerous as a basis for conclusions, as we know from the many examples of profound work which waited for decades or generations before being appreciated. But with due caution, it can be admitted; again because it is frequently the best we have. A case in point here is what might be called the 'Newtonian' programme for explaining the physical and chemical properties of matter in terms of very small particles connected by short-distance forces of attraction and repulsion. The history of this programme is now being traced, through the eighteenth century into the nineteenth.[6, 7] Its last and most explicitly programmatic phase was announced and promoted by the 'Arcueil' group, led by Laplace and Berthollet. Looking back on this programme, we can say that it was 'premature'; not so disastrously premature as Descartes' mechanical physiology, but still lacking so much in the way of information and tools, that it could not succeed in its own terms. And the stock of lasting achievements in physics surviving into later decades of the nineteenth century, is far less impressive than the lustre of the school, in its own time and later, would suggest.[8] Whether an individual is caught up tragically in the failure of a research programme, depends very much on his circumstances and personality. It appears that younger recruits to the Arcueil group, as Gay-Lussac and Arago, got out early enough in their careers to pass unscathed; while such senior men as Laplace and Biot were, each in his own way, sufficiently flexible to survive; but Berthollet and Poisson were both tragic figures.[9]

The Varieties of 'Science'

Up to now I have been arguing implicitly within the framework of a certain conception of 'science', one descended from the German universities of the nineteenth century, in which the attempt was made to assimilate the study of nature to the existing disciplines of academic scholarship. The extension

[6] For Newtonianism in chemistry, see A. Thackray, *Atoms and Powers* (Harvard University Press, 1970).

[7] For general Newtonian natural philosophy in the eighteenth century, see R. E. Schofield, *Mechanism and Materialism, British Natural Philosophy in an Age of Reason* (Princeton University Press, 1970).

[8] The downfall of the 'Laplace school' has been studied by Dr. R. Fox, to whom I am indebted for this information.

[9] The first and standard history of the 'Arcueil group' is M. P. Crosland, *The Society of Arcueil* (London, Heinemann, 1967).

of this idea of 'science' out of its particular historical context, is as common among historians as it is illegitimate. In the present period, when the 'ideology of pure science' is in tatters, it is possible to speak of the many possible functions of a completed research project, without being universally condemned as a traitor to scientific integrity. In the simplistic ideology of which Sarton was a spokesman, the only explicit function of a scientific result is to raise the edifice of scientific knowledge by one more unit, and thereby to serve as a footing for the next brick. Anyone involved in biographical history recognizes that research performs another sort of function altogether in providing benefits of various sorts to the author. The relation of this 'personal' function to the 'objective' one of the advancement of the field, is delicate: only a saint could ignore it altogether, while if it preponderates in choices of problems, it easily leads to fashion-chasing and eventually to corruption.

Once they are achieved and made public, the results of research can be made to perform other functions, quite independently of the original intentions and subsequent wishes of their author. In the present period, the possible functions of any scientific result in the investigation of technical problems (roughly, the manufacture of things for commerce or war) present pressing and difficult moral problems to the community of scientists. And in earlier times, the natural sciences, at least as much as the sciences of man, could be used as weapons in ideological debate, dealing with questions both fundamental and urgent about man and his relation to Nature and God. The period during which the study of nature was 'ideologically sensitive' is roughly that during which it was called 'natural philosophy' rather than 'science': in modern Europe, from the beginning until at least the very late nineteenth century.

To ignore the ideological sensitivity of modern natural science through its great formative period is very convenient for a simplistic defence of the autonomy of science and the integrity of scientists. But the price to be paid, in the understanding of science, is heavy. It is not merely that the presence of tragedy in the history of science is denied, by making the antagonists in any conflict to be plaster saints on one side and tin devils on the other. It also deprives scientists of the present time of a historical perspective on their own moral problems, with the result that the illusion of suddenly lost innocence makes their dilemmas seem unprecedented and hence worse than they really are.

To put these ideological functions in perspective, we can consider two classic versions of the situation. The first derives mainly from the Enlightenment, when certain aspects of the thought of Descartes and Bacon were developed and applied for ideological functions quite foreign to the tastes of their authors. It affirms an ideological function of science, as a weapon in a struggle against a variety of material and spiritual ills, most of their causes capable of location in the policies of the Catholic Church of the time. Both in its knowledge and in its methods, science was bearer of truth and reason, against dogma, superstition and oppression. By its very nature, science could not produce either error or evil; and so it had a privileged position among all sorts of ideologically engaged activities. This conception of science persisted, in a variety of currents, through the nineteenth century; in the English-speaking world, it embraced such liberal Christians as A. D. White, author of *The History of the Warfare between Science and Theology in Christendom*,[10] and rationalists, and also eminent late-Victorian scientists of the 'X-Club' group.[11] It is clear that such an extended set of claims for Science is vulnerable to refutation, not so much by the ideological enemy as by events; and hence it contains within itself the seeds of tragedy.

A position which in intention is diametrically opposed to this can be found in the 'ideology of pure science', which started in nineteenth-century Germany, and was explicit in the eloquent writings of its last flowering in England in the late 1930s.[12] Here, science is totally inward-looking; its only offerings to the outside world are general contributions to knowledge and culture, unpredictable technological applications, and the example of its endeavour. Any direct ideological or political engagement by scientists, except in self-defence against direct attacks on their autonomy, would be a contamination of the ideals of science. The folk-memory of scientists is no more extended or accurate than that of any other group; and so the recollections of 'Little Science' (perhaps better described as 'Our Town' science) are of a golden age of simple living and noble innocence. This happens to have been true for many fields in many centres, and it is also a very useful *prisca* myth to invoke in the very real battles for the autonomy of universities from vulgar or reactionary external attacks. But as a description of the normal or essential state of science as a whole, during the nineteenth century or at any other time, it simply will not do. Also, as we shall see later, even this 'quietist' position is not a guarantee against the development of the insoluble problems that constitute tragedy.

One reason why these two principled positions have not been distinguished clearly up to now, is that they have frequently been muddled,

---

[10] First published 1896, as the culmination of a stream of such literature, reprinted New York, Dover, 1960.

[11] See R. M. Macleod, 'The X-Club; A Social Network of Science in Late-Victorian England', *Notes and Records of the Royal Society*, **24** (1970), pp. 305–22.

[12] See G. H. Hardy, *A Mathematician's Apology* (Cambridge University Press, 1940).

both in practice and conception, and also mixed with yet a third interpretation of the social rôle of science. This is that of 'useful' science, where the results and methods of science are applied directly to technical and practical problems; and these external tasks provide stimulus, goals, and partial justification for scientific work. An extreme example of this confusion of conceptions is provided by the movement for 'Freedom in Science' of the late 1930s and the war years. They reacted to the crude political pressures on science exerted by the Nazi and Soviet regimes, and also to the theoretical arguments for 'planning' in science advanced by English Marxists; their defence was an assertion of the purity and autonomy essential for genuine scientific creation.[13] But their campaign for purity led them not only into ideological polemics, but also into organizational work through the 'Congress for Cultural Freedom', whose political affiliations were only belatedly revealed.

Tragedy from Ideological Engagement

We can now consider the ways in which tragedy can enter into the life of science, because of its ideological connections. Let us take the classic case of 'martyrdom for science', that of Galileo; it is rich enough to be worth re-examining every time our understanding of these problems develops. Concentrating at first on the principles involved in his confrontation with the Church, we must first ask whether he was an honest research-worker being persecuted, whether by bigots or by a State machine;[14] or whether he was from the outset engaged in an ideological struggle analogous to that of the Enlightenment albeit on more restricted ground. My view is that the latter interpretation comes much closer to the truth of a very complex situation. If this is the case, then the issues in the conflict were much deeper than the correctness of Aristotelian physics or Ptolemaic cosmology. What was the fundamental objection from the other side? For me again, the crucial evidence is the letter from Ciampoli of February 1615, conveying the warning of Maffeo Babbenini that Galileo's conclusions were ideologically sensitive;[15] and that he would do well to present them in such a way as to minimize their dangerous consequences. For the primary task of the Church is not the establishment of philosophical truths, but the salvation of souls. And when interesting hypotheses, presented by however brilliant, pious,

13 Michael Polanyi argued this on many occasions; see his *The Logic of Liberty: Reflections and Rejoinders* (London, Routledge, 1951).

14 The former interpretation is traditional in the history of science; the latter was introduced by G. de Santillana, in *The Crime of Galileo* (London, Heinemann, 1958).

15 Galileo *Opere*, Vol. XII, pp. 145–7, translation S. Drake, *Discoveries and Opinions of Galileo* (New York, Doubleday, 1957), p. 158.

and well-connected a philosopher, are capable of being used for mischievous ends, the duty of responsible members of the Church is to issue a warning and then to call a halt.

Viewed at this abstract level, the tragedy of Galileo is quite Hegelian. The conflict is between right and right: on the one hand, the duty to pursue and announce the truth; and on the other, responsibility for human welfare. I doubt that I am the first to see it this way; in his public account of his own reaction to the news of Galileo's condemnation, Descartes refers to persons 'whose authority governs my actions no less than my own reason governs my thoughts'.[16] Here we have the problem of 'social responsibility in science', with a vengeance. It is interesting to remark that the problem first appeared with such tragic intensity, in association with the revolution in natural philosophy of the seventeenth century. I do not believe this to be a coincidence, for this 'new philosophy', denying as it did the possibility of magical powers over man and nature, cut itself off from the ancient traditions which recognized the category of knowledge too powerful to be published.[17] The reappearance of this category since 1945 is of the greatest significance for science, destroying as it does the basis of an ideology of beneficence starting with Galileo and Descartes.[18]

Of course, life is never as neat and symmetrical as art. Whether this rational reconstruction of the Church's motives in the Galileo case is correct, the fact remains that personal spite and nasty methods were involved in the campaign to cut him down to size. And Galileo himself showed a less than ideal appreciation of the situation, not only in his failings in personal tact, but also in his propagandist writings and actions. He was aware, at the intellectual level at least, of the problems of conveying truths to the subliterate masses; indeed, he invoked the old principle of allegoric expression as a means of explaining away any inconvenient passage in Scripture. But he seems to have shown no awareness of the principled case on the other side, at least before the composition of a revealing passage in the *Discorsi*. There he commented on the attempts of certain people to 'lower the esteem in which certain others are held by the unthinking crowd'

[16] See the *Discours de la Méthode*, sixième partie, first paragraph.

[17] Although in the received doctrine of the history of astronomy Tycho Brahe is seen as primarily an observer, cautious on cosmological questions, he was in fact an astrologer, famous as such and committed to the science. In his *Astronomia Pars Mechanica* (1598), he offered confidential information on his astrological and alchemical discoveries to men of position and wisdom. See the English Translation, *Tycho Brahe's Description of His Instruments and Scientific Work* (ed. H. Raeder, E. Strömgren, B. Strömgren) (Copenhagen, Munksgaard, 1946), pp. 118–19.

[18] See N. Wiener, *God & Golem Inc* (M.I.T. Press, 1964), Chapter V for a discussion of 'sorcery' in the contemporary period.

('solo per tener bassa nel concetto del numeroso e poco intelligente vulgo l'altrui reputazione').[19] Since Galileo had never previously shown any concern with that audience for his own works, it seems likely that only at this late stage did realize its relevance to the struggle.

Indeed, if we choose to look at Galileo in a particularly uncomplimentary way, we may say that he claimed the right to power over men's minds, through his teachings, while ignoring his ordinary responsibility for the spiritual and social consequences of his actions. Such a position can be maintained only by someone who is above ordinary morality, as a prophet, or someone below it. Galileo doubtless considered himself a prophet, merely uttering God's simple mathematical truths about Nature. But he did not submit himself to the personal abstinences of the traditional prophet; he enjoyed the good things of material life, and never lost a chance to convert his inventions to cash. Also, it can be argued, somewhat speculatively to be sure, that he was aware of the contradiction inherent in the *Dialogo*, of conveying a message that the Truth is clear and simple, while formally obeying the instructions of the Pope to show the opposite; and that he failed to resolve this contradiction, perhaps because he could not face it clearly. Hence we can see Galileo not so much as a Hegelian tragic figure, as a Shakespearean one: a great man with a fatal flaw.

Such classic tragedies are rare in science, partly because such great men appear only rarely, and then can usually pass through life without the contradictions latent in their position becoming manifest and destructive. There was a long relatively peaceful period when natural science had neither such crucial ideological sensitivity in the face of an effective machinery of thought-control, nor enormous powers in its application for good and evil. In the present period, as is quite well known, the moral problems of science in its technological functions, are as real and deep as these earlier ones of natural philosophy in its ideological functions. In the 1950s de Santillana wrote a biography of the honest liberal Galileo on the inspiration of the Oppenheimer case; perhaps in the 1970s there will be a biography of the flawed philosopher Oppenheimer, on the analogy of the Galileo tragedy.

Another source of tragedy in ideologically engaged science can be discerned in the history of Galileo: the loss of simplicity and certainty, on which so much of the programme depends for its foundations. Just as every successful social revolutionary finds new, unexpected, and sordid problems, which becloud and confuse the vision conceived in Yenan or the Sierra Maestra, so does the prophetic philosopher discover a growing thicket of nasty special problems, when he attempts to articulate his general scheme of things into its particular details. This progression is quite clearly seen in the

[19] Galileo, *Discorsi*, edition No. 2, p. 204.

career of Descartes: from the complete dedication and confidence up to 1633; through the period of moral confusion but continued philosophical optimism in the *Discours*; on to the last decade of his life, marred by unseemly squabbles with Dutch theologians and former disciples, by the cool reception of all his works, and by continued difficulties and degrading priority disputes in his scientific investigations; until his final retreat to tuition in moral philosophy to princesses. Galileo would of course never admit failure or defeat; but his theory of the tides, designed both to demonstrate a particular cosmology and to illustrate a 'mechanical' method in natural philosophy, was a disaster from the outset. It contradicted 'sense experience' by its prediction of a single daily tide; and its 'necessary demonstrations' violated every principle of simplicity and comprehensibility. Yet by his vehement adherence to this theory, which was after all the only evidence he had that was not simply destructive of the old system, he tended dangerously close to that subtle boundary which demarcates the genius from the crank.

This process of erosion of the early, revolutionary simplicities can sicken the soul. If the early ideological struggle has also lost its intensity in the meantime, then the philosopher can fall back into some minimalist or 'probabilist' interpretation of his works, and suffer only the private hurts of lost ideals. But if the battle is still on, then a public admission of defeat on this front compromises and betrays the self that has been created and exists in its context. For although the other side has spokesmen who can discourse on methodology and truth, they are but servants to an enemy whose fundamental beliefs are as simple as those of the philosopher, and utterly opposed in detail and in principle. The principled debate between competent scholars is only one front of a fierce struggle for power, in which the other side possesses and uses weapons of mental and physical coercion. Any concession on the philosophical front is bound to be seized and used in an appropriately distorted form by the hired intellectuals of the enemy, for rallying their own followers and demoralizing the philosopher's ranks. Had Galileo sincerely admitted the slightest doubt about his conclusions and methods, he would have suffered the collapse of his lifelong campaign against the claims of institutionally certified authorities to decide what could and what could not be true; hence the Pope's speech had to be put in the mouth of the simpleton. But even Simplicio could not be allowed to raise the long-standing objections against Galileo's theory of the tides; this cut too close to the heart of the struggle. Thus when the problems become so complex, the cause of intellectual integrity requires defences by methods through which intellectual integrity is stretched and even compromised. Whether this is done deliberately or not, does not matter; the contradiction is there, and, with it the tragedy.

The False Refuge of 'Neutrality'

It might be argued that my analysis of Galileo's tragedy, whether correct or not, is irrelevant to science. For Galileo's misfortunes did not result from his scientific discoveries and theories, or even from his philosophical interpretations of them; but only because he persisted in playing the *idéologue* even when the game had become one which could not be won. Had he restricted himself to a statement of his results with a clear assessment of their reliability, so goes the familiar story, he would have suffered nothing worse than the inconveniences of respectable renown. Indeed, the Jesuit scholars themselves hinted as much at the time.[20]

We can expose this interpretation by asking, where should he have drawn the line? Perhaps he should have suppressed that first analogy between the changing pattern of circular shadows on the moon, and the phenomena of sunrise over a terrestrial valley; for this alone was sufficient to alert the thought-police. Even the most suffocating positivist admits that there is more to science than the collection of data; and we might ask, what sort of unifying hypothesis and arguments could Galileo have offered, without being cast under suspicion of meddling in matters that were officially excluded from his competence? An example is provided by his rival and archenemy Christoph Scheiner, himself an unjustly neglected experimenter, observer and mathematician. For in his *Rosa Ursina*, he finally accepts the Galilean hypothesis that the 'sunspots' are indeed moving configurations of fluid on the sun; and the last third of the book is devoted to a scholastic balancing of accepted authorities, showing that the rigidity of the heavens is only a minority view among the Fathers and Christian scholars.[21]

Thus, Galileo could not begin to do science in any sense conformable to our own, without becoming involved in the whole gamut of debates ranging through Scriptural exegesis to epistemology. In this respect his situation was analogous to that of the 'orthodox' Soviet geneticists when attacked by Lysenko. It was not merely their particular doctrines that were denounced, but their methods and the very principles underlying them: the very elementary procedures for control of evidence, that distinguish the scholar from the crank and charlatan. In the context of Stalinist forced marches to industrial-

[20] de Santillana, op. cit., p. 290. The remark of Father Grienberger was, 'If Galileo had only known how to retain the favour of the Jesuits. . . .'; this would involve not only personal relations but also an approved style in ideologically sensitive matters.

[21] See G. MacColley, 'Christoph Scheiner and the Decline of Neo-Aristotelianism', *Isis*, 32 (1940), pp. 63–9.

ization and collectivization, the basic principles of intellectual integrity, in this field of science as much as in politics and history, were ideologically sensitive. They were, and remain, subversive of the Stalinist concept of Socialist construction.

Viewed in this context, the ideological neutrality of science is not a prerequisite for scientific work and scientific integrity; rather it is luxury, dependent for its possibility on vagaries of the social and ideological context of science. Of course, it is a very precious luxury, one of those that make the difference between a society with potentialities for peaceful change and improvement, and one either rent by conflict or turned into a sophisticated concentration-camp. But when the ideological involvement of science is real, the mere assertion of its neutrality cannot accomplish its ends. Just as the undergraduates' arguments about the futility of philosophy are themselves philosophical, so is any answer to an ideological attack on science itself a statement of an ideological position. This will not be so clear to the defenders of science when their own ideology is one of the autonomy and neutrality of science, supported not by rationalistic theory but by a 'conservative' argument compounded of tradition and a quasi-religious personal commitment. But to assume that one is in this respect essentially more holy and pure than one's enemies, can involve its own threats to integrity, and its own characteristic tragedies.

Like all other creative endeavours, science has the paradoxical property that its greatest achievements belong to all mankind for all time, but the work leading to those achievements is done by particular individuals living in, and conditioned by, particular milieux. In the history of any effort relating closely to human affairs, this double character is recognized, and the tradition of later assessments of any great figure cannot escape from judging him by the good and evil consequences, intended and unintended, deriving from the circumstances of the achievement and reception of his work. Such a critical awareness has hitherto been foreign to the history of science; its categories would have been incomprehensible to such as Sarton. The faith in the essential 'progressiveness' of science was supported by some generations of experience, as reinforced by the assumption that the effects of science were necessarily beneficial, and diffuse in their workings. Even the aggressive campaigns of ideologically engaged scientists could be brought under this comforting assumption, for they too stood for Truth and Reason against false ideologies. But the growing malaise of science deprives both the ideologists of science, and the scientists themselves, of protection against the moral judgements, and moral problems, to which the rest of humanity has always been subject.

A scientist, be he the most pure and abstracted puzzle-solver, is involved in a community whose influence on his work is all the more pervasive because it is informal and frequently tacit and unselfconscious. The criteria of adequacy applied to his results, and the criteria of value applied to his problems, are social possessions, which he absorbs from his environment and accepts as natural and obvious unless he is involved in an immature or contentious field. When the terms of exchange between some field of science and its patrons are vague and generous, these governing principles may develop nearly autonomously, conditioned only by the general criteria of fruitfulness and success within the discipline. This may well have been the situation in many subjects over long stretches of time in many places. But in the present period of 'science policy' and 'criteria of scientific choice', the terms of exchange, however generous on occasion, are now explicit. The existence of conscious choices, favouring some projects at the expense of others, makes us aware that unselfconscious choices, with the same effects, have been present all along. The sudden increase in research funds for the study of the 'environment' does not result from new possibilities for pure and applied research; but from the belated recognition that this area had previously been systematically starved. The ideological bias inherent even in the most excellent and scholarly political and social history has long been known to operate through the process of selection. By concentrating on kings and wars, the scholar and teacher implicitly credit these phenomena more value, and assign them more reality, than say technology and exploitation; and the converse is also true. We are now seeing that in its own priorities, implicit or explicit, natural science makes and has always made judgements of reality and value in the world it explores and tries to comprehend.

In itself this is no criticism; for the work could not proceed otherwise. But to the extent that the conception of its tasks implicit in the direction of scientific work, is tied to the conceptions of reality and value held by one section of a divided society, then to that extent is science involved, consciously or not, in the struggles within that society. To defend the integrity of science and scientists by pretending that the whole enterprise is somehow in an aetherial world of its own, is to invite defeat and tragedy.

There are only a few instances recorded so far in the history of science, when the relation of natural science to its base of recruitment and support in the ruling sections of society, became an explicit focus of debate and struggle. One such is the controversy over the 'academies' in England in the 1650s;[22] another is the episode of 'Jacobin' science during the Terror in

---

[22] See A. G. Debus, *Science and Education in Seventeenth-Century England: The Webster Ward Debate* (London, Oldbourne, 1970).

the French Revolution.[23] More recently we have had Lysenko,[24] and certain aspects of the Cultural Revolution in China.[25] The different controversies occurred in very different contexts, and gave different emphases to the three elements in dispute: the criteria of value, relating to the social functions of the body of doctrine; criteria of adequacy, relating ultimately to the style and the class basis of the science; and the objects of enquiry, relating to the conception of reality presupposed by the science. The general tendency of these radical attacks is to advance the claims of a science directly serving the interests of (some at least of) the lower, less educated orders of society; shunning and condemning the abstracted experimental and mathematical approach which is necessarily a cultural monopoly of the educated élite; and finally invoking a more rich and lively conception of the natural world, participating in or descended from the tradition of alchemy and magic.

It is significant that in every case their direct antagonists were not the patent reactionaries, but individuals and institutions that themselves were committed to, and had been engaged in struggle for, their own version of the ideals of the radicals. Whether these more moderate forces felt the pangs of guilt on the inevitable discovery of enemies to the Left, is not known to me. But the campaigns of the sectarian radicals of science have always, up to now, ended in failure and tragedy for themselves. This was probably inevitable, and, many will add, richly deserved. For their programmes and style of debate were a betrayal of the principles of science and of rational discourse; and to the extent that they achieved political influence, they only succeeded in damaging the life of science. Indeed, the phenomenon of populist demogogues is known in all other spheres of social life; and it is not surprising that such types have occasionally chosen science for their target.

It is quite legitimate to make a condemnation, on ideological grounds, of these frequently unpleasant people. But it would be an error of historical judgement to dismiss them as accidental, insignificant phenomena. For they are a sign of an unhealthy, contradictory situation of science, which cannot be completely cured so long as society is divided into sections of which one possesses a monopoly of literate culture and the power that it yields. C. C. Gillispie provides a vivid picture of the effects of the perception of this monopoly by those outside. Given what I have said already, I need only

[23] See C. C. Gillispie, 'The Encyclopédie and the Jacobin philosophy of science: a study in ideas and consequences', in M. Clagett (ed.), *Critical Problems in the History of Science* (University of Wisconsin Press, 1959), pp. 255–89.

[24] See Z. Medvedev, *The Rise and Fall of T. D. Lysenko* (Columbia University Press, 1969); the 'romantic' aspect of Lysenko's thought is indicated on p. 168.

[25] For science in the Cultural Revolution, see C. H. G. Oldham, 'China Today: Science', (Canton Lecture), *Journal of the Royal Society of Arts*, 116 (1968), pp. 666–82.

quote: 'Early in 1789, Laplace, impatient at the quantity of chimerical projects (of inventions), had proposed that every applicant be subjected to a test in geometry before his designs could be considered.'[26] Gillispie analyses the whole episode of Jacobin science in great depth, and explains the inevitable failure of the romantic, populist sectarians in terms of the essential neutrality of scientific knowledge: the deepest ideological conflict is between genuine science and any 'naturalistic moral philosophy', be it that of Diderot, Comte or Marx.[27] But this is wisdom after the event; he observes that none of the victims of the Jacobins made this simple point, but rather defended themselves clumsily and inappropriately in terms of the same set of 'Baconian' slogans.[28] And they had to do so, for none of them were nineteenth-century German academics; they were heirs to another tradition of 'naturalistic moral philosophy', as developed by d'Alembert and continued by Condorcet.[29] However we view it *sub specie aeternitatis*, at that time the issue was not of the essential neutrality of science from ideological needs and desires, but the explicit involvement of different sorts of science with the different sides of a political struggle at least partly related to ideology and class.

Contemporary Roots of Tragedy

It is just possible that during the relatively brief period when the ideals of 'pure science' were strong and socially effective, there could have been a viable defence of science from external attacks, on the grounds of its essential neutrality with respect to ideology and politics. But pure science has not yet recovered from the tragedy of the Bomb, and will not do so for a long time to come. It was the most aristocratic, philosophical and pure branch of science that was converted into a new technology of mass destruction. The tragedy was inherent in the original decision to press for an anti-Fascist 'deterrent' bomb; it was realized in the bombing of Hiroshima, compounded in the assault on Nagasaki, and completed in the development of the H-Bomb, the ballistic missile systems, and the obscene gematria of 'nuclear strategy'. It is no longer possible for anyone to sneer at the lines of Brecht:

[26] Gillispie, op. cit., p. 272.
[27] Ibid., p. 281.
[28] Ibid., pp. 277–8.
[29] The earlier history of this tendency, developed as one wing of the Enlightenment, is indicated in T. L. Hankins, 'Jean d'Alembert', in *Science and the Enlightenment* (Oxford, The Clarendon Press, 1970).

> Out of the libraries strike the slaughterers.
> Mothers gaze numbly at the skies
> For the inventions of the scholars.[30]

The resulting moral crisis of science works itself out in many ways. The 'swing' of potential recruits away from science is dismaying to those who had built their careers on the expectation of ever-increasing numbers of students providing jobs and research assistance; and disheartening to those who sense the condemnation of their own existence, implicit in this quiet defection. It is partly science itself that is rejected; and partly also the bureaucratic-technical society in which it is now embedded and assimilated. Within science, there are symptoms of a loss of coherence and direction within the most sensitive fields, as mathematics[31] and physics[32]; as well as a proliferation of 'shoddy science' in journals whose only functions are to provide profits for the publisher, and titles counting as publication-points for the author. Voluntary organizations based on the image of science as beneficial and uplifting, as the British Association, now find themselves dying of ennui.

In the generation since the war, science has grown very rapidly, and progressed magnificently. But this resulted in its being in a dangerously bloated condition, unable to comprehend or to cope with the crisis resulting from the sudden manifestation of the contradictions within its situation. The various ideologies by which science has traditionally justified itself to its practitioners, recruits and supporters, have simultaneously collapsed. The struggle against clerical dogma and superstition is now receding from memory; and science itself is under attack as a vehicle of bureaucratic obscurantism. The ideals of pure, academic science become increasingly remote from the practical experience of planned, mission-oriented research. And the 'usefulness' of science is now increasingly identified with the technomania that is the curse of our civilization. There is a focus for a new idealism and commitment within science, in the 'critical science' devoted to the protection of the environment and mankind from the depredations of the technocrats. But this is as yet small and marginal, and at risk of being bought and corrupted by the more enlightened sections of the machine before it comes to maturity.

It is not the case that tragedy, for individual scientists or for science as

---

[30] I am indebted to Dr. C. Pedler for this quotation; it was seen by him on a wall at Warwick University.

[31] See W. G. Spohn Jr., 'Can Mathematics be saved?', *Notices of the American Mathematical Society*, **16** (1969), pp. 890–4.

[32] See J. M. Ziman, 'Some Pathologies of the Scientific Life'. *The Advancement of Science*, **27** (September, 1970), pp. 7–16.

a whole, is inevitable. But to avert tragedy, something more will be required than the very straightforward intellectual integrity which had sufficed as a basis for the ethics of science in simpler times. The problems confronting any scientist who is not merely a white-collar worker, are compounded of aspects of nature, of society, and of the commitments and values that are best described as 'religion'. Considered as an institution with its own life-style, science has enjoyed an extremely long period of adolescence, rejoicing in its growing strength but not yet encountering the problems of responsi-bility. During those generations, the possibility of tragedy could be excluded from the self-consciousness of science; and the cases where it did occur, ignored in its history. But within the last quarter-century, maturity, in the form of present and future tragedy, has been thrust upon science; and it is an open question, whether it will survive in a form recognizable to ourselves.

Conclusion

The idea of tragedy that I have implicitly and informally used, is that of an insoluble problem: when an agent is in a situation where he can neither evade nor resolve a moral or existential problem. If the person involved is a patient rather than an agent, we have misfortune, pathos, sadness; but not tragedy. The type of the tragedy will vary, depending on the genesis and outcome of the insoluble problem. Most of us go through life with insoluble moral problems latent in our existence; when they become manifest and urgent, then if we are still agents, we find ourselves in tragedy.

In retrospect, the illusion that science could escape from tragedy, and that the life of science could somehow prevent it, appears as shallow as the vulgar optimism of the high Victorian age. I have indicated its roots, and given a preliminary discussion of some ways in which tragedy can strike in science. I am keenly aware of the crudities of my analysis, and of the paucity of historical evidence for it. But I am convinced that from this analysis one must either go forward to a deeper study of the problem, bringing forth the historical evidence illuminated by it, or backwards to the simple vision of Sarton, itself the bearer of the deepest tragedy of all in the science of our time.

XIII

TIME AND MOTION: REFLECTIONS ON THE NON-EXISTENCE OF TIME*

Maria Luisa Righini Bonelli

In the natural sciences time is not explicitly defined so much as assumed as a primitive concept. It appears as an abstract parameter in all the functional relationships that measure or characterize the concrete and real entity: motion. Accordingly, time is normally regarded as an empirical reality. In what follows, however, I wish to present some speculations which argue that time is better thought of as a purely rational entity.

Time, it would seem, is necessary for the creation of kinematics out of geometry. When in geometry we compare two configurations that occupy different spaces, we need the notion of local displacement, an essentially static or atemporal concept. In kinematics, however, when we transform a point or configuration A into a point or configuration B we need a velocity, the operator which intimately links a change of place with an interval of time. Hence, if in geometry displacement is defined with regard to the initial and final states of configuration, i.e. with respect to position, shape and size, in kinematics the displacement is properly the particular modality called velocity, the velocity with which the system functions and which distinguishes and defines the trajectory of the configuration, i.e. its initial, intermediate and final positions.

Now in kinematics the parameter time is normally introduced right from the first, in a proposition on uniform motion. We accept the definitions given without presuming to demonstrate them because to try at this stage would result in being caught in a vicious circle. But if we reflect on the case, it is clear that the objective phenomenon of time is none other than a body in

* The author wishes to thank Professor Thomas B. Settle, Polytechnic Institute of Brooklyn for translating this article.

some motion (the hands of a clock, for instance). Thus a kinematic calculation, the calculation of a space covered in that motion, is comparable to a
geometric calculation of length. The first depends on the unit of velocity
just as the second depends on the unit of length. If we again think of a clock,
the motion of a hand could be considered a sample or a standard of uniform
motion.

Even Newton exhibits a certain ambiguity about time, as when he says:
'Possibile est, ut nullus sit motus aequabilis quo, tempus accurate mensuretur.
Accelerari & retardari possunt motus omnes, sed fluxus temporis absoluti
mutari nequit'.[1] In fact, while initially he seems to make time depend on
motion, he then introduces an extra or independent time to which movement is referred. But how else do we detect the flux of time if not with a
moving body? The speed of a moving body is its basic identifiable characteristic, and the only way we have of measuring speed is with reference to
another motion, conveniently assumed to be standard.

Hence, in order to define time it is necessary first to come to terms with
real, physical motion, and it is vain to search for a non-existent uniform time
outside actual motion. And to define any motion whatever it is necessary to
postulate a conveniently chosen standard velocity. One could adopt as a
standard, for instance, the speed exhibited by a body occupying successive
positions with reference to fixed markings on a linear scale, itself presumed
to be invariable. One widely used procedure is to observe the apparent
displacement of a star with a transit instrument and, supposing the entire
angular motion between transit sightings to be uniform, to accept the
interval between successive sightings as a basic unit of velocity. Next we
divide the entire circuit into 86,400 parts, and in so far as we can confirm
with our instrument the uniformity of motion among the fractional sightings
(the motions between appropriate cross-hairs in the same instrument, for
instance), we make these fractions of the celestial course coincide with the
successive displacements of the second hand of a clock or the excursions of
a pendulum. These kinematic units we call sidereal seconds, and with them
we introduce a time number. Thus astronomers take the motion of the stars
as a basis for time. 'Dixit autem Deus: Fiant luminaria in firmamento coeli,
et dividant diem ac noctem et sint in signa et tempora et dies et annos',
months, seasons, days and years.

Clearly, then, we observe time by observing motion, and not the reverse.
And if we are to have an even flow of time, we must have a continuous
motion. In reality, however, we perform the empirical measure of time and
motion with discontinuous methods, with a sequence of instantaneous

[1] J. Newton, 'Philosophiae Naturalis Principia Mathematica' (Londini, Jussu Societatis
Regiae ac Typis Josephi Streater, Anno MDCLXXXVII), p. 7.

photographs spaced at short intervals, a sequence our mental processes merge into a uniform flow which gives rise in turn to something we can call psychological time. This takes hold of us in such a manner that, if we were not provided with this subjective time (that is, if memory did not exist), we should even affirm that in the interval between one action and another nothing exists, because the only existing things would be the instantaneous flashes, and there would be no occasion for the notion of duration between the flashes.

But what is this psychological or memory time? External facts give rise in us to two classes of the facts of consciousness which are very distinct. The first is that of the immediate interpretation of the direct, sensory image, the pattern or projection of the external object, in a reduced scale perhaps, but conforming identifiably to the real external object. Thus we say that we 'see' or 'follow' a phenomenon in all the successive phases of its reality. The second class is that of the ideal and subjective construction that looks more to internal, self-consistent, aesthetic and rational desiderata in its judgements than to exact correspondences with elements of the external world. Among these subjective constructions are the rational entities such as mathematics and psychological time. And as a rational entity this time is easily made compatible with mathematics, becoming almost a mathematical entity, and easily enters the functional relationships of physics.

In a sense, however, this psychological or memory time does not stay inside. We have a long history of trying to project it back out into the external world, of trying to make time an independent entity actually existing in the external world, of wanting to construe phenomena as dependent on time instead of time as a consequence of phenomena. In fact, what we know of actual physical processes we know by remembering a succession of projections of the external world, i.e. in psychological time. But we try to justify the notion of a real, external and objective time by emphasizing the striking correlations between the projections in our memory and the functional fellowships we have developed to represent physical processes. And so successful have these correlations been that it now is next to impossible to eliminate the notion of a real objective time from physical calculations. We even want time to determine our own material existence; we attribute to it an irresistible destiny, our own decay and death.

We adopt, in short, a convention for a uniform succession: an ordered whole, constituted of the local positions of a point always moving in the same sense and without discontinuities. If we assume this succession to be a standard of constant velocity, then every succession with elements having a one to one correspondence with it has the same uniform velocity. If the correspondence happens to be between an element of one succession and a group of elements of another, the relative velocity will be different; it will be uniform if the groups succeed themselves with a constant numerical index of the relative velocity, and it will be non-uniform if the groups succeed themselves with a variable index. That is, the numerical index associated with a given succession of groups (a series) is the defining characteristic of that motion. But the concept of an ordered succession (the relation of ordering in which a fact A is always prior to a fact B, and this in turn is prior to a fact C, which thereby is after B and even more remote from A) does not in fact depend on the concept of a pre-existent time outside this succession or movement. Properly speaking it is the succession that constitutes time. Since we can take any succession for time, there are infinite times, because there are infinite successions. In this context one can avoid considering duration as an absolute length of time between two points of a series; it must be a relative concept referring to the relative positions of two points in each of two ordered series; and thereby one has no need of velocity of succession or time in an absolute sense.

Then if we observe a pendulum clock, we find ourselves using our hearing and our vision together to associate the sound of the 'tick', say, with the series of odd markings on its face and the sound of the 'tock' with the series of even markings. Thus the tick-tock represents the swinging pendulum and refers it to an alternating numerical order, and we say that the motion of the pendulum is alternating. But we cannot state that the tick precedes or follows the tock. Likewise other periodic phenomena seem to offer similar ambiguities with regard to the future and the past. If memory did not intervene, there would not exist a determinate sense in any cyclic event. For instance, it would be the same thing to say that five in the afternoon is after midday of the same day or before noon of the following day as to say that tomorrow precedes today, because there is no clear temporal succession; there are only coincidences in stars. Without remembering what has gone before, one coincidence can be another and there is no way whatever to differentiate among them. So one cannot find absolute time in such events; it is psychological time that distinguishes the past from the future.

What I have suggested amounts to seeing the usually assumed, real, objective time as only a projection into the external world of a psychological time, and in turn reducing this psychological time to a product of our memory. While I would hope that this might solve some problems for physicists and philosophers, in so far as I have not attempted to clarify the problems around memory itself I do not pretend to have resolved the purely subjective problem of time. Here my concern has been to show that an objective time can be systematically replaced by an objective velocity.

Simultaneity and succession can be thought of as coincidence and distance in our kinematic sense. In the first instance simultaneity is identified as spatial continuity or coincidence. But by extension one can construct simultaneity by imagining two phenomena or points transported so as to be contiguous or superposed. Alternatively one could imagine the two events connected by a rigid rod; in a rigid rod what happens at one end is signalled instantaneously to the other. But if the universe were such that all its parts were rigidly interconnected then there would not even be subjective time; past and future would coincide with the present. Therefore, though conceptually intelligible, absolute rigidity does not exist. In practice we make approximations by correcting for the phenomenon of propagation, and we assert as contemporary two phenomena not otherwise connected by causal succession.

Occasionally we misuse the notion of simultaneity in referring to phenomena really connected by ties of causality. When I lift a chair from the ground by grasping its back, do I really simultaneously both lift its back and detach its feet from the ground? Or if I lift a rope by one end, do I simultaneously lift the whole rope? As simultaneity supposes rigidity, propagation supposes deformation. And though rigidity and absolute simultaneity are understandable constructs, they are incompatible with real physical space, and the degree of simultaneity is always relative to the lack of rigidity of the system that serves to unite the phenomena under observation. We could say, therefore, deformability is connected to the determination of succession.

In a limited sense, the problems arising out of the necessity of resorting to a finite velocity of propagation in deformable bodies for any communications, both in everyday life and in more scrupulous scientific determinations, are not new. The ancients had already noted, in fact, that two contemporary phenomena, the impact of a hammer and the contact with the body hit, were acoustically and optically differentiated in a greater degree the greater the distance of observation. The 'thud', coming after the 'contact' should have introduced to us, so to say, the concept of an acoustical time.

By analogy, once we have determined that the velocity of light is finite and also have determined a value for it, we introduce the notion of an optical time. We are able, for instance, to discriminate the apparent positions of bodies as given by optical data (especially in astronomical observations) from their absolute, 'unknown' positions. In a sense Römer has resolved already, as a result of his calculations concerning the eclipsing of the moons of Jupiter, the problem of an 'optical time base' in our theoretical accounts of phenomena. Unfortunately this has not yet brought about a revolution in our conception of the relation between space-time and empirical reality, and perhaps we should not place too much hope in the thought expressed by Perrin: 'It is natural that the development of the theory of relativity should oblige us to rework our concept of a mechanics dependent on the notions of time and space. The replacement of absolute time with optical time should modify it profoundly'.[2]

To clarify the concept optical time (which is very like acoustical time and any other 'elastic' time) we need only refer to the particulars of the propagation of signals in a fixed, homogeneous and isotropic space and at a constant velocity. This will in turn allow us to interpret physically the question of simultaneity. Let us first imagine light being emitted from a single, fixed source into space. If the light is of constant frequency, circular wave-trains, concentric to the source, will expand into space, and the distances between successive waves will be the same everywhere. We could call such a situation a 'Huygens configuration'. Such a configuration in principle, if not in experimental practice, is in itself sufficient to determine all the distances and positions of the source and of the observer with respect to the physical field (the waves from a fixed point in space being sufficient to constitute an absolute system of co-ordinates). And if we postulate a constant C, we come *ipso facto* to be able to discriminate simultaneity and succession. If, however, we were to have to deal with a system of two sources at least one of which was moving, the problem would be complicated by the Doppler effect. At least one set of waves would be non-concentric. But every wave in such a case would have its centre in the one of the successive positions occupied by the source. And as every wave would still be characterized by a centre of action or source, we could still say that there are points of true simultaneity, those points in space which are at the same distance from an original source of propagation, the surface of a given wave-envelope.

In this physical Huygens-space there would be no need for two observers, who wish to co-ordinate observations between themselves, to erect two sets of Cartesian orthogonals and either determine the locus of one with respect to the other or determine the angles which their axes make among themselves. The two observers are uniquely determined in a universal field and every 'extra' system is in principle superfluous.

Clearly all this depends on being able to isolate single waves and determine their radii of curvature, a task quite beyond what we are capable of empirically. Even so, such conjectures can cast light on contemporary problems.

---

[2] F. Perrin, 'La dynamique relativiste et l'inertie de l'énergie', *Actualités Scientifiques et industrielles*, XLI Ed. Hermann (Paris, 1932), p. 3: 'Il est naturel que le développement de la théorie de la relativité nous oblige à remanier notre conception de la mécanique, qui dépend des notions de temps et d'espace. Le remplacement du temps absolu par le temps optique doit la modifier profondément.'

For instance, the notion of 'local time' has occupied modern writers, and there has been a tendency to put it in opposition to the idea of a 'universal time'. But it would seem that the determination of time depends on the determination of motion and not on the real existence of either of the above 'times'. Hence we should only talk of the computation of signals in physical space with kinematic properties of its own.

When we say, then, that a phenomenon has a certain relation to a spatial order, it is superfluous to add also 'in time', which we have not been able to find in objective reality. To say that a phenomenon is such that it is ordered in space and time would be to suppose that the spatial order is insufficient to determine it unequivocally. On the contrary, putting a phenomenon in a spatial order relative to a reference system removes any and all ambiguities of succession in so-called 'space-time'.

Minkowski has contributed not little to our modern conception of time, but perhaps we should argue with some of his views, informed as they are by an almost mathematical mysticism. He says, for instance:

> From now on space in itself and time in itself must fall into shadow and only a type of union of the two should retain its individuality.[3]

He adds:

> The object of our observation is always, and only, space and time considered together. No one has ever observed a place if not at a certain time, nor a time if not at a determined location. I still respect the dogma that space and time have their own proper, independent significance. A space-point along with a time-point, i.e. a system of co-ordinates x, y, z, t, I call a *universal* point (Welt-punkt).[4]

It seems to me that a basic confusion stems from the assertion that the object of our observation is always space and time considered together. This could be true, but only in so far as one intends to refer to the results of the interference of the observer with the object. By actually observing that which is in space, a particular phenomenon, we weld to our memory, as it

[3] H. Minkowski, 'Spazio e tempo'. Traduzione di G. Gianfranceschi. Nuovo Cimento. Serie V, Vol. XVIII, fasc. 11–12, Nov.–Dic. 1909, p. 334: 'Da quest'ora in poi lo spazio in se stesso, e il tempo in se stesso, debbono piombar nelle tenebre e soltanto una specie di unione dei due deve serbare la sua individualità.'

[4] H. Minkowski, op. cit., p. 335: 'Oggetto della nostra osservazione sono sempre, e soltanto, spazio e tempo insieme considerati. Non ha mai alcuno osservato un *luogo* se non ad un certo *tempo*, ne' un *tempo* se non in un luogo determinato. Io rispetto ancora il dogma che spazio e tempo hanno ciascuno una significazione propria indipendente. Un punto-spazio, in un punto-tempo, ossia un sistema di valori x, y, z, t, lo chiamo *punto universale* (Weltpunkt).'

were, sensations and thereby external facts. This is accomplished by means of that vague psychological time that neither is observed nor exists in the object, but which is in us and which we unthinkingly presume to be reflected into us from the outside. We only observe spatial configurations and their relative changes with respect to one another, not space-time.

From this should be obvious the artificiality of those modern constructions that introduce false entities into physical speculations. Perhaps there was an attempt to create a vision of harmony between pure mathematics and physics, but it was forgotten that in nature there are neither numbers, nor triangles, nor circles; these are our own creations.

Without a doubt Descartes gave to scientists a marvellous analytical instrument. But we must not forget that the reduction of a figure to its components in a set of Cartesian orthogonals is an arbitrary act, a conventional unreality; helpful as it is, it is extrinsic to the phenomenon that is, in itself, and is independent of any system of reference or analysis. A velocity is what it is, and it has nothing to do with its components. Certainly we cannot easily analyse a problem without the artifice of a system of axes, but if we consider a problem in itself, we ought to be able to eliminate any extraneous system of references. With the creation of the world-line Minkowski introduced a representation in space-time, an ideal creation of a systema of four co-ordinates. In turn some have been convinced that the formal graphical representation has a real content, i.e. that the unusual marriage of space and time actually exists as an entity. This tendency to confound a graphic representation with the very phenomenon represented is more or less unconsciously followed by anyone who uses mathematical symbols with familiarity. But the physicist should put himself on guard and remember that the same real trajectory of a moving body can have for a logically correct representation a multiplicity of traces, all different among themselves and not necessarily congruent to the actual form of the trajectory; all depends on the choice of co-ordinates.

It would seem evident, then, that 'physical time' is none other than 'active physical space', that is, the enumeration of displacements that have occurred in that space. If we do not postulate the existence of a universal physical space, we cannot conceive of a universal time. Physical space is the absolute reference for motion and thus for time, without the addition of any relative co-ordinates whatever. As Minkowski wrote in his memoir:

> The equations of Newtonian mechanics present a double invariance: the first case, in which their forms remain invariant, is when a system of spacial co-ordinates has undergone any displacement whatever; the second is when (such a system) becomes modified in its state of motion, that is, when there is impressed on it any other uniform motion of translation; the origin of the times does not

have any bearing. It is usual to consider the axioms of geometry as established when one feels ready for the axioms of mechanics, and therefore these two invariances are rarely named together. Each one of them represents a certain group of transformations in themselves through the differential equations of mechanics. The existence of the 1st group is regarded as a fundamental characteristic of space. The 2nd group is usually treated contemptuously or passed by lightly on the grounds that from the physical phenomena one cannot decide whether the space, initially supposed to be at rest, is actually in uniform translation.

Thus these two groups have an existence totally different from one another. Perhaps their character of complete heterogeneity has deterred bringing them together. But in the total ensemble, considered as a whole, there is material for consideration.[5]

Here location and motion are the prime objects of our interest; time plays no independent rôle.

Moreover, it should be clear that the concept of time alone is insufficient to furnish us with a metric criterion of continuity or uniformity of movement. Measures themselves are discontinuous and could never justify in themselves the idea of continuous motions, whether uniform or variable. To have continuity we must resort to an assumption from outside the discontinuous measures of changing things. If we postulate the continuity of space, we thereby gain continuous motion and also time by derivation; there is no need to postulate an independent time continuum.

If time is a *sine qua non* for the representation of the external world, it is so only in the sense of subjective or psychological time. The representation comes from the projection of external phenomena on to our memory screen. But when we follow, analyse and interrogate the elements of that picture, we do so instantaneously and fleetingly, without time. We follow spatial dislocations, not an objective time. There is no real counterpart of this

[5] H. Minkowski, op. cit., p. 335: 'Le equazioni della meccanica newtoniana presentano una doppia invarianza. Un primo caso in cui rimane invariata la loro forma è quando si sottopone ad uno spostamento qualunque il sistema fondamentale di coordinate dello spazio; il secondo quando esso viene modificato nel suo stato di moto, cioè quando gli si imprimeun qu alsiasi moto uniforme di transazione, inoltre l'origine dei tempi non ha alcuna influenza. Si è soliti di considerare come stabiliti gli assiomi della geometria quando ci si sente maturi per gli assiomi della meccanica, e perciò quelle due invarianze vengono ben di rado nominate insieme. Ciascuna di esse rappresenta un certo gruppo di trasformazioni in se stesse per le equazioni differenziali della meccanica. L'esistenza del primo gruppo si riguarda come un carattere fondamentale dello spazio. Il secondo gruppo si suole punirlo col disprezzo, sorvolando con leggerezza sul fatto che, dai fenomeni fisici non si può mai decidere se lo spazio supposto dapprima come in quiete si trovi alla fine in una transazione uniforme. Così quei due gruppi conducono una esistenza totalmente separata l'uno dall'altro. Il loro carattere del tutto eterogeneo ha forse distolto dal comporli. Ma appunto il gruppo totale composto come un tutto ci dà materia a considerazione.'

psychological time of our receiving screen in the outside world, the objective source of the projection. If elements of reality have their exact counterparts on the screen, the reverse is not time. There is no externally real element which would function as time independently of motion.

It would seem, then, that space, matter and motion are the sole entities necessary and sufficient for empirical reality, even if for subjective representation they need to be ordered in psychological time. Perhaps I could rephrase the comment of Minkowski noted above—

Time in itself must fall into darkness, because it cannot claim any right to real physical existence, and in the place which it has till now unjustly occupied we must put movement, the only entity that has the right to preside over the markers of succession.

XIV

A NOTE ON HARVEY'S 'EGG' AS PANDORA'S 'BOX'

I. Bernard Cohen

I

William Harvey's *De Generatione Animalium*[1] contains a striking frontispiece (Fig. XIV.1). On a large pedestal, the figure of Jove may be seen, accompanied by an eagle. In each of his hands, he holds one part of a large bisected eggshell, opened as if it were a box, from which a cloud of living things emerges: including animals, plants, large insects, etc., and at least one human figure. On the upper part of the eggshell, there may be plainly seen the word *Ex*, and on the lower the words *ovo omnia*, together forming the motto: *Ex ovo omnia*.[2] These words are noteworthy, as Joseph Needham has remarked, since they embody 'a conception which Harvey is continually expounding (see especially the chapter, 'That an egg is the common original of all animals'), but which he never puts into epigrammatic form in his text.' Hence, Needham quite correctly concludes, 'the saying, *Omne vivum ex ovo*, often attributed to him, is only obliquely his.'[3]

[1] See Geoffrey Keynes, *A Bibliography of the Writings of Dr William Harvey, 1573–1657* (2nd edition revised, Cambridge, University Press, 1953), pp. 45 ff.

[2] According to Keynes (op. cit., p. 52), in the original London edition 4° ('Typis Du-Gardianis; impensis Octaviani Pulleyn in Cœmeterio Paulino. M.DC.LI.') there is a 'representation of Jove seated on a pedestal and holding in his hands an egg inscribed: *ex ovo omnia*. He has lifted the upper half of the egg with his right hand and animals, insects, and plants are springing out of the lower half. At his side is an eagle with thunderbolts. A pediment below is inscribed: *Gulielmus Harveus/de/Generatione Animalium*. In the background is a landscape with buildings. The plate-mark measures 21 × 16 cm.' (A minor correction: the egg is inscribed *Ex* . . . and not *ex* . . . .)

For a description of the forms of life emerging from the egg, see n. 37 below. There may be identified a bird and a stag, a fish, a lizard, a snake, a spider and grasshopper, a butterfly, and a human figure. Additionally, there appear to be one or two plants hanging from the lower half of the egg shell.

[3] Joseph Needham, *A History of Embryology* (Cambridge, University Press, 1934), pp. 112–13; and second edition (1959), p. 133.

The first edition of this work (in Latin) was published in London by
Pulleyn in 1651, and an English version followed—without the frontispiece
—two years later.[4] Three other editions or printings of the Latin *editio
princeps*, all dated in the same year, 1651, contain slightly variant versions of
the original frontispiece; in all three Jove is seated, holding one half of the
bisected eggshell in each hand, and the various forms of life are let loose
into the world. In only two of the three does the eggshell contain the motto,
Ex ovo omnia.[5]

Although Harvey's book has been the subject of study by such men as
F. J. Cole, H. P. Bayon, Sir Henry Dale, Adolph Meyer, Joseph Needham,
and Walter Pagel,[6] and others, the frontispiece has not been examined
notably—save in relation to the motto *Ex ovo omnia*. Some consideration has,
of course, been given (for instance by Cole, Meyer, and Needham) to the
degree to which the allegorical representation is or is not truly expressive of
Harvey's thesis. But I have not been able to find any discussions of two
themes, each of general interest. The first is the iconographic origin of the
personage opening an egg from which a variety of living things fly out.
The second is the possible reason for the use of this particular representation
in Harvey's book.

The purpose of the present note is primarily to suggest that the origin—
conscious or unconscious—of the frontispiece may be found in artists'
renderings of a variant form of the myth of Pandora and her box. In particular, attention is called to a printer's mark of the sixteenth century,

[4] For this version (8⁰, 'Printed by James Young, for Octavian Pulleyn, and are to be
sold at his Shop at the Sign of the Rose in St. Pauls Church-yard. 1653.'), see Keynes,
op. cit., no. 43. The frontispiece is now an engraving 'by [William] Faithorne of a bust
of the author'.

[5] In the Elzevir edition (12⁰, Amsterdam, 1651, 2 issues), according to Keynes (op. cit.,
nos. 35, 36), there are some alterations in detail in the frontispiece, which is reduced:
'Jove is seated on a pedestal beneath a classical archway and has the same attributes as
before. At the foot of the pedestal on either side is a bird sitting on a basket of eggs. The
top corners of the archway are ornamented with eggs in the act of hatching.' In another
Amsterdam edition of 1651 (12⁰, 'Apud Joannem Janssonium. CIƆ IƆ C LI'.), according to
Keynes (no. 37), the frontispiece—reduced, as in the Elzevir editions—varies as follows:
'Above Jove's head hangs a curtain, and the egg in his hands bears no inscription;
otherwise the attributes are as in the first edition'. In the third Amsterdam edition of
1651 (12⁰, 'Apud Ioannem Ravesteynium'), according to Keynes (no. 38), 'The engraving
is a partial copy of those in the Elzevir and Jansson editions of the same year. Jove is
seated on a pedestal with a curtain overhanging him on his right and with three pillars
beyond him on his left. Between the pillars is seen a distant landscape and a crocodile.
On the pedestal is inscribed the title given above and the publisher's imprint is at the
bottom of the plate. On the ground on either side of the pedestal are birds on baskets of
eggs with numerous hatching chicks and reptiles and a tortoise scattered beyond them'.

[6] A brief discussion of the comments on the frontispiece and its motto by these scholars
(and some other writers on Harvey) is given in note 37 *infra*.

Figure XIV.1 Frontispiece to Harvey's *De Generatione Animalium* (courtesy of the Francis A. Countway Library, Boston, Mass.).

Figure XIV.2 (a, b) Gourbin's printer's mark, from the title page of Ononce Finé's *Practique de la Geometrie* (courtesy of the Houghton Library, Harvard University).

PRACTIQVE
DE LA GEOME-
TRIE D'ORONCE, PROFES-
feur du Roy és Mathematiques , en laquelle
eſt comprins l'vſage du Quarré Geometrique,
& de pluſieurs autres inſtrumens ſeruans à meſ-
me effeĉt : Enſemble la maniere de bien meſu-
rer toutes ſortes de plans & quantitez corporel-
les : Auec les figures & demonſtrations.

Reueuë & traduiĉte par Pierre Forcadel , le-
ĉteur du Roy és Mathematiques.

A M. le Duc de Guyſe.

A PARIS,

Chez Gilles Gourbin , à l'enſeigne de l'Eſperance,
deuant le college de Cambray.

M. D. LXXXVI.

showing a spheroidal box or casket of about the same size (relative to the personage holding it) as the egg in the Harveian frontispiece; it too is bisected so that the upper half is held in one hand and the lower half in the other. This printer's mark displays Pandora and her 'box', from which there emerges—as from the Harveian Jove's egg—a cloud of living creatures. These similarities suggest more than a mere coincidence of motifs; they hint at the possibility that either Harvey or the artist who designed the frontispiece might have seen this printer's mark or could have drawn inspiration from (and used actual elements of design taken from) the tradition out of which this mark emerged. Since the mark was used by a well-known printer of medical and other scientific books, Gilles Gourbin, it is not far-fetched to suppose that Harvey or his illustrator may have been familiar with this representation of a cloud of living creatures emerging from an opened spheroidal box, laterally bisected. We shall see below that a shift from Pandora to Jove (and hence from 'Pandora's box' to Harvey's egg in Jove's hand) is easy to conceive, since at that time there was a common association of the images of Jove and his jar or jars and Pandora and her 'box'.

A second suggestion, to be made below, consequent upon the proposed identification of the source of the subject of the frontispiece, is that the particular representation of Jove and the egg may have a sense understood at that time and not now. And thus the present inquiry may be of more than a purely iconographic interest. It is especially appropriate for a comment on this particular subject to appear in a volume in honour of Joseph Needham, since his writings contain a most profound and perceptive analysis of Harvey's ideas on generation and the motto, *Ex ovo omnia*.[7]

II

As to Pandora and her box, the versions in general circulation go far beyond the original sources. The legend of Pandora has been studied by the late Dora and Erwin Panofsky,[8] who endorse the suggestion that 'the original

[7] In his bibliographical preface to the account of the edition of *De Generatione Animalium*, Keynes (op. cit., p. 47) found Needham's summary of Harvey's views so clear and sound that he quoted *in extenso* the ten-point 'titles to remembrance' from Needham's *History of Embryology* (op. cit., pp. 129–30).

[8] Dora and Erwin Panofsky: *Pandora's Box: the changing aspects of a mythical symbol* (2nd edition, revised, New York, Harper & Row [Harper Torchbooks, The Bollingen Library], 1965)—an 'exact reproduction' of the 2nd edition (Princeton University Press) of 1962 (1st edition 1956) with a correction and a new preface, and with all the illustrations.

The information which follows concerning Pandora and Epimetheus is based on the Panofskys' book, but presented so as to have a possible relevance to Harvey's *De Generatione Animalium*, which is not mentioned by the Panofskys.

sense of the myth' may be more accurately presented in the Fifty-eighth Fable of Babrius than in 'the version forced upon posterity by Hesiod'.[9] According to Babrius, the famous 'vessel contains goods rather than evils' and 'man as such ($\check{\alpha}\nu\theta\rho\omega\pi\sigma$) takes the place of Pandora'. In the more usual accounts, deriving from Hesiod,[10] Pandora—whether formed of earth and water by Prometheus, or by Hephaestus (at the instigation of Zeus)—is supposed to have received gifts from all the gods, and hence was named 'Pandora'. Taken as wife by Epimetheus, the brother of Prometheus, Pandora 'became the mother of all women'. The famous vessel (containing all the evils, according to Hesiod and others; but possibly containing all goods, according to Babrius) was said to have been opened by Pandora, whereupon all its contents save Hope flew out into the world. In early versions of this fable, the vessel itself was 'invariably designated as a $\pi\iota\theta\sigma$ (*dolium* in Latin), a huge earthenware storage jar used for the preservation of wine, oil, and other provisions, and often large enough to serve as a receptacle for the dead or, later on, a shelter for the living'.[11] With the sole exception of Babrius, the motive for opening the box is not specified.[12]

The frontispiece to Harvey's *De Generatione Animalium*, showing Jove opening a small casket (or box) in the shape of an egg—as large as the egg of an ostrich—will hardly seem to have been inspired by the concept of 'the mother of all women' removing the heavy lid of a huge[13] earthenware vessel, a *pithos* or *dolium*. But in the sixteenth century, as a result of a 'philological accident', *pyxis* was substituted for *pithos*. This transformation to 'box' has been pinpointed for us by the Panofskys as due to 'none other than Erasmus of Rotterdam, and the pulpit from which he preached his heresy was his *Adagiorum chiliades tres* (first edition, 1508), one of the world's

[9] Op. cit., pp. 6 ff.

[10] The relevant texts of Hesiod's *Works and days* (lines 57–101) and *Theogony* (lines 570–90) are given in the original Greek and in English translation by the Panofskys, pp. 4–6.

[11] Op. cit., pp. 7–8.

[12] The Babrius version, quoted in English translation by the Panofskys (p. 8) reads: 'Zeus assembled all the goods in the vessel and gave it sealed to man; but man, unable to restrain his eagerness to know, said, "What in the world can be inside?" And, lifting the lid, he set them free to return to the houses of the gods and to fly thither, thus fleeing heavenwards from the earth. Hope alone remained.' The Panofskys observe that this version is just as logical as Hesiod's is not. With Babrius, that which is inside the vessel (the goods) remain available to man, while that which escapes from it is lost to him; by opening it, he forfeits the goods but retains hope. With Hesiod, only that which escapes from the vessel (the evils) gains power over man; so that hope, remaining within, could not have any influence on him. This is obviously the opposite of what Hesiod meant to convey, no matter whether he conceived of hope as evil, good, or neutral.'

[13] The 'very lid that prevents Hope from escaping is described as "big" '.

most popular and influential books'.[14] Furthermore, a close grammatical analysis of Erasmus's Latin text indicates the possibility that Erasmus supposed that Epimetheus, and not Pandora, had opened the 'box'.[15] This appears to be the source of 'the otherwise inexplicable fact that several poets and artists from the sixteenth to the nineteenth century depict Epimetheus opening Pandora's box'.[16] And, indeed, from this 'box', a small receptacle, 'now shaped like a box in the proper sense of the word, now like a largish pomander, now like a little vase (as in Raphael's famous fresco in the Villa Farnesina and its derivatives, among them a beautiful bronze group by Adriaen de Vries. . .), now like a goblet',[17] it is a simple step to an egg-shaped vessel in two halves, one of which is held in each hand of Jove in the frontispiece to Harvey's *De Generatione Animalium*. Furthermore, it is easier to envisage the artist introducing Jove as opener of the egg-shaped box as a replacement for the male figure Epimetheus, than for the female Pandora.

III

In the frontispiece to Harvey's *De Generatione Animalium*, the egg is opened as a box—not broken, as would be the case when a bird is actually hatched and emerges from a shell. The eggshell thus appears as an egg-shaped box, like the curio-boxes of more recent times made from ostrich eggshells sawed in half. The artist's representation of the creation of all life *ex ovo*, in an allegorical association with Epimetheus and Pandora, does not appear overly strained when we recall that not only was Pandora considered 'the mother of all women', but—as the Panofskys have shown—Pandora often tended

[14] Op. cit., p. 15. The story of Pandora occurs twice in this work, and each time the *pithos* is transformed into a *pyxis*.

[15] 'Moreover, Erasmus' narrative could seem to imply, by virtue of a grammatical ambiguity, which has been indicated in our translation, that it was Epimetheus rather than Pandora who committed the actual offence.' Op. cit., p. 17.

[16] Op. cit., p. 17. Chapter VII (pp. 79 ff.) is devoted to sixteenth- and seventeenth-century representations of the myth, in which the box is opened by a male and not a female figure, but not in each case Epimetheus.

[17] Op. cit., p. 21. The Panofskys suggest (pp. 18–19) that Erasmus may have 'fused— or confused—the crucial episode in the life of Hesiod's Pandora with its near duplicate in Roman literature, the last and equally crucial episode in the life of Apuleius' Psyche.

'Having acquitted herself of three other seemingly impossible tasks imposed upon her by Venus, Psyche is finally handed a pyxis, which she has to carry down to Hades and to bring back, filled with "a little bit [*modicum*] of Persephone's beauty". Psyche succeeds, against all odds. . . . But on her way back—and this is a motif obviously borrowed from the myth of Pandora—she cannot resist the temptation of opening' the 'sealed' *pyxis* she had been given by Persephone.

to be equated with Eve and, hence, Epimetheus with Adam. For example, a rather remarkable pair of wood statuettes in the Prado, formerly in the collection of the Conde de las Infantas, attributed to El Greco, show Epimetheus and Pandora as an Adam and Eve, and are 'obviously inspired by Dürer's famous engraving'.[18] Significantly, in this pair it is Epimetheus and not Pandora who holds the small casket or box in his hand. Relative to Epimetheus, this box is of about the same size as Harvey's egg is relative to Jove.

One or two further points may be made. First, in at least one of the artistic representations of the sixteenth century, by Giulio Bonasone,[19] the word 'SPES' is actually printed on the *pithos* or *dolium*; thus the placement of words on the egg-shaped box in the preface to Harvey's book has a counter-part in the Pandora-Epimetheus tradition. Second, some further confusion appears to have existed in relation to Pandora's 'box' (and Epimetheus) and a pair of vessels at the Gates of Jupiter, which are represented as 'flanked by two gigantic vases inscribed *Boni* and *Mali*'. The theme is derived from Homer's *Iliad, xxiv,* 527, and 'became extremely popular in emblem books' of the sixteenth and the early seventeenth centuries.[20] Jupiter is shown between two vessels, marked 'BONI' and 'MALI' or 'ΚΑΛΟΝ' and 'ΚΑΚΟΝ', and in one representation he is actually seated on a throne-like cloud (reminiscent of the frontispiece to Harvey's book).[21] There is even a representation (by Giulio Bonasone) in which there are two main figures—one for each 'pithos' —Jupiter, together with his eagle, alongside the 'good', and Saturn along-side the 'evil'.[22] The Panofskys adduce evidence for the correlation of 'Jupiter's two pithoi, one containing good and the other evil, with Pandora's

[18] Op. cit., Addenda to the Second Edition, 17, pp. 150 ff. These two figures were obtained by the Prado after the book had been written; they may now be seen on exhibition among the El Greco paintings.

[19] This is shown as Fig. 39 in the Panofskys' book. The figure in question, opening the vessel, is male.

[20] Op. cit., pp. 48–54. See especially the discussion of the fresco of Rosso. The Homeric lines in question, cited by the Panofskys in a seventeenth-century English translation, read:

'Two Tuns with Lots stand at Jove's Pallas Gates,
From whence he draws our good and evil Fates.'

The Panofskys (p. 50, n. 24) leave no doubt that despite the antiquity of the correlation between Jupiter's two *pithoi* and Pandora's single *pithos*, this correlation is 'frowned upon by modern scholarship'. But see n. 34 *infra*.

Of course, in Homer there is no direct reference to gates as such (in the passage trans-lated by these two lines).

[21] Op. cit., Figs. 25, 26, 27.

[22] Fig. 27; here we may see Jove portrayed with his eagle and a lettered vessel—three components of the frontispiece to Harvey's book.

one pithos, containing evil alone' going way back to antiquity,[23] and especially as a theme for the sixteenth and seventeenth centuries.

IV

Let me summarize the above results of interpreting the findings of the Panofskys in the light of Harvey's frontispiece. We see that in the late sixteenth and early seventeenth centuries there were current on the Continent a number of associated iconographic elements which appear in the frontispiece, of which the major ones are:

(1) the *pithos* being held by Pandora (or Epimetheus); we note the change from *dolium* or *pithos* to *pyxis*, a vessel of the size (and even roughly the shape) of the egg-casket in the frontispiece,

(2) the representation of a mass of living creatures streaming out of the opened box or casket in a cloud (of about the same shape and size as the cloud in the frontispiece),

(3) the male figure: we have seen that it was not uncommon to substitute Epimetheus for Pandora (so that the use of a male figure in the frontispiece is not bizarre),[24]

(4) Jove–Pandora: there was a correlation of Pandora and her box with Jove and his two *pithoi*[25] (and hence it would not have been an extraordinary feat of the imagination to have substituted Jove or Zeus for Pandora [Epimetheus]),

(5) the representation of Jove on a throne-like cloud with his eagle and a jar with a label (thus requiring only the substitution of the box of the associated figure of Pandora–Epimetheus, together with a cloud of creatures, in order to achieve the frontispiece).

Even if we grant that the substitution of a male personage for Pandora (3, 4, above) is not far-fetched, and that the seated figure on the throne in Harvey's frontispiece may resemble Jove on a cloud with eagle and jar (5, above), there remains the question of the vessel itself. We have seen that it was common enough practice to show a small vessel (1, above) with a cloud issuing from it (2, above), but this is hardly a bisected egg-shell. In the early seventeenth century, however, there appeared a number of representations of Pandora in which the box became a spherical casket, or even a spheroidal casket (both oblate and prolate), bisected just like Jove's egg in

[23] Op. cit., p. 49, n. 23, and pp. 50–2.
[24] Since Epimetheus was associated with Adam, just as Pandora was with Eve, this male Adam-figure would obviously be an appropriate symbol for a book on generation.
[25] For further evidence on this point, see note 34 below.

Harvey's frontispiece. In an etching made about 1625 by the Frenchman Jacques Callot, called *La Création de Pandore*, a female central figure holds such an ovoid casket at eye level, contemplating it 'as if in doubt whether she should open it'.[26] In Michel de Marolles' *Tableaux du Temple des Muses tirez du cabinet de feu M. Favereau* (published in Paris in 1655), there is an engraving by Cornelis Bloemaert, showing Pandora holding the box against the lower part of her body; it is now a spherical casket, bisected just like Harvey's egg. This engraving follows a drawing by Abraham van Diepenbeeck. The Callot etching of 1625 was thus available in 1651, when Harvey's book was published, but Bloemaert's engraving did not appear until 1655, and so is of interest here only as an example of the spread of the figure of Pandora with a small spherical box, of the same size relative to the main figure as Jove's egg, and bisected in exactly the same manner.[27]

V

There was, furthermore, an additional source of this representation which was available in Harvey's day. It goes back to a pen drawing made by Il Rosso Fiorentino (Giovanni Battista de' Rossi, 1494–1541), now in the École des Beaux-Arts in Paris, and probably drawn after 1534 or 1536. In this drawing, Pandora has just lifted the lid off a small 'round, flat, metal box—the Erasmian pyxis, here making its first appearance in art'.[28] This drawing shows a blinding flash of light spreading out from the opened box, and the evils are full-scale human-like figures, representing the Seven Deadly Sins, and so directly related to Christian theology. There is no direct representation of Hope, but there is a limp crow hanging from the rim of the box.[29]

Il Rosso seems to have been alone in representing the evils as the Christian Vices, showing these evils as full-scale figures or as human types. At least one of his followers rather used 'diminutive demons' which, as they escape

[26] See the reproduction in the Panofskys' book, fig. 33. In size and shape it resembles the egg of Jove in Harvey's frontispiece.

[27] Bloemaert's engraving and van Diepenbeeck's drawing are reproduced by the Panofskys, Figs. 35 and 36. The existence of the Bloemaert engraving, published in the same decade as Harvey's *De Generatione Animalium*, shows the currency at that time of the concept of a spherical (or spheroidal) bisected box in association with Pandora.

[28] Quoted from the Panofskys' book, p. 34. The drawing is reproduced as Fig. 16.

[29] On the crow motif, see the Panofskys' account on pp. 28–33, 140–1. Presumably, this 'harsh-voiced bird' was connected with Hope and Pandora because it 'always holds out a promise' with its *caw-caw* or 'Cras, cras' ('Tomorrow, tomorrow').

from the small receptacle, swarm 'about like so many mosquitoes'.[30] And it is this very swarm of diverse but diminutive creatures, streaming out from the box as a cloud, that interests us, because of the great similarity to the swarm of creatures in Harvey's egg: this motif was widely diffused in Harvey's day.

This is, in fact, the representation that occurs in a French printer's mark, used by Gilles Gourbin (or Gorbin), who was active as a printer in Paris from 1555 to 1587. The use of Rosso's Pandora was symbolic for him of Hope, and he selected as his motto the phrase SPES SOLA REMANSIT INTUS, from Niccolò della Valle's translation of Hesiod's account of Pandora. It was an appropriate device, since he worked out of a house near the Collège de Cambrai, named *À l'Espérance*.[31]

Gilles Gourbin's printer's mark occurs in (at least) four major versions;[32] two are rather large woodcuts, the other two small; all have the above phrase in Latin. In these there are different versions of the landscape background, and two different representations of Pandora. In one she holds an elongated vase-shaped or pitcher-shaped box, but in the other she holds a box which has the shape of an oblate spheroid, standing on a small pedestal. As may be seen in Fig. 2, the bottom half of this spheroidal box is held in the left hand and the top half (or cover) is held in the right hand. Various creatures emerge in a cloud from the opened spheroidal box, and a crow[33] stands within the box—plainly suggesting a bird hatching out of its egg. This portion of the printer's mark is strikingly similar to the representation in the frontispiece to Harvey's book.

We may observe here the way in which the numbered iconographic elements (listed above) are presented. Pandora's *pithos* has plainly become —thanks to Erasmus—a *pyxis*, and is no longer 'a big, practically immovable storage jar', but rather 'a small, portable vessel' of spheroidal shape, laterally bisected in the manner of the Harveian egg. The bird (crow) makes the suggestion of an egg even stronger. From this *pithos* emerges a cloud of varied forms of life. But of course these are monstrous, and not the plants,

---

[30] Rosso, therefore, according to the Panofskys (p. 34), 'had the courage to represent the Vices as full-scale human figures' and in 'contrast to all his followers' he saw that a swarm of diminutive figures 'are apt to strike the beholder as a nuisance rather than a tragedy'.

[31] On Gourbin and his printer's mark, there are works by L. C. Silvestre and P. Delalain; and a valuable study by Renouard, cited in the Supplementary Note to the present article. The Panofskys refer to Silvestre and Delalain, but not Renouard. They observe that Delalain (p. 40) 'correctly' explains 'Gourbin's reasons for adopting Pandora as his printer's mark' but add that he had mistaken 'the crow for a dove'.

[32] The Panofskys, Renouard, and others cite only three versions, but there are at least four; see Supplementary Note at the end of this article.

[33] See note 29 above.

animals, insects, and man himself that we see emerging from the Harveian egg. The artist would have had to use his creative imagination to get from one to the other.

If the artist who designed the frontispiece to Harvey's book had been familiar with the versions of the Pandora legend then current, he would have been aware of the correlation of Pandora with Eve, which might well have suggested a transformation of the Pandora 'box' into an egg from which all life comes (*Ex ovo omnia*). It may be noted that in Gilles Gourbin's printer's mark (especially the version reproduced in Fig. 2), the central figure is not very feminine—the apparent baldness is indeed more suggestive of a male than a female. In any event, since often it had been Epimetheus, rather than Pandora, holding and even opening the vessel, the artist could without difficulty have substituted in his mind's eye the male Epimetheus for the female Pandora, which would have made a shift to Jove much easier, since it would not have entailed a change in sex. If this were the train of suggestion, the then-current correlation of Epimetheus and Adam would (as mentioned above) at once further suggest generation—the subject of Harvey's book. The artist represented Jove in a seated position on his throne, accompanied by his eagle. And this transformation, too, would have readily come to mind, since (as we have seen) there was current a correlation of Pandora and her 'box' with Jove and his two vessels, Jove being accompanied by his eagle, and having one or two vessels with lettering.[34]

VI

The attempts to reconstruct the creative processes of the artist who designed the frontispiece to Harvey's *De Generatione Animalium* are necessarily inconclusive, since we do not know either who made the drawing and engraving, or even whether they were the same individual. Geoffrey Keynes has made the following suggestion:

Harvey's *De Generatione Animalium*, clearly a work of very great importance, was first printed and published in London in 1651. The first edition is a handsome quarto with a curious allegorical frontispiece representing Jove seated on

---

[34] The Panofskys (pp. 52–3) quote a poem of Joachim du Bellay (first printed in 1549), in which he 'automatically associated Pandora's box with the two vessels in front of the *Ostium Iovis*', or Gate of Jupiter. The Panofskys point out that 'the motif of Jupiter's pithoi was current in France as late as the eighteenth century and was still automatically associated with the Pandora myth'. A sample of the evidence in support of this assertion is a letter from Voltaire to J-B. de Laborde. See the Panofskys' note 29, p. 53, and note 6, p. 120; they observe that 'du Bellay follows, of course, Babrius'. On the Gates, see n. 20 *supra*.

a pedestal and holding in his hand an egg inscribed *ex ovo omnia* from which are springing animals, insects and plants. The plate is anonymous as regards both designer and engraver, but the name of Richard Gaywood suggests itself as that of a contemporary craftsman who might well be the author of this plate. Gaywood was a pupil of Hollar, the well-known Czech artist who was closely associated with Harvey, both having accompanied Lord Arundel on a diplomatic mission to Vienna in 1636. His name has also arisen in connexion with an etched portrait of Harvey in his old age which has been attributed to Hollar, but which has seemed to me, on the grounds of style and competence, to be more likely to be the work of Gaywood. Furthermore, it has recently been discovered that this plate was done from the life and was intended to be inserted in the first edition of *De Generatione*. This information is derived from a letter from Dr Jasper Needham, F.R.S., to John Evelyn, dated Covent Garden, 5 April 1649, as follows: 'Dr Harvey's picture is etcht by a friend of mine and should have been added to his work, but that resolution altred: however I'l send you a proof with your book that you may bind it up with his book *De Generatione*. I'm sure 'tis exactly like him, for I saw him sit for it.' Both the letter and the book with the etching inserted are still in the Evelyn collection formerly at Wotton and how housed at Christ Church, Oxford. The portrait, reproduced here from an impression in the British Museum, is unflattering, but appears to be an honest and lifelike representation of Harvey in 1649. The attribution of both etchings to Gaywood must remain for the present conjectural.[35]

Clearly, Hollar would have been in touch with the Continental traditions of emblems and representations of Pandora, Epimetheus, and Jove (with the vessels). There is no evidence available as to whether Harvey himself 'was in any way responsible for the frontispiece',[36] as Arthur Meyer reminded us. But it is not likely that he would have permitted a frontispiece of which he did not approve.

A possible means of substantiating the suggestions I have made would be to see whether anything in the life or works of Gaywood or Hollar shows an acquaintance with the emblems, devices, or artistic representations I have mentioned. I believe, however, that anyone who looks carefully at the illustrations in the Panofskys' book cannot help being convinced that

[35] Keynes: *Bibliography* (cited in n. 1 *supra*), p. 48.
[36] In his *An Analysis of the* DE GENERATIONE ANIMALIUM *of William Harvey* (Stanford, Stanford University Press, 1936), p. 73. This whole statement reads:

'It will be recalled that the phrase *ex ovo omnia* is found on the bisected eggshell in the hands of Jupiter in the frontispieces of the four editions of Harvey's *De Generatione* which appeared in 1651. As far as I know, the origin of these illustrations remains unaccounted for, and since this large treatise is otherwise wholly without illustrations, and apparently purposely so, and especially since the *De Motu* contains only one plate of four modified illustrations from the *De Venarum Ostiolis* (1604) of Fabricius, it seems highly unlikely that Harvey was in any way responsible for the frontispieces, although he may have approved their use.'

C.P.H.S.—9*

Gaywood (or whoever did the frontispiece) was in contact with, and influenced by, the Pandora–Epimetheus tradition.

VII

These notes may perhaps indicate no more than that the frontispiece to Harvey's *De Generatione Animalium* has elements in common with the story of Pandora (or Epimetheus) and the 'box', notably as seen in pictorial representations of the sixteenth and seventeenth centuries. I would suggest, however, that if this interpretation has any validity, it would imply that this frontispiece may have a significance not otherwise detectable. For if the figure of Jove in Harvey's book were intended as an artistic fusion of Jove and Epimetheus–Pandora, then the allegory represented would be the one deriving from Babrius: a letting loose into the world of 'goods' and 'evils', in short, *omnia*. In this case, Harvey's egg would be a conjunction of Jupiter's two *pithoi* into one *pithos*, transformed into a single *pyxis*. This interpretation accords with the fact that everything ('all things', good and evil, *omnia*) would come out of the single *pyxis*; just as in the Harveian frontispiece, 'all things' (*omnia*) come out of the egg—not merely mammals, fish, reptiles, and birds—but even plants. This general concept of 'ovum' is in harmony with the *pyxis*, but not with a narrow interpretation of the 'ovum' in a present-day embryological sense[37] (i.e. the sense in which we speak of Karl

---

[37] In addition to Keynes' *Bibliography*, and the works of Needham and Meyer cited above, note may be taken of the following studies concerning (or related to) Harvey.

Kenneth J. Franklin, in his *William Harvey, Englishman, 1578–1657* (London, MacGibbon & Kee, 1961), pp. 112 ff., merely summarizes the successive views on *De Generatione Animalium* of Andrewes, Dale, Meyer, and Bayon, without contributing anything of his own, nor discussing the frontispiece. D'Arcy Power, in his *William Harvey* (London: T. Fisher Unwin ['Masters of Medicine'], 1897), pp. 89, 147–54, 238–63, alleges that Harvey's 'treatises on development are so full of detail that it is impossible to give an exact notion of their contents in a popular work' and refers to neither the frontispiece nor the motto, *Ex ovo omnia*. Geoffrey Keynes, in his *The Life of William Harvey* (Oxford, The Clarendon Press, 1966), devotes a whole chapter (38) to *De Generatione Animalium*, analysing the main features of this work in full. He devotes one paragraph to the frontispiece, as follows: 'The frontispiece of 1651 provides a rather undistinguished figure of Jove seated on a pedestal with his eagle beside him. He holds an egg from which he lifts the upper part with his right hand to allow the escape of a variety of animal forms, including a tiny human being, a bird, a stag, a fish, a lizard, a snake, a grasshopper, a butterfly, and a spider. On the two halves of the egg is the inscription: *Ex ovo omnia*, "everything from an egg". This legend must be presumed to have been approved by Harvey, although it is most unlikely that he had any hand in the design of the emblems. The earlier commentators accepted the legend as his dictum; in the middle of the eighteenth century, however, it was quoted as *Omne vivum ex ovo*, "every living thing comes out of an egg", and during the nineteenth century currency was given by many writers to this "famous

Ernst von Baer's discovery of the mammalian ovum), or even in any sense restricted to the subject of Harvey's book, on the generation of *animals* (i.e. not of plants, or of life in general).

VIII

Of course, it is always possible that the illustrator or Harvey conceived the egg of Jove without any knowledge of the current (or recent) representations of Pandora's box or of Jove and his vessel or vessels. I find it more likely that Harvey's egg in the hands of Jove was in large measure a fusion, transformation, and adaptation of existing designs rather than a radically new creation. The similarities I have pointed out are too great and too many to allow a high probability for this frontispiece to have been wholly independent of the Pandora tradition in post-Erasmian art.

Furthermore, there is a strong possibility that either Harvey or his illustrator might have been acquainted with Gourbin's printer's mark, and hence that the concept of the frontispiece was not derived independently

---

misquotation", which was sometimes repeated in the twentieth, even by the great medical historian Sigerist, who quoted it as *Omne animale ex ovo*. The misquotation gives a more precise meaning to the idea than Harvey ever intended.'

F. J. Cole, in his *Early Theories of Sexual Generation* (Oxford, The Clarendon Press, 1930), pp. 137–8, observed that 'Wahlbom (1746), in commenting on this passage says: *Harvaeus etiam, omne vivum ex ovo, olim exclamavit.* This appears to be the first use of the famous misquotation, and the first attribution of it to Harvey. It was not, however, adopted by other writers until later. The modern vogue of the error is perhaps due to Oken. . . .' The passage in question is from Linnaeus, who 'in his *Fundamenta Botanica* of 1736, uses the expression *Omne vivum ex ovo provenire datur,* but does not associate it with Harvey'.

Walter Pagel, in *William Harvey's Biological Ideas: selected aspects and historical background* (Basle, New York, S. Karger, 1967), pp. 45, 274, 295, 314, while discussing fully Harvey's own views or doctrines concerning generation, barely mentions the frontispiece, only referring (p. 45) to 'the popular parlance of the frontispiece to Harvey's book: *Ex ovo omnia*'. H. P. Bayon: 'The lifework of William Harvey and modern medical progress: an essay to commemorate the tercentenary of the publication of *Exercitationes de Generatione Animalium*', *Proceedings of the Royal Society of Medicine*, 1951, **44**, pp. 213–18 (Section of the History of Medicine, pp. 13–18), does not refer to the frontispiece as such, but merely mentions that 'a further example of misunderstanding is the often repeated misquotation of the dictum: *Omne vivum ex ovo*—which Harvey did not state with such lapidary brevity'.

Sir Frederick W. Andrewes, in his *The Birth and Growth of Science in Medicine, being the Harveian Oration* . . . (London, Adlard and Son and West Newman, 1920) barely touches on *De Generatione Animalium*. Nor does Sir Henry Dale have much to say concerning this work in his *Some Epochs in Medical Research* (London, H. K. Lewis & Co., 1935). H. R. Spencer, in his *William Harvey, Obstetric Physician and Gynaecologist* (London, Harrison and Sons, 1921) and Elizabeth Gasking, *Investigations into Generation 1651–1828* (Baltimore: The Johns Hopkins Press, 1966), ch. 2, discuss Harvey's *De Generatione*; Spencer deals with the motto of the frontispiece, but not with the iconography.

from the same tradition that produced Rosso's drawing. We shall probably never know for certain about such matters. But I would suggest that Harvey could very well have seen a book printed by Gourbin, and he might even have had such a work in his own library.[38]

Gourbin was an active printer of scientific books, especially in the medical field. By referring to the catalogue of a single collection, I have been able to list 21 titles of medical books or pamphlets which appeared in Paris under his imprint between 1555 and 1587.[39] The authors were Hippocrates, Pierre de Gorris, Giovanni Battista da Monte, Giovanni Argenterio, Mesuë, Étienne Gourmelen, Remaclus Fusch, Thierry de Héry, Jean Martin, and Jacques Dubois [or Jacobus Sylvius, or Silvius]. Gourbin published at least ten works by Sylvius.

Harvey's own library has disappeared, 'presumably mostly lost when the College of Physicians was destroyed in the fire of London in 1666, though a few books are said to have been rescued by Dr. Christopher Merrett'. There appears to exist no catalogue of this library, nor even a partial inventory of its contents. Only three books are known today in copies that belonged to Harvey. One is a 'fabulous' book, Sylvius's *De febribus commentarius ex libris aliquot Hippocratis & Galeni* . . . Venice, 1555, 'which carries annotations by Fabricius ab Aquapendente, and the signatures of Harvey, dated 1621, and of Dr. Richard Mead. . . .' The other survivors are Falloppius's *Opera* (Frankfurt, 1584) and Galen's *Opuscula varia* (ed. Goulston: London, 1640).[40] Since Harvey owned one book by Sylvius, an important author and commentator, he might very well have possessed (and would certainly have seen or have read) still other works by Sylvius, possibly even one printed by Gourbin. Here then is a plausible link between Harvey and Pandora's box, a possible source of a central part of the frontispiece to *De Generatione Animalium*.[41]

---

[38] Another possibility to be considered is that the choice (or development) of the subject of the frontispiece may have been due (at least in some measure) to George Ent, who saw this work through the press. A physician of Dutch parentage, Ent would have easily had contact with the Continental tradition of Pandora (especially notable in Holland, see n. 27 *supra*), and might easily have been familiar with one or more medical books published by Gilles Gourbin.

[39] The catalogue of the U.S. National Library of Medicine (Bethesda, Md.) was chosen purely for convenience. See Supplementary Note at the end of the article.

[40] Keynes, *Bibliography* (cit. n. 1 *supra*), pp. vi–vii.

[41] I should like to thank Mr. Richard J. Wolfe, Rare Books Librarian in the Francis A. Countway Library of Medicine, for his kindness in helping me to identify, locate, and study the books printed by Gourbin which are in the Countway Library.

Supplementary Note on Gilles Gourbin

I have referred above to 21 medical works printed by Gilles Gourbin; these are listed in Richard J. Durling (compiler) *A catalogue of sixteenth century printed books in the* [*U.S.*] *National Library of Medicine* (Bethesda, Maryland: National Library of Medicine [U.S. Department of Health, Education, and Welfare: Public Health Service], 1967). I give below the dates, the names of authors or commentators, and the numbers identifying the works in the above-mentioned catalogue:

1555

2140: Pierre de Gorris
3259: Giovanni Battista da Monte

1556

3279: Giovanni Battista da Monte

1557

262.1: Giovanni Argenterio
2403: Hippocrates, commentary by Adrian Aleman

1561

1234: Jacques Dubois [= Jacobus Sylvius], commentary on Galen
1236: *idem*, commentary on Galen
1242: *idem*, commentary on Hippocrates & Galen
1252: *idem*, commentary
1255: *idem*, commentary on Hippocrates
1259: *idem*, commentary on Hippocrates & Galen (corr. Alexandre Arnaud)
1269: *idem*, commentary on Galen
1279: *idem*, commentary on Hippocrates & Galen
3146: Mesuë, int. by Dubois

1562

1246: Jacques Dubois

1566

2142: Étienne Gourmelen

1569

1729: Remaclus Fusch
2282: Thierry de Héry

1572

2141: Pierre de Gorris

1578

2974: Jean Martin

1587

1260: Jacques Dubois, commentary on Hippocrates & Galen.

Of these I have seen nine, all of which are at present in the Francis A. Countway Library of Medicine, Harvard Medical School, Boston. One of these, *1269* (1561), has a small printer's mark like that shown above in Fig. 2; another, *2282* (1569) has almost the same printer's mark, a variation occurring primarily in the landscape background. A larger printer's mark, in which the figure holds a vessel shaped more like an elongated vertical urn, occurs in *1242* (1561), *1259* (1561), *1236* (1561), *1234* (1561), and *1279* (1561): here the oval is enclosed in a rectangular frame. Yet another form of the oval within a rectangle is shown as Fig. 18 in the Panofskys' book: this woodcut differs in the frame, the landscape, the figure, and the sky—the 'box' is much like the one in our Fig. 2, but a larger number of emerging creatures may be distinguished.

Yet additional books printed by Gourbin are listed in Ph. Renouard: *Les marques typographiques parisiennes des XVe et XVIe siècles* (Paris: Librairie ancienne Honoré Champion, 1926), pp. 114–15. Renouard identifies four non-medical works, for each of which I give the author's name, date of publication, and catalogue number:

1556

375:   Oronce Finé

1561

376:   Joan. Sambucus

1567

377:   Ja. Carpentarii . . . oratio

1580

378:   Gilb. Genebrard

The work by Oronce Finé is entitled *La composition et usage du quarré geometrique* . . . and seems not to be the work from which our Fig. 2 is taken. It has, however, the same printer's mark.

In the Harvard College Library, there are three further books printed by Gourbin:

Oronce Finé: De re & praxi geometrica libri tres . . . (1556)
idem . . . (1586)

Oronce Finé: Practique de la geometrie . . . reveuë & traduicte par Pierre Forcadel . . . (1586) [the source of our Fig. 2]

The title-page of the first of these three works is shown (no. 230) in Ruth Mortimer: *Harvard College Library, Department of Printing and Graphic Arts, Catalogue of Books and Manuscripts. Part I, French 16th Century Books* (Cambridge, Mass.: The Belknap Press of Harvard University Press, 1964), vol. I.

XV

✿

THE ✡, ✦, AND ⬡, AND OTHER GEOMETRICAL AND SCIENTIFIC TALISMANS AND SYMBOLISMS

Derek de Solla Price

The unspeakability of the title of this piece is an attempt to exemplify its thesis. There exists a type of human mind to which the three symbols in the title speak without the intervention of words and in the absence of direct pictorial representation. Such non-representational iconography, it will be shown, forms a long and honourable figurate tradition. It is a fellow to the more familiar literate tradition, common to many cultures and subjects, and the numerate tradition which stands as a characteristic of the quantitative sciences. It is a vital component of the aesthetics of scientific theories, both ancient and modern, communicating a sense of interrelationships amongst a complex 'Gestalt' and embodying the principles and the results of theories based on such relationships.

Curiously enough the figurate tradition seems never to have been discussed in general although specific instances abound of descriptions of particular diagrams and their uses for magical or scientific purposes. A great deal of confusion arises from the circumstance that the preservation and transmission of the tradition has depended upon manuscript scribes and copyists who may have been amply competent in literate qualities but deficient in the numerate, as historians of astronomical tables know only too well, and in the figurate, as is also attested by many blanks in texts where the pictures should be. Even when such diagrams appear, they are often hopelessly garbled by being misunderstood and left uncorrected, and by veiling them in a secrecy appropriate to their valuable magical content as an embodiment of potent theoretical understanding. The consequence is that most understanding has vanished and the modern scholar is unable to develop a history which is more than a flat statement of instances of the various diagrams. Even then they appear to be little more than arbitrary emblems that appear and disappear through the pages of history as, for example, the well-known and surprisingly recent history of the six-pointed Star of David as a symbol of Judaism,[1] the five-pointed figure which attains significance as the pentacle of witches and the Pentagon Building in Washington, D.C., and such curious symbolisms as the forms of the alphabet letters.[2]

The fundamental quality of a geometric symbol of this sort is that it gives at a glance a reminder of a theory whose very elegance is displayed by the form of the lines. A trivial example can be found in the famous incident of the discovery of the forgotten tomb of Archimedes by Cicero in 75 B.C. when he was quaestor in Syracuse.[3] The tomb, unmarked by surviving literate description or name, bore as legend the simple diagram of a cylinder enveloping a sphere. As such it was immediately obvious to the educated discoverer as a depiction of the Archimedean rectification of the spherical surface—unquestionably the most powerfully elegant product of the methodology of Archimedes, and a precursor of the integral calculus. The whole method, the proof, and the results, were keyed to this non-obvious construction, whereby the sphere, whose surface area was to be found, was encased in a cylindrical surface that just touched it and could be compared with it, infinitesimal element by element. That diagram spoke for the scientific personality and achievement of only one man, Archimedes.

A more modern instance might be seen in a popular book by Nobel Laureate Chen Ning Yang[4] in which the content and the elegance of symmetry principles in the physics of fundamental particles is conveyed in terms of simple diagrams that 'speak louder than words'. Even the book jacket is a symbol of this sort; it reproduces one of the cleverest and most mind-bending illustrations by the modern Dutch artist M. C. Escher, showing a tessellated formation of mounted horsemen moving in a contrary direction. My point in citing this example is to explain that it is not only the content of modern theory in fundamental particle physics that requires the use of diagrams that would obviously and trivially show the same symmetry as the theory and indeed of Nature herself. The diagram goes beyond this in assuming a form of such inner elegance and economy whereby a few lines or simple forms imply a much greater amount of communication than could

[1] Gershom Scholem, 'The Curious History of the Six Pointed Star', *Commentary*, **8** (1949), pp. 243–51.

[2] S. Goudsmit, 'Symmetry of Symbols', *Nature*, 6 March 1937.

[3] Cicero, *Tusculanae Disputationes*, V, 23.

[4] Chen Ning Yang, *Elementary Particles, A Short History of Some Discoveries in Atomic Physics* (Princeton University Press, 1962).

otherwise be made. Indeed it would appear that the amount of symmetry and the ingeniousness of its interrelation is virtually an argument for the assumption that this particular theory or set of theories must be true. They must be true because they are so neat and so cleverly interwoven. We shall maintain furthermore that when a scientific theory has been developed on such a basis, the diagram tends to take on a life of its own, not just as a representation of the theory or as an *aide-mémoire*, but as a magical talisman and an object of contemplation and speculative philosophy.[5]

What we have here is an historically important principle of elegance which acts not just as an aesthetic criterion, but as a guide to philosophic truth of scientific theories. Everyone is familiar with the test of Occam's Razor; all other things being equal we should prefer the theory that is most simple, the one that involves least by way of assumptions and postulates. Now we have in addition to simplicity a second test that, all other things being equal, we shall prefer the theory which displays most of this elegance, this interlocking Gestalt which seems to force a feeling of necessity and can apparently in many cases only be conveyed in the figurate mode. There would seem to be many strands in the history of scientific thought where an obscure but powerful literate tradition is in fact just such a figurate mode; the obscurity creeps in only through the difficult process of attempting to translate (as I do now) from the figurate to the literate. It is perhaps worth noting that a similar difficulty seems to attend the translation of numerate to literate. The main threads of Greek mathematics are literate, but the Babylonian tradition is almost exclusively numerate in its very sophisticated armoury of higher mathematical astronomy.[6] Whenever historians of mathematics have sought to explain the ways of thought that seem to pervade Babylonian methods they are forced to rely on a method of communication which is that of the wrong blood-group. Babylonian astronomers seem to have thought of their theories in purely numerate terms like a stockbroker knowing the state of the market from the ticker-tape alone without the intervention of graphical methods or statements in words. It seems very likely that the obscure Pythagorean tradition of pre-Socratic Greek philosophy may in fact be at least partially due to a poor literate translation from the numerate astronomical science of the Babylonian contemporaries.

[5] For general history but little by way of rational explanation of derivation see:
 Sir E. A. Wallis Budge, *Amulets and Talismans* (University Books, New York, 1961).
 Jean Margues-Rivière, *Amulettes, Talismans et Pantacles* (Payot, Paris, 1950).
 Kurt Seligmann, *The History of Magic* (Pantheon Books, New York, 1948), pp. 154, 194, 296–9, 354, 355.
[6] See Derek J. de Solla Price, *Science Since Babylon* (Yale University Press, New Haven, 1961), Chapter 1.

It is also remarkable that the few Babylonian tablets containing figures seem to bear just the type of diagram we shall discuss in which the connected polygonal and star-shaped 'talismans' play a special rôle.

The ultimate foundation for this entire tradition in East and West seems to be the concepts of an Element Theory. What is at stake is not the predecessor of our modern chemical elements but rather a theory which relates the various forms of substances to all the forces and changes which may be wrought with them and upon them. Thus element theory contains the rationale of physics, astrology and alchemy, not just the nature of substance. In particular it should be noted that the concept of atoms is in a quite separate department in the history of ancient science. It seems to derive rather unexpectedly not from chemistry or physics at all but rather from a preoccupation with the discovery of mathematical irrationality. The easy proof that $\sqrt{2}$ could not be expressed 'rationally' as a number, p/q had a disastrous effect upon early logicians who were forced to conclude that the integral numbers caused a certain graininess of the universe and forced the abandonment of such intuitive devices as the use of similar triangles in geometrical argument. The style of Euclid's Elements is not so much a pedagogic device of inexorable logical steps, as a successful hunt for a way round the unfortunate hiatus of the irrationality of the real world and its 'mathematical atomicity'.

The concepts of elements then had nothing to do with atoms or other units of substances which could be mixed and compounded like medicinal or culinary ingredients. The element theory had to contain a rationale of forces or qualities which would change and transform one substance to another. The central concept of the four-element theory, the *tetrasomia*, was that the set of basic modalities of matter were produced by the working of two pairs of qualities that acted, so to speak, at cross-purposes to each other.[7] One pair consisted of the opposed qualities of hotness and coldness, the other of wetness and dryness, each set therefore containing a positive and a negative manifestation of a principle that seemed part of the essential character of all substances and all change.

From this central concept a whole theoretical structure could now be erected. The two pairs cross with each other to form the four possible combinations, the four elements of air, earth, fire and water, each of these terms being taken with the greatest of generality. Air is the symbol and support for all vapours and volatility, earth for solidity, water for all fluids and liquidity, water and earth are visible substances, air and fire invisible. The four elements are necessarily arranged by the crossed principles into a square in which each side corresponds to one of the four periodic exchanges which

[7] Serge Hutin, *A History of Alchemy* (Walker and Co., New York, 1962), p. 80.

together comprise a Platonic cycle;[8] fire condenses into air, air liquifies to water, water solidifies to earth, earth sublimates into fire. In the reverse order, fire condenses to earth, earth dissolves into water, water vaporizes into air, and air becomes rarified into fire again.

This doctrine of Aristotelian elements lends itself very easily and naturally to the geometrical symbolisms of figures composed of a cross or of crosses within squares, or of squares set diagonally within squares (see Fig. XV.1).

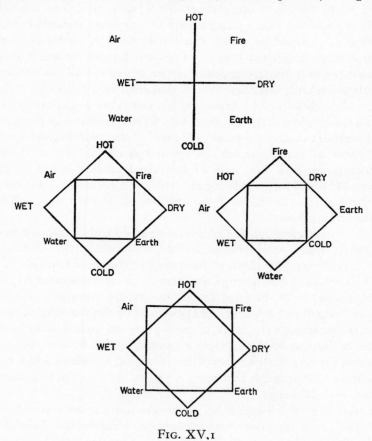

FIG. XV.1

The antiquity of the figures themselves is indisputably great, but at what period they become associated with element theory is a matter for conjecture. The square figures with diagonals are common decoration found in incised pattern and in tessellations in antiquity. One presumes that the

[8] Maurice P. Crosland, *Historical Studies in the Language of Chemistry* (Heinemann, London, 1962), p. 29.

Aristotelian text must have been illustrated originally with some such diagram, and of course innumerable versions exist from the later medieval and renaissance manuscripts. The whole issue takes on a new significance through the recent identification of the Tower of Winds, built in the Roman Agora of Athens by Andronicus Kyrrhestes *ca.* 50 B.C., as an architectural exemplification of the octagonal form of the symbolism resulting from the square-set-diagonally-within-a-square form of the element diagram.[9]

In the original archaeological examination it had been determined that this building, perhaps the only surviving classical structure known to have been designed by a mathematician, was an exercise in drawing-board geometry. The orientation along the meridian and a certain determination of the form were essential if the tower was to be used for mounting a wind-vane above, and a set of panels depicting the gods of the eight cardinal winds. Joseph Noble and the present author have been able to make a plausible reconstruction of the water-clock within the tower and show that the entire structure seems to be intended as a giant cosmic model rather than as a utilitarian combination of a timepiece and wind-vane. There seems good reason to suppose that the form of the building was intended to demonstrate that there must be indeed eight winds and not four or twelve as had otherwise been suggested by rival philosophers.[10] In the same spirit we suggested that the use of water to turn a sky disc (a star map in projection) behind an earth grillwork net, probably lit by flames of fire and decorated with playing fountains was all part of a symbolism of the elements.

Thanks to the publication of an account of the Tower of Winds written by a Turkish traveller in 1668 we are now able to confirm and extend this view on the symbolism and use it as a fixed point in the general history of this figurate mode of thought.[11] The traveller, Evliya Çelebi, though full of fanciful tales and dubious interpretations, indicates quite clearly that the tower also contained some sort of zodiac ceiling, now lost, depicting the twelve constellations and within them representations of the planets set in various named signs. The names all agree completely with the standard

[9] John V. Noble and Derek J. de Solla Price, 'The Water Clock in the Tower of Winds', *American Journal of Archaeology*, 72 (1968), pp. 345–55.

[10] Note especially that in Vitruvius I vi, 4, it is stated that Andronicus built the Tower at Athens as an exemplification (*qui etiam exemplum*) of the eight wind theory or system. Homer and the Bible use the four cardinal winds only, but Hippocrates has a six wind system and Aristotle uses a zodiacal division into twelve winds. This latter system is exemplified in a stone table of the second to third century, A.D. found in 1779 at the foot of the Esquiline and now on the Belvedere Terrace next to the Museo Clementino at the Vatican. On it the twelve winds are named in both Latin and Greek. See Wood, James G., *Theophrastus of Eresus on Winds and on Weather Signs* (Edward Stanford, London, 1894), page facing p. 89.

[11] Pierre A. MacKay, *American Journal of Archaeology*, 73 (1969), pp. 468–9.

convention of planetary houses given in Ptolemy's *Tetrabiblos* I.17. The traveller then goes on to speak of a mirror of the world that was once there but now missing, originally set on a pivot—this may well be some misunderstanding of the star map disc of the water clock[12] and adds that there were also 366 talismans, one for each day of the year, and a set of stones such as Yemeni alum and blue vitriol eye-stone which were related to the black and yellow bile and other humours of the body and thus of great effect in curing and preventing diseases.

We thus learn that in addition to the octagonal element symbolism the tower contained the twelve-sided divisions of the zodiac and a set of associations with planets, humours and lapidary talismans. Some of this theory is well attested by medieval texts; we know for example that conventionally in astrology the element of earth was associated with melancholic humour, fire with choler, air with blood, and water with phlegm. We know moreover that the zodiac cycle began with Aries and springtime and was aligned with air and blood to the South point of the compass and the corresponding wind, as well as to youth. The choice of alignments is not at all arbitrary, but certain key points are obvious choices and these being made the rest of the cycle falls into place naturally and uniquely determined so as to form an interlocking set of theories covering virtually all creation and comprehending cosmology, chemistry and physics, meteorology and medicine. Such was the ambitious burden of the Tower of Winds.

The method of aligning the square and octagonal symmetry of the element theory with the twelve-sided division of the zodiac has a special historical interest. It is not attested in detail by any surviving evidence at the Tower nor indeed in any literary text. Nevertheless the general method by which it must have been achieved has been preserved in the traditional forms of the horoscope diagram, this significance of them never having been noted before. All three early forms of astrological horoscope diagram are formed on the basis of a square intersected either by a cross or by another square placed diagonally over it in a manner very similar to that of the element diagram and quite compatible with it (see Fig. XV.2).[13] Once the general principle has been stated it becomes quite obvious that such a diagram has been used as a basis or rationale for much of the underlying theory of astrological science and previously obscure alignments and

---

[12] For the tradition of 'mirrors of the world' and their identification of star maps see F. N. Estey, 'Charlemagne's Silver Celestial Table', *Speculum*, 18 (1943), pp. 112–17.

[13] For the 'modern' form see Frederick H. Cramer, *Astrology in Roman Law and Politics*, American Philosophical Society (Philadelphia, Pa., 1954), pp. 20, 21; for ancient forms see Cramer, op. cit., p. 165 and O. Neugebauer and H. B. Van Hoesen, *Greek Horoscopes*, American Philosophical Society (Philadelphia, Pa., 1959), p. 156.

associations may be seen as necessary results of two cycles being aligned from other elements.

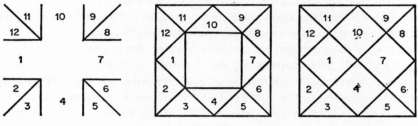

FIG. XV.2

It is, I believe, also significant in this figurate scheme that so much of the rest of astrological theory depends on the aspects, particularly those relating one sign of the zodiac to another, where the original text appears to have been illustrated with diagrams (see Fig. XV.3) that serve not so much as illustrations but as figurate theories in this tradition. The figures referred to

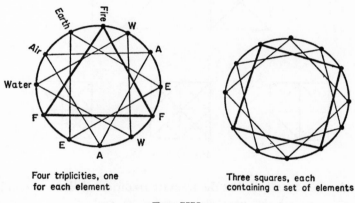

Four triplicities, one
for each element

Three squares, each
containing a set of elements

FIG. XV.3

are those of the triplicities and the squares linking sets of signs distant from each other by a right angle so that they form a square, or by 120 degrees so that they form an equilateral triangle. There are necessarily four of the triangular triplicities, one corresponding to each element,[14] and there are three squares where each square contains a set of elements. Again the alignments come naturally so that Aquarius for example must be in the watery

[14] Such a diagram of four triplicities is attested in a Babylonian tablet from Uruk, see F. Thureau-Dangin, *Tablettes D'Uruk*, Musee du Louvre, Department of Oriental Antiquities, VI (Paul Geuthner, Paris, 1922), plate XXVI.

triplicity. It seems quite plausible that much of astrological theory may rest on just such a basis of figurate rationality rather than upon empirical or special omen lore. In this sense astrology, quite apart from its utter falsity in the light of modern knowledge, developed on a very rational basis with a figurate theory and the associated symbolism at its centre.

In view of the ingenuity of this matching of the twelve-fold division of the zodiac and horoscope with the fourfold symmetry of the element diagram it is especially interesting to find that amongst the relatively few diagrams occurring in the corpus of Old Babylonian mathematical texts we find an entire collection of squares divided in this fashion and accompanied by a text that seems quite enigmatical.[15] Although the text is usually interpreted as pertaining to area calculations for the figures given, I think it may be more reasonably viewed as an exercise in what was peculiarly difficult for the Babylonians, an interpretation of a written text in pictorial form (see Fig. XV.4).

FIG. XV.4

It may also be remarked that the figurate tradition of the cross and square in element theory has also been elaborated to several other well-known and attested magical forms. The standard magic square of the third order clearly has some of the cross-like symmetry of the element diagram and can be

$$
\begin{array}{ccc}
4 & 9 & 2 \\
3 & 5 & 7 \\
8 & 1 & 6
\end{array}
$$

forced into various sorts of agreement with it. With the numbers transposed into alphabetic numerals it was taken as the source of magical nonsense words

---

[15] H. W. F. Saggs, *A Babylonian Geometrical Text*, in *Revue Assyriologique*, **54** (1960), pp. 131–46.

in Arabic and in Greek, and it may well be that the famous acrostic word square

S	A	T	O	R
A	R	E	P	O
T	E	N	E	T
O	P	E	R	A
R	O	T	A	S

has been designed with the same symmetry and figurate significance in view.[16] In another variation it may be seen that if one starts from the third order magic square numbers and draws lines joining the triads of numbers as follows: 1, 2, 3; 4, 5, 6; 7, 8, 9 the resulting figure is the mystic 'demon' of the planet Saturn. Very likely many of the other weird signatures and demons have similar origins in squares of other order. Unfortunately for the four-element theory there is no possible magic square of the second order in which the totals of rows and columns and diagonals is constant. If there had been it would doubtless have become a central object in mystic symbolism. The very absence may have indeed some indication that the four-element theory could not be a sufficient and complete explanation of all substance and change in nature. It seems, however, more likely that the ingenuity of the explanation was an indication that the theory was on the right track, but in all explanations it became clear that just some little modification would be necessary to make it just perfect.

For this reason it seems evident that the four-element theory was followed during antiquity and the middle ages with an elaboration designed to bring it to perfection. I suggest now that there were in fact two rather different sorts of attempts to improve the valuable figurative core of the theory and that these resulted in the symbolisms of the pentagram and of the hexagram respectively.

In the first modification the theory is improved simply by increasing the number of elements from four to five by the addition of a 'quintessence'. The problem then is to determine what in this new scheme can correspond to the neat double duality of principles that was built into the figurate structure of the old Aristotelian theory. By using the complete pentagon, the pentagram taken as their emblem by the Pythagoreans, occurring naturally as a knotted strip, linked to the essential and perfect 'fiveness' of the Platonic solids, one could show that the new scheme also had a natural beauty and perfection. If, for example, each side of the pentagon is made to correspond with one of the five elements, the five external and five internal

---

[16] Charles Douglas Gunn, *The Sator-Arepo Palindrome: A New Inquiry into the Composition of an Ancient Word Square*, dissertation (Yale University, 1969), p. 235.

vertices represent all the combinations of elements taken two at a time, and just four such combinations are grouped on each of the lines. Alternatively the points of the pentagram may be taken to represent elements and the lines then become relations between them.

In the second modification of the theory, the improvement is obtained not by adding a new element but by adding a third duality to the original two principles. An obvious way of symbolizing all the possible combinations of three intersecting dualities would be by means of three circles in the customary representation of a Boolean diagram of formal logic. It does not seem to have been previously noted that the hexagram or Star of David or Seal of Solomon is formally identical with the three circle diagram. If three alternate vertices are taken to represent the three principles, then the other three vertices represent the combinations of the principles two at a time, and the central hexagonal area represents the combination of all three principles. Furthermore the sides can also bear interpretation in this way and the whole symbolism can be suitably embroidered and elaborated with the greatest ease.

The possibility that these familiar talismanic diagrams are part of this figurate tradition of an element theory naturally leads one to ask if there are other figures that can be so generated. The figures sought are those formed by the joins of n points equally spaced around the circumference of a circle. The system in which each point is linked to the next gives only a regular polygon, an n-gon, which appears as a trivial solution. For three points there exists only this solution, the regular equilateral triangle, common enough in the figurate language of mysticism, but not readily bearing any sophisticated interpretation of this sort. For four points the only solution apart from the square is the cross formed by its diagonals and already described as the Aristotelian element diagram that stands near the heart of this tradition.

For five points the only possibility apart from the pentagon is the pentagram which has been discussed as a Pythagorean symbol, perhaps illustrating a five-element theory. For six points, again there exists apart from the hexagon, the hexagram which is famous as the Seal of Solomon and Star of David. There exists also the degenerate cross-like diagram formed by the three diametral lines of the hexagon, a sort of set of snow-flake axes, but that seems again without any significant symbolical properties.

For seven points, apart from the regular convex polygon it is possible to form two distinct types of heptagram; one in which each point is connected to the two vertices distant from it, and one in which each point is connected to the third distant therefrom (see Fig. XV.5). The first of these variants never seems to have been used as a mystical or magical diagram. This is

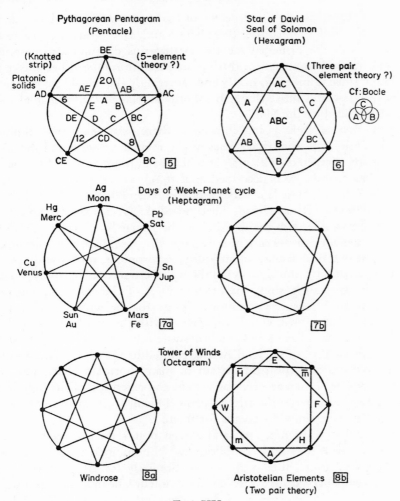

FIG. XV.5

strange, for the second variant is one of the more frequently occurring such instances of the figurate tradition. It is attested on a Babylonian tablet from the Khabaza Collection now in the Philadelphia University Museum[17] in which it is said to represent the 'seven regions' or *heptamychos* of the philosopher Pherecydes of Syros. Astrologically it is very familiar as the heptagram of the weekday gods[18] in which a diagram containing the planets placed in their astronomical order of distance from the earth is made to yield

[17] See Robert Eisler, *The Royal Art of Astrology* (London, 1946), Plate XVIa and p. 273.
[18] See for example Cramer, op. cit., p. 20.

by jumping three places at a time the order of planets in the days ruled by them in the week. Of course the planets and their gods are also found associated with the principal metals, so that this diagram also assumes a special alchemical significance; this figurate tradition indeed became central in alchemy since it linked so neatly and temptingly the metal lead designated by the heaviest and most sluggish outermost planet with the goal metal, gold, symbolized by the Sun.

For the case of eight points in a circle there exist again two significant forms in addition to the trivial cases of the regular octagon and the star of four crossing diameters. The case in which each point is joined to the next but one has already been described as that on which the structure of the Tower of Winds is based; a version of the Aristotelian two-pair theory of the four elements. It has already been noted that it has special significance as being compatible with the division of the zodiac into twelve parts using one of the versions of the square horoscope diagram. It has also been noted that at least one philosopher of antiquity, Andronicus, took the diagram as indicating a basis for the eight-wind theory of classical meteorology. This is particularly interesting since the other variant of the eight-point diagram occurs in many places as a traditional design for the windrose or compass card which is, of course, closely associated with the winds. I do not think that this association has previously been noted. It may be seen, for example on the compass card of Cecco d'Ascoli, printed in 1521[19] and also as a basis of the windrose and the grid system of many portolan charts and other antique maps. It is such an obvious variant and extension of the other eight-point figurate representation that it seems difficult to separate the traditions and establish independent lineages for them.

Diagrams based on nine, ten and eleven points do not seem to occur, probably because they add complications without increased insight when compared with those already discussed. Similarly for all greater diagrams we find no evidence except for what is undoubtedly the most famous tradition of all, the duodecimal division of the zodiac and the associated astrological theory replete with trines and sextiles, squares and triplicities and other such alignments and correspondences. It may well be that just such a technique of skipping around a circle, well known from Seleucid astronomical mathematics, may be at the origin of this entire corpus of figurate methods though, as has been remarked, the evidence concealed by mysticism and bad copying is too difficult to follow at this stage.

The figurate tradition of all these related polygonal diagrams having now

---

[19] Silvanus P. Thompson, *The Rose of the Winds: The origin and development of the Compass-Card* [Read at the International Historical Congress, April, 1913, from the *Proceedings of the British Academy, Vol. VI*] (Oxford University Press, London), p. 11.

been explored we must turn finally to what appears to be a relatively small collection of other varieties, including some from other cultures. The existence of one series points to the possibilities of others. Is it entirely capricious to see some association between the Yin and Yang diagram symbolizing the paired principles of Chinese elemental philosophy, the three-legged *triskelion*, and the four-legged swastika? These each occur in left-and right-handed varieties, and they can be set in a series as curvilinear partitions of a circle.

Quite different is the case of the mystic Hebrew figure known as the Sefirotic tree of the Cabbalists.[20] Although the diagram contains correspondences between the letters of the Hebrew alphabet, the elements, seasons, parts of the body, days of the week, months of the year, etc., it seems evident that the system is based not so much on the shape of the diagram as upon the sequence and significance of the letters of the alphabet; the tradition is indeed much more literate and perhaps numerate than figurate.

Lastly, and most diffidently as a contributor to this volume, I must consider the Chinese tradition. The essentials of the five element theory are well known and I can add nothing to the historical evidence.[21] For five elements there must be $4 \times 3 \times 2 \times 1 = 24$ different arrangements around a circle or a pentagon, but of these half are mirror images of the rest and there are therefore no more than 12 basically different arrangements of this sort (not 36 as maintained by Eberhard and followed by Needham, op. cit. p. 253). It is interesting that only three of these twelve seem to attain considerable importance as a sequence of physical significance. Perhaps more significant from the figurate point of view is the tradition that comes near to the Aristotelian four-element theory and may well be the origin of the elaboration of this to include a quintessence. In this the element Earth is placed at the centre of the (square) diagram, and the familiar pair of elements, Water and Fire, occupy the North-South axis. On the East-West axis however, instead of the Air/Earth combination of the Aristotelians is the peculiarly Chinese duality of Wood and Metal. As before the set of elements is aligned with several other sets of properties and objects. The zodiac is presumed to run from Aries in the East, clockwise via the South, planets and colours, and as a very Chinese touch, tastes are given their alignments (see Fig. XV.6). In quite another version there is a Chinese figurative scheme which seems to be in the same tradition, the hexagrams, a set of eight triplets of whole or broken bars—essentially a set of three-place binary numerals. These are associated with the eight compass directions, similar to the Western diagram

[20] See Seligman, op. cit., figs 155, 156.
[21] Joseph Needham, *Science and Civilization in China*, Vol. II (Cambridge, 1956), especially section 13.d, pp. 253 ff.

of the Winds associated with the Tower of Andronicus at Athens. In this
set however we have associated the set of five elements, augmented by other
things like mountains and wind, thunder and lightning, and with water
occurring as fresh and salt varieties. I think it is likely that the specific
association of particular trigrams with their designated elements follows a

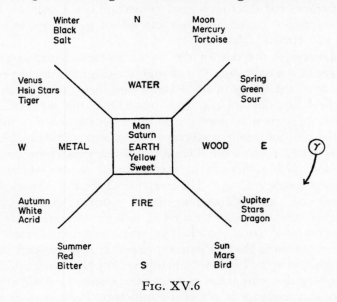

FIG. XV.6

rational and figurate scheme, probably through a topological correspondence
which must exist between the trigrams and the six-pointed Seal of Solomon
figure already discussed; each of them is merely a formalized version of a
Boolean logic diagram showing the overlapping of three logical classes. It
seems very likely too that the alchemical symbols of both East and West
may draw quite heavily on this sort of figurate tradition, the relevant portion
of the element diagram standing as a symbol for a particular element or
combination of them.

XVI

BOTANICAL NOMENCLATURE AND ITS TRANSLATION*

André Haudricourt

I

In order to study what happens within the language of a science at the time of a scientific revolution, such as that which occurred in botany in the seventeenth and eighteenth centuries, let us examine first the question of the choice of language in which the science is expressed.

In periods before the birth of modern nation-states, each with its language serving every purpose, it could be said that as a rule there was a multiplicity of functional languages. Alongside the vernacular language of the illiterate masses there were written languages for administrative, religious or literary purposes. The classic example of this differentiation at its maximum is eighteenth-century Mongolia, where the religious language was Tibetan, the administrative language Manchu, the commercial language Chinese, the literary language Classical Mongol, and the vernacular language the Khalka dialect of Mongol.[1] This linguistic situation corresponded to the social specialization of functions: religion, administration and commerce were the occupations of an insignificant minority in relation to the mass of the population.

In Renaissance Europe the situation was analogous: Latin combined the functions of a religious, administrative and learned language. In France, for example, it was in 1539 that the Edict of Villers-Cotterets imposed French in place of Latin as the administrative language, and it was in 1541 that Calvin translated his plea in favour of the Reformation from Latin

*The editors wish to thank Dr. M. J. S. Rudwick, University of Cambridge, for translating this article.
[1] See my article, 'Note sur la formation de langues nationales (A propos de l'Asie du Sud-est)', *Tiers-Monde, Problèmes despays sous-développés*, II, **8**, pp. 479–84, (Paris, PUP, 1961).

into French. But it was not until the following century, with Descartes, that philosophical and scientific discussions began to be treated in French.

In botany, interest in plants being above all practical, the gap between learned and vernacular languages was narrower than elsewhere. Even in Latin itself, Greek had been the learned language, and the Latin botanical vocabulary is full of more or less assimilated Greek words.[2]

Since European languages are written alphabetically, the difference between terms in the learned language and those in the vernacular are obvious. In the Far East, on the contrary, Chinese writing did not take account of the evolution of pronunciation in the course of time, but only of lexical additions. Written words remained unchanged and did not allow any distinction between learned and vernacular vocabularies.

II

The principles of the nomenclature and traditional classification of plants were summarized well by J. B. Monet de Lamarck, in the preface to his *Flore françoise*[3]—

> Need, so to say, was the first guide which led man to a knowledge of the vegetable kingdom. The foods that plants offered him, the remedies that he discovered by lucky chance among some of them; these made him look at them with more or less interest, according to the more or less striking benefit that he drew from each. He named them after their virtues and properties, and drawing from this the division appropriate to his own advantage, he distributed them according to the different services they gave him and the diverse kinds of disease against which they offered him help; so that the earlier works on these matters were really Treatises on Practical Botany.
>
> Next it was noticed that certain plants favoured particular climates; that in the same climate, damp places, dry or hilly ground, woods and fields, each displayed a distinct scene, which changed a little from one season to another. Some observers classified plants from this general point of view of Nature, and their treatises are like the accounts of their travels.
>
> It was then felt that neither the properties of plants—which to some extent are only revealed by the very destruction of the individual—nor purely local circumstances, could furnish any exact and methodical distribution. Thus divisions of plants were suggested on the basis of whatever was most striking to the eye—their size, consistency or duration. Consideration was given to roots, stems and leaves, sometimes even to flowers and fruits. These attempts,

---

[2] Jacques André, *Lexique des termes de botanique en latin* (Paris, Klincksieck, 1956), p. 343.
[3] De Lamarck, *Flore françoise ou description succincte de toutes les plantes qui croissent naturellement en France* (Paris, 1778), 3 vols. (the passage cited is in Vol. I, pp. xxxiii–xxxv).

although very imperfect at first, were improved little by little, and prepared the way, by degrees, for the happy resolution that has occurred in botany in about the past century.

In fact these three principles of classification had been used simultaneously ever since the first herbals (*herbarium, herbier*, or in Chinese *Běn-cǎo*); and the apparent incoherence of this classification for a modern European is well shown in a text by the Argentinian writer José-Luis Borges, which is cited by Michel Foucault in the preface to his *Les mots et les choses*. Thus in the *Běn-cǎo* of Lǐ-shǐ-zhen, there is a chapter on plants that have a name but no uses (*zhòng yǒu mìng wèi yòng*).

Let us see now how Lamarck describes what he rightly calls a happy resolution—

> It was then that certain celebrated men, convinced of the inadequacy of all the characters employed by their predecessors, turned their whole attention towards the parts of the fructification, and believed indeed that in the importance of these organs, serving for the reproduction of individuals, they perceived the tokens of Nature. They grouped together different plants that seemed to them to have several of these characters in common, and formed, as I have already said, small distinct families, known by the names of *genera*. The smallest difference that appeared constant in the plants composing a genus served to form species, while accidental and inconstant differences were to form, or at least should form, varieties.

This revolution took place in Europe, and in the learned language—Latin. It was in this language that the *genera* were designated and defined, with Caspar Bauhin and above all Pitton de Tournefort in the seventeenth century; then the *species* likewise with Linnaeus. In 1753–8 Linnaeus invented the binomial nomenclature which bears his name, that is to say the designation of the species by two Latin words: a substantive representing the genus, followed by an adjective (or a genitive) denoting the species.

It was indeed a revolution in the manner of understanding plants when, instead of describing their usefulness, abundance, size, smell and colour, attention was given exclusively to the disposition and form of the parts of the flower and seed.[4] The result was a new semantic field with a structure totally different from the old. Botanists decided that the Latin nomenclature begun in 1753 (and still used today in the international rules) was indispensable, because the change in semantic field had not been accompanied by a total reconstruction of vocabulary. The Latin words used by Linnaeus and

---

[4] It was only at the end of the seventeenth century that it became possible to name the parts of the flower in French: 'pistil' appeared in 1685, 'étamine' in 1690, 'pétale' in 1698. Before that time it was not possible for poets to speak of 'pétale de rose'.

his successors in a new and precise sense had previously been used—but in a different sense—within the older traditional semantic field; and it was therefore necessary to abolish all pre-1753 meanings.

Let us see more precisely, for the sake of clarity, how the old vocabulary appeared in relation to the new nomenclature: for cultivated plants it defined varieties, for many useful plants it defined species, more rarely genera, while lastly several words such as 'moss' or 'mushroom' were applicable to some thousands of species.[5]

But the traditional semantic field, which was eliminated by scientific Latin, had consisted of the vernacular language; yet from the eighteenth century onwards there was a desire that these languages, now raised to the dignity of national languages, should be capable of being used for all subjects. The problem of translating botanical Latin into the national languages was resolved in very different ways in French—where there was an attempt to establish a word-for-word translation of the Latin words for genera and species—and in English and Chinese, where the old vocabulary was retained.

III

To illustrate this point, I will take a concrete example of four plants which were distinguished from each other at an early date, but which today are classed as four species of the same genus. This is a particular case where there is an isomorphism between the two semantic fields, the old and the new, so that divergence in translation does not arise from differences of meaning but appears above all as a cultured trait.

The example involves two plants indigenous to Eurasia: A nightshade (la morelle) and B bittersweet (la douce-amère); and two introduced cultivated plants: C potato (la pomme de terre) and D egg-plant (l'aubergine).

The elder Pliny calls A *Solanum*, and writes about it in chapter 27, in which medicinal plants are listed alphabetically; and he calls B *salicastrum* and deals with this liana in chapter 29 after the vine. The sixteenth-century authors are in the same position. Leonhard Fuchs (1501–66) calls A *Solanum*

---

[5] The lack of isomorphism between the old semantic field and the new has been neglected as much by botanists as by linguists. For example the botanists Max Walters and David Briggs, in *Les plantes, variations et évolution* (Univers des connaissances) (Paris, 1969), p. 14, write, 'the commonly accepted idea of a type of animal or plant corresponded more to a modern genus than to a species'. As for the linguists, G. B. Milner, for example, in his *Samoan Dictionary* (Oxford University Press, 1966) gives only the generic name for Samoan plants: see my review in *Bulletin de la Société de Linguistique de Paris*, **62** (2), pp. 218–20 (1967).

hortense, Solatrum; B *vitis sylvestris, Atrigena*; and D *Mala insana*.[6] Rembert Dodoens (1518–85) terms A *solanum hortense* and B *Dulcamara*.[7] Johann Bauhin (1541–1613) calls A *solanum hortense*, but B is called *glycipicros* and D *Solanum pomiferum*.[8]

In the Běn-căo of Lĭ-shĭ-zhen (1518–93) the situation is similar.[9] A is called *lóng-kúi*, in the chapter on medicinal herbs found in cool places; it follows the mallows termed *kúi*, and could be translated as 'dragon's mallow'. B is called *bái-ying*, in the chapter on lianas;[10] and D is termed *jié* and is the first plant in the chapter on gourds and melons.

Until there was a correlation between learned and common names, the difference between A and B was that the first had retained its ancient Latin name, while the second was no longer identifiable, so that it was necessary to create a new name on the basis of the common name.[11] In the Far East, the same Chinese characters were employed in China and in Japan for A but not for B.[12]

The 'revolution' which was to group A and B in the same genus, and consequently under the same name, was begun by Caspar Bauhin (1550–1624), who termed A *Solanum officinarum, ascinis nigricantibus et fuscis*, B *Solanum scandens seu dulcamara*, and C *Solanum tuberosum esculentum*. It was continued by J. Pitton de Tournefort (1656–1706); and it only remained for Linnaeus to simplify the epithet and so obtain his binary nomenclature, namely, A *Solanum nigrum*, B *Solanum dulcamara*, C *Solanum tuberosum* and D *Solanum melogena*. Then, with Bernard de Jussieu (1699–1777) Adanson, (1727–1806) and Antoine-Laurent de Jussieu (1748–1836), the genera were grouped in families, each family being named after a genus chosen to typify it. Thus the genus *Solanum* was the type of the family of *Solaneae* (later, *Solanaceae*).

[6] *Historia plantarum* (Lugduni, 1567), see pp. 506, 623, 514.

[7] *Stirpium historiae pemptades* (Antuerpiae, 1166), see pp. 454, 402.

[8] *Historia plantarum uniuersalis* (Ebroduni, 1650), see Vol. 3, p. 608; Vol. 2, p. 109.

[9] *Běn-căo gang-mù*, appeared in 1590, republished *xin-húa shu-diàn běi-jing*, 1965.

[10] *bái-ying* means 'white *ying*'. *Ying* no longer has a precise meaning: dictionaries give it as 'flower', but the occurrence of this syllable in plant names shows that in fact it involved a definite type of flower, namely a gamopetalous tubular or tunnel-like corolla. It should also be noted that the wild plants A and B are not identical in Europe and the Far East: B normally has violet flowers in Europe, but in the Far East they are white and the plant is distinguished by botanists as *Solanum lyratum* Thunb.; similarly for A, the Far Eastern varieties are often raised to specific rank as *Solanum nodiflorum* Jacq.

[11] *Dulcamara* is the Latin equivalent, and *glycipicros* the Greek, of bittersweet, douce-amère, bittersüss, etc.: at this period scholars likewise translated their own names into Latin (Bauer: Agricola, Kaufman: Mercator) or into Greek (Schwartz-erd: Melancthon).

[12] Japanese scholars believe they can recognize B in another solaneacean: shu-yang-quan, p. 685 of the Běn-căo gang-mù, in the same chapter as A and separated from it by two other Solaneaceae. The beginnings of a natural grouping will be noticed in these works.

The problem of translating the results of this revolution into the vernacular languages were posed at once, for example by Jean-Jacques Rousseau in his famous botanical letters (1771–3) to Madame de Lessert. Let us see how Lamarck discussed this problem in the preface to his *Flore françoise*: 'As for the names I have given to the plants that are described in the analysis, I have used most often those of M. Linné, which I have translated into French, my work being written in that language'; and later on: 'I have adopted M. Linné's specific names as far as possible; but I have been forced to change some of them, which seemed to me too defective, and which I was unable moreover to describe in French because they expressed no meaning to the mind'.[13] The result, for our plants, was as follows:

A	Morelle noire	*Solanum nigrum*
B	Morelle grimpante	*Solanum scandens*
C	Morelle tubéreuse	*Solanum tuberosum*

Lamarck give no other French name beyond his word-for-word translation of the Latin nomenclature; he changed the name of B, doubtless because *dulcamara* is not classical Latin.

The French botanical vocabulary thus created persisted for more than a century. In 1889, in the *Petite Flore*[14] by G. Bonnier (1853–1922), there are no Latin names, but scientific names in bold type and common names in italics, thus:

A	**Morelle noire**	*Tue-chien, raisin de loup*
B	**Morelle Douce-amère**	*Douce-amère, Vigne de Judée*
C	**Morelle tubéreuse**	*Pomme-de-terre*

There were at the same time Floras without translation of the Latin nomenclature, such as the *Flora of Paris*[15] by J. L. de Lanessau (1843–1919); but it was the *Bonnier Flora* that continued to be reprinted, even to our own day.

The contrast with English nomenclature can be seen in the *Flore-manuel de la Province de Québec, Canada*, published in 1959 by Père Louis-Marie, in which we find:

A	Morelle noire (Crève chien)	*Solanum nigrum* L.	Black Nightshade
B	Morelle Douce-amère	*Solanum dulcamara* L.	Bittersweet
C	Morelle tubéreuse (Vulg. pomme-de-terre ou patate)	*Solanum tuberosum* L.	Potato
D	Morelle aubergine	*Solanum melogena* L.	Egg-plant

---

[13] Lamarck, op. cit., I, p. lxxxii, II, p. ii.
[14] G. Bonnier and G. de Layens, *Petit Flore*, 144 pp. (republished 1969).
[15] J. L. de Lanessan, *Flore de Paris*, xlii + 902 pp.

The Canadian author explains that in French the Latin family name is simply gallicized, thus 'Solanacée (famille de la pomme-de-terre)', whereas in English it is translated, 'Nightshade family'.

Seeing this, one might assume that the term 'Nightshade' corresponds to '*Solanum*' in the name of A and in the family name, and that it is in fact a translation of *Solanum*. But if we look at the *Flora*[16] by J. Hutchinson, one of the most distinguished English botanists, we find that while he retains some 'Nightshade' names they are used in genera of *Solanaceae* other than *Solanum*, for example 'Deadly Nightshade', *Atropa belladona* L., and even in one plant, 'Enchanter's Nightshade', *Circaea lutetiana* L., which is not and never has been classed among the *Solanaceae* and which belongs to the *Onagrariaceae* (Evening-primrose family). The only attempt I have noticed in English to translate the Latin nomenclature word for word, namely Loudon's *Hortus Britannicus*, does not avoid this ambiguity, for one finds there '*Solanum*: Nightshade' and '*Circaea*: Enchanter's Nightshade'.[17]

Other European languages are intermediate between the two preceding cases. In some German authors the same systematization can be found as in France:[18]

A	Schwartzer Nachtschatten	*Solanum nigrum* L.
B	Rankender Nachtschatten, Bittersüss	*Solanum dulcamara* L.
C	Knolliger Nachtschatten, Kartoffel	*Solanum tuberosum* L.
Family	Nachtschatten gewächse	

Other European languages retain the common name for C, thus:

	Finnish[19]	*Danish*[20]	*Czech*[21]
A	musta-koivo	sort Narskygge	lilek černý
B	puma-koivo	bottersød Nakskygge	potměchuť
C	peruna	kartoffel	brambor
Family	koivo-kasvit	Natskyggefamilien	lilkovité

In the Far East the situation is analogous: the current names are used to

---

[16] J. Hutchinson, *Common Wild Flowers* (Penguin Books, 1945).

[17] Loudon's *Hortus Britannicus, A catalogue of all the plants indigenous, cultivated in, or introduced into Britain* (London, 1830).

[18] O. Wünsche, *Die Pflanzen Deutschlands* (Leipzig, 1897) (7th edition of *Schulfora von Deutschland* (Leipzig, 1871)).

[19] Ilmari Hiitonen, *Otavan värikasvio* (Helsinki, 1956).

[20] E. Rostrup and C. A. Jørgensen, *Den danske Flora* (Copenhagen, 1961).

[21] Ladislav Čelakovský, *Analytická květena česká* (Prague, 1879).

translate the Linnaean binomials, without trying to indicate the generic name in each species. Thus in Japanese:[22]

	pronunciation (or Kana script)	characters used (Chinese pronunciation)
A	inu-hoozuki	lóng-kúi
B	hiyodori-joogi	shŭ-yáng-quán
C	jaga-imo	mǎ-líng-shŭ
D	nasu	jié
generic name	nasu-zoku	jié-shŭ
family name	nasu-ka	jié-ke

It will be noticed that in the Far East it is no longer A, but D, which is the type of the family, and that the use of a suffix to indicate the genus avoids all ambiguity. On the other hand, it is only the generic and family suffixes that correspond, and then only in the learned pronunciation of the characters; while the name for D is different in the two languages. Furthermore, if we translate the terms for A, B and C, we find the following:

	Spoken Japanese (or written in Kana)	Written Japanese (in Chinese)
A	Dog's Winter-cherry[23]	Dragon's Mallow
B	Bulbul's Funnel	Shu's Sheeps'-trough
C	Djakarta Taro	Horse-bell's Yam

In the face of such discordance, it is hardly surprising that the Japanese tolerate the discordance between Latin scientific nomenclature and its translation into current terms. It is the same in China, where the creation of a new name only takes place if the plant has no current name.

In South Vietnam, a French-trained botanist, M. Pham-hoàng Ho, gives in his Flora:

A cà den (nomen nobis)
C khoai tây
D cà
Family họ cà

---

[22] Works consulted: *Vegetation of the National Park for Nature Study*, (Tokyo, 1965); Lemaréchal, *Dictionnaire Japonais-Français* (Tokyo, 1904); *Kenkyusha's new little japanese-english Dictionary* (Tokyo, 1960).

[23] The Winter-cherry is a solaneacean of the genus *Physalis*: in Chinese, *suan-jiàng*; in French, *coqueret*; in German, *Schlutte*; in American English, *ground-cherry*; in Quebec, *Cerise-de-terre*.

The author, not knowing an indigenous name for A, creates one—'Black Egg-plant'—which is not very felicitous,[24] but which cannot be confused with the indigenous name for C (Western tuber). The syntax of the language not allowing the use of a suffix for the genus, the family is indicated by a prefix.

In some works the translator, being in a hurry and not knowing the true indigenous name, simply makes a copy. In the *Russo-Chinese Dictionary of Sylviculture*,[25] for example, we find:

	Russian	Chinese
A	cërny paslën (pozdnika)	lóng-kúi
but B	paslën sladko-gorkij	gan-kŭ-jié (= bittersweet/egg-plant)

In conclusion: every scientific revolution involves the creation of a new terminology. How are new terms to be formed; how are they to be trans-ferred to other languages, foreign or dead; and does this change the meanings of words already employed in the language? In France and Germany many botanists, from Lamarck onwards, have tried to introduce the whole Linnaean nomenclature, by using copied or borrowed words; but it can now be said that this has been a failure. In English, on the contrary, there has never been an attempt to modify the common names of plants, but instead an attempt to standardize them,[26] that is to say, to urge that the same name should always be used for the same species, to avoid confusion. It is the well-known difference between English empiricism and the French systematic spirit[27]. And at this point the Far East is like the English.

Instead of studying four species very briefly, hundreds of common plants ought to be considered;[28] but I think that the same conclusions would be reached.

[24] For nothing then distinguishes the nightshade from the 'black egg-plant', Vietnamese being unaccented (compare English, in which 'Blackbird' is distinct from 'black bird'). But I should mention that in the *Háng-zhou yào-yòng zhí-wu zhì*, (Shanghai, 1961), *hei-jié* (black egg-plant) is in effect given as a synonym of *lóng-kúi*.

[25] *-è-hua lín-yè cí-diǎn, Russko-kitajskij slovarj*, Bei-jing, 1959, 727 pp.

[26] For example for America: Kelsey and Dayton, *Standardized Plant Names*.

[27] This can usefully be compared with what happened with the metric system, in French, with the creation of new terms: *mètre, gramme, litre, centime*. In Chinese the old terms (already decimal, it is true) are used with the (provisional) prefix *gong*. The same thing is happening in England with the old and the new *penny*.

[28] It would be necessary to consult a large number of Floras and of popular works; and for the two introduced plants the history of the current names in the European languages is related to their cultivation and their consumption: see for example, on D. R. Arveiller, 'Les noms français de l'aubergine', *Revue de linguistique romane*, 33 (1969), pp. 131–2, 225–44 (1969).

XVII

THE RÔLE OF SCIENCE IN THE INDUSTRIAL REVOLUTION:

A STUDY OF JOSIAH WEDGWOOD AS A SCIENTIST AND INDUSTRIAL CHEMIST

Neil McKendrick

Much has been written, in the recent controversy over the rôle of science in the Industrial Revolution, of the pervasive popularity of science in the late eighteenth century. As James Keir wrote in 1789 in his *Dictionary of Chemistry*, 'The diffusion of a general knowledge, and of a taste for science, over all classes of men, in every nation of Europe, or of European origin, seems to be the characteristic feature of the present age.'[1] And a powerful array of detailed evidence has been marshalled to support this view.[2] By this period, science, or rather the vocabulary of chemical processes (and it is an instructive distinction), had certainly entered the English language. Even Burke took metaphors from chemistry and when writing of 'the spirit of liberty in action' quite naturally adopted a chemical analogy. 'The wild *gas*, the fixed air, is plainly broke loose: but we ought to suspend our judgement until the first effervescence is a little subsided, till the liquor is cleared, and until we see something deeper than the agitation of a troubled and frothy surface.'[3] So marked and so powerful was this influence that Professor Davie has written, 'The influence of science upon literature in this period is so pervasive and so protean that the only way to grasp and handle it is through

---

[1] Quoted by A. E. Musson and Eric Robinson in *Science and Technology in the Industrial Revolution* (1969), p. 88.

[2] Ibid., *passim*.

[3] *Reflections on the French Revolution*, pp. 6–7 [Everyman edition]. Quoted in Donald Davie, *The Language of Science and the Language of Literature, 1700–1740*, (1963), pp. 76–7. As Professor Davie commented 'in Burke . . . the scientific analogy is never far off', op. cit., p. 76.

close and particular analysis of the vocabulary.'[4] Equally clearly it is only through a close and particular analysis of individual industrialists' knowledge and use of science that we can finally answer the important questions about the rôle of science in stimulating or facilitating economic growth. Nowhere was this interest in science more intense or more purposeful than in the Midland society of industrialists and scientists characterized by the Lunar Society. It can be observed in a particularly well evidenced and well recorded way in the social, intellectual and business milieu of Josiah Wedgwood. In this society there most certainly was 'a general knowledge and . . . a taste for science'. It intrudes even into their poetry. Erasmus Darwin has suffered the satire and the parody of posterity for his heavy handed enthusiasm for science and the technical achievements he envisaged for 'UNCONQUER'D STEAM', and Anna Seward has understandably not escaped critical mockery for writing of tribes of 'fuliginous chemists' and their 'troops of dusky artificers' invading Colebrookdale with 'their pond'rous engines'.[5] To them the current interest in science was unmistakably economically purposeful. Its purpose was 'to glut [Britain's] rage commercial'. In Birmingham 'that town, the mart Of rich inventive Commerce', it was 'Science' which in Anna Seward's words 'Leads her enlighten'd sons, to . . . Plan the vast engine'.[6]

Science not only invaded their poetry, it invaded their painting. For the most vivid pictorial record of this new found enthusiasm for, and belief in, experimental science still survives in the paintings of Joseph Wright of Derby.

The fascination exerted by science and the thrill of scientific exploration have perhaps never been captured more brilliantly than by Wright in his two famous paintings entitled *A Philosopher giving a lecture on the Orrery*[7] and *An Experiment on a Bird in the Air Pump*.[8] The march of industrial technology was equally faithfully portrayed by Wright with his paintings of forges, mills, kilns and glass ovens—like his *Iron Forge* of 1772 or his *Interior of a Glass House*.

That he should so accurately mirror the twin interests of industry and science is scarcely surprising for the leading artist of what has been called that remarkable 'syncretism of pure science and advancing industry'—the

[4] Donald Davie, op. cit., p. 64. See also W. K. Wimsatt, *Philosophic Words* (1948), John Arthos, *The Language of Natural Description in Eighteenth Century Poetry* (1949), Marjorie Nicolson, *Newton Demands the Muse* (1946).

[5] Jeremy Warburg, *The Industrial Muse: The Industrial Revolution in Poetry* (1958), pp. 5–8.

[6] J. Warburg, op. cit., p. 5, quoting from Anna Seward's poem 'Colebrook Dale' written in 1785.

[7] Painted circa 1763–5.

[8] Painted in 1768.

Lunar Society of Midland England in the 18th Century. The contrast between this and the rural lyricism of a Gainsborough, or the cultural preoccupations of the metropolitan world of a Horace Walpole or a Boswell is stark and significant. For this bourgeois group of midland manufacturers, merchants, lecturers and scientists was characterized by a remarkable enthusiasm and seriousness of purpose. They passionately believed in the benefits of industry, commerce, and experiment. They passionately believed in progress. The self-doubt bred by the defeats, disappointments and disasters of a later factory age was largely unimagined as yet. They did not envisage defeat. Science and industry would march in triumphant unison towards the solution of all the world's problems—social, political and material. Joseph Priestley was voicing no lonely belief in the practical objectives and future achievement of science when he wrote in 1768 in his *Essay upon the Principles of Government* that: 'Nature, including both its materials and its laws, will be more at our command; men will make their situation in the world absolutely more easy and comfortable; they will probably prolong their existence in it, and will daily grow more happy.'[9] This atmosphere of optimism glows out from Wright's paintings. He was painting for a group of scientific 'philosophers' struggling to get their ideas to work, of inventors struggling to get their discoveries applied, of entrepreneurs struggling to maximize the profits of their innovations, and of manufacturers struggling to match their output to demand. All were buoyed up by the belief that they would succeed. The seriousness of purpose is not surprising: nor is the optimism. In an economic climate of expansion, growth, and rampant demand, the proffered rewards promised to be great. Fame and profit beckoned encouragingly. The hopes, the aspirations, the optimism gleam in the rapt, candlelit faces of Wright's paintings: the concentration on a job of work and the satisfaction derived from its successful conclusion are unmistakable: 'Sir Richard Arkwright is resting his hand proudly on a model of the machine which crowns his life's work and will change the world; a learned philosopher is lecturing on the orrery to a spellbound audience; that picturesque figure the alchemist is discovering, not a nebulous mediaeval philosopher's stone, but the substance phosphorus which had for Wright's contemporaries the same ring as the word "penicillin" for us; men are beating out iron by the light of the moon, into undreamt-of-shapes which will revolutionize the pattern of private life: others are uncovering the secrets of the laws of nature by torturing a bird in an air pump. All . . . is justified in the name of progress.'[10]

[9] Quoted by Sir Vincent Wigglesworth in 'The Religion of Science', Address of the President of the Association of Applied Biologists delivered to the Annual General Meeting, 14 April 1967, cf. *Ann. appl. Biol.*, 60, pp. 1–10.

[10] Cf. the brilliant introduction by Benedict Nicolson, *Joseph Wright of Derby, 1734–1797* (The Arts Council, 1958), p. 5.

It was Wright's contribution to historical evidence to capture on canvas the busy glare of the blacksmith's shop, the intense preoccupation with the hard facts of the physical world displayed in the fascinated faces of his candlelit pieces. It was this and his paintings of 'cotton mills humming by night along the Derwent valley, each window pricked with hard work'[11] which has won him the title of 'the first professional painter to express the spirit of the Industrial Revolution'.[12] What he portrayed so vividly and exactly some two hundred years ago, has recently been painstakingly re-created by the work of modern scholars. With perhaps less panache and excitement, but with impressively deep reserves of historical scholarship and detailed evidence, scholars like Musson, Robinson and Schofield have sought to illustrate the fruitful inter-relationship of science and industry. They have uncovered what amounts to a 'subculture' of interest in science and industry in late eighteenth-century England, but their problem is an even more demanding one since they seek not only to illustrate the interconnections, but to *prove* the value of the contribution of science to industry.

Their critics have made much play of the point that mere association between scientists and industrialists is not proof of fruitful interdependence. Professor Mathias' very pertinent remark on this controversy that 'the assumption of a particular causal relationship by association is as conceptually injudicious a conclusion in history as guilt by association in law'[13] is an echo, albeit a more eloquently turned one, of D. W. F. Hardie's complaint that 'In democratic societies at least to assert guilt by association is generally considered bad law; finding significance by association, and association alone, is equally bad historical procedure. While it is possible to document the association and personal inter-relations of the members of the Lunar Society it is only possible to assert—as Dr. Schofield repeatedly does—the significance of that association.'[14] More pointedly Hardie queries the description of Wedgwood as a 'scientific potter', asking 'what significance has "scientific" when applied to the great potter's preoccupation with canals, clays and jasper ware?'[15]

It is with seeking an answer to that question in particular, and in contributing to the debate over the significance of the rôle of science in the industrial revolution in general, that this paper will be mainly concerned.

[11] Benedict Nicolson, op. cit., p. 6.

[12] F. D. Klingender, *Art in the Industrial Revolution* (1947).

[13] Peter Mathias, 'Economic Growth and Economic History', *Cambridge Review*, 30 January 1970, p. 90.

[14] D. W. F. Hardie's review of Robert E. Schofield, *The Lunar Society of Birmingham: A Social History of Provincial Science and Industry in Eighteenth Century England* (Oxford, 1963), in *Business History*, Vol. VIII, no. 1, January 1966, p. 74.

[15] Op. cit., p. 73.

In view of the complaint of a causal relationship claimed by association alone, it is a pity that the pictorial evidence should consist simply of the proof by 'association' of *The Orrery* experiment with the industrial mill, or the *Air Pump* with the industrial kiln. The transfer of formal scientific knowledge through this kind of tenuous connection is clearly nil. Proof by contiguity is perhaps even less satisfactory than proof by association. In view of this it is even more regrettable that historians were robbed of 'the perfect illustration of the scientific interests of Midland manufacturers'[16] when Josiah Wedgwood's plan to have Wright paint the Wedgwood boys performing scientific experiments failed to materialize. Here the visual evidence of association would have been more specific, the links more clearly underlined, the belief at least in the power of experiment more vividly secured. For the painting was to be of 'Jack standing at a table making fixable air with the glass apparatus &c, and his two brothers accompanying him. Tom jumping up & clapping his hands in joy & surprise at seeing the stream of bubbles arise just as Jack has put a little chalk in the acid. Joss with the chemical dictionary before him in thoughtful mood, which actions will be exactly descriptive of their respective characters.'[17]

Unfortunately for historians of science and industry this family portrait was never painted. George Stubbs wished to 'do something for us by way of setting off against the tablet'[18] which Wedgwood had made for him and Wedgwood decided to take 'payment in pictures'.[19] So the painting of the scientific experiment was taken out of 'Mr Wright's hands', and Stubbs, not surprisingly, preferred to exercise *his* 'scientific vision' on his own aesthetic preoccupations: he preferred animal anatomy to glass retorts and fixed air experiments and so the Wedgwood children were immortalized on horseback rather than in the laboratory.

That the perfect *symbolic* illustration of the scientific preoccupations of Midland industrialists should never have materialized must have an attractive irony for many historians of the period. For its failure to emerge from suggestion into fact might seem to offer splendid ammunition for the current debate, a perfect piece of evidence of the insubstantial and nebulous nature of the claimed inter-relationship between science and industry. When the evidence in support of a strongly-asserted argument melts away in such a dramatic fashion—and it must be allowed that it not infrequently does—then it is easy to understand the impatience of the more demanding critics

[16] Neil McKendrick, 'Josiah Wedgwood and George Stubbs: An Essay on the Interaction of Art and Industry', *History Today*, VII, no. 8, August 1957, pp. 107–8.

[17] *WMSS*. May 1779.

[18] See note 16.

[19] *WMSS*. Op. cit.

of a science-caused—or even science-oriented and science-influenced—Industrial Revolution.

Wedgwood provides a splendid example of a scientifically-absorbed industrialist whose career seems to bristle with examples of the significant economic influence of science, many of which prove to be less than wholly convincing when examined in detail. In fact the whole controversy over the contribution of science to economic growth in the late eighteenth century can be studied in microcosm in the career of Josiah Wedgwood. So many of the ingredients of the case for scientifically-induced economic growth are present in this case study—the industrialist's friendship with scientists, the mutual discussion of scientific and industrial problems, the unquestioned belief in experiment and the apparent familiarity with theory, the contribution of itinerant scientific lecturers, the central contribution of scientific societies, the purchase of scientific publications, the correspondence with scientists abroad, the publication of scientific papers, the production of scientific instruments—these and many other of the characteristics identified by Musson and Robinson as indicative of the fruitful rôle of science in industry are all prominently displayed. Yet so much of the seemingly powerful evidence in favour of the important contributions of science to Wedgwood's industrial success seems to fade, on detailed investigation, into slight and insubstantial suggestions. Often what looked to be so easily provable becomes unsubstantiated assertion, what looked like concrete evidence becomes intangible suggestion, and what seemed to be highly probable becomes only a possibility. Propositions often fail to become concrete realities, suggestions remain obstinately unrealized. For just as Wedgwood's vision of his children being painted whilst reading about and practising chemistry never materialized, so his projected co-operative research organization never got beyond the proposal stage. It can, of course, be argued—and it has been—that the proposals are more significant than their failure to materialize. Perhaps they are if one is aiming to establish an *interest* in science. They most definitely are not, if one is aiming to establish an achieved economic rôle for science.

When what is proposed is not achieved, and when what is suggested is not implemented, doubts as to the importance of the suggestions and the real significance of the proposals cannot easily be brushed aside. When, with disturbing frequency, doubts about other apparently powerful evidence of Wedgwood's scientific interests and achievements can be raised, it is easy to understand the controversial nature of the case.

At first sight there seems to be a mass of evidence clamouring to support the case for Wedgwood's scientific knowledge and interests, his scientific techniques and achievements, his scientific distinctions, his scientific vision.

At first sight there is also a mass of evidence testifying to his belief in the value of applying science to industry and of his ability effectively to do so. He was after all a Fellow of the Royal Society, he was after all a familiar of the Lunar Society, a member of the Society of Arts and Manufactures. He did invent the ceramic pyrometer. He did deliver and publish three papers to the Royal Society. He did possess an impressive chemical library. He did compile volumes of scientific notes. He did employ a trained chemist. He certainly conducted endless experiments. He believed—and he acted on that belief—that 'All things yield to experiment'. He certainly corresponded with many other scientists, he certainly encouraged the itinerant scientific lecturers (both arranging and attending their lectures), he certainly endeavoured to base his experiments on their theoretical findings, and he certainly borrowed their techniques in order to improve his experimental skill. And he did produce original scientific ideas of his own—such as his ingenious approach to the development of chemical symbolic formulae.

Again, at first sight there is ample evidence of Wedgwood's industrial application of science and his belief in the importance of science's contribution to industry. Did he not propose a co-operative scientific research organization? Was he not deeply involved with societies devoted to the industrial application of scientific principles and scientific discoveries? Was his pyrometer not inspired by industrial need, and was it not industrially effective in the accurate measurement of heat in his kilns? Did not his knowledge of the chemistry of clay allow him to produce an unrivalled series of ceramic inventions and improvements? Even when the answer is unambiguously in the affirmative, there are still certain problems to be overcome before Wedgwood can be accepted as a scientist, and before the rôle of science in the intellectual origins of economic growth can be affirmed.

Doubts about the scientific value of his papers to the Royal Society are as well founded as is scepticism of the scientific significance of being a Fellow in the eighteenth century. Doubts about the importance of proposals which do not develop are equally justified. And such doubts can be multiplied. Wedgwood's pyrometer was certainly industrially useful, but his empirical approach to the problem and his failure to calibrate its scale with that of Fahrenheit raises doubts over the level of scientific achievement involved in its invention. Wedgwood was certainly familiar with the language of chemists but that does not necessarily invest his use of it with any scientific significance. Wedgwood certainly learnt and borrowed experimental techniques from the chemists he corresponded with and read, but that does not necessarily raise his experimental practice above the level of empiricism. Wedgwood certainly produced a novel and ingenious solution to the problem of rendering chemical properties visible, by giving them pictorial symbolic values, but he did not develop the ideas beyond its use in his children's education.

It is this contrast between promise and performance, between suggestion and reality, which has led to such firmly contradictory views about Wedgwood's status as a scientist. It is the same contrast which has fed the general controversy over the rôle of science in the Industrial Revolution. For just as Sir Eric Ashby and Professors M. M. Postan, D. S. Landes and A. R. Hall have doubted that 'science was an important element in the general stimulus to the improvement of techniques',[20] whilst Professors T. S. Ashton and W. W. Rostow, along with Messrs. Musson and Robinson have held that it was; so the biographers of Wedgwood have divided fiercely on this issue. The eighteenth century adopted the unsophisticated assumption that Wedgwood's claims as a scientist were too obvious to be questioned.[21] The mid-nineteenth century shared the assumption, Meteyard never questioning it,[22] but by the late nineteenth century opinions had divided. For although in 1894 Smiles supported the traditional view, even inventing 'the "Transactions" of Josiah Wedgwood' in support of his 'great . . . scientific . . . reputation',[23] in the same year Professor Church (significantly a professor of chemistry) wrote that 'it is impossible to argue . . . that Wedgwood was a great chemist. . . . Chemistry, in Wedgwood's day, was a science in the very early stage of development; even had he mastered all that was known of it, the aid it could have afforded him in his practical enquiries would have been comparatively insignificant.'[24] The progress of the Wedgwood bibliography has not, however, been one of growing scepticism, and the twentieth century has seen a revival of his scientific reputation. For although William Burton echoed Church when he wrote 'he was no chemist as has been popularly supposed—indeed, such chemistry as was known in his day, would have helped him singularly little in his business as a potter',[25] many other historians,

[20] P. Mathias, 'Economic Growth and Economic History', *Cambridge Review*, 30 January 1970, p. 90. See also 'Who Unbound Prometheus?', *Yorkshire Bulletin of Economic and Social Record*, XXI, no. 1 (May 1969).

[21] *The Gentleman's Magazine* of 1795 printed an obituary of Wedgwood [Vol. 65, no. 85] which stated 'His communications to the Royal Society shew a mind enlightened by science, and contributed to procure him the esteem of scientific men at home and throughout Europe'.

[22] Eliza Meteyard, *The Life of Josiah Wedgwood*, 2 vols. (1865-6).

[23] Samuel Smiles, *Josiah Wedgwood F.R.S.* (1894), pp. 270-1. In fact these volumes were made by Alexander Chisholm for Dr. William Lewis who employed Chisholm. After Lewis's death Chisholm was employed by Wedgwood who bought the notes.

[24] A. M. Church, *Josiah Wedgwood* (1894), pp. 84-6.

[25] William Burton, *English Earthenware and Stoneware* (1904), pp. 121-51.

including A. T. Green,[26] E. Bostock,[27] J. W. Mellor, [28] John Thomas,[29] Robert E. Schofield,[30] in particular, and Archibald and Nan L. Clow,[31] together with Musson and Robinson,[32] have argued either explicitly or implicitly, that 'Wedgwood's reputation as an industrial chemist be re-established'.[33]

Professor Schofield has gone further and suggested that 'the eighteenth-century pottery industry of Josiah Wedgwood demonstrates early deliberate use of science in the industrial revolution':[34] a view categorically denied by Professor A. R. Hall[35] and D. W. F. Hardie.[36]

The survival of the controversy and the existence of much unpublished manuscript material in the Wedgwood archive, perhaps justifies a fuller and more systematic examination of the evidence.[37] Especially when Wedgwood presents such an important case-study with which to exemplify the general controversy. There are after all some formidable denials of any significant rôle for science in the Industrial Revolution barring the way to an easy accept-ance of Wedgwood's scientific knowledge and industrial exploitation of it.

A prime and influential example of such denials is the trenchant general-ization of Professor A. Rupert Hall and Marie Boas Hall that 'the beginnings of modern technology in the so-called Industrial Revolution of the eighteenth and early nineteenth century owed virtually nothing to science, and every-thing to the fruition of the tradition of craft invention'.[38] They saw the

[26] A. T. Green, 'The Contributions of Josiah Wedgwood to the Technical Side of the Pottery Industry', *Transactions of the English Ceramic Society*, Vol. XXIX (1930), pp. 5–35.

[27] E. Bostock, 'Contributions of Josiah Wedgwood to the Technical Side of the Pottery Industry', *Transactions of the English Ceramic Society*, Vol. XXIX, (1930), pp. 36–58.

[28] J. W. Mellor, 'Wedgwood as an Industrial Chemist', postscript to Julia Wedgwood's *The Personal Life of Josiah Wedgwood* (1915).

[29] John Thomas, 'Josiah Wedgwood and his Portraits of Eighteenth Century Men of Science', *Royal Society of Arts*.

[30] R. E. Schofield, op. cit.

[31] A. and N. L. Clow, *Chemical Revolution* (1952).

[32] Musson and Robinson, op. cit., *passim*.

[33] Robert E. Schofield, 'Josiah Wedgwood, Industrial Chemist', *Chymia,* 5 (1959), p. 192.

[34] Schofield, op. cit., p. 192.

[35] A. R. Hall, 'What did the Industrial Revolution in Britain owe to Science?', *Historical Perspectives, Studies in English Thought and Society in Honour of J. H. Plumb* (ed. Neil McKendrick) (1973).

[36] D. W. F. Hardie, op. cit., *Business History*, 1966, p. 74.

[37] For despite the dogmatic assertions and well known contradictions, there is no exhaustive survey of the problem. Professor Schofield has done most to establish a clear picture, but he has not devoted himself to the whole field of Wedgwood's scientific activities, and there is still a great deal of unexamined evidence.

[38] A. Rupert Hall and Marie Boas Hall, *A Brief History of Science* (1964), p. 219. It is fair to add that the Halls then conceded that 'the chemical industries from the last years of the eighteenth century fall into a different category, for in them scientific knowledge was conspicuous'. Ibid., p. 219.

characteristic features of the Industrial Revolution (including the steam engine) as 'the results of empirical experiments, products of craft skill and large quantities of hard labour'.[39] In Professor Hall's view 'virtually all the techniques of civilisation up to a couple of hundred years ago were the work of men as uneducated as they were anonymous'.[40] Others have gone further in their denials of any meaningful economic rôle for science. Professor Ashworth wrote that 'the heroic inventions of the eighteenth and early nineteenth century . . . were mostly the work of practising craftsmen and enthusiastic amateurs with, at best, a very modest knowledge of scientific theory'. Although he allowed that 'in . . . fields, such as chemistry, purely empirical experiments, undertaken out of idle curiosity or in response to some pressing economic difficulty, led to a few very important industrial innovations', he added that 'In all fields, except mechanics, technological progress was retarded because of lack of knowledge of basic scientific principles, without which the relatively small amount of experiment was bound to be somewhat haphazard and its occasional success too heavily dependent on chance coming to the aid of persistence and shrewd observation'. Even chemistry is allowed very little contributory rôle in economic growth by Professor Ashworth, for he wrote 'the creative contribution of chemistry to economic life remained small until the last third of the nineteenth century'.[41] Even more sceptical of the effective economic rôle of science is Sir Eric Ashby. In his view 'the Industrial Revolution was accomplished by hard heads and clever fingers . . . [by] men [with] no

[39] Ditto. Even in the early nineteenth century, in their view, 'technological development was still dependent upon chance, empiricism and craftsmanship'. Ibid., p. 219.

[40] A. R. Hall, *The Historical Relations of Science and Technology* (inaugural lecture) (1963), p. 10.

[41] William Ashworth, *An Economic History of England, 1870–1939* (1960), pp. 27–33; see also pp. 311–12. However, although Professor Ashworth believed that 'even the numerous economically important discoveries of the laboratories came mainly from empirical programmes of research without much concern to establish or confirm general principles', he does add the rider that 'it was empiricism made fruitful by a new, rational conception of the subject. The establishment of the necessity of quantitative methods of investigation and the formulation of several empirical laws concerning the combination of different substances all came in the closing years of the eighteenth century and the first few of the nineteenth, and they led to a great increase in the scope and reliability of chemical analysis'. In my view this 'new rational conception of the subject', this 'establishment of the necessity of quantitative methods of investigation', and this 'formulation of . . . empirical laws' were of considerable significance. It was this changed attitude to the conduct of experimental research which made the empiricism of a man like Wedgwood more informed, more persistent, more exhaustive and more rational than the empiricism of the uneducated untutored craftsmen. It also made it more economically purposeful and more commercially productive. The succession of effective inventions and improvements which resulted from Wedgwood's experiments indicates something more than the haphazard investigations and accidental discoveries.

systematic education in science or technology. . . . In this rise of British industry . . . formal education of any sort was a negligible factor in its success. . . . There was practically no exchange of ideas between the scientists and the designers of the industrial processes'.[42] Others, like Professors J. D. Bernal, M. M. Postan, E. E. Hagen and D. S. Landes, have found the link between the Scientific Revolution of the sixteenth and seventeenth centuries, and the Industrial Revolution of the eighteenth, to be, at best 'an extremely diffuse one'.[43]

In sharp contrast, Professors T. S. Ashton, W. W. Rostow, R. E. Schofield, and, of course, Messrs. Musson and Robinson have seen 'the stream of English scientific thought' as 'one of the main tributaries of the Industrial Revolution'.[44] Dr. Samuel Lilley has claimed even more for science, writing 'science, at the highest level then available, was the key to techno-logical innovation'. Of a list of 'recorders and projectors' intimately con-nected with industry he writes they 'made full use of all that science could offer', and he adds 'Science played a similar, if not always so central a part in all the chemical industries of the eighteenth century, and in other industries that rely heavily on chemical transformations, such as pottery, glass and paper making'.[45] Rostow, in enumerating his pre-conditions for economic growth, singled out as one of the two essential features 'the initially slow but then accelerating development of modern scientific knowledge and attitudes';[46] and, in enumerating the causes of economic growth, his first two are 'the propensity to develop fundamental science and to apply science to economic ends'.[47] But perhaps the best known statements in favour of a more central rôle in the Industrial Revolution for scientists and scientific knowledge occur in the work of Messrs. Musson and Robinson whose 'findings necessitate', they have argued, 'considerable modification of the traditional view of the Industrial Revolution as being almost entirely a

[42] Sir Eric Ashby, *Technology and the Academics: An Essay on Universities and the Scientific Revolution* (1958), pp. 50–1.

[43] D. S. Landes, *Cambridge Economic History of Europe*, Vol. VI, Pt. 1 (1965), p. 293 and pp. 550–1; J. D. Bernal, *Science in History* (1954). Others have been more certain in their judgements like E. E. Hagen who has written that the technical advances of the eighteenth century 'were the products of technical ingenuity far more than scientific logic'. He dismisses the suggestion that these advances were the result of a combination of economic incentives and the accumulating progress of scientific knowledge, writing quite categoric-ally, 'They were not'. 'British Personality and the Industrial Revolution: The Historical Evidence' in *Social Theory and Economic Change*, ed. by Tom Burns and S. B. Saul (1967), p. 41.

[44] T. S. Ashton, *The Industrial Revolution* (1948), pp. 15–16.

[45] Samuel Lilley, 'Science in the Early Industrial Revolution', ch. 12, pp. 49–51; Vol. III, ch. 3 of *Economic History of Europe*, ed. Carlo M. Cipolla.

[46] W. W. Rostow, *The Stages of Economic Growth* (1960), p. 31.

[47] W. W. Rostow, *The Process of Economic Growth* (1952), p. 23.

product of uneducated empiricism'.[48] Their massive scholarship has perhaps received less praise than its due;[49] and their conclusions have been less than universally applauded, partly because of an essential difference between their criteria for what was truly scientific, and the more stringent criteria adopted by, say, Professor Hall.[50]

The controversy has remained irritatingly unresolved partly because, despite the pioneer work of Dr. and Mrs. Clow, Professor Schofield and Messrs. Musson and Robinson, 'much more research is required', as the last two admit, 'on particular industries, industrialists and scientists'. Musson and Robinson have made full use of the Boulton and Watt papers. and Schofield has dipped productively into the Wedgwood papers, but even for the activities of industrialists so closely associated with Priestley, Keir, Black, and the Lunar Society, there is little agreement on the extent to which they were inspired, informed and influenced by science. Certainly there is little sign of an agreed consensus on the importance of science in Wedgwood's commercial activities.

With Wedgwood part of the problem is that there is so much surface evidence of his apparent scientific interests and achievements—F.R.S., publications, scientific research scheme, scientific library, scientific correspondence, membership of scientific societies, etc.—and yet so much evidence that both his mental processes and his industrial processes were based on a deeply empirical approach to problem solving.

Sir Kenneth Clark, in his great work on Leonardo da Vinci, wrote, 'To the end of his life he continued to draw patterns of squares, circles, and arcs, trying to exhaust every possible combination, rather as a chemist might try every possible combination of fluids in order to discover the elixir of life'.[51] Josiah Wedgwood showed a similar restless preoccupation with the possibilities of clay—and a similarly exhaustive experimental approach to his search for solutions. The evidence of the ten thousand trial pieces, which he produced on his route towards the perfection of his

[48] A. E. Musson and Eric Robinson, *Science and Technology in the Industrial Revolution* (1969), p. vii.
[49] See Professor D. C. Coleman's review in *Economic History Review,* 2nd series, Vol. XXIII, no. 3 (December 1970), pp. 375–6.
[50] A. R. Hall, 'What did the Industrial Revolution in Britain owe to Science ?', *Historical Perspectives, Studies in English Thought and Society* (ed.) Neil McKendrick. (Forthcoming.) A. R. Hall, *The Historical Relations of Science and Technology* (1963), pp. 127–8, 'it is easy to overlook the smallness of the scientific community of the past . . . the impact of any question of abstract science upon a human brain was exceptionally infrequent—it could only happen to, say, one individual in a hundred thousand . . . virtually all the basic techniques of civilization up to a couple of hundred years ago were the work of men as uneducated as they are anonymous'.
[51] Sir Kenneth Clark, *Leonardo da Vinci* (1948).

invention of jasper, still survives as physical proof of his persistence and his belief in repeated trial and experiment; and jasper was only one of many major discoveries. From his earliest apprenticeship he had struggled to wrest every possible new glaze, and new body, from the simple clays at his disposal. Every conceivable mixture was tried, every possible combination tested, all the techniques known to the Staffordshire potters, and many unknown to them, were utilized, scrutinized, improved or rejected. His imagination was fertile and fortunately his industry seemingly inexhaustible. Even when he anticipated defeat, his intellectual curiosity and obsessive concern drove him on. 'I have too much experience of the delicacy and unaccountable uncertainty of some of these fine bodies to be very sanguine in my expectations', he wrote, but he felt that the only solution to 'unaccountable uncertainty' was remorseless experiment. Even when he saw the prospect of himself 'ever pursuing—just on the point of overtaking—but never in possession of his favourite object!' he resigned himself to this Tantalus-like existence, writing 'Fate . . . has decreed that we must go on. We must have our Hobby Horse, & mount him, & mount him again if he throws us ten times a day. This has been my case, & I suspect it is pretty general amongst the fraternity'.[52]

In fact few of his rival potters could compete with Wedgwood's devotion to the experimental method. When industrial secrets could be stolen so quickly and so cheaply the Staffordshire potters could afford to allow Wedgwood his unquestioned technical superiority in invention and improvements. Perhaps it was an acceptance of this (and an acceptance of their very different price and profit policy from that of Wedgwood),[53] which explains *their* apathy and Wedgwood's willingness to set up a co-operative experimental research scheme in 1775.

This proposed organization is perhaps the best and most quoted piece of evidence in favour of Wedgwood's belief not only in scientific experiment, but also in the industrial application of that experiment. It is not surprising that it should excite attention, for a public research organization with the stated aim of carrying out scientific investigations for a whole industry has a startlingly anachronistic ring for the eighteenth century,[54] and Wedgwood

[52] 6 August 1775.

[53] At the prices they sold at they could afford to lag behind Wedgwood in novelties and fashionable appeal and yet still find an ample market for their cheaper products. See Neil McKendrick, 'Home versus Foreign Demand in the late 18th Century: The Legend of Wedgwood's Exports', *Economic History Review* (forthcoming).

[54] E. Bostock drew attention to this scheme in 1930 in his article on the 'Contribution of Josiah Wedgwood to the Technical Side of the Pottery Industry', *Transactions of the English Ceramic Society*, Vol. 29 (1930), pp. 39–41; see also R. E. Schofield, 'Josiah Wedgwood and a Proposed Eighteenth-Century Industrial Research Organization', *Isis*, Vol. 47, Part I, no. 147 (March 1956), pp. 16–19, but there is further evidence about the scheme in the Wedgwood papers which Schofield did not use.

first mooted the idea at 'a general meeting of the Potters . . . at Moreton on the Hill',[55] as early as 22 June 1775. He 'propos'd a Public experimental work, after cautioning them against a too precipitate change from a branch of business they were well acquainted with, to one untried by anybody, and quite unknown to themselves. They seem'd to take the caution, and the proposal very kindly, & we are to meet again in a fortnight to try if we can bring our plans to some tolerable degree of maturity, in which I foresee many difficulties will occur'.[56] Bentley foresaw further difficulties, and Wedgwood replied 'Your idea of the Experimental work I am afraid is too true, but the chief objection some of our principal Potters have to the Plan is, that all our improvements when they are known to 100 Members of an Experimental work will instantly be carried out of the Country & out of the Kingdom, for we have some People who now make a trade out of carrying our present improvem[ts] to distant works & receiving sums of money for the purpose. This will make me a little cautious how I proceed in this business'.[57] Despite his caution, Wedgwood was sufficiently convinced of the future benefits to press on with 'a plan for an Experimental work—Wish us good success for it is a most intricate business I assure you'.[58] Five months later the difficulties seemed to be on the point of being overcome, for Wedgwood reported to Sir John Wrottesley that 'we have made some progress in our improvements with the Cornish Materials, but not much, as we wait for the Establishment of a Public Experimental work, which I hope will take place soon. We had a meeting of the Potters yesterday upon the subject, when I deliver'd in a Plan, which was approv'd of, & it was agreed to put it into execution as soon as proper Buildings and other conveniences cd. be prepar'd for the purpose'.[59] Three days later Wedgwood enclosed this plan, headed 'Proceedings in the Scheme of an Experimental Work', in a letter to Bentley,[60] and four days later Thomas Byerley,[61] Wedgwood's nephew and later partner, sent Bentley 'a copy of the agreement as it now stands, and as, I suppose it will be actually entered into'.[62] Schofield by combining these two copies[63] produced his own

[55] *WMSS*, J. W. to T. B., 23 June 1775.
[56] *WMSS*, ibid.
[57] *WMSS*, J. W. to T. B., 3 July 1775.
[58] *WMSS*, J. W. to T. B., 6 July 1775.
[59] *WMSS*, J. W. to Sir John Wrottesley [a copy] dated, 29 November 1775.
[60] *WMSS*, J. W. to T. B., 2 December 1775.
[61] Mysteriously rechristened Richard by Professor Schofield.
[62] *WMSS*, Thomas Byerley to T. B., 7 December 1775. In the postscript he asks Bentley '. . . will (you) choose to become a Member of this Experimental Company, shd. it be determined that every partner of a joint concern must sign and pay *severally*'. This 'difficulty which was started towards the close of the last meeting and was the cause the articles were not then signed, will be talked over—the question is, whether those who are engaged in Partnership shall sign and pay for the whole concern *jointly*, or whether they shall sign or pay for themselves severally'.
[63] *WMSS*, E.18628–25 and E.12426–13.

composite version, an unnecessary, if effective, act of historical splicing, for in fact an actual copy of the final version is neatly recorded in Wedgwood's Commonplace Book,[64] copied out by Alexander Chisholm, Wedgwood's

[64] *WMSS*, E.28408–39, Josiah Wedgwood's Commonplace Book, I, pp. 215–17.

'Heads of an Agreement intended for An experimental work in partnership—1775.

(1) We the underwritten potters and subscribers to the Opposition to Champion's porcelain bill, & the journey to Cornwall, do mutually & severally agree to establish & carry on an experimental work for the purpose of trying the materials lately brought from Cornwall, as well as those which may in future be imported, from that county or any other place, into this manufactory; in order to improve our present manufacture, and to make an useful *White porcelain body*, with a colourless glaze for the same, and to paint the same with Blue under the glaze: and to prevent our being lost in too wide a field at first setting out, we likewise agree that the Experim^{ts} shall at the beginning be confined to these objects only.

(2) That the company shall be established and carried on in all respects as a partnership, and that there be a joint stock.

(3) That

(4) That a deposit of twenty-five pounds be paid down by each of us, on the signing of these articles, into the hands of the treasurer; which deposit shall be our joint stock. But if this joint stock should not be immediately wanted for carrying the plan into execution, it shall be placed out to interest in such a manner as shall be directed by a majority of the company.

(5) That there shall not be any transferring of shares; nor shall any heirs (except a son, or heir who continues the business of the deceased member) any executors, administrators or assigns, become partners, without the consent of a majority of the C^o; but the property of a deceased member, leaving neither son nor heir who continues the business, shall be valued, and satisfaction made.

(6) If any member withdraw himself, and the company be in debt, he shall first pay his share of the debt; but if the company be not in debt, he shall forfeit his whole share as a punishment for deserting the company.

(7) If any member shall refuse or neglect to pay into the hands of the treasurer his proportion of such calls as may from time to time be made on the company, such member shall forfeit 10 per cent of his respective share if he pay not within ten days after the call shall become due, & of which he has had notice; and if he refuse or neglect to pay within twenty days after another notice given him, he shall forfeit his whole share.

(8) No member shall disclose the experiments made by this company, or the knowledge obtained by them, to any person or persons not of the company, under pain of forfeiting his share in the joint stock, and of incurring a penalty of one thousand pounds.

(9) No member shall take advantage of the knowledge acquired by the experiments of this Society, by employing any of the improvements made hereby, in his own manufactory or otherwise, until the time and plan of generally adopting & receiving such improvements into all our manufactories be agreed upon by the Society—under a penalty of one thousand pounds.

(10) That the time & manner of adopting & removing such improvements shall be determined by a number of the proprietors not less than two-thirds of the whole; but if two-thirds of the proprietors should not attend the meeting for this purpose, then a General meeting shall be called, and two-thirds of such meeting be sufficient.

amanuensis, after the scheme had fallen through. For despite Byerley's optimism,[65] Wedgwood had to report to Bentley before the year was out that 'our experimental work is over'.[66] Wedgwood had done all he could to plan and promote the scheme but it finally 'expir'd in Embrio [sic]'[67] because 'we could not settle the question whether the Partners in Co.ʸ shd. pay separately, or jointly . . . I consented to agree to either plan, being determin'd it should

---

(11) That after the first signing of these articles, & the C°. being formed, no person be admitted into the company without the consent of all the members.

(12) That the business of the company be done by general meetings of the partners, or by a committee of them to meet weekly in a morning, at a room in the works when they are fixed.

(13) That five or more in number shall constitute a meeting or committee, for making orders & calls for money, and for doing all other the business of the company, subject however to the following provisos:

1ˢᵗ That more than £5 shall not be called for at one time, nor shall the money for any call be demanded within ten days after the notice given.

2ᵈ That no new call shall be made within one month after a former one.

3ᵈ That notice be given of an intention to make a call, at a meeting that in which the call shall be made.

4ᵗʰ That no order made at a former meeting shall be reversed at a subsequent one, unless there be as great, or a greater, number present at the latter.

5ᵗʰ That if any questions cannot be determined without taking the opinions of the partners severally, that shall be done by a ballot, with black balls & white ones.

(14) That the expence of carrying the plan into execution shall not exceed £50 for each share; and when the expence amounts to that sum on each share, the partnership shall be dissolved, but not before.

(15) That the first meeting for proceeding to carry the plan into execution, be held on the first Friday after lady-day next ensuing, at the house of John Moreton, known by the Sign of the Queen's Arms on Man's hill, at 10 o'clock in the morning.

(16) That on a dissolution of the company, the effects be disposed of in manner following, viz.: Three or five persons shall be chosen by ballot, to allot & value the effects (the Experiment-books, and collection of proofs or results, excepted)—The partners shall have the option of purchasing these effects, balloting for the first choice of a lot, & for the subsequent ones in order: but if there should be any articles left, which the partners do not chuse to purchase, they shall be sold to the best advantage, and the produce of the whole placed to the joint stock, as in other partnerships.

(17) That the experiment-books, and the results to which they refer shall be put up by auction to the Members of the Company *only*, and sold to the best bidder; any member having first had the liberty of copying the experiments from the book.

That then the accounts shall be closed, and general releases given by the respective partners, as at the expiration of any other partnership.'

[65] *WMSS*, Thomas Byerley to T. B., 7 December 1775. Wedgwood, himself, was always less confident, writing to Bentley, 'Do not give yourself any trouble with the papers Mr Byerley & I have sent you about the experimental work. I do not know if it will be worth it'.

[66] *WMSS*, J. W. to T. B., 30 December 1775.

[67] *WMSS*, J. W. to T. B., undated, but written before 30 December 1775, because on that date Wedgwood was asking Bentley to inform Mr. Horn of the scheme's collapse with the clear implication that Bentley already knows about it.

not fail on my account . . . as I heartily wish a general improvement to the Manufacture—But it seems that it cannot be in this way, & having done my duty I am contented, & shall take my own course quietly by myself as I can, & may perhaps have it in my power to serve the trade some other way. . . . I shall now begin with the materials in earnest. I scarcely thought myself at liberty to do so whilst the Partnership plan was in agitation'.[68] As Professor Schofield understandably commented 'the significant point . . . is not that the scheme failed, but that it should have been proposed at all. It indicates a recognition of the possible fruitful inter-relation of science and technology in the eighteenth century, the kind of recognition that current historical opinion has been too inclined to assign to the nineteenth century'.[69] And as 'the first organization of this type ever to have been proposed'[70] it deserves its honourable, if modest, place in the history of relationship between science and industry. Its scope in terms of numbers—a point which Schofield failed to notice—was very considerable, for to combine '100 Members [in] an Experimental work'[71] was a startlingly ambitious project for 1775: but its scope in terms of what was to be investigated—a point which Schofield failed to stress—was distinctly limited, for the proposals were restricted to making 'an useful *White porcelain body*, with a colourless glaze for the same, and to paint the same with Blue under the glaze . . . and to prevent our being lost in too wide a field at first setting out, we likewise agree that the Experim⁵ shall at the beginning be confined to these objects only'.

That in conception it was remarkably advanced cannot be denied. It is yet further evidence of the originality of Wedgwood's thinking, for it is clear that the idea, the proposal, the initiative, the energy and the detailed plans all originated with him. It was his belief in experiment together with his massive powers of organization which kept it alive until it reached 'Embrio' form, but it was clearly ahead of its time. It is more a tribute to Wedgwood's innovative zeal than to the readiness of eighteenth century industry to invest in a co-operative research scheme. Wedgwood never had great confidence in its successful operation, for whereas he could, of his own volition, be the first potter to introduce steam power into the Potteries, be the first potter to introduce a clocking-in scheme into the Potteries, be the first potter accurately to cost his accounts, be the first potter to allow his customers to 'serve themselves', be the first potter to take a successful merchant into partnership, be the first potter to organize a training school for artists, be the first potter to introduce a satisfaction-or-money-back policy, be the first English potter to

[68] *WMSS*, J. W. to T. B. (late December 1775).
[69] *Isis*, Vol. 47, Part I, no. 147 (March 1950), p. 19.
[70] Ibid., p. 16.
[71] *WMSS*, J. W. to T. B., 3 July 1775.

dominate the European market, be the first potter to employ famous artists on an industrial scale, be the first potter to introduce the substantial use of female labour into the potbank, be the first potter to gear his prices to the fluctuating nature of demand, be the first potter to harness the fashionable taste for antiquities to mass production, and so on to a whole series of innovative 'firsts',[72] he could not, in this instance, prevail over the suspicious doubts and hesitations of his contemporaries. So for all its undoubted significance in the history of the changing relationship between science and industry, for all its importance in the history of the birth of industrial belief in experimental research, its actual contribution to economic growth in the eighteenth century was nil. It failed. And when one is attempting to measure the rôle of science in industry—leaving aside for a moment the more stringent interpretations of whether this scheme would have involved science or empiricism—one has to concede that, in this instance, it remained merely hypothetical. It was a lonely forerunner of later schemes which were actually implemented. Unlike so many of Josiah Wedgwood's other ideas it never bore commercial fruit, and one cannot overlook the significance of the fact that it *did* fail. It is a superb example of Wedgwood's passionate belief in the power of experimentation, but as an example of the direct contribution of science to the Industrial Revolution, it is a non-starter. Interpretations which rely on abortive schemes serve only to invite the scepticism of evidentially rigorous scholars. As Professor Hall has written, 'Discussion of the intellectual origins of the Industrial Revolution has, like medieval philosophy, provoked a

---

[72] Neil McKendrick, 'Josiah Wedgwood and George Stubbs: An Essay on the Inter-action of Art and Industry', *History Today*, VII, no. 8, August 1957, pp. 504–14.

Neil McKendrick, 'Josiah Wedgwood: An Eighteenth Century Entrepreneur in Salesmanship and Marketing Techniques', *Economic History Review*, 2nd series, XII, no. 3, April 1960, pp. 408–33. (Reprinted in *Essays in Economic History* (ed. E. M. Carus-Wilson), Vol. III, 1962, pp. 353–70.)

Neil McKendrick, 'Wedgwood and his Friends', *Horizon*, Vol. 1, no. 5, May 1959, pp. 88–130.

Neil McKendrick, 'Josiah Wedgwood and Factory Discipline', *The Historical Journal*, IV, no. 1, 1961, pp. 30–55. (Reprinted in the *Rise of Capitalism* (ed. D. S. Landes) (New York, 1966), pp. 65–80.

Neil McKendrick, 'The Discovery of Pompeii and Herculaneum and the Neo Classical Revival', *Horizon*, Vol. IV, no. 4, March 1962, pp. 42–70.

Neil McKendrick, 'Josiah Wedgwood and Thomas Bentley: An Inventor-Entre-preneur Partnership in the Industrial Revolution', *Transactions of the Royal Historical Society*, 5th series, Vol. 14, 1964, p. 1–33.

Neil McKendrick, 'The Story of the Portland Vase', *Horizon*, Vol. V, no. 8, November 1963, pp. 63–66.

Neil McKendrick, 'Josiah Wedgwood and Cost Accounting in the Industrial Revolu-tion', *Economic History Review*, 2nd series, XXII, no. 1, April 1970, pp. 45–67.

Neil McKendrick, 'Home versus Foreign Demand in the late Eighteenth Century: The "Legend" of Wedgwood's Exports', *Economic History Review* (forthcoming).

division between nominalists and realists. One school of historians seeks to apply (in one formulation or another) the word *scientific* either to the process of technical change in the eighteenth century or at least to the mental habits of those who effected these changes; others, the realists, search without success for precise examples of a technical innovation being derived consciously from pre-existent theoretical knowledge of a non-trivial character'.[73] The nominalists might claim Wedgwood's 'public experimental work' as clear evidence of one of the leading entrepreneurs of the Industrial Revolution thinking like a scientifically-orientated industrialist, and they would be right to do so, but the realists would continue to search in vain for any sign of the successful application of scientific theory to a technical problem.

Much of the evidence of Wedgwood's experimental zeal would be more persuasive to the nominalists than to the realists. His restless desire to quantify led him into elaborate measurements not only of such vital questions as the temperature of his kilns or the cost of his labour, but also of such transient, even trivial, concerns as the numbers of bricks in a given area of wall or the exact distance covered during a complete tour of his ground floor. Pages of elaborate calculations occur over and over again in the Wedgwood archive: calculations recording the cost per yard of printing the different patterns on his plates, the differential costs for male, female or apprentice labour, the cost for different sizes, the cost for different qualities of decoration or different qualities of paint, and all bear testimony to a belief that only by discovering and measuring the facts would one make progress possible and probable. The restless itch to know and to quantify found ample scope for exercise. Casually observed phenomena are characteristically reduced to, and recorded in, numerical terms. A lime work at Linsell seen in 1776 yields the jotting '9 kilns—100 strike per week at each kiln—at 4½″ pr strike'.[74] 'A Railway in Cumberland' seen in the same year elicits the brief but pertinent comment to a man interested in the economics of transport: 'near two miles—To support the railway with timber and workmanship, 1 shilling for each carriage going up and down—the owners of (the) colliery are obliged to carry 33 waggons (about 100 Cumberland tons, 1400 cwt) per day'.[75] Should Wedgwood ever wish to consider such a method of transport the salient facts were there recorded, indexed ready for use. Where the facts were for immediate, not future, use everything that measurement can possibly squeeze from them is extracted. At times the obsession with exact

---

[73] A. Rupert Hall, 'What did the Industrial Revolution in Britain owe to Science?', *Historical Perspectives, Studies in English Thought and Society in Honour of J. H. Plumb*, ed. Neil McKendrick (forthcoming).

[74] *WMSS*, E.28408–39, p. 36.

[75] Ibid.

knowledge and precise quantification seems to border on mania. A single example should suffice. From 1767 to 1784 his concern with the fineness of the sieves so necessary for the control of the quality of his clay led him to a constant search for improvements. A Mr. Mills of Manchester was applied to in 1767. When he failed 'an ingenious wire worker in London, Mr. William Panton of Turnmill Street' was tried. That he too failed, along with Mr. Edward Smith of Manchester and Mr. Dominick Rouse of Shoreditch, is not surprising in view of Wedgwood's requirements which he recorded in his Commonplace Book.[76]

The Lawns used for sifting the clay for the Thermometer pieces of which an account is given in the following paper, appeared, upon examining several pieces, to contain 200 by 180 threads on a square inch, that is 36000 partitions of thread, & consequently the same number of interstices, in an inch square. If therefore the intervals between the threads be equal to the thickness of the threads themselves, each interstice must be $\dfrac{1}{72000}$th part of a square inch: if the threads are double in thickness to the interstices, $\dfrac{1}{144000}$dth part.

In the above computation, the number of threads and interstices being surrounded by 4 threads, the number of threads in the square inch will be near double—— If the inch contains 200 threads by 152, as counted by Mr. Smith, page 25, then the square inch will contain, of thread partitions

$$200 \times 152 = 30400$$
$$\text{and } 199 \times 151 = 30049$$
$$\text{of Interstices,} - 199 \times 151 = 30049$$

$$\overline{}$$

90498th of a sq^r inch

So that if the threads be equal in thickness to the interstices, each interstices will be $\Big\}$ one

and if the threads are double in thickness 180996th.

He carried his computations to such lengths because of his conviction that without such constant and precise attention to detail, many of the startling advances in technique and in quality which he made would not have been possible. Quantification became a habit of mind with Wedgwood with as valuable consequences for his experimental method as for his costing techniques. Inevitably not all of his calculations proved useful: many were wholly unnecessary; many potentially useful measurements were never exploited, for his belief in the careful quantitative method and his almost automatic pursuit of evidence in quantifiable form and his compulsive need to record it, gave his experiments far greater chance of success and a far

[76] Ibid., p. 82.

greater possibility of being successfully repeated over and over again on the production line. When, in his suggestions for improving the quality of 'Achromatic Glass', he proposed a process which 'will divide it into 11 thousand billion of laminae' and felt the need to record the figure (11,251,338,370,842,624),[77] he displays the lengths to which he carried his beliefs. On many occasions his suggestions proved abortive. Unfortunately 'all things did not yield to carefully quantitative experiment' but an impressively large number of Wedgwood's problems did.

With the preoccupation that I have detailed, the belief in the necessity of quantitative methods of investigation is of far greater significance than the subjects these exact methods were practised upon. For it is in this quantitative method (the belief in it and the undeviating practice of it) which (together with the belief that one should always make whatever use one could of the available scientific method and knowledge) gives a different dimension and a different quality to Wedgwood's empiricism—a quality very different from that of the uneducated craftsman. Almost inevitably it also gave it a far greater chance of arriving at a successful conclusion.

In much the same way his eagerness to submit unknown materials to the test led to experiments equally distant from anything which could reasonably be called scientific. No subject was safe from Wedgwood's experimental approach. When he wished to find the best fertilizer for his garden and the best way to use it, he automatically adopted an experimental approach in his pursuit of the answer. He conducted the trials himself. He headed a new page in his journal for August 11th, 1781,[78] 'Some Experiments for increasing the effects of a given quantity of Lime as a manure, and likewise of Dung, by dissolving them in water, and watering the land with their solution'. He proceeded carefully to record the various stages of his experiments: dung alone was tried and dung with water, dung in lumps and in solution, and finally all the available varieties of dung were tested, neatly headed 'Dung of different kinds, but chiefly Cow'.[79] Much of these trials is described with the formality of a mathematical proof, the instructions are written with the absorbed care of a French recipe, for Wedgwood was wholly absorbed by such problems, and wholly confident that his methods were the only route to a successful discovery. Experiment, together with the knowledge that could be derived from a scholarly knowledge of all available sources, was, he argued, the proper approach to all problems. Nothing could better illustrate Wedgwood's attitude of mind than such examples, but it was an

[77] Ibid., p. 113.
[78] *WMSS*, E.28408–39.
[79] Ibid. Quoted in Neil McKendrick, 'Wedgwood and His Friends', op. cit., Note 72, p. 94.

empirical attitude, not a truly scientific one. Here again the significance clearly lies in the belief in the experimental method as the key to problem-solving rather than in the intellectual content or the economic results of the actual experiment. It is evidence for the 'nominalists', not the 'realists', but it should not be too lightly disregarded on that account. Informed empiricism, logically conducted and remorselessly pursued, may not be science, but it is something more than the product of 'uneducated craftsmanship' undertaking 'haphazard experiments out of idle curiosity'. Such an approach certainly existed and produced notable discoveries. As Joseph Black said, in a letter to James Watt, of Henry Cort, pioneer and inventor of the puddling process, 'he is a plain Englishman without Science but by dint of natural ingenuity and a turn for experiment has made such a Discovery in the Art of making tough Iron as will undoubtedly give to this Island the monopoly of that Business'.[80] But that approach, which fits so well with the traditional explanation of 'an Industrial Revolution . . . created by unlettered craftsmen of genius fumbling in the dark but divinely inspired',[81] co-existed with an 'empiricism made fruitful by a new rational conception of the subject'.[82] For Wedgwood's empiricism *was* rational, *was* logical, *was* aware of the need for quantitative methods. It was persistent, exhaustive, precise *and* economically motivated. And it was as informed as contemporary scientific knowledge could make it.

That Wedgwood was an assiduous experimenter is impossible to deny. That his empiricism was informed and knowledgeable can also be easily proved. That his experiments were backed by a familiarity with a wide selection of contemporary scientific literature is also well attested. His notes contain references to a rich army of such sources, and his knowledge of them was sufficient for him to be able to furnish his correspondents with precise page references to scientific problems which they discussed. On 2 December 1793 Wedgwood wrote to Mr. Gisborne, in answer to a suggested method of increasing the intensity of fire by applying dephlogisticated air, pointing out at great length the faults of this method, and advising Gisborne to read Lavoisier's *Elements of Chemistry* and to look particularly at p. 474 of the English translation.[83] There are many such examples, but a clear distinction between Wedgwood and the empirical craftsman of Industrial Revolution text-book fame is, perhaps, best recorded in Wedg-

[80] Eric Robinson and Douglas McKie (eds.), *Partners in Science. Letters of James Watt and Joseph Black*, 1970, p. 140.
[81] D. C. Coleman's review of E. Robinson and D. McKie, op. cit., in *Econ. Hist. Rev.*, 2nd series, XXIV, no. 2, May 1971, p. 300.
[82] W. Ashworth, op. cit., pp. 27–33, 311–12.
[83] *WMSS*, J. W. to Mr. Gisborne, 2 December 1793.

wood's commonplace book when Wedgwood describes his conversation
with a Liverpool glassmaker on experimental problems of interest to both.[84]
'Mr. Knight, who keeps the only working Glass house now in Liverpool,
the others having all given over, on account of the last heavy duty' was
according to Wedgwood 'a sensible man, & a *working* (Wedgwood's
emphasis) master in a small way of business'. He was an experienced man
who had served his time with Mr. Grazebrook of Stourbridge[85] and
Wedgwood particularly noted that he 'has made many experiments himself,
& is very willing to make more for me'.[86] Yet although Wedgwood found
the conversation fruitful and recorded his intention to try[87] several of
Knight's ideas, he added the very significant remark '*To make myself understood I did not find so easy as one would imagine—owing chiefly to my not understanding
sufficiently the technical, nor he the philosophical terms, so that in fact we spoke two
languages*'.[88] With Mr. Knight Wedgwood had to spell out his questions in
the simplest form, and on some queries he realized that he would do better
to consult written authorities, as his curt self-instruction, 'Examine Neri,
Kunkel &c.'[89] clearly indicates. Mr. Knight was, like Henry Cort, one of
those 'plain Englishmen without science' but with 'a turn for experiment' of
textbook fame. Wedgwood had no more difficulty in categorizing Mr. Knight
than Black had in categorizing Henry Cort. The economic value of their
empiricism obviously varied with their 'natural ingenuity' and perseverance.
Both Black and Wedgwood recognized the achievements of such men, both
clearly recognized the difference between the craftsman's approach and their
own, and both had a greater ultimate faith in their own more 'philosophical'
methods and sources.

In fact in view of the frequent references to the work of other scientists
it is difficult to doubt that Wedgwood was well acquainted with a very wide
range of chemical literature. Certainly he made enthusiastic references to the
help which they provided, writing of James Keir's translation of Macquer's
Dictionary of Chemistry, 'I have bought a Chemical Dictionary Translated by
Mr. (late Captn) Keir with which I am vastly pleased . . . I w^n not be without

[84] J. W. Commonplace Book. p. 185 all quotes.
[85] Possibly the glass works at Stourbridge briefly managed by James Keir, F.R.S.
It is interesting to note that Wedgwood had earlier (1777) consulted with Keir on one of
the same problems he raised with Mr. Knight. See *WMSS* E.28408–39, p. 29. Keir
published some reflections on this problem in *Philosophical Transactions*, lxvi (1776),
pp. 530–42, 'On the Crystallizations Observed on Glass'.
[86] P. 185.
[87] P. 186. 'Mr. Knight shewed me . . . a Bead made for the African trade, in imitation
of Opal' and explained the necessary technique. Wedgwood resolved to 'try it', p. 186.
A further example of his constant willingness to experiment.
[88] P. 185, my italics.
[89] *WMSS*, E.28408–39. J. W. Commonplace Book, I, p. 185, 30 June 1783.

it at my elbow on any acct. It is a chemical Library!'[90] He eagerly bought Priestley's scientific publications. He subscribed to Priestley's *History of Optics* in 1772, and he bought, and extensively annotated, both volumes of Priestley's *Experiments and Observations on Different Kinds of Air* which were published in 1774 and 1775. He had always been keen 'to contribute in any way you can point out to me, towards *rendering Doctr. Priestley's very ingenious experiments more extensively usefull'*.[91] At this time Wedgwood and Bentley were interested in 'experiments relating to gilding by Electricity',[92] but Priestley's work does not seem to have been commercially fruitful on this occasion, and when years later he was credited by the *Cambridge Intelligencer* with having invented 'a valuable improvement in the method of gilding', Priestley denied having done so and gave the credit for the invention to Boulton.[93]

Wedgwood, however, continued avidly to explore Priestley's works for useful techniques or discoveries. His Commonplace Book simply bristles with references to such problems. The first pages of the index to Volume I of his Commonplace Books gives some indications of his overriding pre-occupation with the problems of ceramic chemistry, and its industrial application.

Achard, Mr. his experiments on Earths, by strong fire, p. 123, 152
Acid
    of Blacklead, its general properties, 129
    of Heavy Spar——ib.
    Phosphoric, found in the mineral kingdom, 129
                lead found mineralised by it, 131
Air
    Observations on the generation of air from various bodies, glass &c. by heat; & the means of obviating its ill effects in pottery—43, 54
    In weighing gravity of airs in a bladder,—a plaster mould proposed for the bladder, that it may be distended equally (by filling the mould) in all the exp^to. 34
    Alabaster & light ashes proposed for confining nitrous air, 40, 41
    Hint for applying Priestly's production of pure air, &c. to improving the air of sick rooms—105
    Inflammable: The more slowly heat is applied to phlogistic bodies, the less inflammable air is produced—37
    Fixt: Copper & lead found mineralised by it, 131
Alcali, fossil—method of extracting from Sea Salt, 59
Ale, See *Vinous* liquors

---

[90] *WMSS*, J. W. to T. B., 30 November 1771.
[91] *WMSS*, J. W. to T. B., 9 October 1776 [My italics].
[92] *WMSS*, J. W. to T. B., 2 March 1767.
[93] R. S. Schofield, *The Lunar Society of Birmingham*, 1963, pp. 92–3.

Alum earth
> To obtain pure—130
> The most refractory of earths—104
> Hints for using it to give opacity to glaze. 104 for getting out grease spots, ib. imbibes lime from water—130 (if the alum earth has been precipitated by *caustic* alcali)

Aqua tint Engraving, process of—173
Arsenic, peculiarly disposed to vitrefy with clay, not with other earths, 103
Arkwright, Mr.—Proposal for spinning wool by machinery, 325.[94]

But his detailed comments on Priestley's publications give an even clearer indication of how he constantly tried to apply what he read to his own immediate industrial problems, how he tried to resolve Priestley's teaching on Phlogiston with his own observations and other theories known to him; and how his respect for his scientific friends did not exclude his ultimate belief in an experimental check on their findings. Emphatic comments like 'According to Mr. Crawford's theory it is so'[95] display a confident belief in his own judgement of the value and conclusions of contemporary theorists. Further comments show that he compared Priestley's findings with other theorists and with his own experience. 'The Experiments of Mr. Walsh with a more perfect Vacuum seems to be in favor of Crawford's theory, and the whole section appears to me consistent with it. Some facts I have lately been furnished with by a philosophical friend have a tendency to confirm the same theory . . . when the air is highly dephlogisticated, and consequently will not receive the Phlogiston so rapidly from bodies in combustion, the works depending upon fire & which require a given degree of heat to bring them to perfection, take more time and fuel to bring them to that point than is necessary in colder weather, and this he instances in Glass Houses and Iron Furnaces. The same is observed by our Workmen in the cooling of heated Masses of Matter.'[96]

But more pertinent than either of those comments, is the one that Wedgwood attached to p. 6, vol. 3 of Priestley, headed '*Calces of Metals perfectly dephlogisticated*'.[97] Wedgwood obviously scented the possibility of a further improvement in his enamel colours and was eager to experiment, writing 'These Calces may possibly improve Enamel Colors, and I wish to have a small specimen of the Calces of Iron and Copper to make a few Experiments upon, and to try whether they produce a finer, or more or less color than the Calces we use, which are slightly calcined in the common way'.[98] The next paragraph indicates that whatever his respect for theory,

[94] *WMSS*, E.38408–39, p. 1 of Index.
[95] Ibid., p. 30. Commenting on Vol. 1, p. 280 of Priestley.
[96] Ibid, p. 30.
[97] Ibid., p. 31.
[98] Ditto.

and his attempts to make his observations conform to theoretical requirements, theory was for him only a guide to be tested by experiment

> *According to the Theory of Phlogiston being the cause of color* [he wrote], *they ought, perhaps, to produce less of it than Calces not so completely dephlogisticated.*[99] I say, *perhaps,*[100] because *admitting the Theory, the inference may not follow;*[101] for, in order to produce any vitreous color, the glass, with the Calx united, must pass through a red heat, which may furnish an opportunity of acquiring any quantity of Phlogiston it is disposed to receive; and whether that disposition is *increased* or *diminished*, or what other dispositions it may have acquired from the dephlog-isticating, *I would rather trust to Experiment than to Theory alone to determine*[102, 103].

Not only was Wedgwood willing to test Priestley's theoretical expectations and experimental findings by subjecting them to his own experimental enquiries, but he was very quick to fasten on to any possibility of applying Priestley's findings to his own industry. In his Commonplace Book quotations from Priestley are followed by Wedgwood's industrially purposeful com-ments and queries: 'Vol. 2nd. p. 126 "Oily matters become extremely viscid by mixing with Spirit of Nitre" Qn: May not this quality be rendered useful in Cements, Glues or Varnishes?'[104]

Another example shows how a simple statement of scientific fact by Priestley triggers off an excited industrial response from Wedgwood. Even more significant, perhaps, the response requires a quick perception of the chemical methods involved.

[99] My italics.

[100] Wedgwood's italics.

[101] My italics.

[102] My italics.

[103] Ibid., p. 31. Wedgwood's faith in empiricism is frequently attested in his letters—he could be as sceptical of scientific theory as he was impatient of Dean Tucker's economic theory on the relations of a poor country to a rich one. Indeed, in his reply to Dean Tucker's pamphlet, *Reflections on the present matters in dispute between Great Britain and Ireland* (1785) he wrote sceptically of both, 'The author's theory is certainly very ingenious, but this is a province in which theory is at best but a blind guide. The theorist may put cases, assume data, and upon those data he may settle the fates & fortunes of empires—with the same demonstrative certainty that Archimedes cd. have lifted the earth . . . all the postulates of the mathematician, cd. have been given to him, he wd. have found himself miserably decieved in practice, because he had not taken into his calculation even the properties of that very power by which the effect was to be produced; and the commercial theorist will find himself equally deceived, from neglecting circumstances as essential in his case as gravitation was in the other.

'Instead of supposing *fictitious* cases, let us briefly state the *real* one.'
Having done so, Wedgwood felt constrained to add that 'one would have imagined' these arguments to be 'too obvious to escape a writer of the dean's penetration and acuteness, and which indeed most possibly woud not have escaped him *if he had not been wedded to a theory, which, when once espoused, must be maintained in all events*'. (My italics. *WMSS.*)

[104] Ibid., p. 36.

'Vol. 2ⁿᵈ: p. 197. "Fluor Acid Air dissolves Glass very freely." *This may be a valuable discovery for bringing Enamel Colors to their utmost degree of Perfection.*[105] At present when the Colors are prepared in the form of colored Glasses, the next process is to reduce them to a very fine powder by pounding & grinding upon a Stone or Glass, and their beauty depends in great measure upon the degree of this Levigation. *But every Chemist will perceive the difference there is—* one may almost say infinite—*between the degree of fineness procured by Levigation, and the chemical solution of any body.*[106] To say nothing of the unavoidable adulteration of such color by the substance upon which it is ground nor *the saving wᶜᵇ· may perhaps be made in time & expence by this method.* [107], [108] The techniques are simple enough, but here surely is an example which leads one to question Professor Hall's disparaging comment on Wedgwood's chemical knowledge: 'being familiar with chemists' terminology, he used it, but it had no technological (or logical) significance'.[109] Much of Wedgwood's parade of chemical terminology *was* doubtless no more than a way of rationalizing what was physically observable, and it is undeniable that ultimately he preferred observation to theory. But it seems arguable at least that here his familiarity with chemical terminology allowed an immediate realization of the technological possibilities of Priestley's statement which would not have been open to those lacking what Wedgwood called 'a knowledge of the philosophical terms'.[110] Indeed, without such knowledge he would not have been reading, annotating and criticizing Priestley's chemical writings. As Wedgwood noted *'every Chemist will perceive the difference . . .* procured by levigation, and the chemical solution of any body'. But it is to be doubted whether the uneducated craftsman would have been so quick to recognize the possible industrial significance of Priestley's brief sentence, or, indeed, the possible industrial economies. Familiarity with chemists' terminology was not always without its technological significance.

Wedgwood was as alive to the possibilities of technical advance and industrial economy[111] as he was to new commercial opportunities and few would now doubt that he was acutely responsive to those.[112] But he proceeded in this—as he did in response to other 'hints . . . for the extension and improvement of my manufacture'[113] from his scientific friends like Dr.

[105] My italics.
[106] My italics.
[107] My italics.
[108] P. 36.
[109] A. R. Hall, 'What did the Industrial Revolution owe to Science?' (forthcoming).
[110] *WMSS*, E.28408-39, p. 185.
[111] Neil McKendrick, 'Josiah Wedgwood and Factory Discipline', op. cit., Note 72.
[112] Neil McKendrick, 'Josiah Wedgwood: An Eighteenth Century Entrepreneur in Salesmanship and Marketing Techniques', op. cit., Note 72.
[113] *WMSS*, E.28408-39. Josiah Wedgwood's Commonplace Book, I, p. 240.

Priestley,[114] Dr. Fothergill[115] or James Watt[116]—'as sound theory and experience shall dictate'.[117] All that he read was subjected, if possible, to Wedgwood's own experiments. Alexander Chisholm, Wedgwood's scientific amanuensis, recorded one such occasion—

Extracts from Musschenbroek's Tentamina experimentorum naturalium. 1731. [This work consists of Experiments made by the Academy *del Cimento* at Florence with a Commentary & additional experiments by Musschenbroek—In the following extracts, those of the Academy are distinguished by F. Mussenbroek by M.

De Congelatione vol. 2[d].

Let a vessel be taken full of snow, into which let another be put containing water; let the whole apparatus be set over the fire, and as soon as the snow begins to dissolve, the water will freeze: which happens sooner when the vessel is set over the fire, than if left to itself—This proves, that the fire drives something *out* of the snow *into* the water, by which the freezing of the latter is occasioned—The freezing cannot be owing to want of heat, for it is an addition of heat that produces the effect. M. page 190.

[April 3, 1784. Mr. Wedgwood tried this experiment with snow which fell yesterday—which has been thawing ever since its fall, at first slowly, now rapidly —A large saucepan was filled with it, a small vial with a little water immersed in the snow to the top, & a thin tin-ware cup with water placed in a cavity made in the snow—The Sauce pan was set over, the fire till the snow melted away both from the vial & cup, but no appearance of congelation could be observed in either during the whole time, which was about a quarter of an hour.

Another vial, corked, was buried in some of the snow, in a room whose temperature was a little above 50° Fahr.—also an open vial, & a tin cup, both with boiled water—but though kept in the snow till next morning, there was not the least appearance of congelation.

It may therefore be presumed, that in Musschenbroek's experiment the snow was colder than the freezing point, and that the freezing of the water was owing simply to this greater degree of cold.][118]

Such evidence is a tribute to Wedgwood's independent mind, to his intellectual curiosity, and his willingness to offer an alternative explanation to recorded phenomenon. Above all it is testimony to his endless belief in the experimental method. But it is scarcely testimony to any advanced scientific knowledge. Yet Wedgwood was never afraid to offer comment and advice to his scientific peers in the light of his own experiments. For Wedgwood's willingness to offer aid to his scientific friends was not limited

[114] Ibid., pp. 30–8.
[115] Ibid., p. 33, on Borax and p. 240.
[116] Ibid., p. 35.
[117] Ibid., p. 35. This in response to the experiments occasioned by several specimens of Scotch clays sent to Wedgwood by 'Mr. Watt of Birmingham' in 1779.
[118] *WMSS*, E.28408–39, p. 171.

to his well-known provision of crucibles and retorts. Priestley's gratitude for such apparatus,[119] and his gratitude for Wedgwood's 'serious contribution to the expences of my experiments',[120] was matched by his gratitude for Wedgwood's criticism of his experiments: 'Your observations on zinc, arsenic and antimony, are certainly very just, and account very well for the failure of my experiments with those semi-metals'.[121] Nor was he afraid to draw their attention to experiments which he had read about elsewhere. 'Refer Dr. Priestley to No. 3, vol. I of the Philosoph. Transact. for some curious experiments' reads one entry into his Commonplace Book.[122]

Indeed the references in his Commonplace Book to the work of other experimentalists reveals a familiarity with a wide range of scientific sources. Constant references to such sources pepper his notes: such as 'See Memoirs of the French academy for the year 1737 the article M. Hellots experim^{ts} upon the solution of cobalt & the color's liquors from the same';[123] or 'Mr. Smeaton upon the expansion of metals: Phil. T. for 1754';[124] or 'To discover lead in wine or any liquors. See Neumans Chemistry P. 155. line 30';[125] or 'Hints upon Lime as a cement from Anderson, Higgins and my own experiments';[126] or see 'Lewis's Commerce of Arts, p. 617 where copper is also said to be more fusible than gold';[127] or 'Mr. De Luc says very justly in his observation upon physical resources, Phil. Tr. Vol. 68, p. 483';[128] or 'Dr. Priestley on Air etc . . . Vol. 2, p. 229. An excellent method of concentrating oil of vitriol';[129] or 'Extracts from the Memoirs of the Royal Academy of Sciences at Berlin for the year 1779—Printed in 1781';[130] or 'See a detail of this manufacture by Mr. Cockshutt. Collect Chem.ii. 290';[131] or 'Macq.^{rs} memoirs on clays. M.S. p. 179. Collect. Chem. vol. vii. Allum earth most refractory. Does not vitrify with glasses—not even glass of lead. Renders

[119] Joseph Priestley to Josiah Wedgwood, 8 December 1782, quoted by R. S. Schofield, *A Scientific Autobiography of Joseph Priestley (1733–1804)* (1966), p. 215.

[120] Joseph Priestley to Josiah Wedgwood, 21 March 1782, ibid., p. 206.

[121] Joseph Priestley to Josiah Wedgwood, 10 October 1782, ibid., p. 213.

[122] *WMSS*, E.28408–38, p. 29.

[123] *WMSS*, E. 28408–39, p. 9.

[124] *WMSS*, E.28408–39, p. 11.

[125] *WMSS*, E.28408–39, p. 9.

[126] Ibid., p. 47.

[127] Ibid., p. 56. This is a footnote to an entry which reads: 'Fusibility of Copper and Gold—Mr. (Hancock*) who made a visit at Etruria recommending Dr. White, April 1787, bears testimony from his own experience to the greater fusibility of copper than of gold. . . .', see *WMSS*, E.7984–10.

[128] Ibid., p. 11. J. A. de Luc was a contemporary Swiss scientist.

[129] Ibid., p. 57.

[130] Ibid., p. 123.

[131] Ibid., p. 184.

the glasses opaque. Try it in a glaze for that purpose';[132] or 'Wallerins, in the Swedish Memoirs, given an Acco^t. of Exp^ts. on the earths of different animal and vegetable substances';[133] or 'Magnesia, its general properties' and 'Manganese, properties of its metals' recorded in 'Extracts from Bergman Sciagraphia regni mineralis 1782'.[134] Quotations in one source are quickly followed up. A crisp reminder that 'Baumé Memoire sur les Argiles (is) to be procured' results from discovering that Baumé is 'quoted by Bergman' to the effect that 'clay is obtainable from animal substances'.[135]

When on the track of information about the chemistry and the behaviour of clay Wedgwood's appetite was voracious, and with the help of Alexander Chisholm he gutted the work of chemists, mineralogists, meteorologists, geologists, vulcanologists and natural historians in his pursuit of knowledge which would help his experiments. Successive pages in his Commonplace Book yield notes on du Halde's History of China,[136] Faujas de Saint Fond's *Recherches sur la Pouzzalane, sur la Theorie de la Chaux et sur la cause de la dureté de Mortier* (Paris & Grenoble, 1778),[137] and De la Faye's *Recherches sur la preparation que les Romains donnaient à la Chaux* (Paris 1777)[138]. This was clearly no 'unlettered craftsman', no 'plain Englishman without science', no 'uneducated tinkerer'.

My purpose at this stage is merely to indicate the range and variety of sources that Wedgwood consulted, and although many of the authorities were recommended to him by his scientific friends and correspondents, and many more were the fruit of Alexander Chisholm's scholarship—the fact that he had such friends and followed up their suggestions, and the fact that he employed a trained chemist like Chisholm, are surely powerful evidence of his interest and his belief in the help that science could provide to an industrialist.

When he was conducting his experiments to find a reliable instrument with which to measure the heat in his kilns, he went systematically through the Philosophical Transactions in order to take advantage of any earlier published work on thermometers.[139] When he recorded his 'thoughts upon

[132] Ibid., p. 104.
[133] Ibid., p. 124.
[134] Ibid., index and pp. 129–32.
[135] Ibid.,
[136] Ibid., pp. 134–41.
[137] Ibid., pp. 141–4.
[138] Ibid., pp. 144–9.
[139] *WMSS*, E.28404–39, p. 10.
'Thermometers, accounts of them in the Philosophical Trans.'

Vol.	Page
17	650. Halleys standard for them+
39	297, 221+ ....

the improvements of flint glass', (ibid., p. 108) he did so only after first discussing Macquer's views on this process with his friend Keir, and thereafter having Alexander Chisholm, his amaneuensis, record the views of the major written authorities: quotations from 'Macquer, Dict. de Chimie, art. Vitrification', 'Baume, Chimie experimentale', and from 'Keir's notes on Macquer' recorded in Chisholm's neat hand are appended on a separate sheet before Wedgwood's own suggestions. It is interesting to note that although he liked to exhaust the scholarly sources available to him for a clarification of a problem and suggested causes, he very often submitted his proposed cure to a technical craftsman.[140] He did not accept the statements of others uncritically, and a reference to another scientist's work is often accompanied by a sceptical note of his own.

He noted for instance that 'according to the opinion of most Chemists . . . Color is . . . supposed to depend upon Phlogiston', but although as a disciple of Priestley he accepted the phlogiston theory, it was not an uncritical acceptance, for he added the query 'from what then proceeds the very full color of those Calces, Iron, Lead &c., after they have undergone the dephlogisticating process?'[141] Of other published work he was even more sceptical, and often very rightly so. '2u. If mercury boils (or nearly boils) at 600 how did Doct^r. Fordyce measure 700 by this instrum^t. The heat at which he says bodies begin to be luminous in the dark. Phil. Tr. Vol. 66. P. 504. Upon asking the D^r. this question at the Royal Societies rooms in the spring of 1780 he told me he did not mean to be accurate as it was of no great consequence in that instance & he guess'd at it as near as he could.'[142]

Vol.	Page
44	686, 672, 675, 693. This is a short history of Thermometers to the year 1746, & an acc^t of a thermometer for heats sufficient to melt metals. It contains some usefull hints.
45	259, 261, 338—130-1201. A very simple metal Thermometer, & may be made more so.
46	1 20B+
47	4+
48	107+
50	300+
51	823+
52	146. A curious but intricate thermometer.
64	214 to 300+Nothing new in the construction.
65	202+
66	174, 320, 370, 588.
67	350 to 382, 816.
68	419 to 553.
	Fahrenheits Thermometer is in No. 381.

[140] Ibid., pp. 108–14 and unnumbered appendage.
[141] *WMSS*, E.28408–39, p. 39.
[142] *WMSS*, E.28408–39, p. 10.

Clearly here the critical note was well justified, but the pursuit of the detail which seemed to contradict Wedgwood's own findings was quite fruitless—one of many such false trails which littered the conscientious experimenter's path in the eighteenth century. An amateur enthusiasm for science clearly had its pitfalls as well as its advantages when a man like Doctor Fordyce[143] recorded his 'guesses' without admitting that they were no more than that.

Not all the victims of Wedgwood's criticism turned out to be so unhelpful. For although he pointed out in his notes on the measurement of temperature that 'The want of known, establish'd measures for heat is nowhere more conspicuous than in Mr. Pott's very valuable experiments in his Lithogeognosie [sic], & must be severely felt by all who attempt to follow him in any of them',[144] he was very ready to admit his debt to Pott on other occasions. In 1767 he wrote to Bentley: 'I could wish to have a translation of Potts Lithogeognosia if you could find out a hand for that purpose, who wo^d not be too expensive'.[145] He was already pursuing his ideas for a new use of 'the *Spath Fusible* or fusible spars which he mentioned in that letter, and later in 1767 when he was planning to visit the Lead Mines near Matlock 'being in the course of Experiments which I expect must be perfected by the *Spath Fusible*, a substance I cannot at present meet with, but I will bring *Pott* along with me, *who will direct us in the pursuit of it*'.[146] Before his experiments were successfully concluded in 1774 with the triumphant production of jasper, Wedgwood consulted Dr. Erasmus Darwin, Dr. Priestley, Dr. John Fothergill, Dr. William Wittering, James Keir and James Brindley, the canal engineer, in his pursuit of what he called *terra ponderosa*. As Professor Schofield commented, the route which led to the invention of jasper is 'one of the best examples of Wedgwood's chemical work, combining science and empiricism, reading and help from his friends'.[147] The combination was often repeated, for although Wedgwood's ultimate faith was always in his own experiments—and it should not be forgotten that the number of experimental pieces of jasper was said to have far outnumbered the famous ten thousand that were preserved—he believed in tapping every

[143] Dr. George Fordyce was a friend of many of the members of the Lunar Society—see Schofield, *The Lunar Society*, p. 142. Both he and Wedgwood met with other 'men of real science, and of distinguished merit' at *Young Slaughter's* Coffee House, see Richard Lovell Edgeworth, *Memoirs* (1820), Vol. I, p. 188.

[144] *WMSS*, E.28408–39, p. 12.

[145] *WMSS*, J. W. to T. B., 5 August 1767. He was at this time experimenting on a new ceramic body which was to culminate in his most famous and most characteristic invention —jasper. In this letter to Bentley he asked him to 'let me know in half a dozen lines what your Dictionary says concerning the *Spath Fusible* or the French making use of that Fossil in their China'.

[146] *WMSS*, J. W. 1767. [My italics.]

[147] R. S. Schofield, *The Lunar Society*, p. 93.

available scientific source for helpful leads, explanations and techniques. Indeed, to judge by his Commonplace Book his belief in the value of other scientists' work increased with the passing years. Not satisfied with personal sources of information which he enjoyed, he scoured every scholarly source on the chemistry of clay that he could obtain as well. Although he recorded a defence of 'the illustrious Pott' from the 'several chemists' who reproached him for 'want of accuracy' being 'persuaded they do him wrong. Baumé and Macquer repeated together most of his experiments, & found the results sufficiently conformable to what he describes', he did allow that 'his work has become much less useful, since the discovery that the earths then reckoned pure, are compounded of others'.[148] Having conceded that there were other more reliable sources he proceeded to annotate them.

His notes and those of Alexander Chisholm were not allowed to lie idle. Wedgwood believed that industrial innovation should ride close on the heels of scientific discoveries, and any new idea which came his way was quickly scrutinized for its potential usefulness. The value of such a quick entre-preneurial response was often as valuable—some would argue more valuable —than the original discovery. Schumpter stated firmly, 'Invention only makes progress possible: innovation makes it real'. One feels that Wedg-wood's more optimistic assessment would have been, 'Invention makes progress probable, innovation makes its fulfilment certain'.

Few potentially useful ideas were ignored. For Wedgwood was as quick to exploit the possibilities of scientific and technological developments as he was to exploit the potentialities of fluctuations in demand. A man who could market black basalt to exploit the female fashion for bleached white hands, a man who could produce piecrust ware in response to the Flour Tax of the Napoleonic Wars, was not a man to miss a commercial opportunity provided by a new technique or a new scientific theory.

Obviously such knowledge was not always commercially helpful in any direct sense, but a specific example of Wedgwood benefiting from the scientific work of others can be found in his Commonplace Book where he records his method of purifying cobalt.[149] Cobalt was a vital colouring agent

---

[148] *WMSS*, E.28408–39, p. 123.

[149] *WMSS*, E.28408–39. Josiah Wedgwood's Commonplace Book, I, p. 33. 27 April 1782.
'Cobalt, To purify—
   Dissolve 1 (illegible) of Cobalt in Aqua-regia. Dilute the solution with water— Mix it with 1^{oz} of Oil of tartar, and stir them well. Filter. Precipitate the filtered solution with another ounce of the oil of tartar & filter again, and so on for five or six times.
   The arsenic, nickel, iron, and a small portion of the cobalt itself precipitate first. The last precipitations are fine cobalt, of which one part will stain 300 parts of flint glass of so deep a blue, that in the lump it is black, in the thinnest strings or hairs it is a fine blue.'

in Wedgwood's famous blue jasper, and a reliable method of procuring it in pure form was an important step in Wedgwood's ability to control the colour of his wares. It is difficult to believe that he could have overcome the technical problems posed in copying the Barberini vase without a mastery of such basic techniques. And he acknowledged his debt when he jotted down, alongside his record of the technique, the words 'given to me as a very valuable process by Mr. Peter Woulfe'.[150] Peter Woulfe was a chemist, who was awarded the Copley Medal for 1768 for his experiments on the distillation of acids, and volatile alkalies.[151]

Wedgwood often acquired new techniques from the chemical journals which provided him with technical short cuts in his experiments, but it is not always explicitly stated that the new process proved economically valuable at the production state. When he read M. Achard's 'On the changes which different earths undergo, when mixed with metallic calces, and exposed to a fire of fusion' which was printed in 1781 in the *Memoirs of the Royal Academy of Sciences at Berlin*, he noted 'I separated the earths by precipitation with alcaline lye, from Solutions of the Salts which I had previously purified by repeated crystallisations'.[152] But whether the new method proved worthy of adoption on a more permanent basis is not recorded. Problems of security, and the fact that Wedgwood confided most of his private thoughts to the commercially-minded Bentley and not to, say, the scientifically-minded Alexander Chisholm has probably robbed us of certain knowledge of exactly what was profitably absorbed from his scientific sources into his manufacturing process.

What is certain is that the possibilities of any new scientific technique or instrument were eagerly taken up and submitted to experiment: when he learnt of William Parker's burning lenses (which Priestley used in his famous experiment in which he isolated 'dephlogisticated air'[153]) he was quick to examine their potentialities for his own experiments and soon recorded a variety of 'Clays tried by Mr. Parker's Burning Glass'.[154]

What is equally certain is that the possibilities of any experiment were optimistically explored for any useful application either to Wedgwood's business or to the world 'May not Dr. Priestley's Experiments in producing pure air be applied to the improving or changing the air of sick rooms

---

[150] Ibid., p. 33.

[151] R. S. Schofield, *A Scientific Autobiography of Joseph Priestley (1733–1804)* (1966), p. 118, n. 2.

[152] *WMSS*, E.28408–39, p. 124.

[153] Douglas McKie, 'An Unpublished Letter from Priestley to John Parker', *Archives Internationales d'Histoire des Sciences*, No. 35, 1956, pp. 117–24.

[154] *WMSS*, E.28408–39, p. 32.

or hospitals ? Possibly the fumes of vitriolic acid might answer this purpose, and precipitate the phlogistic matter in the state of sulphur'.[155]

It is clear too that Wedgwood benefited industrially from the accurate control of temperature made possible by his pyrometer. Indeed, he clearly stated that his pursuit of an accurate method of measuring temperatures above the range of the mercury-in-glass thermometer—a pursuit which continued the noble tradition of the earlier efforts of Newton and Musschenbroek—was inspired by industrial need. 'In a long course of experiments for the improvement of the manufacture I am engaged in, some of my greatest difficulties and perplexities have arisen from not being able to ascertain the heat to which the experiment pieces had been exposed.'[156]

Inspired by industrial need, Wedgwood's criteria for success were not surprisingly based on *industrial utility* not *scientific perfection*, and it is instructive to read his notes on the criticisms of Sir Joseph Banks, the President of the Royal Society, when Wedgwood read to him a copy of his paper on 'the construction of a thermometer for measuring high degrees of heat'.[157] 'Sir Joseph desired to have a copy of the paper just as it was, praised the idea, &c, but . . . objected first to the scale being composed of different shades of colour; in the precise distinction of which, few people were very accurate. I admitted the objection as far as it went, that is, against the *perfection* of this measure; but not against its utility, as being the *best in use* (Wedgwood's emphasis), or indeed the only one offered to the publick.

'The next objection was, if I understood it right, its being united to Fahrenheit's scale upon *Crawford's principles of heat* (original emphasis), which have by no means been established, or allowed of by our present philosophers.

'It was answered to this, that I had endeavoured to point out *a method of uniting them upon the best principles I knew, or which I believed had hitherto been* discovered (my italics), but I was willing to omit that part, as not being essential to the measuring of the high degrees of heat'.[158]

The scientific reception of Wedgwood's invention echoed both the doubts and the recognition of its utility. William Playfair wrote in congratulation on 'your newly invented thermometer' saying 'I have never conversed with anybody on the subject who did not admire your thermometer and consider it as perfect as the nature of things will admit (sic) of for great heats, but I have joined with severall in wishing that the scale

[155] *WMSS*, E.28408–39, p. 105.
[156] *WMSS*, E.28408–39, p. 84.
[157] Ibid., p. 81.
[158] Ibid., p. 82.

of your thermometer were compared with that of Fahrenheit (so universally used for small degrees of heat) that without learning a new signification of affixing a new idea to the term *degree of heat* we might avail ourselves of your useful invention'.[159] Like so many other admiring critics, Playfair joined chemists of the calibre of Priestley and Lavoisier in seeking a supply of Wedgwood's invention. Indeed its usefulness, if not its scientific sophistication, can be judged from the eagerness of Lavoisier and Armand Séguin to acquire it. 'Mr. Lavoisier has sent for two of my thermometers which I have accordingly forwarded to him. Mr. Séguin says "We find this instrument of the greatest use and at the moment feel more than ever its indispensability; because we are occupying ourselves, Mr. Lavoisier and I, in completing the theory of furnaces of fusion".'[160]

Indeed few contemporaries denied its utility. As respected a geologist and vulcanologist as Faujas de Saint Fond sang its praises, writing of Wedgwood and his invention, 'This able artist having daily occasion to study the action and different modifications of fire, has made himself, so to speak, master of that element, taking it captive and dissecting it at pleasure. The course of his enquiries led him to invent a graduated instrument for ascertaining various degrees of heat, which bears his name, and does honour to his genius. The *pyrometer of Wedgwood* has a place in all the laboratories of chemistry and physics'.[161] To indicate its versatility he added in a footnote that 'the celebrated Spallanzani has very happily applied it to determine the degree of fire necessary to fuse the lavas of volcanos'.[162] Later, Guyton de Morveau, who himself devised an apparatus for measuring high degrees of heat, paid tribute to Wedgwood's contribution when giving an account of the methods employed by Newton, Musschenbroek, Mortimer and Wedgwood.[163]

In fact, although doubts concerning the reproducibility of the Wedgwood thermometric scale, which continued to be voiced throughout the nineteenth

[159] Rylands English MSS. 1110. f. 9 (formerly Joseph Mayer collection of Wedgwood Correspondence, now in the John Ryland's Library, Manchester. (R.93406)), William Playfair to Josiah Wedgwood, 12 September 1782. In this letter Playfair suggested a method of overcoming the difficulty, which Wedgwood clearly followed up, to judge from the comment in a letter to him from Priestley, 15 October 1782 (see R. E. Schofield, *A Scientific Autobiography of Joseph Priestley (1733–1804)* (1966), p. 214). 'I am exceedingly pleased with Mr. Playfair's method of reducing the scale of your thermometer to that of Fahrenheit. I think it cannot fail to answer, provided due precaution be taken in conducting the experiment, and those you mention are certainly very important ones.'

[160] *WMSS*, J. W. to Joseph Priestley, 2 September 1791.

[161] F. de Saint Fond, *Voyage en Angleterre, en Écosse et aux Iles Hebrides,* in 1784 (Paris, 1797), translated and edited by Sir Archibald Greikie (Glasgow, 1907), p. 95.

[162] Ibid., p. 95, note 1.

[163] Repository of Arts, Ackermann (December 1809) p. 445. Proceedings of the French National Institute.

century, were subsequently proved to be justified, Wedgwood's pyrometer continued to serve several generations of potters, and in a modified form it is still used in one large pottery firm today. Wedgwood's confidence in its utility proved to be as justified as his willingness to accept that it was not perfect. Perhaps more significant is that it derived from Wedgwood's wide reading of the earlier scientific work on the subject, that it originated in Wedgwood's industrial needs, and that it was sufficiently effective not merely to serve Wedgwood's production purposes, but also to meet the needs of chemists of the calibre of Priestley and Lavoisier.

That it successfully emerged from a background of correspondence and discussion with scientists, that it was indebted to much antiquarian philosophical literature, that it stemmed from a prolonged course of carefully quantified experiments, would be as persuasive to the 'nominalists' of its links with science and its debt to science, as its obvious scientific shortcomings would be convincing to the 'realists' of the meagre rôle which science played in its invention. In much the same way, the fact that what emerged was economically beneficial and industrially useful is as persuasive to the 'nominalist' of the contribution that such scientific interchange and such a scientific background could make, as its lack of scientific sophistication and lack of command of theory would be damning to the 'realist'. As in the whole debate, much hangs on the stringency of the criteria which one adopts.

Where science and empiricism co-exist, and where the intellectual calibre of the science is in such doubt, and the empiricism seems to be so clearly dominant, it is tempting to belittle, or at least to underestimate, the contribution of science. If the contribution, although small and intermittent, were on occasion crucial to the rate of growth, yielding to such temptation could be a serious mistake. If one can show that science made a significant contribution to the bleaching process in the textile industry, if one can show that science made a significant contribution to Wedgwood's method of purifying cobalt, if one can show that carefully quantified experiments derived some impetus from a pervasive belief in science and were helped on their way to a successful conclusion by a knowledge of scientific techniques, if one can show even that industrialists' wide reading in scientific periodicals was directed by a belief that the knowledge thus acquired would aid their industrial and commercial activities, if one can show that these things occurred—and I think that one can—then one cannot entirely dismiss the rôle of science in economic growth. Obviously many of the major characteristics of the Industrial Revolution—the canals and the textile inventions—owed nothing to science; and science's contribution to the steam engine is still very much in doubt. But that does not invalidate a science-aided Industrial Revolution. Few serious scholars would now expect any *single* factor—certainly not science—to be a sufficient explanation of Britain's rapid economic growth. All models of growth—whether sectoral or aggregative—now include a multiplicity of causes, and the catalytic action of science could play a vital contributory rôle in that complex of causes—even if the science involved made itself felt less in terms of the straightforward transfer of scientific knowledge and more in terms of a growing interest and popularity of science, and a growing belief in the contribution that science could make to industry. And it would be difficult to argue against the evidence in favour of such growing interest and growing confidence in the benefits of applying scientific knowledge to industry. The links between the worlds of science and the worlds of industry are now well established and richly evidenced—Sir Eric Ashby's much-trumpeted view that 'there was practically no exchange of ideas between the scientists and the designers of the industrial processes'[164] would find few supporters today, thanks to the work of Musson and Robinson.[165] That these links 'did not all operate in a one-to-one, direct linear fashion'[166] does not invalidate their importance—nor indeed do the facts that in the interchange science often gained as much from industry as it contributed, or that there were timelags between scientific discoveries and industrial innovation, or that the determinants of what industry took from science were economic not intellectual. Such facts are important blocks to a belief in a science-caused industrial revolution—they form a formidable barrier to an acceptance of the idea of economic growth controlled exogenously by the level of scientific knowledge, or initiated by the acquisition of new scientific ideas. But they need not subvert a belief in a science-*influenced* industrial revolution. They need not invalidate a belief in the importance of the contribution of scientific attitudes and scientific methods to economic growth. This contribution can be detected first—and most easily—in the language of assertion, motivation and endeavour: in the pervasive interest in science, in the attempt to harness it to industry, in the belief that 'all things yield to experiment'. It can be measured in the spread of the experimental method, and the growing acceptance of the idea that knowledge and progress would stem from research, logically organized and ruthlessly pursued and empirically tested. Wedgwood's career offers striking proof of the existence of eighteenth-century business men who passionately believed that 'progress would come by experiment, by discovering more facts, by measuring, by analysing; spurred on by the faith that progress *would* come, that the unknown would become knowable'.[167] The transference of such a

---

[164] Sir Eric Ashby, op. cit., pp. 50-1.
[165] Op. cit., *passim*.
[166] Peter Mathias, 'Economic Growth and Economic History', *Cambridge Review*, 30 January 1970, p. 91.
[167] Ibid.

mental attitude from the laboratory to the business world could, when it occurred alongside a greater diffusion of existing scientific knowledge through the new educational channels of itinerant lecturers, provincial scientific societies, dissenting academies, new syllabuses, and scientific journals, be of greater importance than new scientific breakthroughs.

If one accepts such an argument then the intellectual calibre of the science available to eighteenth-century entrepreneurs matters less than the depth of their belief in the pursuit of knowledge through a programme of experiments based on precise measurements, existing philosophic literature, and using the best available techniques. A belief in scientific method and scientific techniques would well prove to be more important than the quality of their scientific knowledge. And it is very easy to over-intellectualize the mental processes by which scientists achieve their results. Empiricism still plays a major part in the scientific methodology in the twentieth century; *a fortiori* it played a major part in late eighteenth-century Europe. Much contemporary science was naïve, unsophisticated and insufficiently critical. Much that passed for science was error-ridden, unscholarly, and at best mere antiquarianism. But it is important that an awareness of the limitations of many eighteenth-century scientists should not lead to a wholesale dismissal of the value of the science which impinged on an industrial manufacturer like Wedgwood. A recognition of the dominating rôle of empiricism, and a realization that the background of theory was very variable in quality, need not necessarily push one into the 'realists' camp. Some of those who do not measure up to the standard tests of scientific knowledge and behaviour set up by the 'realists', look even less like the unlettered craftsman of textbook fame.

Wedgwood was clearly nor unlettered, nor ignorant of the knowledge, methods and ideas of the leading scientists of his day. Indeed the strict criterion, which would deny him the title 'chemist', would one feels, equally deny it to Richard Watson, and come dangerously close to denying it to Joseph Priestley. And insistence on criteria as strict as that can be historically unhelpful. It can hide important distinctions, and conceal important changes in motivation, interest and endeavour. When in 'their hunt for the residuals' economic historians now eagerly examine psychological theories of achievement motivation and sociological theories of status reorientation by way of economic achievement for disprivileged minority groups; when in their effort to explain the changing psychology of the entrepreneur they have enquired minutely into the impact of different methods of child-rearing, into the measurement of need achievement as expressed in contemporary children's literature, into the ethical precepts of different minority sects, into the value and aspirations of different social groupings, into the belief in self-help or the success ethic; when research is expanding into such sophisticated problems and into such subtly distinguished groupings, it seems unnecessarily restrictive and unjustifiably over-simplified for historians of science to set up such crude alternatives for identifying the human agents of technological invention: between on the one hand theoretical practitioners wrestling with the problems of abstract science, and on the other 'men as uneducated as they were anonymous'. One can only feel that the insistence on such demanding and exclusive criteria for establishing the existence of a man of science, or for awarding the label scientist, exacerbates this problem. If the scientific controversy is not to escalate, like so many academic controversies, into a barrier to further historical advance, strict definitions are obviously very necessary, but if they are to do justice to the historical reality of the eighteenth century they must allow for a greater variety in the gradations of scientific knowledge and expertise, and they must not, in assessing the impact of science, exclude a consideration of levels of scientific interest, endeavour and motivation, or an examination of the adoption and spread of scientific method and techniques, or an acceptance of the importance of growing confidence in, and growing expectations from, science.

The cloak of inventive enterprise covered a vast range, from a chemist of the calibre of Lavoisier, through the informed empiricism of Wedgwood, to the unlettered experiments of Mr. Knight. In identifying the position in this intellectual spectrum one finds it difficult not to feel that the influence of science penetrated at least as far as Wedgwood. According to the strictest definitions there were very few scientists in the eighteenth century. As Professor Hall has said 'the impact of any question of abstract science was exceptionally infrequent—it could only happen to, say, one individual in a hundred thousand'[168]—which in the Britain of Wedgwood and the newly founded Etruria would mean only about a hundred men. Even such a tiny nucleus could, if linked, through the agency of provincial scientific societies, with industrially active entrepreneurs, have a disproportionately large economic impact.

If Wedgwood would not qualify for that exclusive group, the evidence of his proposed research organization, his pyrometer, his symbolic formulae, his reading, his experiments, his methods, his techniques, his correspondence, his friends, his employment of Chisholm and his purchase of Dr. Lewis' library,[169] his financial support of scientists like Priestley, and the scientific education he planned for his sons would testify—amply enough—to his

[168] A. R. Hall, *The Historical Relations of Science and Technology* (1963), p. 10.

[169] See F. W. Gibbs, 'A Notebook of William Lewis and Alexander Chisholm', *Annals of Science*, Vol. 8, 1952, pp. 202–20, and F. W. Gibbs, 'William Lewis, MB.., F.R.S. (1708–1781), *Annals of Science*, Vol. 8 1952, pp. 122–51. 'The experimentalist's laboratory described by Lewis was unique in being the first designed specifically for research in applied chemistry. . . .'

belief in science and his close connections with it. If there is little evidence to support any claim for Wedgwood as a pure scientist, there would seem to be plenty of evidence of applied science being actively pursued at Etruria. His experiments alone, however numerous, would not be sufficient support for this view. It has been cogently argued that 'experiments, however cleverly performed, are not science unless they are guided by some sort of theoretical structure. Many things have been discovered by a process of empirical testing, but empiricism is not science'.[170] Fortunately there is supplementary evidence which strengthens his scientific claims. For as Schofield wrote, 'the collection of a library of scientific works shows that Wedgwood's experiments were not simply chains of empirical tests undirected by study of contemporary scientific discoveries . . . there is sufficient proof in Wedgwood's commonplace books of a theoretical direction to his experimentation'.[171]

In fact Wedgwood's scientific writings reveal (like the other evidence in the Wedgwood archive) a similar pattern of empiricism co-existing with a knowledge (if no profound grasp) of other scientists' theoretical findings, and with the empirical approach always dominant.[172]

That the empiricism, and the pursuit of scientific knowledge to aid his experiments, were economically purposeful can be seen from his published papers as clearly as it can be discerned in his experiments. He firmly stated, in an address to the Royal Society, that 'the progress which the manufacturer makes in the improvement of his manufacture will generally be in proportion to his knowledge of the nature and properties of his raw materials, singly and compounded together'.[173] Again in his 'Observations upon Coloured Stones'[174] he began with the view that 'It would be of great importance in several arts and manufactures to find means of introducing beautiful colours into hard and durable materials' and he stressed that his 'attempt to investigate the means by which . . . Nature . . . fixes both plain colours, and infinite variegations, in bodies of the very hardest clay that she produces . . . [was] more than a matter of *mere* curiosity'.[175]

[170] R. S. Schofield, *Chymia*, p. 181.

[171] Ibid., pp. 186–7.

[172] They provide further evidence of the familiarity which Wedgwood had gained with laboratory techniques. His paper on the semi-qualitative analysis of a manganese dioxide ore (published in the *Philosophical Transactions* for 1783—one of the five papers he published in the *Transactions*) records his use of a fractional precipitation with alkali, to separate the iron and the manganese.

[173] *WMSS*, E.19112–26. 'Analytical Experiments on Steatites or Soaprock' by Josiah Wedgwood, F.R.S. Written by Alexander Chisholm. Qualified and extended in a later version—see *WMSS*, E.31346.

[174] *WMSS*, E.19111–26. 'Observations on Coloured Stones', by Josiah Wedgwood, F.R.S. and Alexander Chisholm, pp. 1–23.

[175] Ibid., p. 1.

If his aim was clearly to further the progress of his business, his modest disclaimer on the scientific value of his work reveals equally clearly the limits of his scientific ambitions. 'I am far from attempting any theory of the coloration of stones. A valid theory, on such a subject, must be built upon facts too numerous, and too extensive in their nature, to come within the observations of any individual.'[176]

He did not, however, class this kind of work as scientifically valueless. His observations, 'however inadequate . . . and however inconsiderable they may appear . . . in their detached state', could make, he felt, their humble contribution to what he called '*the chain of science* which . . . had begun to form', by '*supplying . . . some links that were wanting*'.[177] Many scientists of serious repute regarded his assessment as excessively modest. The respectful correspondence between Wedgwood and Priestley, Wedgwood and Lavoisier, Wedgwood and Watt, Wedgwood and Playfair, Wedgwood and Watson, Wedgwood and Warltire, Wedgwood and Seguin, Wedgwood and Keir, Wedgwood and Woulfe, Wedgwood and many others has been well established by Schofield,[178] Musson and Robinson, [178] and McKie.[178] And the traffic in scientific information was not all one way. The help which Wedgwood received from the scientists ('Dr. Fothergill invited me to Lea Hall in order to communicate some hints to me for the extension and improvement of my manufacture'[179]) can be readily matched by the technical advice which flowed back to the scientists from Wedgwood. An unpublished letter from Richard Watson,[180] Professor of Chemistry at Cambridge, is typical of many. Watson's original contribution to science may have been small, but his belief

[176] Ibid.
[177] Ibid.
[178] R. S. Schofield, op. cit., Musson and Robinson, op. cit., E. Robinson and D. McKie, *Partners in Science* (1970).
[179] *WMSS*, E.28408–39, p. 240. See also p. 33.
[180] Watson was elected to the Cambridge chair having never, on his own admission, 'read a book on chemistry or seen an experiment performed' [R. Watson, *Anecdotes of the Life of Richard Watson, Bishop of Llandaff*, pp. 29–30]. His honesty has won him undeserved notoriety as the textbook caricature of scientific ignorance. In fact his *Chemical Essays* (Cambridge, 5 vols., 1781–7) did a great deal 'to bring to the notice of a wider public the new techniques of careful quantitative experiment that we find in the work of Black, Cavendish and Lavoisier'. See L. J. M. Coleby, *Richard Watson, Professor of Chemistry in the University of Cambridge, 1764–71*, Annals of Science, Vol. 9, no. 2, June 1953, pp. 101–22. His critics have argued that his 'experiments . . . are chiefly concerned with manufacturing processes than with the advancement of pure science' [Sherwood Taylor, *Philosophical Magazine*, Commemorative Number, July 1948, p. 163], but there are impressive and authoritative tributes to his rôle as a popularizer of chemistry—'Sir Humphry [Davy] said [in 1813, after the "new" chemistry had been generally accepted in England] that he could scarcely imagine a time, or condition of the science in which the Bishop's "Essays" would be superannuated'. *The Collected Writings of Thomas de Quincey*, Vol. ii (Edinburgh, 1889), p. 197.

in the contribution that it could make to industry matched Wedgwood's. He proposed that an Academy of Applied Chemistry should be founded at public expense to promote the 'proper application of chemical principles'[181] to industrial problems. It is little wonder that such a man should invite Wedgwood to Cambridge, writing 'for tho' I have resigned my Professorship of Chemistry I still retain a great love for the study, and *want to have some information from you upon that branch of it in which you are engaged with so much advantage to yourself & credit to our Country*'.[182]

Scientists of far greater intellectual stature than Watson also wrote to Wedgwood for advice, even men, like Lavoisier and Seguin,[183] whose concern was the advancement of pure science rather than applied. Mutual help and advice (or more often the latest news of scientific research and controversy from the scientists in exchange for technical help and 'a couple of crucibles' from Wedgwood) bound the industrialist to the scholar. A typical example was a letter from Richard Kirwan in 1783 informing Wedgwood, among other things, that 'you may see in the Paris Memoires of 1778 that M. Lavoisier also found that excellent & almost infusible Biscuit may be made of Steatites',[184] and promising to inform him 'of Scheele's process as soon as it is published'.[184] Sandwiched between this helpful information sits a request for a quid pro quo: 'would welcome a couple of crucibles & a retort the same size as you make for Dr. Priestley'.[185]

One could multiply the detailed evidence of such interchanges between industrialist and scientist almost endlessly, but I trust I have examined sufficient to establish Josiah Wedgwood's position as a vital link between the worlds of science and industry, and as a crucial bridge between the two. To do this is to do more than re-evaluate Wedgwood's position as an industrial chemist whose productivity and commercial success were increased and stimulated by the research endeavours of himself and his many scientific contacts. To do this is to do more than claim that science

---

[181] R. Watson, *Chemical Essays*, Vol. i, pp. 39–48. See L. J. M. Coleby, op. cit., pp. 115–16.

[182] *WMSS*, E.4825–6, Richard Watson to Josiah Wedgwood, 9 July 1774. [My italics.]

[183] *WMSS*, J. W. to Joseph Priestley, 2 September 1791, quoting their request for information. It is true that the pure scientist's major interest was in his pyrometer, and their main concern was with the problem of reproducing Wedgwood's thermometric scale. It is also true that Wedgwood could not always satisfy their needs as he freely admitted to Priestley, writing 'Mess: Lavoisier & Seguin have requested some information from me which I do not find myself very capable of supplying out of my own funds.' [*WMSS*, J. W. to J. Priestley, 2 September 1791.]

[184] Richard Kirwan to Josiah Wedgwood, 3 August 1783. John Rylands Library, Manchester. [Wedgwood Correspondence. 1101–1110], 1110, f. 12. [Joseph Mayer collection R.93406.]

[185] Ibid.

influenced the success of a single famous industrialist. Because if one can establish that an industrial leader like Wedgwood benefited in his ceramic inventions from both the stimulus of scientific theory and the short cuts of improved scientific techniques, then one can confidently claim that science influenced hundreds of his industrial competitors. For Wedgwood's industrial secrets—despite all his efforts to preserve them—were quickly stolen and successfully imitated. By such a process of linkage and diffusion through espionage and imitation, the contribution which chemistry made to Wedgwood would be transmitted throughout the potteries of Stafford-shire—and with a growing timelag—eventually to the potteries of the rest of England and of Europe. What Wedgwood owed to science, was very quickly owed by potters who aspired to neither science nor empiricism, who knew nothing of chemistry, geology or carefully quantified experiment, who attended no scientific societies, who read no scientific or even anti-quarian literature, who advised no scientists and received the advice of none in return, who listened to no scientific lectures and who wrote no scientific papers. Such was the dominance of the market of Wedgwood's Queensware, jasper and black basalt, that anything they owed to science was inevitably owed as well by the booming Potteries. Anything that the success of the Etruria production machine owed to science was owed by a whole industry experiencing unprecedented growth—the value of production increased by a factor of thirty in little over half a century.

One of the reasons why the British Industrial Revolution occurred without significant research costs, was that industrial espionage could so quickly carry a new technique or invention to the eagerly awaiting horde of followers and imitators. Men whose knowledge of science was minimal, whose interest was scarcely more, and whose capacity for sustained experi-ment (if it existed at all) was never exercised or explored, flourished because the fertile mind of Josiah Wedgwood kept them supplied with perhaps the most concentrated list of major ceramic breakthroughs ever achieved by one man. It would border on nihilism to suggest that Wedgwood could have achieved so many new ways of exploiting his native clays—green glaze, cream-colour, black basalt, jasper, to mention only the most famous—without the aid of either scientific methods or scientific knowledge.

In a craft where so many had laboured for so long with such erratic and thinly spread advances, such a concentration of invention and improvement suggests something very much more than the result of 'experiments under-taken out of idle curiosity'[186] or 'in response to some pressing economic difficulty'. Wedgwood's experiment books alone are evidence of a different order of enquiry: it was experiment number 411 which produced Queens-

---

[186] W. Ashworth, op. cit., pp. 27–33.

ware; it was experiment number 3681 which yielded in Wedgwood's view the best formula for jasper; it was number 3839 which produced a good yellow jasper.[187] Experimental enquiry carried to this level of persistence, backed by scientific advice, sometimes made possible by scientific method, and rendered industrially reliable by accurate measurement, not surprisingly produced a high yield in successful innovations.

Wedgwood stood alone amongst Staffordshire potters in the quality of his experimental work, but the results were shared. So if one accepts that Wedgwood alone was influenced or aided by science, one can quickly multiply that scientific contribution to the economy on behalf of the whole industry. The queue of Wedgwood imitators ensured that what science had spawned would soon be utilized by the Potteries as a whole. They needed to know nothing of science, nothing of scientific method and technique. With the identification and proportion of the ingredients, and a bribed workman for the method, they could reproduce an approximate copy of Wedgwood's results.

Wedgwood's career, in fact, not only exemplifies the routes by which scientific knowledge was diffused by means of scientific societies, itinerant lecturers, and scientific correspondence; it also exemplifies the means by which the results of science were spread and released and multiplied throughout the industry as a whole. Yet Wedgwood—understandably in view of the ambiguity of much of the evidence—is often quoted as a prime example of one for whom science could have afforded 'insignificant...aid'.[188] If this were so, then as shrewd an entrepreneur as Wedgwood devoted a remarkable proportion of his time to the pursuit of such insignificant benefits.

Just as the vocabulary and metaphor of science was entering the English language (subtly inserting itself into common usage), so the attitudes and techniques of science and in particular of chemistry were beginning to impinge on the world of industry and to influence the rate of economic growth. The most powerful evidence for an interest in science comes from the chemical branches and it was chemistry which was the most influential industry. Wedgwood's career provides individual case-history support for what the Clows have argued in general for the chemical industry,[189] and what Schofield has argued for its influence on the Lunar society—'strong currents of scientific research underlay critical parts of this movement'.[190]

---

[187] *WMSS*, J. W.'s Experiment Books. These four volumes merit a separate article. I have made further use of them in a forthcoming paper, 'Science and Industrial Innovation in the 18th Century England'.

[188] A. H. Church, op. cit., pp. 84–6.

[189] A. and N. L. Clow, *The Chemical Revolution* (1952).

[190] R. S. Schofield, *The Lunar Society of Birmingham* (1963), p. 436.

But in attempting a re-evaluation of the rôle of science in his career, and in trying to redress the balance in favour of a significant scientific contribution, one must be careful not to overestimate science in comparison with more important variables. To argue in favour of an empiricism made fruitful by the quantitative methods of investigation, made more successful by the availability of the best available scientific techniques, and usefully directed by the best endeavour of contemporary scientific advice, is not to argue any initiating or dominating or even central rôle for science in Wedgwood's career.

In Wedgwood's career, as in the economy as a whole, the major pull came from the demand side of the economy rather than from a push from scientifically-induced advance on the supply side. Indeed in the hierarchy of causal significance, science would not rank very high, but that does not mean that it would not rank at all. As a dependent variable, the latent potential of which was released by other more commanding variables, it played a necessary but not sufficient rôle in easing the path of industrial success and economic progress.

XVIII

THE INDUSTRIAL REVOLUTION AND THE RISE OF THE SCIENCE OF GEOLOGY

Roy Porter

'There rolls the deep where grew the tree.
O earth, what changes hast thou seen!'

The earth has always charged human emotions, imagination and intellect. In some Chinese myths and fables, art and science, in Near and Middle Eastern Creation stories, or in the work of a Renaissance genius such as Leonardo, such varied responses fuse in integrated harmony. But not until, say, Tennyson, does an artist deeply defer to the existence of *geology*. Yet, why a self-aware science of geology should have emerged in Europe in the second half of the eighteenth century and the early part of the nineteenth is a question largely ignored by professional historians of science, who have mainly treated of the history of geology from an internal, intellectual standpoint—an approach which may possibly be adequate to explain the succession of paradigms, the dialectic of concepts within the science, but which cannot conceivably account for the existence or nature of the science itself.[1]

1780 to 1830 is the critical period for the development of geology in Britain. Before that, study of the earth, though often acute in its theory and accurate in observation, was nevertheless an incoherent jumble of cosmogony, fossil-collecting, Biblical theorizing, topography and so on, lacking

---

[1] Noting that geology developed much later than mathematics or astronomy, Robert F. Legget at least asks, 'Could it be that, unlike the case with these other sciences, the awakening of modern geological study had to await the start of the industrial age and the practical demands of modern engineering for information about the earth's crust and the materials in it?' Unfortunately, he directs his questions only to a few individual cases. 'Geology in the Service of Man', in C. C. Albritton, *The Fabric Of Geology* (1963), p. 246.

even the name 'geology'. After about 1780, devotees revolted against their 'unscientific' past, and within half a century had united themselves in broad agreement over the principles and methods of their science, and had achieved a uniform nomenclature and an institution, the Geological Society of London, founded in 1807, which guaranteed the standards of the science and the direction of its future progress. Geology had essentially become stratigraphy. Its focus was the delineation of the strata of the British Isles, through detailed local studies.

But this half century is also the epoch of the Industrial Revolution in Britain. 'Next to its constitution, England owes its prosperity very largely to its coal deposits. Without its coal, England would not have one thousandth of the factories that she now possesses', was the very typical comment of Hans Caspar Escher when he visited Britain in 1814.[2] The notion was already commonplace at home: 'No country in the world depends so much upon the productions of the mineral kingdom for the means of comfortable accommodation, wealth and power, as the island of Britain'.[3] Indeed, United Kingdom coal output had leapt from two and a half million tons a year in 1700, and six million in 1770, to ten million in 1800, and thirty million by 1830. But as well as using more fuel, the Industrial Revolution required new minerals and metals. Josiah Wedgwood's quests for china-clays and decorative spars are famous. Furthermore, in addition to mining, other forms of exploitation of the earth—canals, quarrying, road-building—also demanded a greater practical knowledge of the resources and structure of the British Isles.

The urgent new insistence that geology must become scientifically precise, and the switch from 'speculative' theories of the earth to detailed local stratigraphical studies both seem so well adapted to these economic ends that they seem a function of the pressure of industrial needs. Particularly so as William Smith, 'the father of British stratigraphy', was himself a 'practical man', mining engineer, canal surveyor and agricultural improver, who put his own geological ideas to constant use, and always insisted on 'the PRACTICAL ADVANTAGES of Geology and its beneficial application to Agriculture, Mining, Coalworking and Commerce'.[4] 'A knowledge of our Subterranean wealth' declared the President of the Board of Agriculture, Sir John Sinclair, 'would be the means of furnishing greater opulence to the country than the acquisition of the mines of Mexico and Peru'.[5] Did,

[2] In a letter from Sunderland, printed in W. O. Henderson, *Industrial Britain Under the Regency* (1968), p. 46.
[3] John Williams, *The Natural History of the Mineral Kingdom* (1789), i, p. ix.
[4] Summary of a lecture given by Smith to the Yorkshire Philosophical Society in 1824, in John Phillips, *Memoirs of William Smith* (1844), p. 109.
[5] This was adopted as the epigraph of the Royal Geological Society of Cornwall.

then, the Industrial Revolution create scientific geology to discover this wealth, and was geology its loyal servant in maximizing the exploitation of mineral wealth in the industrialization of Britain?

Authoritative recent work argues that the growth of certain manufacturing industries in Britain in the Industrial Revolution (notably those involving chemicals and power), the developing technology of these industries, and the progress of certain sciences, especially chemistry, mechanics and the physics of heat, stand in relation to each other rather as the Three Persons of the Trinity.[6] Here I am investigating the relations between the vast expansion of mining, the 'art' (i.e. technology) of mining, and the science of geology. Clearly, the issues are enormously complex, especially as both mining and geology developed distinctively region by region. But two questions seem basic: firstly, how far did mining (and similar enterprises) create or stimulate geology? And how far did geology aid mining, either directly through the participation of geologists or the diffusion of geological know-ledge, or more indirectly through encouraging a better technology of mining?

Geology and Utility: the Scientific Élite

Certainly two great 'truths', repeatedly asserted by the leading geologists of the period, seem to bespeak a close and fruitful interplay: that geology, as other sciences, should and did progress through co-operation between 'persons attached to various pursuits and occupations . . . the Miner, the Quarrier, the Surveyor, the Engineer, the Iron-Master' and 'philosophers';[7] and that geology was 'of the greatest utility to mankind in a civilised state'.[8] Thus Lyell, Buckland and Sedgwick praised the proposal to establish a Geological Survey as 'a work of great practical utility, bearing on agriculture, mining, road-making, the formation of canals and railroads and other branches of national industry'.[9] Such a justification of the science was thought too obvious to need to be developed: 'It would be superfluous to enumerate the many advantages which may be derived from Geology' the Geological Society of London laconically announced: 'practically considered, its results

[6] A. and N. Clow, *The Chemical Revolution* (1952); R. E. Schofield, *The Lunar Society of Birmingham* (O.U.P., 1963); A. E. Musson and E. Robinson, *Science and Technology in the Industrial Revolution* (Manchester U.P., 1969).

[7] *Geological Inquiries*, a set of queries sent out by the Geological Society, reprinted in *Philosophical Magazine*, xlix (1817), pp. 421–2.

[8] Westgarth Forster, *A Treatise on a Section of the Strata*, 2nd edition (1821), p. vii.

[9] Charles Lyell, 'Address to the Geological Society', *Geol. Soc. London Proc.* ii, no. 44, 1838, p. 247.

admit of direct application to purposes of the highest utility'.[10] Hence, it was a commonplace to claim that the foundation of a geological society in the important mining area of Cornwall was certain to lead 'as well to the progress of geological knowledge as to the advancement of the mining resources of the County' through 'THE DISCOVERY OF NEW FACTS TO ENRICH SCIENCE AND THE APPLICATION OF SCIENCE TO IMPROVE ART'.[11]

These assumptions undoubtedly cohere; but they reveal more about the ways in which geologists thought, than about their activities in the rise of the science. For, except William Smith, not one of the intellectual and social élite of British geology at this time was by birth, by profession, or even by proprietorship importantly involved in mining or similar enterprises relating to geology. Their interest in geology did not derive from mining. Mostly they were gentlemen of leisure—Hutton, Kirwan, Greenough, de la Beche, Murchison etc.; several followed the liberal professions, chiefly medicine, as Babington and Fitton; some like MacCulloch and Arthur Aikin were professional men of science; and a large proportion resided in the universities, usually with professorships, such as Kidd, Jameson, Playfair, Buckland, the two Conybeares, Daubeny and Sedgwick. They clustered in London and Edinburgh, Oxford and Cambridge. They did not make systematic tours of British (or continental) mining areas—the British Mineralogical Society regretted that 'Englishmen should be almost wholly endebted to strangers for an acquaintance with their own mineral treasures, and that the names of Ferber, Klaproth, Raspe, Jars, and Faujas de St. Fond should stand the foremost among those who have illustrated the mineralogical geography of Britain'.[12]

On the whole, these men were not the great publicists of industry; they were country-men, nature lovers. Buckland preferred geologizing where he could ride, Murchison abandoned field-sports for field science. Of course they visited coal and metallic regions, descended mines, talked to miners; not to help discover coal-seams but to delineate the general geology of the area in question, as Buckland and Conybeare did with the South-Western Coal District. They argued, violently and at length, about coal seams and mineral veins—both issues the source of some of the bitterest differences between the Huttonians and the British Wernerians—but their disputes concerned the nature of coal, the origin of mineral veins, not where to find them. The context of discussion was a scientific problem about the relative plausibility of rival explanations of the earth. In fact, Britain had specialists

[10] *Trans. Geol. Soc. London*, i, 1811, p. viii.
[11] *Trans. Royal Geol. Soc. Cornwall*, i, 1818, pp. v–vi.
[12] Notice of the British Mineralogical Society in *Phil. Mag.* xii, 1802, p. 284.

on the Cambrian and Silurian Systems—Sedgwick and Murchison—and at the other end of the scale on tertiary formations—Webster, Lyell, Fitton— before she had an expert on the Carboniferous Series.[13] No distinguished British geologist of the period devoted himself to coalfields. None wrote a popular applied geology.

And if these geologists derived little from personal involvement in practical mining and industrial pursuits, neither did the geological ideas and theories of the age originate from such a source, if we again except William Smith. It is true that developments in the glass and porcelain industries certainly offered useful experimental corroboration of Vulcanist and Plutonist positions on the origin of rocks, but they neither provided the first inspiration for these theories, nor can they account for their growing purchase.[14] The mechanical analogies found in much geological theory almost certainly owe more to the general background of the mechanical philosophy than to industrial machinery. James Hutton may have had some aspects of his theory suggested to him by his friend, James Watt's steam engine; but Hutton wanted to show that the earth's processes were not merely mechanical but organic.

The foundation of the Geological Society of London in 1807, so that 'above all, a fund of practical information could be obtained applicable to purposes of public improvement and utility'[15] provides a test of this philosophy of utility, the more significant because the Society effectively directed the pattern of English geology for the next half century. The founders had before them the experience of the great mining schools of Germany and Hungary, where geology and 'the arts of mining, surveying, working and smelting are taught, by able masters, upon scientific principles'[16]—two of the able masters, Werner and Charpentier, being among the greatest geologists in Europe; and also the success of the *Agence, École,* and *Journal des Mines* in France, where a similar programme, uniting science with technical education, was followed. Hence, dedication to the progress of mining was a real option for the Geological Society. Instead, a gentlemen's dining club was originally created, its fees, dining charges, and atmosphere all so elevated as *de facto* to exclude practical men. Although Leonard Horner thought a publication possible 'like the *Journal des Mines*', and even the Society's trustees recommended that 'to render it more generally acceptable [sc. than straightforward Transactions] to the numerous class of

---

[13] John Phillips—significantly William Smith's adopted nephew. Yet even he did not make a comprehensive study of the coalfields of the British Isles.

[14] C. S. Smith, 'Porcelain and Plutonism', in C. J. Schneer, ed., *Toward a History of Geology* (1969), pp. 317–38.

[15] *Geological Inquiries*, in *Phil. Mag.,* xlix, 1817, p. 422.

[16] R. E. Raspe, translation of J. J. Ferber, *Travels Through Italy* (1776), p. vi.

practical miners and others concerned in the several branches of art in any degree connected with this department of science, it is proposed by means of correspondents to render the Journal a medium of communication respecting the actual state of British mines',[17] in fact mere *Transactions* were published, in an extremely lavish edition, which include, at most, nine papers of practical interest out of eighty-eight published in the first decade. 'Considering the great importance of the coal and lead mines and of the quarries of Northumberland and Durham, and of the opportunities which they offer to geological research, it is rather singular that no history of the physical structure of these counties has yet been laid before the public' complained the north-eastern naturalist, Nathaniel Winch in 1817, and another local man, Westgarth Forster, noted that the same area had been 'neglected by scientific men'.[18] The point is that the same could have been said about all mining areas: a story of neglect by the 'scientific men' of the Geological Society.

The nature of the commitment of the geological élite towards furthering local mineral exploitation is well illuminated by the great jealousy which arose between surveyors and engineers such as William Smith, John Farey, and Robert Bakewell, who were provincial, practical geologists, and a leading group of the Geological Society. These provincial geologists claimed that if geology was to advance in Britain, it must chiefly be through the work of provincial practical men, involved in engineering, mining enterprises, etc., who would only be able to delineate the strata in rough, local terms; and that only if geology retained such a language could practical men understand and deploy it. 'Some of my Readers may perhaps be disposed to complain of the want of precise Mineralogical Descriptions and Terms in various parts of this Report' Farey apologized at the beginning of his *General View of the Agriculture and Minerals of Derbyshire*; 'to such I beg to observe that the great object in this Volume has been to state and shew the economic purposes to which the various Mineral Products of these Districts are or may be applied'—so he had used as far as possible 'such terms as are known and understood by *practical* Miners, Quarrymen etc.'[19] Hence, they accused the Geological Society, led by the 'worthy Geognost', Greenough, and the 'dupes of a foreign Geognostic faction' of imposing the alien 'new language' of the 'Wernerian Geognosy' on British geology, which,

[17] M. J. S. Rudwick, 'The Foundation of the Geological Society of London', *Brit. Jnl. Hist. Sc.,* i, no. 4, 1963, p. 353. I am endebted to Dr. Rudwick for all I know about the Geological Society, and much more besides.

[18] N. J. Winch, 'Observations on the Geology of Northumberland and Durham', *Trans. Geol. Soc. London,* iv, 1817, p. 1; Westgarth Forster, op. cit., p. viii.

[19] John Farey, *General View of the Agriculture and Minerals of Derbyshire* (1815), i, pp. vii–viii.

because unintelligible to most people, 'terrified men of plain understandings from continuing or from publishing their own observations'[20] and which consequently made Establishment geology practically useless. Obviously, if geological science was to flourish as a discipline, it needed to 'adopt one nomenclature', as the Geological Society urged; but once more the stance of the Society shews little central regard for the practical application of its geological work. The isolation of William Smith from the Geological Society élite throughout his working life exemplifies the same point.

Was utility then a false banner, raised by leading geologists perhaps unavoidably in an age of militant utilitarianism and 'vulgar Baconianism'? In some cases; but more frequently it is our reading of utility that is too facile. For when contemporaries used utility in an unqualified way, they envisaged intellectual, religious and moral benefits quite as powerfully as material advantage. And the practical utility of science applied to the arts was often seen not in concrete applications but in general access of understanding, 'giving clearness and consistency to a knowledge before obscure and confused, and by giving general and comprehensive views instead of local and confined ideas'.[21] For since within a progressive Baconian metaphysics and sociology of knowledge, knowledge *is* power (as a consequence of destroying the reality of distinctions between experiments of fruit and light), the advancement of learning, simply *qua* knowledge, must itself be useful.

Geology and Utility: the 'practical provincial men'

But perhaps British geology progressed through its ranks as well as its officers, through the many men in an industrial society whose work required some geological knowledge, however empirical; who also cultivated geology as a science, and who ploughed back the profits of study into their professional work. Men such as John Farey, and Robert Bakewell, mineral surveyors; Edward Martin, coal engineer; John Mawe, mineral dealer; Westgarth Forster, lead-mine agent; White Watson, 'sculptor, marble-worker, mineral-dealer' and 'estate mineral surveyor'. As also perhaps it progressed through the efforts of the many societies and institutions established in Britain at this time expressly to forge co-operation between the arts and sciences.

Brilliant as these individuals often were in their professions, the effective

[20] Robert Bakewell, 'Observations on the Geology of Northumberland and Durham', *Phil. Mag.*, xlv, 1815, p. 95.

[21] Frederick Burr, *Elements of Practical Geology*, 1838, p. 251.

scope for applying their geological wisdom practically was of course highly personal and local. And many factors conspired to disqualify them from contributing in proportion to their talents to the growth of geological science. Schofield has rightly insisted on some of the advantages for scientific and technological endeavour of living in the dynamic industrializing areas of Britain at this time; but these men often felt isolated and suffered for it— few became honorary members of the Geological Society. 'It is difficult in these days to conceive of such insulated and independent research as that into which the young philosopher entered' wrote John Phillips in 1839 of William Smith.[22] They complained of 'distance from books and those assemblies of the learned who had turned their studies into the same chanel', 'a discouraging and in some ways an insuperable disadvantage'.[23] Isolation, and lack of formal education, meant that their concepts were often uncritical and antiquated. Throughout his life, William Smith held to a theory of the earth akin to that formulated by William Stukeley in the 1720s; John Farey's belief that the strata had been disrupted by a comet dates back to William Whiston's *New Theory* of 1696. Yet theories such as Whiston's and Stukeley's were being mocked by Geological Society contemporaries as, in Fitton's words, 'a species of mental derangement'.[24] But as well as their theories, even their patient and detailed stratigraphy was in many instances vitiated by the extreme localism of their knowledge and experience. Because William Pryce knew only the geology of Devon and Cornwall at first hand; John Whitehurst and White Watson only that of Derbyshire—there is no record of Watson's ever travelling more than twenty-five miles from his native Bakewell; Westgarth Forster only that of Northumberland and Durham, the generalizations they offered on the order of strata, the location of mineral deposits, the relations between veins and beds of rock, were frequently erroneous. They also experienced great difficulties in literary composition, and in financing publication of their writings. Hence, given the narrow circulation of books published locally by subscription, it is not surprising that these men failed to shape the dynamics of national geological development. In two respects, however, local practical experts played a key rôle: by acting as popularizers of the science at a local level, through giving lectures, patronizing societies, personal contacts, etc.; and by drawing and colouring each local piece of the great jig-saw puzzle of the stratigraphy of the British Isles, waiting to be fitted into place by members of the Geological Society.

[22] John Phillips, 'Memoir of William Smith', *Mag. Nat. Hist.*, iii, 1839, p. 215.
[23] William Borlase, *Natural History of Cornwall* (1758), p. vii.
[24] (W. H. Fitton), 'Transactions of the Geological Society of London', *Edin. Rev.*, xix, 1811–12, p. 207.

Geology and Utility: the Provincial Societies

Societies deliberately set up to encourage co-operation between men of science and practical industry clearly did not face the same problems of isolation, lack of funds, or the sheer vastness of the labours confronting the individual. Did they fare more successfully? Metropolitan societies can largely be discounted at the start. The Society of Arts never devoted much attention to geology or mining. The Royal Institution failed to raise a subscription for its planned mineralogical and assaying laboratory. Its original plans for training artisans (which were soon dropped anyway) did not include mining. The Board of Agriculture reports were almost silent on the relations between geology, soils, and agriculture, and apart from Farey's, none of the originally planned mineralogical surveys—'which series has been promised ever since 1793'[25]—was undertaken. Admittedly, the British Mineralogical Society, founded in 1799, sought to be useful by offering free analysis of mineral samples, but failed utterly in its far more important project of making a collection of 'specimens, with their provincial names, as well as their scientific, of each county in this kingdom', and the whole society disintegrated within a decade.[26] None of this is very surprising; London could well be a centre for industrial chemistry and engineering, but without artificial—i.e. state—stimulus, scarcely of mining technology.

Their provincial equivalents never did less than fulfil two functions: disseminating geological knowledge generally through their libraries, papers, cabinets, lectures, etc., and facilitating the actual interchange of mineral specimens and expertise between mine-owners, manufacturers, chemists and mineralogists.[27] Obviously, some societies did no more than this, sometimes because, like the Manchester Literary and Philosophical Society, they were situated in districts where mining was not the crucial industry, sometimes because the dominant élite of the society was either mercantile or largely composed of the liberal professions, as seems to have been the case with the Derby Philosophical Society, or the Leeds Philosophical and Literary Society.

By contrast, the Royal Geological Society of Cornwall, 'promising great practical as well as scientific ability'[28] achieved a highly successful fusion of

[25] John Farey, 'Notes and Observations on Mr. Robert Bakewell's "Introduction to Geology" ', *Phil. Mag.*, xliii, 1814, p. 335.

[26] Report of the British Mineralogical Society, in *Phil. Mag.*, xix, 1804, p. 85.

[27] This—rather than any advances in geology or in mining technology—seems to have been the rôle of the Lunar Society as a group. For evidence of this, see also the mineralogical correspondence between James Watt, Boulton, Wedgwood and Black, in E. Robinson and D. McKie, *Partners in Science* (1970).

[28] Davies Gilbert to the Rev. D. Lysons, cited A. C. Todd, 'The Origins of the Royal Geological Society of Cornwall', *Trans. Roy. Geol. Soc. Cornwall*, xix, pt. 3, 1959–60, p. 179.

geology and mining in its papers, its Transactions balancing pure geology, mineralogical analysis, and mining technology in nearly equal proportions, and including detailed statistics on the economy of the county's main industry. Much of the dual success of this society was due to the local social identity of cultured gentry and mine-adventurers, and to the energy of such adventurers as Henry Vivian and John Hawkins, who had themselves studied at, or visited Freiberg or Schemnitz.[29] Yet even this society was not sufficiently active and united as a body to achieve the fruition of its major geological and mining projects, the former, to complete a full map and geological history of the county, the latter, to establish a records office and a school of mining, the plan for which met with a lukewarm response from mine-owners.

On the same hopes of 'ingenious persons who are employed as viewers . . . supplying better information' to geologists, and such 'speculative philosophers . . . returning the obligation'[30] by aiding the miner, the Newcastle Literary and Philosophical Society was also set up, in 1793. Yet even this society, at the heart of the country's chief coal-field, illustrates the almost insuperable problems involved in fulfilling such a programme. Certainly the need for some such body was plain. The frequency of colliery disasters and the loss of rich seams on encountering faults and dykes both bore witness to the contribution a flourishing local geology could make to improve the art of mining. And the attempt to trace the geology of the area had scarcely yet begun. The society made great plans towards both these ends. Monthly papers were given to the society, on geology and mining, as well as other topics; the relevant parts of Jars' *Voyages Métallurgiques* and Faujas de Saint Fond's *Journey* were translated and read; a mineral collection was formed. Lecture courses on or including geology were given in the town by Henry Moyes and Robert Bakewell. The New Institution, a permanent scientific lectureship established by the society 'through the aid and encouragement' of which it was intended that 'several of the younger pupils of our mining department should in time be distinguished by the same stores of science' as mechanical engineers then possessed, included geological and mineralogi-

---

[29] Hawkins declared his geological interests were the 'practical result of a much longer attention to the subject, begun in Saxony, under the guidance and instruction of Werner, and continued in Bohemia and Hungary'. 'Queries', *Trans. Roy. Geol. Soc. Cornwall*, i, 1818, p. 243. Vivian wrote for the society a 'Sketch of the plan of the Mining Academies of Freyberg and Schemnitz', *Trans. Roy. Geol. Soc. Cornwall*, i, 1818, pp. 72–9.

[30] *Prefatory Observations on the Propriety of Establishing a Literary Society in Newcastle* (1793), pp. 5–6, bound up in Anthony Hedley's *Reports of the Literary and Philosophical Society of Newcastle* (1823), i, in the library of the Lit. and Phil. Society, Newcastle-upon-Tyne. I am very grateful to the Society for allowing me to use these collections.

cal courses.[31] The society acted in 1813 and subsequently as the focus of experiments on safety lamps. And most important of all, the society tried to conduct a full enquiry into the 'natural history of coal' and the geology of Northumberland and Durham, which 'must certainly be an object of much consequence to the curious enquirer' and of 'great advantages' to 'future naturalists',[32] by sending out a lengthy questionnaire to local mine-owners, managers, viewers and engineers. In addition, from 1796, it promoted plans for the establishment of a Newcastle Mines Record Office 'for collecting and recording authentic information relative to the state of the collieries', again to increase available information about the geological structure of the North East, but above all to provide a complete body of data about coal workings, for the sake of safety and to promote future finds, by rendering 'unnecessary' 'a considerable expense in the article of Boring'.[33]

In practice, however, little was achieved in any of these aims. Speaking generally of the society, it fell far short of hopes, as most others did, as an active, co-operative venture. Most members used it only for its library, and factions attempted to terminate its scientific and educational functions 'for diffusing useful knowledge' as 'chimerical and impracticable schemes', doomed to an 'inevitable fate' of failure by 'human nature' and 'local circumstances'.[34] For long periods few papers were given, and they were often letters from outsiders. The mineral cabinet remained unorganized and 'in every department very imperfect'. More specifically, the society, run by the town gentry, merchants and members of the liberal professions, in fact failed to enlist the support of mine-owners and managers (except during periods of public outcry, for example, the succession of great colliery disasters in 1813 to 1815). Thus, only two local replies were received to the society's Queries, which therefore 'failed of the desired success';[35] and, equally negatively, mine-owners resolutely declined to co-operate in setting up a mines record office, which was not in fact established until 1840 in London. The failure of the Literary and Philosophical Society was only highlighted when its offshoot, the Newcastle Natural History Society felt that one of its first vital tasks was to draw up a great geological map of the North Eastern Coalfield, with seams, dykes, faults, etc. all marked in; an

[31] Thomas Bigge, *On the Expediency of Establishing in Newcastle-upon-Tyne a Lectureship On Subjects of Natural and Experimental Philosophy* (1802), Hedley *Reports*, ii.

[32] *Queries*, drawn up by the Lit. and Phil. Society, Newcastle-upon-Tyne, 20 January 1795, Hedley *Reports*, i.

[33] *Hints For Establishing an Office in Newcastle*, read to the society, 13 September 1796, printed 1815.

[34] *A Brief Account of the Literary and Philosophical Society And of the New Institution of Newcastle-Upon-Tyne* (1809), Hedley *Reports*, iv.

[35] *Sixth Year's Report to the Society* (1799), Hedley *Reports*, i.

undertaking still necessary because 'the importance of Mineralogical and Geological knowledge to all persons concerned in mining operations, will be apparent when we consider the immense sums of money which have been lost even in this district, in a fruitless search for coal and metallic ores, in situations where a slight Geological knowledge would have taught that none was to be expected'.[36]

Hence, as with provincial individuals, the main function of these societies within the development of geological science in Britain seems to have been to act as general repositories of local knowledge, which facilitated the work of neighbouring men of science and was ready to be used—plundered often— by outsiders. They clearly failed to stimulate advance in the geological aspects of mining technology, but not, unlike the Geological Society, for want of genuine convictions and initiatives. The failure lies less with them, than in the nature of the various sections of the mining industry in Britain.

Mining, Geology and the Technological Gap

'I do not know of any striking improvements in mining, which may be said to have been derived from the progress that geological science has made in later years', the distinguished mining engineer, John Taylor, admitted in 1837.[37] But this was true, not because geology could be of no use to mining, but because, on the whole, mining in Britain never strove to create geological knowledge, experts and expertise appropriate for its own use, and remained blind to the science that was coming independently into being. British mining remained 'a system of working . . . pursued in each locality that has probably been handed down from time immemorial'.[38]

To pinpoint mining's disregard for geology might appear inappropriate, on the grounds that at this stage, when most mining was small-scale, 'yet shallow', and profitable, and when no metal or mineral was in short supply, little geological expertise was needed by the industry. Again, to emphasize backwardness in nursing a geological technology might seem misplaced, in view of the admittedly highly developed state of *mechanical* technology in Cornish metal mining and North Eastern coal-mining in particular. Yet the steam engine was not essentially a development by mines management, and

---

[36] *Report of Subcommittee to the Natural History Society of Northumberland, Durham and Newcastle* (1831), cited in T. Russell Goddard, *History of the Nat. Hist. Soc. of Northumberland, Durham and Newcastle* (1929), p. 30.
[37] John Taylor, 'On the Economy of Mining', an address to the Society of Arts, printed in *Quarterly Mining Review*, (1837), pp. 261–72, reprinted in R. Burt, *Cornish Mining* (1969), p. 44.
[38] Matthias Dunn, *Treatise on the Winning and Working of Collieries* (1848), pp. 1–2.

undoubtedly the bad safety records of British mines, the 'too great apathy shown by many of the proprietors to the waste of human life which annually takes place';[39] the failure to exploit seams systematically, and for long-term advantage; the leaving of lethal flooded 'wastes'; refusal to keep, and make public, detailed records of workings, and the frequency of complaints of 'mismanagement', 'the bad administration of the mines of England and the defective method of working them',[40] all argue a highly customary and empirical art of mining rather than a rationalized, modern technology. And more specifically, there is massive evidence of valuable minerals being constantly squandered—roads repaired with copper ore—and trials being made for coal and shafts sunk in hopeless locations 'where nature, interpreted by science, forbids the discovery of that inestimable treasure',[41] for the want of the most elementary, and available, geological and mineralogical knowledge. Even half a century after Smith's demonstration of strata identified by organized fossils, Sir Henry de la Beche felt the need to insist 'Did time permit, it would be very easy to show how many mistakes have been made, especially in seeking for coal, thousands having been thrown away in a vain search for that which a very moderate acquaintance with palaeontology would have prevented'.[42]

Why then was British mining so unresponsive to the growth of this potentially useful science? Part of the answer lies within the structure of skills in the industry. Labouring colliers and miners of course had had no education in their art except pure pragmatic experience and intuition, and so were 'wholly ignorant of the principles and facts of mineralogy and geology'.[43] Surveyors such as Smith and Farey regularly pointed out how often practical colliers' judgements 'on matters of inference', such as the continuation of seams after dykes, were in error, because of 'the instances being so extremely rare of any general or extended idea of Stratification with practical Miners'.[44] None of this is very surprising: what is, perhaps, is that the opinions of these 'ignorant coal-finders' were misguidedly followed 'in hundreds of instances' in developing mining adventures, 'attended with heavy and sometimes almost ruinous expences'.[45]

[39] Robert Bakewell, op. cit., p. 96.

[40] John Williams, op. cit., i, pp. 399 f.; note in the French translation of J. J. Ferber's *Essay on the Oryctography of Derbyshire*, reported in J. Pinkerton, *A General Collection of Voyages* (1808), ii, p. 483.

[41] John Phillips, op. cit., p. 66.

[42] Sir Henry de la Beche, 'Inaugural Discourse at the Opening of the School of Mines', *Records of the School of Mines*, i, pt. 1, 1852, p. 8.

[43] William Phillips, 'On the Veins of Cornwall', *Trans. Geol. Soc. London*, ii, 1814, p. 141.

[44] John Farey, *General View*, i, p. 163.

[45] John Farey, article 'COAL', *Rees Cyclopaedia* (1819).

Unlike contemporary entrepreneurs, many of whom were famous for their technological skills, mine-owners were not noted for their geological expertise, as many critics observed, from Sir John Clerk—a notable exception—condemning Sir James Lowther, owner of the great Whitehaven pits, as 'an indolent old man, and knows nothing about coalworks'[46] to Matthias Dunn's finding as late as 1848 of owners and managers that 'a great proportion of them are deficient in those enlightened and scientific acquirements which are known to be so necessary to a suitable management of mining property'.[47] But the scientific ignorance of manual colliers was inevitable and expected; that of owners not an unsurpassable obstacle. The crucial structural reason why mining failed to deploy geology is that the industry did not generate an extensive class of technologists, educated and trained surveyors, assayers, prospectors, managers. William Phillips castigated the 'total ignorance of almost everything relating to the sciences of geology and mineralogy and above all chemistry in the conductors of mines and their agents'. 'Even the conductors of the mines, termed captains, are men generally of little or no education, who have risen to that station by a superior attention to their art, in which they have been incessantly occupied from the early age of five or six years'.[48] The fact was even more apparent to foreign observers: Diederick Wessel Linden's reaction to metal mining in North Wales, 'Mining in this part of the globe is in such a perfect decay that it is probably to be feared it will in a few years become totally extinct', chiefly because of the 'unskillfulness of the mine directors'[49] bears all the doom-laden misanthropy of self-interest, but epitomizes an authentic foreign response.

On the continent such a corps of technologists—frequently gentlemen—was created by state mining academies, particularly the Saxon academy at Freiberg, 'supported by the influence and funds of the Government'[50] to provide expert personnel for the management, at all levels, of state-owned and state-controlled mines. It was widely recognized that 'the Art of Mining, and its many subordinate branches, are in Germany and its dependent countries for various reasons so highly improved that for these last ages Germany has been justly considered as the most ancient and best school for miners'.[51] 'As late as the mid-eighteenth century English land-owners and

---

[46] W. A. J. Prevost, ' "A Trip to Whitehaven to Visit the Coalworks There in 1739", by Sir John Clerk', *Trans. Cumberland and Westmorland Antiq. and Archaeol. Soc.*, xlv, 1965, p. 312.

[47] Matthias Dunn, op. cit., p. 4.

[48] William Phillips, op. cit., p. 141.

[49] Diederick Wessel Linden, *Three Letters on Mining and Smelting* (1750), p. 4.

[50] John Taylor, *Records of Mining* (1829), preface.

[51] R. E. Raspe, translation of Baron Born's *Travels Through the Bannat* (1777), p. xiii.

mine adventurers were still turning to German experts for advice and guidance, as they and the Crown had done two centuries earlier', and still German surveyors and prospectors, half expert, half charlatan, such as Linden or Raspe, touted their services as superior scientific miners. 'One condition sufficient for a miner these days, to have been born in Germany or Hungary' was Lewis Morris's bitter, ironical comment.[52]

British engineers and geologists were well aware that 'in mining, Germany is so pre-eminent',[53] that this was due to 'the very regular and scientific system of mining which is there pursued'[54] and that the difference essentially represented a difference of education: 'while in France, in Germany, there are national institutions for the education of those intended to conduct the working of mines . . . in this country all is left to accident, and the rich gifts which nature has bestowed upon us are consequently often neglected or lavishly thrown away'.[55] In view of inefficient exploitation of mines, which constituted an absolute national wastage of resources as well as private loss —and men such as Bakewell forecast total exhaustion of British coal within a few hundred years—spokesmen called for a state mines school in Britain which is 'yet without any school where its miners may find a suitable education'.[56] Some also demanded at least a national mines survey, if not a full mines commission, for as Thomas Thomson pointed out, 'in all parts of Europe where mining has been carried to a great degree of perfection, it has under the inspection and control of Government. That coal-mines should be in this predicament and that exact plans should be preserved of all the excavations and of all the coals left, is too obvious to require any illustration'.[57]

Of course, the politics of British *laissez-faire* ruled out this excessively obvious solution. Mines inspection, when finally established in 1842, was inspection of working conditions for humanitarian reasons, not for technological efficiency, and no state school of mines was set up till the profits of

---

[52] J. N. Rhodes, 'Dr. Linden, William Hooson, and North Welsh Mining in the Mid Eighteenth Century', *Bull. of the Peak District Mines Historical Society*, iii, pt. 5, 1968, p. 259. French governments equally recognized the need to train technologists. The order of Louis XVI authorizing the founding of the School of Mines acknowledged that for mines to flourish 'it is not sufficient merely to give concessions, but it is also necessary to train men to carry on such work with both safety and economy'. Cited by F. B. Artz, *The Development of Technical Education in France* (1966), p. 85.

[53] William James McNeven, translation of Geiss's *Essay on the Construction and Use of a Mine Auger* (1788), p. ix.

[54] John Hawkins, 'Of the Produce of the Copper Mines of Europe and Asia', *Trans. Roy. Geol. Soc. Cornwall*, iii, 1828, p. 258.

[55] William Phillips, op. cit., pp. 142–3.

[56] John Taylor, 'Economy of Mining', p. 47. See Note 37 above.

[57] Thomas Thomson, *Annals of Philosophy*, vi, November 1815, p. 387.

the 1851 exhibition became available, the mid-century scare about foreign technological superiority gathered momentum, and a German directing hand became prominent in influencing policy—Prince Albert.

In view of the acknowledged importance of the superiority of continental mining technology, and the force of jealousy against state patronage, it might be expected that private 'mineral schools' would have been founded, especially as they were so widely recommended. John Taylor in fact drew up and circulated plans for such an academy, for 'we have nothing in the whole kingdom analogous to the schools of mines of Germany and Hungary',[58] even calculating the fair shares of its expense to be paid by each section of the mining community. But no such scheme was acted upon, nor did an education in mining figure in any of the eighteenth or early nineteenth century developments in regular schooling and technical education, nor become available to practical miners through Mechanics' Institutes.[59] Furthermore, outside the Cornish metal-mining industry (which desperately required advanced technical skills, in view of its complex metallurgical processes) few owners or managers of mines, it seems, attended continental schools of mining, or made tours of foreign mines.

No schools: no books. Britain was 'destitute of books that either treat of Mining or describe the actual state of the art among us' for there was no one competent to write them—'the persons who know most of the subjects . . . are from habit and constant occupation unable to write at length or to communicate much of the results of their experience'.[60] So few were the treatises, that an author could claim his to be not merely best, but 'the first of this kind'; 'I have much lamented that no useful Book of Mineing has been printed in the English tongue, at least I have never met with any that was satisfactory'.[61] As late as 1848 Matthias Dunn could justly write, 'It has long been matter of deep regret to persons interested in mining property . . . that in a nation so deeply interested as this is in the study of mining, there should not exist any established elementary work on the science'.[62] Above all, the entire period is devoid of regular published statistics and records of any kind of mining, a want that Taylor hoped to fill with his *Records of Mining*, which, however, ceased to appear after the first volume.

[58] John Taylor, 'Prospectus of a School of Mines in Cornwall', in *Records of Mining* (1829), p. 2.
[59] Mabel Tylecote, *Mechanics' Institutes of Yorkshire and Lancashire* (1957) shewed that in general the institutes failed to teach trades (pp. 107 f.) and did not flourish in mining areas (pp. 66 f.).
[60] John Taylor, *Records of Mining* (1929), preface.
[61] William Hooson, *The Miner's Dictionary* (1747), preface.
[62] Matthias Dunn, op. cit., p. 1.

The Technological Gap and Empiricism

One feature unites all British mining treatises from the smallest and cheapest pocket *vade mecum* of the practical hewer, such as William Hardy's *The Miner's Guide*,[63] to the folio production of a gentleman adventurer, such as William Pryce's *Mineralogia Cornubiensis*. They all show British mining technology totally empirical. 'Poking out his road like a blind man', 'the Empiric gains esteem so far beyond the Rationalist'.[64] Intuition and personal experience—'forty years practice and experience in the mines'[65]—often from a confessedly very narrowly limited geographical area, formed the essence of advice offered on the location, nature, interruptions and complexities of mineral deposits. Instructions for the discovery of minerals and metals invariably involved nothing more 'scientific' than searching out external 'symptoms' on the surface—such as the discolouration of springs and streams, or particles of minerals left exposed by the plough and similar 'accidents', and required 'an acquired habit of judging from particular signs'.[66] Sometimes they included rough and ready generalizations, from the early eighteenth century ''tis a good sign to meet with *dry* earth, if it be *yellow, red, black, green,* or any other extraordinary Colour' to the early nineteenth's 'coal is seldom found under the mountains of solid lime-stone or sand-stone',[67] but they were never derived from a philosophical view of the geological structure of the country or of the formation of coal. Treatises occasionally offered geological speculations (with apologies that such philosophical thought was 'above my capacity'[68]) but without attempting to relate them to the empirical advice, which is quite the opposite from the best German mining manuals, such as those of Lehmann and Delius, in which both aspects are crucially related. But these superior German works were never quoted from in British treatises, nor were they translated into English. In fact, the very few German works of mining technology which did appear in English translation sold badly, and offers to translate more were not taken up. Yet no lengthy account of German and Hungarian mining appeared

[63] First edition (1748), Sheffield.

[64] Diederick Wessel Linden, op. cit., p. 68.

[65] Part of the title page of Hooson's *The Miner's Dictionary*—cf. John Williams's 'more than forty years' experience', *Mineral Kingdom,* i, p. xi.

[66] William Pryce, *Mineralogia Cornubiensis* (1778), p. 112.

[67] Thomas Heton, *Some Account of Mines* (1707), pp. 93–4; J. H. H. Holmes, *Treatise on the Coal Mines of Durham and Northumberland* (1816), p. 72.

[68] J. C., *The Compleat Collier* (ed. 1848), p. 41. The first coal-finding work deriving practical advice from adequate geological principles is Farey's Smithian article 'COAL', in *Rees Cyclopaedia* (1819).

in a native English work between Edward Brown's *Account of Some Travels Through Germany,* 1677, and Robert Townson's *Travels in Hungary,* 1797.

So empirical were British works on the geology of mining that even the one highly-educated man who wrote such a *Treatise Upon Coal Mines or an Attempt to Explain Their General Marks of Indication*[69] on self-consciously scientific principles, and who recognized the need for an overall view of British coalfields, merely attempted, following a purely mechanical method, *a posteriori* statistical generalizations of such symptoms, especially superincumbent rocks. Because his experience was limited to one coalfield, and so little information was available about others, the attempt proved worthless. But not only was mining advice empirical in fact, but was positively justified as by necessity an empirical art and not a rational and theoretical science. It was to be 'learned by practice, by experience and masters; not from books'.[70] Because 'we do not know of anything material or useful that has been found for the better or more easier to the Discovery of lead ore than what has been left to us by our forefathers, but rather much impaired by neglect and idleness'[71] mining instruction could only be an apprenticeship in acquired customary skills, argued William Hooson, to which the German, Linden, retorted, 'Every common Miner here is already rivetted in that Opinion, and if your Knowledge extends no further than theirs, you might as well have been silent'.[72]

Paucity of enlightened professional miners, lack of formal technical education or books, and other factors which hindered a fruitful rapport between mining and geology, are all symptoms of a traditional industry, still chiefly looking within its own natural processes for talent and expertise. In these respects, the trade seems conspicuously backward in a country increasingly dedicated to the diffusion and application of useful knowledge. Why mining should have had this structure is not altogether clear. Certainly, few well-educated, professionally competent outsiders seem to have taken up mine-engineering, viewing or managing as a choice of career (unlike mechanical engineering, or industrial chemistry). If Scotland is a fair index,

[69] The treatise is anonymous. The coalfield with which the author was familiar was the Somerset.

[70] William Borlase, *Natural History of Cornwall* (1758), p. 168.

[71] William Hooson, *Miner's Dictionary,* appendix.

[72] Diederick Wessel Linden, *A Letter to William Hooson* (1747), p. 10. The one exception to the unhelpful empiricism of the mining guides may be John Williams's *Natural History of the Mineral Kingdom,* which through sheer thoroughness almost bridges the gap between a mining treatise and a work of geology. Yet still Williams relies purely on his individual experience, which is essentially restricted to Scotland, and he has no clear idea of the order of strata or general geology even of that country.

mining was far from a prestige profession, and in addition paid badly for technical skills.[73]

The Technological Gap and Capitalism

But as well as the structure of mining, attitudes within the industry were unfavourable. All industries in Britain at this time were to some degree hostile to outside interference, from governmental inspection down to the casual visitor. From their unwillingness to co-operate in developing safety lamps, refusal to set up a records office, diehard opposition to Parliamentary inspection, and much contemporary evidence of the 'jealousies and prejudices of mine-owners', it appears that the belief that secrecy served self-interest was far stronger amongst the coal trade than elsewhere. Geologists frequently noted a 'shyness'[74] in offering information on the part of owners, or found an 'unworthy prejudice to combat',[75] so much so that Farey dubbed mine-sections 'the grand arcana of the coal trade'.[76] 'The proprietors of the collieries, from mistaken views of self-interest, are anxious to conceal every fact which they observe from the public',[77] was Thomas Thomson's dogmatic judgement, and it was to such an attitude that William Turner, Secretary of the Newcastle Literary and Philosophical Society, attributed the failure of his *Queries*. Referring to the tiny response when trying on a later occasion to cajole the owners into co-operation, he explained, 'it is presumed that it could not possibly arise from any jealousy of discovering beneficial secrets, because all practical questions are studiously avoided, and no information desired but such as related to the natural history of coal'.[78]

At this highly individualistic, competitive stage of capitalism, suspicion and competition within mining blighted prospects of the liberal and open attitudes necessary for the scientific development and deployment of its own technology. Possibly, a similar hostility towards outside 'philosophers' may help explain why no soil science based on a firm geological foundation

---

[73] Baron F. Duckham, in his *History of the Scottish Coal Industry* 1700–1815 (David and Charles, Newton Abbot, 1970), has an excellent chapter on 'Managing the Mine' which shows that in Scotland owners and lessees were fortunate if they had a literate viewer who could keep accounts competently and honestly. But they seem to have done nothing to remedy this situation.

[74] John Phillips, op. cit., p. 14.

[75] 'Preface', *Trans. Roy. Geol. Soc. Cornwall*, i, 1818, p. i.

[76] John Farey, article 'COAL', *Rees Cyclopaedia* (1819).

[77] Thomas Thomson, *Ann. Phil.*, vi, November 1818, p. 386.

[78] Rev. William Turner, *Historical Sketch of the Transactions of the Literary and Philosophical Society of Newcastle* (1807).

emerged until the late 1830s.[79] And certainly relations were not regularly closer between geologists and civil engineers, as testified by abortive early nineteenth-century Thames tunnels (to which geologists were mere by-standers) and the famous quarrel between Brunel and Buckland over Box Tunnel.[80] Probably not until the 1870s were even the elementary principles of stratigraphy as laid down by Smith regularly used in the large-scale development of a coalfield, against the advice of practical miners,[81] and right until the end of the century, highly sophisticated pure geological research 'on matters of academic interest such as the description of fossils or glacial deposits' continued side by side with, but independently of, the rather empirical search for coal in the same measures.[82] Only in this century have the progress of the most refined techniques of geology and the requirements of mineral exploitation become *unavoidably* utterly mutually dependent.

The Social Development of English Geology

Industry, technology and science share only historical relationships. In France, Germany, Sweden and Russia, mining technology and geological science grew hand in hand as a result of state ownership and exploitation of mines and mineral wealth, state patronage of mining education, and (to a lesser degree) of science in general. In Britain, where industrial capitalism was most advanced, yet within an establishment of power still espousing traditional social and educational values, the economic pressure of industry seems of itself to have created neither a science of geology nor even an advanced technology of mining.[83] If this is true, then the case of mining provides some evidence for two traditional views of the Industrial Revolution, one that it was more generally an economic than a technological revolution; the other that, in so far a technical change was important, it occurred more through rule-of-thumb methods than through scientific or strictly technological advance. But more important, the entire study under-

[79] The first such wrok is John Morton, *On the Nature and Property of Soils, their connection with the Geological Foundation on which they rest*, etc. (1838). See Sir John Russell, *A History of Agricultural Science* (1966), p. 81 f.

[80] L. T. C. Rolt, *Isambard Kingdom Brunel* (Longmans, 1957), pp. 138–9.

[81] T. E. Lones, *A History of Mining in the Black Country* (1898), p. 95 f.

[82] W. H. Wilcockson and R. F. Goossens, 'Geological Research in the Yorkshire Coalfield', *Trans. Instit. Mining. Engin.*, 117 (1957–8), p. 623.

[83] Cf. D. S. L. Cardwell's similar conclusion about the contemporary science of power: 'This predominance of French names suggests that the French lead in technological education was more effective in promoting advanced technology, and applied science, than was the superior economic growth, with all its implied incentives, of England', *History of Science*, i, 1962, p. 41.

lines our need to know so much more about the diffusion of various kinds of knowledge and expertise amongst the varied hierarchy of men concerned in science and industry, and their relationship with one another, before we can seriously think of answering these questions.

But where does this leave the social history of geology? It has narrowed the field essentially to a question about the changing ways in which wealthy, liberally-educated British gentlemen chose to spend their wealth and employ their education and leisure. This is no less a social problem for their lives not being immediately determined by economic needs and pressures. It is by no social accident, unconnected with the Industrial Revolution, that the life-patterns of these men were changing at this time. To the question there is a general and a more specific answer.

After the early flush of interest in practical questions shown by the Royal Society, which included *Articles of Inquiries Touching Mines*, answered, among others, by Glanvill, Beaumont, Locke and Boyle, a great mixture of studies of the earth achieved popularity in eighteenth-century Britain among the social and intellectual establishment—landed gentlemen, clergymen, dilettanti, scholars, collectors. But these studies were usually intellectually desultory. Fossils were widely collected, but their relevance to the distinctions between strata was scarcely investigated. Rocks were listed, but not used as a key to earth history. Theories of the earth were constructed, without being grounded on sound stratigraphical learning. Extensive tours were made, by such men as Charles Mason, Woodwardian Professor at Cambridge, and Alexander Catcott, both fine observers, but neither they, nor anyone else, attempted to construct a clear idea of the structure of the British Isles. Both visited mining areas, but did not see their studies as practically useful. Edward Lhwyd and Richard Richardson carefully studied the vegetable fossils of coal, but only as specimens. The entire study was incoherent and inconsequential and uncritical, as was so much intellectual activity in that 'not serious' age.

As Schofield and others have argued, the English Midlands and North industrialized themselves, while the political, social and intellectual establishment of London, the universities, and rural areas, remained *relatively* static; and hence the industrializing, modernizing parts needed to construct for themselves their own superstructural institutions, industrial, economic, urban, educational, intellectual.[84] One such product was an enormous energy for science, for statistics, for avid local factual enquiry of a kind not previously common in eighteenth-century England; and one expression of this was a great wave of provincial geological activity from about 1770 onwards, incapable of creating and sustaining a *science*, but more detailed,

[84] R. E. Schofield, *The Lunar Society*, pp. 438-9.

factual and purposive than anything previously. The new surge of earnest Establishment activity around the turn of the century was as much a response to the provincial challenge, as to continental developments or the French Revolution. Culture and institutions needed to be modified, to come to terms with, embrace and control new provincial activities and standards. Such a trend is very noticeable in the growth of geology. At the beginning of the nineteenth century the Society of Arts for the first time began to offer premiums for mineralogical histories of counties; the British Mineralogical Society to offer free analysis of specimens sent to London. The *Geological Inquiries,* circulated by the Geological Society of London, 'to facilitate and in some measure to direct general research', patronizingly staked the Establishment's claim to be the brain-centre of the science, for 'to reduce Geology to a system demands a total devotion of time, and an acquaintance with almost every branch of experimental and general science, and can only be performed by philosophers; but the facts necessary to this great end may be collected without much labour and by persons attached to various pursuits and occupations'.[85]

But in order to regain control of earth science in Britain, the intellectual aristocracy needed to accept the standards of sober accuracy and the new quest for minute Baconian details in the 'enumeration of the strata' which provincial science had fostered. It needed to revolt against its own past history of 'mental derangement', 'speculation', and 'fancies', admit that little had been achieved in the last two centuries of 'retardation as well as advance', and claim that only now was the science methodologically pure enough to progress, now that it had a 'tendency in all its branches to assume a character of strict experiment and observation'.[86] Thus it was a significant social rivalry which brought a genuinely accomplished British geology into being; for although it is true that some price had to be paid for the restored kudos of the élite—the reintroduction of Mosaic Geology in Oxford in the 1820s— only such a wealthy, well-educated and travelled, above all leisured, élite could pursue geology with success on a par with the continent. At least that part of the propaganda rings true.

The more specific answer encompasses the *means* by which a reinvigorated social élite generated the rise of geological science in Britain. One of the reasons why eighteenth-century geology was so desultory was that it lacked any clear-cut method: should it follow a natural history, classificatory method, or the model of Newtonian physical science, or work within a historical framework? During the second half of the eighteenth century,

---

[85] *Geological Inquiries, Phil. Mag.,* p. 421.

[86] (William Fitton), 'Transactions of the Geological Society of London', *Edin. Rev.,* xxix, 1817, p. 175.

precise and rigorous systems of mineralogy were developed in Sweden, Germany and France, which offered the internal and external mineral characteristics of rocks as the alphabetical key to the book of nature. The mineralogical approach to the study of the earth achieved great popularity in Britain, being especially authoritative with men who had actually studied in Germany—Kirwan, Greenough, Jameson—and chemists—Babington, Arthur Aikin, Knight, Pepys, etc. Although it had its practical uses, it established itself as a pure science—indeed one of the complaints of practical men such as Bakewell was that mineralogists' attention had been 'too much devoted to the discovery of new species of minerals that possess no importance in nature, and can be of no use in the arts'.[87] At first, the new method was chiefly applied to laboratory analysis and cabinet arrangement, but increasingly mineralogists sought to use it as a means of reading the order of strata of the book of nature, *in situ*. Maton, Aikin, Hatchett, Davy, Greenough and others of the country's leading mineralogists all undertook tours around the turn of the century; and fresh out of their closets, they simultaneously discovered that the simple mineralogical alphabet was inadequate for understanding the complex relationships of the geological facts of nature, and found themselves captivated by 'the grand views of nature'. Hence mineralogy, by a different route, ended up among the same rocks as had provincial studies, but men of mineralogical training could approach them with a superior, precise, scientific set of categories, so sorely missing earlier. By the time of the foundation of the Geological Society—so nearly a Mineralogical society—the geology of nature had become the subject matter, and a distinctively geological method was emerging out of the mineralogical approach.

To understand science we must see it in the context of, and constitutive of, social structure, social change, and social consciousness. But it is simple-minded always to expect to find science *responding*, in any immediate way, to social conditions. For a mere glance will show that because men of science in East and West have always comprised some sort of mandarinate or clerisy, whether as an élite set apart by wealth, talent, privilege, or by state patronage, etc., they have often been cushioned from some of the more obvious and potent social movements of their day, often isolated from economic and technical pressures from below, by systems of cultural mediations. In societies and at times when the cushioning effect and the mediations are sufficiently powerful, the social history of science can graduate insensibly into biography. It is at least a half-truth that British geology, early in the nineteenth century, depended, not just for its chance geniuses but for its very intellectual continuity, upon such social accidents of biography; on a young

[87] Robert Bakewell, *Introduction to Mineralogy* (1819), p. xiii.

lawyer having bad eyes, on the ennui of an ex-soldier, on a 'corrupt' Cambridge professorial election. In such a society, the same distortions which led the mine-owner, with his (rational) need for a geological technology, to despise the man of theory, also encouraged the geologist, who (rationally) needed local expertise, to scorn the 'merely practical' man, to the cost of both. As J. D. Bernal graphically put it, 'the collector in his cabinet could do little but marvel at the odd production of the earth. The miner, on the other hand, was so concerned with the ore and the indications of its presence in the rocks that he had usually neither the inclination nor the learning to formulate any general theories as to the structure and history of the earth.'[88] Geologists and miners alike sought utility; but, as Bentham saw, only a new form of state, acting in the name of reason, could engineer that utility.

[88] J. D. Bernal, *Science in History* (C. A. Watts, 3rd edition, 1965), p. 466.

XIX

THE HISTORIOGRAPHIC AND IDEOLOGICAL CONTEXTS OF THE NINETEENTH-CENTURY DEBATE ON MAN'S PLACE IN NATURE[1]

Robert Young

I

A scientific worker is necessarily the child of his time and the inheritor of the thought of many generations. But the study of his environment and its conditioning power may be carried on from more than one point of view. Joseph Needham (1935).[2]

This essay is about theory and praxis.[3] The theory in question is that of the historiography of science, and the praxis to which it is applied is the attempt to write a book about the debate on man's place in nature in nineteenth-century Britain. The praxis is therefore theoretical praxis. For reasons which reflect both my own ideological position and my personal style of research, I find it meaningless to discuss ideological and historiographic questions in

[1] I should like to acknowledge with thanks the generous help of the following people in providing guidance, criticisms and other contributions toward the conception, production and reproduction of this essay: Tamsin Braidwood, John Fekete, Diana Guyon, Jeremy Mulford, Roy Porter, Martin Rudwick, Mikuláš Teich, Brian Turner, Ingrid Turner, Rita van der Straeten, Margot Waddell, Gary Werskey. Responsibility for the result is entirely my own.

[2] Joseph Needham, *Time: The Refreshing River* (*Essays and Addresses, 1932–1942*) (London, Allen & Unwin, 1943), p. 141.

[3] For a discussion of 'The Marxian Concept of Praxis', see Henri Lefebvre, *The Sociology of Marx* (1966), trans. N. Guterman (London, Allen Lane, 1968), ch. 2. A more concise exposition of the relationship between thought and action occurs at pp. 231–5 in Charles Taylor's excellent discussion of 'Marxism and Empiricism', in Bernard Williams and Alan Montefiore (eds.), *British Analytical Philosophy* (London, Routledge & Kegan Paul, 1966), ch. 10.

the abstract. I shall therefore consider these issues as they have arisen in connection with particular problems in the course of my own research.

The epigraph to this essay is based on a marxist approach to the history of science. It was written at a time when Joseph Needham was just beginning to apply his political orientation to systematic work in the history of embryology, in an essay delivered at Yale in 1935 on 'Limiting Factors in the History of Science, observed in the History of Embryology'.[4] Needham's research since then has continually broken new ground in the field of the history of science. I have chosen the epigraph, not only because this essay is written in his honour, but also, because my main purpose is to attempt to stimulate debate on the requirements of a radical historiography of science in the current period—thirty-seven years after Needham laid out his position and forty-one years after the Soviet delegation to the Second International Congress of the History of Science and Technology came to London and dramatically introduced a version of marxist historiography to the awareness of historians of science in Britain.[5] The attempt to work towards a radical historiography cannot at present be based upon a settled conception of what is meant by 'radical'.[6] However, certain aspects of this conception can be stated. It is concerned with an approach to history which is critical and in the service of transcendence and liberation rather than mere reproduction, one which gives insight into possibilities for achieving a society which is not alienating and repressive.[7]

[4] Op. cit. (note 2), pp. 141–59.

[5] N. I. Bukharin *et al.*, *Science at the Cross Roads: Papers Presented to the International Congress of the History of Science and Technology held in London from June 29th to July 3rd, 1931 by the Delegates of the U.S.S.R.* (1931), reprinted with a new Foreword by Joseph Needham and a new Introduction by Paul Gary Werskey (London, Cass, 1971).

[6] In other branches of historical studies this revival of socialist approaches to scholarship has been under way for some time. In the fields of philosophy and history see, for example, *New Left Review; Theoretical Practice* (13 Grosvenor Avenue, London, N.5.); *Telos* (Dept. of Philosophy, S.U.N.Y. at Buffalo, 4244 Ridge Lea Rd., Amherst, New York 14226); *Radical Philosophy* (c/o R. J. Norman, Darwin College, The University, Canterbury, Kent); *Radical America* (1878 Massachusetts Ave., Cambridge, Mass. 02140)—especially Vol. 4, No. 8–9, November, 1970: 'Special Issue on Radical Historiography'; George Fischer (ed.), *The Revival of American Socialism: Selected Papers of The Socialist Scholars Conference* (Oxford, 1971; also paperback)—especially preface on radical scholarship; Theodore Roszak (ed.), *The Dissenting Academy* (New York, Random House, 1967; also Penguin paperback, 1969); Noam Chomsky, 'Objectivity and Liberal Scholarship', ch. 1 of *American Power and the New Mandarins* (New York, Pantheon, 1969; also Pengiun paperback, 1969).

[7] In this context, a 'critical' approach explicitly includes the use of evaluations both in the definition of the domain and in the analysis of the 'data'. The relevant values are emancipatory, and describe 'the human potentiality for self-reflection and self-determination, that is not yet fully realized and is continuously hindered by the present modes of production'. This conception has been worked out by the Neo-marxist Frankfurt

The argument falls into two main parts and a number of distinct sections. The first section is a brief exposition of my particular problem, written in relatively subjective terms. The second is a discussion of what it meant to be 'a child of one's time' in the history and philosophy of science in Cambridge in the 1960s. The third is an outline of the state of the literature on the nineteenth-century debate on man's place in nature. Taken together, these sections provide a statement of the problem. In the second half of the essay, attention is directed to the available alternative perspectives for doing research in the history of science. It begins with criticisms of the marxist 'base-superstructure' model of interpretation, with particular emphasis on the limitations of what has come to be known as 'vulgar marxism'. This is followed by my own criticisms of the prevailing terms of the historiographic debate in the history of science—the 'internalist-externalist' distinction and the related claims of 'demarcationism'. Within this general context, I go on to consider the scope of, and to make objections to, the positions of those who concentrate on the rôle of external factors in the history and sociology of science—the Weberian-Mertonian tradition—and the debate between that and the internalist historians of ideas. The work of Lakatos, Kuhn, Merton and Hall is briefly evaluated, and the severe limitations of their approaches are pointed out. This section is followed by an analysis of recent treatments of Darwinism by New Left authors and a tentative look at the help which various anthropological perspectives can provide for understanding my problem and, more generally, the study of science and its history. This section refers to the work of Anderson, Thompson, Marx, Lukács, Douglas, Horton and Sohn-Rethel.

In the course of the essay an attempt is made to suggest that we reconsider the marxist 'base-superstructure' distinction, i.e., the view that all intellectual and cultural phenomena are ultimately determined by socio-economic conditions. In so doing, attention should be shifted towards the complex and subtle mediations between social and economic factors and the explicit *content* of scientific findings and theories. It is suggested that the strategy for developing a radical historiography in the current period should have two moments. First, the devotion of serious attention to the dialectic *between* base

---

School of Critical Theory; see Hans P. Drietzel, *Recent Sociology, No. 2* (London, Collier-Macmillan paperback, 1970), p. 209. I have been surprised by the extreme reactions of some scholars to earlier versions of this essay. Some imply that scholarship is—or can be —neutral and objective. Others seem unable or unwilling to grasp that there is a long tradition of philosophy, scholarship and political activism which does not acknowledge that the 'fact-value' distinction can operate with any precision. Similarly, there is an established and growing tradition of writers who explicitly draw on evaluative conceptions of how they would like the world to be to provide a standpoint for criticizing how it is and has been.

and superstructure. Second, the development of a theory of mediations which moves towards a concept of totality in which man, nature and society are seen in fully relational terms. Rather than abandoning the study of the history of ideas, it is important that both ideas and their institutionalizations continue to be given serious attention. But this must be done without losing sight of their historical place in social and economic life, and their ideological rôle in maintaining existing social and economic relations by rationalizing them. A double perspective on science as such and as ideology is advocated. This model is applied to the approach I am attempting to develop for writing about the nineteenth-century debate—a debate which was itself concerned with the philosophies of man, God, nature and society and their interactions. The rôle of science as ideology is relatively over-emphasized in this essay in an effort to counter-balance the existing bias in the literature towards the internal history of ideas. It is hoped that when the existing approach has been sufficiently complemented by studies of the rôle of ideology as a material force in our views of nature, man and society, a more balanced view will develop.

Many would argue that the 'base-superstructure' distinction is beyond redemption, but historians of science may well need to work their way through it before they can achieve a more dialectical perspective. I should like to stress that this essay is offered as a 'working paper' in which I have attempted to bring together in the compass of a single argument a number of related issues, the relations among which have not been worked out at all satisfactorily. My aim is to raise questions and to stimulate debate. The extensive annotations are provided in order to draw attention to the wide spectrum of raw materials for that debate. A great deal of further study of marxist and related writings must be undertaken if the history of science is to take its place among critical studies in the service of liberation.[8]

[8] Most scholars find it very difficult indeed to make any contact at all with marxist historiography. They are thereby precluded from giving serious consideration to its approach. I have found the following works helpful in attempting to relate the assumptions of traditional historiography with those of marxism. On the social determination of truth and the sociology of knowledge, see Karl Mannheim, *Ideology and Utopia: An Introduction to the Sociology of Knowledge*, trans. L. Wirth and E. A. Shils (London, Routledge, 1954; also paperback); Peter L. Berger and T. Luckmann, *The Social Construction of Reality: A Treatise in the Sociology of Knowledge* (New York, Doubleday, 1966; also Anchor and Peregrine paperbacks); James E. Curtis and J. W. Petras (eds.), *The Sociology of Knowledge: A Reader* (London, Duckworth, 1970), includes extensive bibliography; further references are given in the notes at the end of R. M. Young, 'Evolutionary Biology and Ideology: Then and Now', *Science Studies* 1 (1971), pp. 177–206; for a criticism of the assumptions of the sociology of knowledge, see Steven Lukes, 'On the Social Determination of Truth', in Robin Horton and R. Finnegan (eds.), *Modes of Thought: Essays in Honour of Sir Edward Evans Prichard* (London, Faber, 1972, in press). For expositions of marxism which are

II

The following two sections are written autobiographically. This style has been adopted in the belief that others working in the field have had analogous experiences. Thus, although the presentation is personal, one hopes that it is not merely so.

The general problem with which my research has been concerned for the past sixteen years is that of applying the categories of natural science to the study of man. After completing a series of studies on the relationship between psychological and physiological categories—considered in the context of philosophical and biological developments in the nineteenth and twentieth centuries[9]—the scope of the enquiry was broadened to include the intellectual context within which the scientific study of man was undertaken. Thus, although these investigations had originally been concerned with biology (as broadly defined), it soon became apparent that the debate on evolution could not be considered in isolation from theological, philosophical, literary, social, political and economic debates in the same period. It is a commonplace that the evolutionary debate was conducted amid heated controversies in those related areas. However, a close study of the documents has made it clear that it is difficult, and ultimately impossible, to maintain the conventional distinction between the science of the period and the interacting factors which are usually considered to provide its context. This is as true of the boundaries between science and pseudoscience as it is of the putative boundaries between

accessible to scholars trained in the Anglo-American empiricist tradition, see Charles Taylor, 'Marxism and Empiricism', op. cit. (note 3); Henri Lefebvre, *The Sociology of Marx*, op. cit. (note 3); H. Lefebvre, *The Explosion: Marxism and the French Upheaval* (1968), trans. A. Ehrenfeld (London, Monthly Review, 1969; also paperback); UNESCO symposium on 'The Rôle of Karl Marx in the Development of Contemporary Scientific Thought', published as *Marx and Contemporary Scientific Thought* (Paris, Mouton, 1969), especially the essays by E. Hobsbawm, C. Frankel and M. Marković. A most helpful exposition is Bertell Ollman, *Alienation: Marx's Conception of Man in Capitalist Society* (Cambridge, 1971); cf. an illuminating review of Ollman's book, in which the difficulties of Anglo-American scholars are spelled out: Robert L. Heilbroner, 'Through the Marxian Maze', *New York Review of Books*, 9 March 1972, pp. 9–12.

[9] R. M. Young, 'Scholarship and the History of the Behavioural Sciences', *History of Science* 5 (1966), 1–51; 'Animal Soul', in Paul Edwards (ed.), *The Encyclopedia of Philosophy* (New York, Macmillan, 1967), Vol. 1, pp. 122–7; 'The Functions of the Brain: Gall to Ferrier (1808–1886)', *Isis*, 59 (1968), pp. 251–68; 'The Development of Herbert Spencer's Concept of Evolution', *Actes du XIe Congrès International d'Histoire des Sciences* (Warsaw, Ossolineum, 1967), Vol. 2, pp. 273–8; 'Philosophy of Mind and Related Issues', *Brit. J. Philos. Sci.*, 18 (1967), pp. 325–30; *Mind, Brain and Adaptation in the Nineteenth Century* (Oxford, Clarendon Press, 1970).

the internal history of scientific findings and ideas on the one hand, and so-called 'non-scientific' factors on the other.[10]

In the course of the research and in conversations with colleagues working in an earlier period, it became evident that this problem was not at all unusual among historians and in particular among those who found it increasingly difficult to do justice to developments in the sixteenth and seventeenth centuries within a framework which required the separation of 'science' (as retrospectively defined) from social, philosophical and theological issues. They found it reassuring to encounter the same problem in the nineteenth century, since one of the ways in which their critics had dealt with their interpretations was to argue that such 'confusion' was to be expected in the period when the methods and assumptions of modern science were being established.[11] Of course, research in the nineteenth century is faced with a version of the same argument in the assertion that its domain is the period when the methods and assumptions of modern science were being extended to the *biological* sciences and to the study of man and society.

It would be wrong to claim that it was easy to refute the arguments of those who wanted to separate internal from external factors and to concentrate on the advancing edge of objectivity, the history of paradigms or the progressive demarcation of real science from contextual factors and from pseudoscience. Rather, we gained confidence from each others' failure to find the accepted historiography congenial to our own enquiries and felt more at home in the company of political and social historians who thought it odd that historians of science could, with apparent confidence, claim to separate science from other developments in a given period. We were not able to offer a coherent historiography to counter the one which was derived from positivist and sociological models of science. Rather, we called the practitioners of the internalist history of ideas 'Whig' historians and, as our detailed studies found sympathetic readers, we began to refer to our own approach as 'relativist' and 'contextualist', embracing labels which others had used as

[10] R. M. Young, 'The Impact of Darwin on Conventional Thought', in Anthony Symondson (ed.), *The Victorian Crisis of Faith* (London, S.P.C.K., 1970), pp. 13–35; 'Malthus and the Evolutionists: the Common Context of Biological and Social Theory', *Past & Present* No. 43 (1969), pp. 109–45; ' "Non-Scientific" Factors in the Darwinian Debate', *Actes du XIIe Congrès International d'Histoire des Sciences* (Paris, Blanchard, 1971), Vol. 8, pp. 221–6; 'Natural Theology, Victorian Periodicals, and the Fragmentation of the Common Context', *Victorian Studies* (in press); 'Darwin's Metaphor: Does Nature Select?', *The Monist*, 55 (1971), pp. 442–503; 'The Rôle of Psychology in the Nineteenth-Century Evolutionary Debate', in Mary Henle *et al.* (eds.), *Contributions to the History of Psychology* (in press); 'Association of Ideas', in Philip P. Wiener (ed.), *Dictionary of the History of Ideas* (New York, Scribner's, in press).

[11] See, for example, the contributions of Charles Webster and P. M. Rattansi to this volume, as compared to that of M. B. Hesse.

epithets. At the same time we came to ignore the official historiography, since we felt that its strictures did violence to the fine texture of the issues which we encountered in our work with primary sources.

In the same period that a group of professional historians was developing some confidence—or at least mutual comfort—I was conducting research for a book which was tentatively entitled *Man's Place in Nature: The Nineteenth-Century Debate in Britain*. This research—which has extended over seven years with the book in mind—was not being carried out according to a clearly-conceived historiography. However, from the outset its scope was the interpretation of the evolutionary debate in a very wide context of late-eighteenth- and nineteenth-century thought. Since I had come to the research from work in the history of psychiatry, psychology, and neurophysiology, I was more concerned with the question of man than with the development of biological theory *per se*. The research has differed from existing work in the general area in that it has been concerned with geological, biological, theological and social issues from the point of view of what I take to be the central preoccupation of the participants in the debate: a fundamental reorientation of the conception of the relations between man, God, nature and society. As the work progressed, this reorientation became more closely defined as a change from mechanistic analogies employed within an explicitly theistic natural theology to the use of organic analogies based on a secularized, implicit natural theology.

As the research began to lead more directly to a plan for a book, it took the form of a close study of a particular line of central works, those of T. R. Malthus, William Paley, Charles Lyell, the authors of the *Bridgewater Treatises*, Charles Darwin, Robert Chambers, A. R. Wallace, Herbert Spencer, the authors of *Essays and Reviews*, T. H. Huxley, and John Tyndall. The focus of the existing literature on developments in geology and biology was complemented by placing much greater emphasis on the writings of theological, psychological, philosophical and social theorists whose work was an integral part of the debate. As the study developed and as an analysis was undertaken of the reception of the central figures and their works, the crucial rôle of a group of commentators became apparent. These were lesser men in the pantheon of the history of science and were in many cases seen as second-rank at the time. But it became clear that their rôle in the debate—considered as a debate in its own terms rather than as the path to current views—was as important as that of the central figures. They were the critics and interpreters of the new approaches of geology and biology to the intelligentsia, and it was their criticisms which the main writers felt they had to anticipate and to answer. The essays and the reviews of, for example, William Buckland, William Whewell, Adam Sedgwick, Baden Powell, G. H. Lewes, William

Carpenter, and St. George J. Mivart were of considerable importance in this respect. There was also a substantial network of periodicals in which the views of these lesser writers appeared. Each of the main works was reviewed and debated in the Victorian periodicals which approached the issues from identifiable points of view. Finally, one can supplement these public documents with the reactions which were subsequently published in the voluminous lives and letters of nearly all of the significant figures involved. It will therefore be seen that a study of this debate can be based on an intricate network of original contributions, criticisms, interpretations, and reactions extending from the founding of the *Edinburgh Review* in 1802 and growing throughout the century as the periodical press expanded. If one wanted to attach a name to the resulting approach, it might be called 'social intellectual history'. Of course, limits had to be set to any systematic goals in covering the periodical literature, but the extent of the critical discussion on particular works helps considerably in deciding how far to follow a given aspect of the ongoing debate.[12] That is, the danger of retrospective distortion of emphasis is lessened by accepting guidance from the intensity and scope of the contemporary debate as defined by its own organs.[13] This approach had to be tempered by some awareness of 'resounding silences'—problems which were *not* discussed and the reasons for this.[14] Even so, the literature involved is very extensive indeed, and it has seemed prudent to make initial attempts to consider aspects of the debate by writing papers along the way. The approach of each of these is at a tangent to the main line of the proposed book, and an attempt has been made to avoid prematurely codifying one's views on a given theme.

It should be emphasized that the historiographic and ideological questions which are to be raised below are germane to a particular historical problem, one which is concerned with an important issue in the history of science and which is undoubtedly of interest to students in a number of related fields. In attempting to work out a clear approach to the argument of the book, I have found that the prevailing historiographic positions in the field are not

[12] Research in the Victorian periodicals has been made immensely easier by Walter E. Houghton (ed.), *The Wellesley Index to Victorian Periodicals, 1824–1900* (London, Routledge & Kegan Paul, 1966–), which has dramatically lessened the problem of anonymous authorship. Scholars working on these sources also benefit from one another's research through the *Victorian Periodicals Newsletter*.

[13] This research was conducted in conjunction with a Special Subject in Part II of the Cambridge Historical Tripos (1966–70), 'Science and Public Debate in Britain, 1830–1876'. Both the scope and the approach to the research were significantly influenced by working on the same documents with undergraduates in History and in the Natural Sciences.

[14] This problem is discussed in 'The Rôle of Psychology in the Nineteenth-Century Evolutionary Debate', op. cit. (note 10).

adequate for interpreting my findings. Furthermore, the problem should not be separated from that of relating this project to the uncertain rôle of academic research in the current political and ideological situation. This difficulty is particularly acute in research on the nineteenth-century debate on man's place in nature, since the domain of that research is itself a widespread controversy on the relations between the interpretations of man and society and the categories of science which defined the problem for the twentieth century. I was therefore faced with the problem of two levels: (1) the interpretation of the nineteenth-century debate; and (2) the attempt to do this in a period when the relations between scientific and political categories are once again at the centre of controversy among intellectuals and political activists.[15] There is a third, more fundamental, issue: this essay must be seen as the first step towards a critique of the project of writing 'social intellectual history', since a radical historiography must go beyond the history of ideas—however broadly defined—and locate the debate on those ideas in the history of men, events and institutions, of which it forms a constitutive part.

III

The epigraph to this essay should be applied not only to scientific workers but also to workers in the *history* of science: *they* are children of *their* time. That is, *we* are children of *our* time, and the issues raised by that assertion should be applied to the questions being considered here. For an historian of science beginning research in Cambridge (as in most centres in Britain and America) in the early 1960s, this truism would have conveyed no echo of the marxist thesis on which it was based: 'The mode of production of material life conditions the general process of social, political and intellectual life. It is not the consciousness of men that determines their existence, but their social existence that determines their consciousness'.[16] Rather, the apprentice scholar was faced with a set of models which implied the relative autonomy of ideas from their socio-economic bases. The standard texts to which he was exposed in Cambridge (and elsewhere) were all concerned with intellectual history, although there were important differences between those who saw the history of scientific ideas as a relatively autonomous study, e.g. Herbert Butterfield, Rupert Hall, and M. A. Hoskin, and those who were keen to relate them to philosophical themes, e.g. N. R. Hanson,[17] G. Buchdahl, and

[15] The analogies between the debates in the nineteenth century and in the current period are discussed in 'Evolutionary Biology and Ideology: Then and Now', op. cit. (note 8).

[16] Karl Marx, 'Preface' to *A Contribution to the Critique of Political Economy* (1859), trans. S. W. Ryazanskaya (London, Lawrence & Wishart, 1971), pp. 20–1.

[17] N. R. Hanson, *Patterns of Discovery: An Inquiry into the Conceptual Foundations of Science* (Cambridge, 1958).

M. B. Hesse. A. C. Crombie represented an intermediate position, and his *Augustine to Galileo*[18] was one of a relatively small number of works read by students early in their studies in the field which had not been written by someone who was then, or had very recently been, on the teaching staff in the history and philosophy of science in Cambridge (though he had worked in Cambridge for many years as a scientist). Butterfield was still active in Cambridge, although not then teaching history of science. Hall and Hanson had recently left to take up professorships in America. The approaches of the staff members in history and philosophy of science consisted of an amalgam of Hoskin's internalism, Hesse's highly disciplined integration of current philosophy of science and history of scientific ideas, and Buchdahl's rich and allusive studies in the history of the metaphysics of science from Descartes to Kant, coupled with an infectious ability to explore all of the ambiguities in any apparently straightforward statement.[19] The standard introductory texts were Butterfield's *The Origins of Modern Science* (1949) and Hall's *The Scientific Revolution* (1954), both of which were almost exclusively concerned with the internal history of scientific ideas. They were joined in 1960 by a useful collection of short essays, edited by Hall, on *The Making of Modern Science* and Gillispie's synoptic work, *The Edge of Objectivity*. (His essay has broader scope, but its title and its historiography perfectly reflected the prevailing orthodoxy.)[20] These were soon followed by introductory works by Boas and Hall which were more detailed and limited in period but which continued the established tradition.[21]

[18] A. C. Crombie, *Augustine to Galileo* (1952), 2 vols., 2nd edition (London, Heinemann, 1959; also paperback).

[19] See, for example, M. A. Hoskin, *William Herschel: Pioneer of Sidereal Astronomy* (London, Sheed & Ward, 1959; also paperback); *William Herschel and the Construction of the Heavens* (London, Oldbourne, 1963); M. B. Hesse, *Models and Analogies in Science* (London, Sheed & Ward, 1963; 2nd edition, Notre Dame, 1966); *Forces and Fields: A Study of Action at a Distance in the History of Physics* (London, Nelson, 1961, also paperback); Gerd Buchdahl, *The Image of Newton and Locke in the Age of Reason* (London, Sheed & Ward paperback, 1961); 'The Relevance of Descartes's Philosophy for Modern Philosophy of Science', *Brit. J. Hist. Sci.* **1** (1963), pp. 227–49. Buchdahl's fundamental and synthetic monograph appeared at the end of the decade, embracing issues which had been discussed in his seminars at least since 1960: *Metaphysics and the Philosophy of Science: The Classical Origins Descartes to Kant* (Oxford, Blackwell, 1969).

[20] Herbert Butterfield, *The Origins of Modern Science, 1300–1800* (1949), 2nd edition (London, Bell, 1957; also paperback); A. Rupert Hall, *The Scientific Revolution, 1500–1800 The Formation of the Modern Scientific Attitude* (London, Longmans, Green, 1954; also Beacon paperback, 1956); A. R. Hall (ed.), *The Making of Modern Science* (Leicester, paperback, 1960); Charles C. Gillispie, *The Edge of Objectivity: an Essay in the History of Scientific Ideas* (London, Oxford, 1960; also Princeton paperback).

[21] Marie Boas, *The Scientific Renaissance, 1450–1630* (London, Collins, 1962); A. R. Hall, *From Galileo to Newton, 1630–1720* (London, Collins, 1963). Their joint synoptic volume, *A Brief History of Science* (New York, New American Library paperback, 1964) was not widely available in Britain.

Thomas Kuhn's *The Structure of Scientific Revolutions* appeared in 1962 and seemed to many to herald 'a revolution in the historiography of science', but the debate about it in the intervening decade and Kuhn's own replies to his critics and interpreters make it clear that his attacks on the cumulative and positivist approaches do not, in the end, transcend the established view of the history of science.[22] The bridge between historical and philosophical issues was built by the work of Crombie, Hanson and Buchdahl and by Hesse's *Forces and Fields* and Dijksterhuis' *The Mechanization of the World Picture,* both of which appeared in 1961 (the latter in translation).[23] Moving closer to the philosophy of science *per se,* Blake, Ducasse and Madden's collection of case-studies, *Theories of Scientific Method,* appeared in 1960.[24] The standard works in the (relatively ahistorical) philosophy of science were Braithwaite's *Scientific Explanation* (1953, paperback edition 1960) and Popper's *The Logic of Scientific Discovery* (1934, English translation, 1959).[25] These were complemented by a number of volumes of collected readings with a heavy bias towards the philosophy of the physical sciences. Although these volumes included selections which were concerned with the biological and human sciences, the approaches of the editors were towards the physico-chemical sciences as the model for all knowledge.[26]

[22] Thomas S. Kuhn's *The Copernican Revolution* (Cambridge, Mass., Harvard, 1957; also Vintage paperback) was a standard source before his historiographic thesis appeared; *The Structure of Scientific Revolutions* (Chicago, 1962; 2nd edition with Postscript, 1970; also paperback); G. Buchdahl, 'A Revolution in Historiography of Science', *History of Science* 4 (1965), pp. 55–69; Dudley Shapere, 'The Structure of Scientific Revolutions', *Philos. Rev.* 73 (1964), pp. 383–94; Imre Lakatos and A. Musgrave (eds.), *Criticism and the Growth of Knowledge* (Cambridge, 1970; also paperback), includes opening and concluding chapters by Kuhn; cf. David Bloor, 'Two Paradigms for Scientific Knowledge?', *Science Studies* 1 (1971), pp. 101–15.

[23] Hesse, op. cit. (note 19); E. J. Dijksterhuis, *The Mechanization of the World Picture* (1950), trans. C. Dikshoorn (Oxford, Clarendon Press, 1961; also paperback).

[24] Ralph M. Blake, C. J. Ducasse and E. H. Madden, *Theories of Scientific Method: The Renaissance through the Nineteenth Century* (Seattle, Washington, 1960; also paperback).

[25] Richard B. Braithwaite, *Scientific Explanation: A Study of the Function of Theory, Probability and Law in Science* (Cambridge, 1953; Harper paperback, 1960); Karl R. Popper, *The Logic of Scientific Discovery* (1934), trans. Popper, J. Freed, & L. Freed (New York, Basic, 1959; also paperback); a collection of Popper's papers appeared four years later: *Conjectures and Refutations: The Growth of Scientific Knowledge* (London, Routledge & Kegan Paul, 1963), dedicated to the conservative economist F. A. von Hayek.

[26] The most widely used of these were Herbert Feigl and May Brodbeck (eds.), *Readings in the Philosophy of Science* (New York, Appleton-Century-Crofts, 1953); Philip P. Wiener (ed.), *Readings in Philosophy of Science: Introduction to the Foundations and Cultural Aspects of the Sciences* (New York, Scribner's, 1953); Edward H. Madden (ed.), *The Structure of Scientific Thought: An Introduction to Philosophy of Science* (London, Routledge & Kegan Paul, 1960). A collection of the essays of Ernest Nagel appeared in the same period: *The Structure of Science: Problems in the Logic of Scientific Explanation* (London, Routledge & Kegan Paul, 1961), as did the analytic treatise of Arthur Pap, *An Introduction to the Philosophy of Science* (London, Eyre & Spottiswoode, 1963).

Taken together, these works—all of which appeared for the first time or in translation (or were still being widely read) in the late 1950s and early 1960s —made up a formidable orthodoxy in favour of treating the history and philosophy of science in relative isolation from their social, economic and political contexts.[27] Although many of the historical works make gestures towards society, this was their orientation. The classics which somehow came to one's attention were concerned with the history of scientific ideas in the context of philosophy, though they probed further than most of those listed above into the philosophy of nature: A. N. Whitehead's *Science and the Modern World* (1925), E. A. Burtt's *The Metaphysical Foundations of Modern Physical Science* (2nd edition, 1932), A. O. Lovejoy's *The Great Chain of Being* (1936), and the essays of Alexandre Koyré.[28]

It was known that Joseph Needham was working in Cambridge, that he saw himself as both a marxist and a Christian, and that he was engaged in a massive work on *Science and Civilisation in China*.[29] It was also known that he was conducting this distinguished work while continuing to hold the position of Dunn Reader in Biochemistry and that whenever the question was mooted of just recognition of his achievement in the history of science in the form of a Chair, key individuals in the University saw to it that the matter went no further. (He was able to resign his Readership in 1966, when he became Master of Gonville and Caius College. He was awarded the Sarton Medal— the profession's highest honour—at the XIIth International Congress of the History of Science in 1968.) His legitimate preoccupation with his self-imposed task, along with its esoteric subject, meant that although young

[27] A survey of the domain of the history of science was made at a symposium at Oxford in 1961: A. C. Crombie (ed.), *Scientific Change: Historical Studies in the Intellectual, Social, and Technical Conditions for Scientific Discovery and Technical Invention from Antiquity to the Present* (London, Heinemann, 1963). The scope of this overview was far broader than the teaching programmes in British and American Universities. In particular, economic and social issues were not emphasized. The only readily available marxist survey of the history of science was not widely read or discussed in teaching programmes: Stephen F. Mason, *Main Currents of Scientific Thought* (New York, Abelard-Schuman, 1956; reprinted as *A History of the Sciences,* New York, Collier paperback, 1962). Mason returned to research in chemistry when no posts were available to him in the history of science. His survey is, in my opinion, very useful and incisive.

[28] Alfred N. Whitehead, *Science and the Modern World* (Cambridge, 1925; also paperback); Edwin A. Burtt, *The Metaphysical Foundations of Modern Physical Science* (1924), 2nd edition (London, Routledge & Kegan Paul, 1932; also Anchor paperback, 1955); Arthur O. Lovejoy, *The Great Chain of Being: A Study of the History of an Idea* (Cambridge, Mass., Harvard, 1936; also Harper paperback, 1960); Alexandre Koyré, *Newtonian Studies* (1948–1961) (London, Chapman & Hall, 1965); *Metaphysics and Measurement: Essays in Scientific Revolution* (1943–1960), trans. R. E. Madison (London, Chapman & Hall, 1968).

[29] Joseph Needham *et al., Science and Civilisation in China* (Cambridge, 1954–), in progress; Vols. 1–4 (Part 3) published.

scholars perused his volumes and used his *History of Embryology*,[30] his presence in Cambridge did not appear to influence the prevailing orientation of teaching and research in the history and philosophy of science. (It was later possible to discern his rôle in subsequent developments in the subject in Cambridge, but his stature and influence were not directly felt by most students in the 1960s.)

More importantly, the word 'social' was seldom heard, and 'economics', 'politics', and 'ideology' were never uttered with any historiographic implications, although they were obviously employed in conversation. The explicitly marxist writings of Needham and Bernal were seldom referred to. I recall acquiring those of Bernal at bargain prices in 1965, and I sought out Needham's essays only after that. Nor were the names of Marx, Weber, Mannheim or Merton prominent on any reading lists or in the informal reading of graduate students. One was much more likely to study Wittgenstein, Wisdom, Ryle, Austin and Strawson in one's exploratory reading. While there were demarcations between those who were more interested in the history than in the philosophy of science, the continuum led from the history of scientific ideas to analytic philosophy. While relations with the History Faculty were cordial, and history of science lectures were listed in the History syllabus, there were no strong departmental or intellectual connections until 1966, when a Special Subject in the history of science was invited by the History Faculty, and this was an anomaly which caused difficulties on both sides.[31]

The foregoing account would seem to be a necessary background for understanding a series of remarks by Rupert Hall which would otherwise appear bizarre from the vantage point of 1972. In an assessment of the relations between the social and intellectual approaches to the history of science, published in 1963, he drew the opposite conclusion to Marx and wrote, 'Thus recent historians reverse the arrow of economic inference: social forms do not dominate mind; rather, in the long run, mind determines social forms'.[32] The scientific revolution was 'a phenomenon of intellectual history', and interest had been withdrawn from externalist explanations.[33] Modern science 'is the fruit of an intellectual mutation [and] its genesis must be considered in relation to an intellectual tradition'.[34]

Social and economic relations are rather concerned with the scientific movement than with science as a system of knowledge of nature (theoretical and practical);

[30] Joseph Needham, *A History of Embryology* (1934), 2nd edition, revised with the assistance of Arthur Hughes (Cambridge, 1959).

[31] See above, note 13.

[32] A. Rupert Hall, 'Merton Revisited, or Science and Society in the Seventeenth Century', *History of Science* 2 (1963), pp. 1–16, at p. 10.

[33] Ibid.

[34] Ibid., p. 12.

they help us to understand the public face of science and the public reaction to scientists; to evaluate the propaganda that scientists distribute about themselves, and occasionally—but only occasionally—to see why the subject of scientific discussion takes a new turn. But to understand the true contemporary significance of some piece of work in science, to explore its antecedents and effects, in other words to recreate critically the true historical situation, for this we must treat science as intellectual history, even experimental science.[35]

Hall's conclusions were borne out by the contemporary sociology of the history of science:

Even without making a detailed review of the work of other historians of science active at the present time it is clear that the trend towards intellectual history is strong and universal. Since the journal *Centaurus* published in 1953 a special group of articles on the social relations of science no single article that can be judged to represent the sociological interpretation of history has appeared in that periodical, or *Isis, Annals of science, Revue d'histoire des sciences,* or the *Archives internationales.* There has been little discussion of the historiographical issue: indeed, it sometimes seems that the case for setting the development of scientific thought in its broader historical context is condemned before it is heard, though one knows from personal conversations that it is not neglected in pedagogic practice. Clearly, externalist explanations of the history of science have lost their interest as well as their interpretive capacity.[36]

How, in the face of this intellectualist orthodoxy, did the revival of the study of the social dimension of science become attractive? Its relevance began to be felt in Cambridge, not as a result of the intellectual preoccupations of the teaching staff in the history and philosophy of science, but from three main sources. These were: the influence of the Leeds department, the approach of two philosophically-oriented political historians in Cambridge, John Dunn and Quentin Skinner, and the work of a group of Oxford-trained social historians. Something very exciting was going on at Leeds under the catalytic influence of Jerome Ravetz and three young scholars whom he had attracted there: P. M. Rattansi, J. E. McGuire, and Charles Webster. None of these was from the mainstream of British academic life: Ravetz is an American émigré and an ex-marxist whose original training was in mathematics; Rattansi is a Kenya Indian who worked as a journalist and took his first degree in economics; McGuire is an Irish-Canadian of maverick intellectual disposition; and Webster is a highly individualistic British radical who worked in education while doing graduate work in the history of science. All four of the Leeds group were placing the preoccupations of the seventeenth-century natural philosophers in theological and social contexts. They were interested in philosophical issues, but unlike Buchdahl and Hesse, this was not for the sake of the good philosophy to be squeezed out of

[35] Ibid., pp. 13–14.
[36] Ibid, p. 13.

them. Rather, they were shamelessly relativist and contextualist and were far more interested in the social milieux and the philosophies of nature underlying the works of the scientific *virtuosi* than in the so-called mainline of the development of modern science.[37] Their work was consonant with (and attractive to) eminent historians whose writings bore on science in differing ways but were not at the centre of consciousness of the professional group of historians of science: H. Trevor-Roper, Christopher Hill, Joseph Needham, Walter Pagel, Frances Yates, E. H. Gombrich, D. P. Walker.

Although the work of Dunn and Skinner had affinities with the approach of the Leeds group—especially in their studies on Locke and Hobbes[38]—they were also concerned to argue on philosophical and historical grounds against the legitimacy of the history of ideas as traditionally conceived. Both stressed that ideas do not beget ideas but that people do so in particular historical contexts and that the meaning of those ideas is exquisitely bound to the particularity of those contexts. They argued that the study of the genealogy of ideas divorced from close study of their social and political contexts could only lead to elaborate historical punning.[39] Their approaches attempted to

[37] P. M. Rattansi, 'Paracelsus and the Puritan Revolution', *Ambix* 11 (1963), pp. 24–32; 'The Helmontian-Galenist Controversy in Restoration England', ibid. 12 (1964), pp. 1–23; 'The Intellectual Origins of the Royal Society', *Notes & Rec. Roy. Soc.* 23 (1968), pp. 129–143; J. E. McGuire and P. M. Rattansi, 'Newton and the "Pipes of Pan"', ibid. 21 (1966), pp. 109–43; J. E. McGuire, 'Transmutation and Immutability: Newton's Doctrine of Physical Qualities', *Ambix* 14 (1967), pp. 69–95; 'The Origin of Newton's Doctrine of Essential Qualities', *Centaurus* 12 (1968), pp. 233–60; 'Force, Active Principles, and Newton's Invisible Realm', *Ambix* 15 (1968), pp. 154–208; 'Atoms and the "Analogy of Nature": Newton's Third Rule of Philosophizing', *Stud. Hist. & Philos. Sc.* 1 (1970), pp. 3–58; Charles Webster, 'Richard Towneley, 1629–1707, and the Towneley Group', *Trans. Hist. Soc. of Lancashire & Cheshire* 118 (1966), pp. 51–76; 'The Origins of the Royal Society', *History of Science* 6 (1967), pp. 106–28; 'The College of Physicians: "Solomon's House" in Restoration England', *Bull. Hist. Med.* 41 (1967), pp. 393–412; C. Webster (ed.), *Samuel Hartlib and the Advancement of Learning* (Cambridge, 1970).

[38] John M. Dunn, 'Consent in the Political Theory of John Locke', *Hist. J.* 10 (1967), pp. 153–82; 'Justice and the Interpretation of Locke's Political Theory', *Pol. Stud.* 16 (1968), pp. 68–87; *The Political Thought of John Locke* (Cambridge, 1969); Quentin R. D. Skinner, 'The Ideological Context of Hobbes's Political Thought', *Hist. J.* 9 (1966), pp. 286–317; 'Thomas Hobbes and the Nature of the Early Royal Society', ibid. 12 (1969), pp. 217–39.

[39] John M. Dunn, 'The Identity of the History of Ideas', *Philosophy* 43 (1968), pp. 85–104; Q. R. D. Skinner, 'The Limits of Historical Explanation', ibid. 41 (1966), pp. 199–215; 'Meaning and Understanding in the History of Ideas', *History and Theory* 8 (1969), pp. 3–53. Skinner's interests in the sociology of knowledge and philosophy have since led him to become preoccupied with the philosophical analysis of the concept of action, while Dunn has become a conceptual analyst of modern political phenomena. See Skinner, 'On Performing and Explaining Linguistic Actions', *Philos. Quart.* 21 (1971), pp. 1–21; Dunn, *Modern Revolutions: An Introduction to the Analysis of a Political Phenomenon* (Cambridge, 1972; also paperback).

integrate the tradition of analytic philosophy (with its élitist appeal to the intellectual aristocracy) with the investigations of Duncan Forbes on the Scottish Enlightment and Peter Laslett on the Locke MSS. The influence of Dunn and Skinner in the history of science led one to set ideas in their social contexts, using stern criteria of investigation and inference in the close analysis of documents. A contemporary of theirs, John Burrow, shared both their background in the Cambridge History Tripos and the influence of Forbes and wrote an important study in nineteenth-century intellectual history.[40]

The appeal of the approach of Dunn and Skinner to social factors in intellectual history was complemented by the influence of a group of young left-wing social historians based at Oxford who were more concerned with the texture of history than with historiographic and philosophical elegance. Foremost among them was Raphael Samuel of Ruskin College. There is no direct connection between interest in the history of science and in the work of Samuel on nineteenth-century British working class history, that of Tim Mason on the German unions under National Socialism, Gillian Sutherland on nineteenth-century primary education in Britain, or Susan Budd on the Rationalist Press Association. The connection lies rather in the fact that scholars become interested in approaches of colleagues who are doing exciting work, find themselves influenced by their perspectives, and ask how they would see a given problem and how they would treat it. In retrospect, the affinities felt with their work can be seen as an attempt to transcend the orthodoxy of internalist intellectualist historiography of science. Among contemporaries, theirs seemed particularly relevant research in that it was addressed to society.

It was in the atmosphere of a strong orthodoxy in the history and philosophy of science at Cambridge, alongside a network of overlapping intellectual affinities and sympathies, that in 1968 Rattansi and Young brought together a number of the scholars mentioned above, and others, for a seminar (sponsored by the King's College Research Centre) which met monthly to consider the relationship between the history of science and other branches of historical studies.[41] At quarterly intervals senior visitors gave papers to the seminar which represented established points of view in

[40] John W. Burrow, *Evolution and Society: A Study in Victorian Social Theory* (Cambridge, 1966; also paperback). I have commented on this book in *Cambridge Rev.* **89** (1967), pp. 409–11. See also Burrow's lucid Introduction to the Pelican reprint of the 1st edition of *The Origin of Species* (Harmondsworth, Penguin paperback, 1968), pp. 11–48.

[41] Those who attended the seminar more or less regularly were D. P. Walker, G. Buchdahl, M. Teich, M. B. Hesse, G. W. Stocking, Jr., W. Thomas, J. W. Burrow, C. Webster, M. J. S. Rudwick, Q. R. D. Skinner, H. Fruchtbaum, R. MacLeod, G. Sutherland, S. Budd, R. Porter, J. Mulford, P. M. Rattansi, R. Young.

different branches of history: Joseph Needham (history of science in the Chinese culture area), E. H. Gombrich (history of art), Frances Yates (Renaissance studies), Hugh Trevor-Roper (social history), Philip Collins (English literature). The seminar continued for over a year and was very stimulating. There was, however, little pressure or inclination to attempt to integrate the approaches of intellectual history, social history, and the philosophies of nature and society. The dichotomy between internal history of scientific ideas and external or social factors was regularly found to be at odds with our respective investigations, but no framework which was remotely near to being coherent emerged to take its place. On several occasions lists of 'factors' were drawn up from the discussions, but no overall approach to the issues was worked out. Most of the papers presented to the seminar were published in learned journals, but the participants felt that they in no sense formed the basis for a volume with a coherent theme. Indeed, the real work of the seminar was in the discussions, but no one had the inclination to record and edit them. It was decided that the seminar had been very successful, and there was some talk about reconstituting it around the topic of historiography, but other priorities intervened.

The striking feature of this account of the local social history of the history of science is that there was practically no discussion of, for example, economic or political history (as distinct from the history of political theory), or of the writings of Eric Hobsbawm, E. H. Carr[42] or E. P. Thompson. Nor was there any serious consideration of the domain of cultural history, for example the writings of Raymond Williams. This list has not been drawn at random. The surprising point is that there was nothing in the general coinage of intellectual discourse or of the books which were recommended by and to colleagues—and, more particularly, there was little in the deliberations of a seminar explicitly concerned with the relations between history of science and other branches of historical studies—which said, even obliquely: confront marxism. By this I mean that there was no serious pressure to consider the perspective of marxism and its critical approach to the relations between knowledge and the socio-political world in which it exists.

I am sure that most historians of science in Britain and America who were studying or doing research in the 1960s could tell an analogous story. The particular and local influences would be different, but the general ambiance and the conclusions would be the same. Internalist historians of science were reaching out uncertainly for alliances with other approaches in historical

[42] E. H. Carr's Trevelyan Lectures on historiography (1961), were published as *What is History?* (London, Macmillan, 1961; Penguin paperback, 1964) and appeared on undergraduate reading lists in the history of science. I do not, however, recall ever having heard them discussed among senior members.

studies—especially social (and in America, sociological) history. They did this with a sense of growing need for historiographic rigour. However, neither the literature nor discussions with colleagues provide evidence of any awareness that it might be worthwhile to consider the potential relevance of marxism to our problems. Readers who have not lived through this period in the subject can gain a sense of the general atmosphere by considering the analogous situation in historical, economic and social research in America as discussed in William Appleman Williams' *The Great Evasion: an Essay on the Contemporary Relevance of Karl Marx and on the Wisdom of Admitting the Heretic into the Dialogue about America's Future*.[43] It would be an exaggeration to say that there was a conspiracy of silence, but there certainly was silence.

But this is to be unduly behaviourist. I now know that at least one member of the seminar had begun his 'confrontation' with marxism even before the seminar began and has subsequently reached the point where he regards himself as a marxist. That this personal grappling had no significant presence at the seminar can be seen as anecdotal evidence pointing to the magnitude of the task that a radical change in ideological position involves one in. The change has to have resulted in a coherence well above the level of a collection of radical intuitions before it ceases to be readily vulnerable to attack from what it seeks to counter. Another sort of indication of the size of this task is perhaps the nature of this essay—which, though offered as the beginning of a critique, nevertheless is cast to a large extent within the terms of 'social intellectual history' (as defined on pages 350–52) which the critique aims to make constitutive of a totality of relations.

IV

It would be natural for the foregoing discussion of the prevailing orthodoxy in the history and philosophy of science in the 1960s—and its failure even to raise the problem of confronting marxism—to be followed by a discussion of marxist historiography, along with other approaches to the history of science which give weight to the relationship between the internal history of science and social and economic factors. These issues, however, will be postponed

[43] (1964, also Chicago, Quadrangle paperback, 1968); cf. the discussion of Williams' work in *Radical America*, op. cit. (note 6), pp. 1, 74, 76, 82–3, 92–3, 96, 104; and ibid. Vol. 4, No. 6 (August 1970), pp. 29–53. Williams is certainly the most distinguished marxist-influenced historian of the older generation in America. For a longer, though not wholly satisfactory, perspective on the relationship between social thought and marxism, see Irving M. Zeitlin, *Ideology and the Development of Sociological Theory* (Englewood, Cliffs N.J., Prentice-Hall, 1968).

until consideration has been given to the state of scholarship on the nine-teenth century debate on man's place in nature. My purpose is to bring these together. To introduce the specific problem first may help the discussion to avoid departing into generalities.

If we turn, therefore, from the state of general approaches to the history and philosophy of science to the literature on the nineteenth-century debate, the situation is relatively straightforward. The first point is that the topics which were central to the best work in the area were not concerned with 'the debate on man's place in nature' but with particular disciplines—geology and evolutionism—especially the work and influence of Charles Lyell and Charles Darwin. The field was occupied by scholars with an impressive command of the documents. In the case of Darwin, the most precise work was being done by Sydney Smith, whose knowledge of the Darwin archive at Cambridge was (and remains) nonpareil,[44] and Sir Gavin de Beer, who (with others) edited Darwin's *Notebooks* and wrote prolifically on Darwin in the context of the history of biology.[45] Neither had the interest or inclina-tion to set Darwin's work in a socio-political context. Smith concentrated on Darwin's work on classification and on important bibliographical research on the Darwin MSS. To those who came to consult him and seek guidance through the labyrinth, he argued that Darwin's research on cirripedes was the key to his mature work. The most important problems, he felt, lay wholly inside the scientific debates of the time. Similarly, de Beer explicitly opposed the assignment of any fundamental significance to social, political and economic ideas in the development of Darwin's theory of evolution. In his biographical study of Darwin he reviews in detail the evidence for an important rôle for Malthus' theory of population—and *only* in detail, with no reference to broader issues—and draws a narrow conclusion followed by a very general *non sequitur*:

> It is therefore clear that Darwin did not owe Malthus anything on the score of variation or natural selection, but only the realization that the high rate of mortality exacted by nature resulted in pressure, and while Malthus argued that this pressure was exerted against the poor members of the human race,

[44] Sydney Smith, 'The Origin of "The Origin" as Discerned from Charles Darwin's Notebooks and his Annotations of the Books he Read between 1837 and 1842', *Advance-ment of Sci.*, No. 64 (1960), pp. 391–401.

[45] Sir Gavin de Beer (ed.), *Darwin's Journal, Bull. Brit. Mus. (Nat. Hist.), Historical Series,* Vol. 2, No. 1 (1959); *Darwin's Notebooks on Transmutation of Species,* ibid. Nos. 2–5 (1960); G. de Beer and M. J. Rowlands (eds.), 'Addenda and Corrigenda', ibid., No. 6 (1961); G. de Beer, M. J. Rowlands, and B. M. Skramovsky (eds.), 'Pages Excised by Darwin', ibid., Vol. 3, No. 5 (1967); G. de Beer (ed.), 'Some Unpublished Letters of Charles Darwin', *Notes & Rec. Roy. Soc.* 14 (1960), pp. 12–66; 'The Darwin Letters at Shrewsbury School', ibid., 23 (1968), pp. 68–85; G. de Beer, 'The Origins of Darwin's Ideas on Evolution and Natural Selection', *Proc. Roy. Soc.* 155 B (1961–2), pp. 321–338.

Darwin applied the principle to plants and animals and argued that the pressure was exerted against the less well adapted. . . . He had already arrived at the principle of natural selection and had seen how, given variation, it would lead to unlimited change away from the ancestral type, improvement of adaptation, and eventually to the production of new species. Malthus enabled him to see how inexorably nature enforced this principle. The view that Darwin was led to the idea of natural selection by the social and economic conditions of Victorian England is devoid of foundation.[46]

The question of the rôle of Malthus' theory in the actual formative process of Darwin's mechanism of natural selection is very complex and open to a number of interpretations. Nevertheless, de Beer's sweeping conclusion about the irrelevance of social and economic conditions does not follow from his argument, even if it were the case that such questions could be meaningfully discussed in such a narrow context. The approach is simplistic, as is the conclusion. However, de Beer subsequently discovered new evidence which showed that Malthus' theory of population had played a much more intimate part in the process of the formulation of Darwin's theory than had been supposed. A number of scholars who have considered this issue have attempted to reinterpret Malthus' rôle and have pointed out the significance of de Beer's new evidence.[47] He, however, has stood firm in attempting to separate the scientific issues from the socio-economic context and even from direct, acknowledged influences. In a subsequent essay, he cites the relevant passage from Darwin's *Autobiography* (not the more significant passage from the *Notebooks*) and concentrates on making claims for how much Darwin had sorted out before reading Malthus. He once again reduces the issue to whether Darwin got the idea of natural selection from Malthus:

> From this passage some commentators have deduced that it was from Malthus that Darwin derived his principle of natural selection. As I have said above, this is quite false: Malthus had not the slightest idea of natural selection and would have been horror-struck at the notion of evolution. What Darwin got from Malthus was something that Malthus knew nothing at all about, and about which he was not writing: how natural selection is enforced on plants and animals in nature.[48]

It was certainly right to move on from the extreme claims of those who attempted to deduce Darwin from Malthus, but the reaction subtly falsifies both the origins and the context of the evolutionary theory of Darwin and,

[46] Sir Gavin de Beer, *Charles Darwin: Evolution by Natural Selection* (London, Nelson, 1963; also paperback), p. 100, cf. pp. 95–9.

[47] Young, 'Malthus and the Evolutionists', op. cit. (note 10); Peter Vorzimmer, 'Darwin, Malthus and the Theory of Natural Selection', *J. Hist. Ideas* 30 (1969), pp. 527–542; Sandra Herbert, 'Darwin, Malthus and Selection', *J. Hist. Biol.* 4 (1971), pp. 209–17.

[48] Sir Gavin de Beer, 'The Evolution of Charles Darwin', *New York Review of Books*, 17 December 1970, pp. 31–5, at p. 33.

for that matter, of his co-discoverer A. R. Wallace.[49] To move in this argument from the questions of origin and context of the theory, the internalist approach does not address itself to the socio-political context into which the theory was received and which set the terms of the debate on its reception in the periodicals and public debate of the time. To move further, there has been little tendency to consider the various extrapolations which were based on different versions of evolutionary theory.

The point of this extended example is not to assert that it is illegitimate for scholars to concentrate on one aspect of the history of science at the expense of others. Rather, it provides a basis for suggesting that the approach of internalist history impoverishes the understanding of its own subject matter and that its interpretive bias isolates evolutionism from its social, political and ideological context. The most comprehensive study of the reception of Darwin's work in the period 1859–1872 concentrates heavily on philosophical and theological issues at the expense of social and political ones.[50] Similarly, the best work being done on evolutionism by younger scholars concentrates on internal theoretical issues, applications to physiology, problems of demography, and methodology.[51]

The case of Charles Lyell's uniformitarian geology is somewhat different. It would be absurd to argue that the uniformitarian-catastrophist debate could be studied in a purely internalist way. It is just about plausible—though, as I have suggested, ultimately distorting—to concentrate on purely scientific issues in tracing the development of Darwin's theory. But theological issues were explicitly central and crucial in the geological debate. These other aspects have been explored in detail by Cannon, Hooykaas, Haber, Coleman, and Wilson.[52] Rudwick has shown that as time went on,

[49] Young, 'Malthus and the Evolutionists', op. cit. (note 10), pp. 125–34.

[50] Alvar Ellegård, *Darwin and the General Reader: The Reception of Darwin's Theory of Evolution in the British Periodical Press, 1859–1872* (Göteborg, Acta Universitatis Gothoburgensis, 1958).

[51] Gerald L. Geison, 'Darwin and Heredity: the Evolution of his Hypothesis of Pangenesis', *J. Hist. Med.* 24 (1969), pp. 375–411; Richard D. French, 'Darwin and the Physiologists, or the Medusa and Modern Cardiology', *J. Hist. Biol.* 3 (1970), pp. 253–74; Frank N. Egerton, 'Studies of Animal Populations from Lamarck to Darwin', ibid., 1 (1968), pp. 225–59; Michael Ruse, 'Natural Selection in *The Origin of Species*', *Stud. Hist. & Philos. Sci.* 1 (1971), pp. 311–51.

[52] Walter F. Cannon, 'The Uniformitarian-Catastrophist Debate', *Isis* 51 (1960), pp. 38–55; 'The Impact of Uniformitarianism: two letters from John Herschel to Charles Lyell, 1836–1837', *Proc. Amer. Philos. Soc.* 105 (1961), pp. 301–14; R. Hooykaas, 'The Parallel between the History of the Earth and the History of the Animal World', *Arch. Int. Hist. Sci.* 10 (1957), pp. 3–18; *The Principle of Uniformity in Geology, Biology and Theology*, 2nd impression (Leiden, Brill, 1963); 'Geological Uniformitarianism and Evolution', *Arch. Int. Hist. Sci.* 19 (1966), pp. 3–19; Francis C. Haber, *The Age of the World: Moses to Darwin* (Baltimore, Hopkins, 1959); William Coleman, 'Lyell and the Reality of Species', *Isis*

the geological debate became relatively internalist, and he has set new standards of research in studying the positive science of geology in the period.[53] However, all of those who have contributed significantly to the literature in the history of geology have found it necessary—for their own purposes—to understand the theologies and philosophies of nature of the participants in the controversies which raged over the history of the earth and its laws.

Similarly, the authors of the general works which attempt to provide a broader view of the geological and evolutionary debates paid due attention to the interaction between the theological and scientific issues. It can be argued that their perspectives, based as they were on an interactive model, prevented their seeing that theological and scientific issues were constitutive of each other's domain. However, the point to be made here is that none of them looked very far outside science to its social and economic context. The argument of Eiseley's *Darwin's Century* is laid out like a detective story or a jigsaw puzzle, in which the clues or pieces are seen exclusively in the light of their contribution to the 'solution', the picture on the cover of the box.[54] It is very cleverly put together, but the contemporary contexts of issues and the different perspectives in which they were seen at the time— let alone the different interests (social, economic and political, as well as intellectual) that these perspectives served—are regularly sacrificed in favour of their contribution to the view of scientific truth as seen in the light of current science. Gillispie's pioneering study of *Genesis and Geology* gives full weight to the theological context, but his basic approach is to show how the advancing edge of objectivity came to relegate theology to the prefaces and conclusions of geological works as it had earlier done in the physico-chemical sciences.[55] The reader knows from the first page that theology is

---

53 (1962), pp. 325–38; Leonard G. Wilson, 'The Origins of Charles Lyell's Uniformitarianism', in Claude C. Albritton (ed.), *Uniformity and Simplicity—A Symposium on the Principle of the Uniformity of Nature*, Geological Society of America, special paper 89 (1967), pp. 35–62; see also below (note 119).

53 Martin J. S. Rudwick, 'The Foundation of the Geological Society of London: Its Scheme for Co-operative Research and Its Struggle for Independence', *Brit. J. Hist. Sci.* 1 (1963), pp. 325–55; 'A Critique of Uniformitarian Geology: a Letter from W. D. Conybeare to Charles Lyell, 1841', *Proc. Amer. Philos. Soc.* 111 (1967), pp. 272–87; 'The Glacial Theory', *History of Science* 8 (1969), pp. 136–57; 'The Strategy of Lyell's *Principles of Geology*', *Isis* 61 (1969), pp. 5–33; 'Lyell on Etna and the Antiquity of the Earth', in Cecil J. Schneer (ed.), *Toward a History of Geology* (Cambridge, M.I.T., 1969), pp. 288–304; *The Meaning of Fossils: Episodes in the History of Palaeontology* (London, Macdonald, in press).

54 Loren Eiseley, *Darwin's Century: Evolution and the Men Who Discovered It* (1958; reprint New York, Doubleday Anchor paperback, 1961).

55 Charles C. Gillispie, *Genesis and Geology: The Impact of Scientific Discoveries upon Religious Beliefs in the Decades before Darwin* (Cambridge, Mass., Harvard, 1951; also Harper paperback, 1959).

on its way out. Greene's *The Death of Adam* is less elegantly written than the other standard works, but its looser structure makes it more attractive, since the reader can retain some sense of the multiple perspectives in which the issues were viewed in their contemporary context.[56]

It is a commonplace that Darwin's achievement stood above that of the other evolutionists, but even if one considers the nineteenth-century debate on its own terms, the significance of the writings of Robert Chambers, Herbert Spencer, and A. R. Wallace was far greater than has been reflected in writings about the debate. None of them has been given his due. Mill-hauser's monograph on Chambers is significantly entitled *Just Before Darwin*. Its subject is relegated to the subtitle: *Robert Chambers and Vestiges*.[57] The author has obviously done a great deal of research into the contemporary debate, as he did for an earlier study of 'The Scriptural Geologists',[58] but he remains curiously diffident about Chambers' achievement. Chambers leapt over the inhibitions and reservations of his scientific contemporaries and conveyed the whole sweep of naturalism, embracing man, his mind, and society. Like Spencer, he suffers at the hands of scholars from the retrospective judgement that he was a bad scientist. That is a fair judgement, one which was made vehemently at the time by nearly everyone who was sufficiently well-informed to express a sound opinion. Yet the application of current standards to the contemporary situation has helped to obscure the fact that it was Chambers' admittedly speculative theory which provided the basis for the British debate on evolution (or the 'Development Hypothesis') for fifteen years before the theory of Darwin and Wallace was made public.[59] Similarly, Chambers' book helped to stimulate Wallace's enquiries, was the subject of debates between Lewes, Spencer and Huxley, and was important in the thought of a number of other scientists, philosophers and theologians in the period. Indeed, the public controversy over the anonymous *Vestiges of the Natural History of Creation* was far more heated than that over Darwin's *Origin of Species*. Darwin was right to say in a later edition of his book that

---

[56] John C. Greene, *The Death of Adam: Evolution and Its Impact on Western Thought* (Ames, Iowa, 1959; also Mentor paperback, 1961).

[57] Milton Millhauser, *Just Before Darwin: Robert Chambers and Vestiges* (Middletown, Conn., Wesleyan, 1959). In his Introduction to the Leicester University Press facsimile reprint (1969) of *Vestiges of the Natural History of Creation* (London, Churchill, 1844), Sir Gavin de Beer utterly fails to convey the contemporary significance of Chambers' book.

[58] Milton Millhauser, 'The Scriptural Geologists, an Episode in the History of Opinion', *Osiris* **11** (1954), pp. 65–86.

[59] Lyell's treatment of evolution was in a negative sense and in another context—that of geological dating. Spencer's evolutionary speculations were not widely known.

The work, from its powerful and brilliant style, though displaying in the earlier editions little accurate knowledge and a great want of scientific caution, immediately had a wide circulation. In my opinion it has done excellent service in this country in calling attention to the subject, in removing prejudices, and in thus preparing the ground for the reception of analogous views.[60]

Chambers was also like Spencer (and indeed like Wallace) in his overriding interest in the implications of evolutionary theory for social and philosophical issues, yet these have not been explored in depth in the literature. Writings about Wallace have been focussed on his zoogeographical work and theories of the mechanism of evolution, not on his socialism, his work on land nationalization, or his spiritualism, even though these interests were determinate in the course of his changing views on evolution.[61] Once again, restriction of the context of the enquiry impoverishes the elucidation even of the issues which interest internalist historians. By now few historians of science would consider an account of Newton's work to be adequate if it excluded analogous interests which played an important part in his philosophy of nature and his science.

The respective literatures on Chambers, Wallace and Spencer represent three sorts of narrow treatment of the issues. Chambers is neglected because of the weak evidential basis for his theory, leading to ignorance of his influence (both positive and negative) and the sweep of his vision. Wallace is included in the standard accounts, but aspects of his work which were highly integrated at the time are treated in isolation, and the deeper political, economic and philosophical ones are excluded. Spencer is hardly mentioned at all, except by historians of the social sciences. He was neither scientifically reputable, nor politically radical, but his theory, along with the generalizations and extrapolations based on it, was probably more influential in the general debates of the late nineteenth century than those of any of the other evolutionists. *We* make a sharp distinction between the evolutionary theories of Darwin and Spencer, but their ideas were routinely conflated in the public mind. Moreover, the participants in the scientific debate itself considered their ideas to be far closer together than our tidy categories seem prepared to allow.[62]

[60] Charles Darwin, *The Origin of Species by Means of Natural Selection, or the Preservation of Favoured Races in the Struggle for Life* (1859), 6th edition, with additions and corrections (London, Murray, 1895), p. xvii.

[61] Young, ' "Non-Scientific" Factors in the Darwinian Debate', op. cit. (note 10); 'Malthus and the Evolutionists', op. cit. (note 10); 'Darwin's Metaphor', op. cit. (note 10), especially pp. 456–61, 471–3, 489, 495–7; cf. Roger Smith, 'Alfred Russel Wallace and the Evolution of Man', *Brit. J. Hist. Sci.* (in press).

[62] Ibid.; see also 'The Development of Herbert Spencer's Concept of Evolution', op. cit. (note 9); 'The Rôle of Psychology in the Nineteenth-Century Evolutionary Debate', op. cit. (note 10).

Spencer was, at bottom, entirely preoccupied with the social and ethical implications of evolution. Indeed, he became interested in the subject as a result of his failure to find a sound basis for the integration and progress of society. He has received considerable and growing attention from historians of social and political theory, but almost none from historians of science. Two of the most important books which integrate scientific issues with social and economic ones and which have considered the evolutionary debate, are importantly concerned with his work: John Burrow's *Evolution and Society, A Study in Mid-Victorian Social Theory*[63] and John Peel's *Herbert Spencer: the Evolution of a Sociologist*.[64] The former relates Spencer's work to the intellectual tradition of the Scottish Enlightenment and Utilitarianism in the domain of anthropology; the latter provides an excellent study of Spencer's social and sociological theories in their contemporary social, economic and theoretical contexts. Significantly, neither author is centrally interested in the history of science as practised by the specialists or is considered to be a member of 'the profession of the history of science'. Burrow is an intellectual historian of political theory, and Peel is a sociologist. Both works are defective in their appreciation of the narrowly scientific issues, but these defects are far less severe than the failure of historians of science seriously to consider Spencer at all. Their books are symptomatic of the mutual isolation between the study of the history of science and the study of social theory, while their subject is someone who never made that distinction.

It would be wrong to claim that the standard sources utterly fail to mention the rôle of social, economic and political issues in the evolutionary debate. It is nevertheless true that their treatment of these issues would never lead a reader to appreciate the importance of giving due weight to the rôles of, for example, Adam Smith, Malthus, Owenite socialism, the Philosophic Radicals, phrenology, the Positivists, *Essays and Reviews,* Bishop Colenso, G. H. Lewes, George Eliot, the Mechanics' Institutes, the Society for the Diffusion of Useful Knowledge, John Draper, Walter Bagehot, the Metaphysical Society, the X Club. Nor would the standard sources lead him to focus on a number of other figures, works and societies which played important parts in the wide debate on man, nature, God and society which, after all, did lead people in the nineteenth century and in the present to interest themselves in the movement summarized by the terms 'Darwinism' and 'Evolution'. Moreover, in case the reader hasn't noticed (which would be understandable in the circumstances), my list of categories—'figures, works and societies'—is significant for what it leaves out. It is at most

[63] Op. cit. (note 40).
[64] J. D. Y. Peel, *Herbert Spencer: the Evolution of a Sociologist* (London, Heinemann, 1971).

tangential to the concerns of, for example, Thompson's *The Making of the English Working Class* or Hobsbawm's *Industry and Empire*.[65]

Although it is not central to my present purpose, it should be mentioned that, conversely, the authors of standard works in the social, political, economic and literary history of the period make little or no attempt to include the influence of science and scientific naturalism in their accounts.[66] In the history of theology the issues are considered, but at a relatively superficial level. Even the question of the relationship between science and technology in the economic and social history of the period has been sorely neglected.[67] The gap between the internalist history of science and the perspective of writers in other branches of historical studies means that there is little impetus to investigate their relations. The thesis that they were part of a single debate does not even arise if one works backwards from the secondary literature to nineteenth-century documents. There are, however, two other paths to that thesis. The first is to place oneself in the midst of the periodical literature of the period and to discover the highly integrated network of issues in all these spheres. The second lies in applying certain fundamental assumptions of the sociology of knowledge or one of its parent traditions—marxism.

The existing literature in any field sets very strong constraints on how one finds it possible to conceive of a problem. There is no neutral naturalism in historical research any more than there is in science itself. Science is a social activity, born of society, and mediating its structures and values, at least as much as it is born of nature. The same is true of the history of science.

---

[65] Edward P. Thompson, *The Making of the English Working Class* (London, Gollancz, 1963; also Penguin paperback); Eric J. Hobsbawm, *Industry and Empire: An Economic History of Britain since 1750* (Harmondsworth, Penguin paperback, 1968; reprinted with corrections, London, Weidenfeld & Nicolson, 1969).

[66] It would be tedious and pointless to list the standard works on Victorian history, from G. M. Young to the present, which do not include any serious consideration of science and its ideology. Two recent works are typical. In J. F. C. Harrison's *The Early Victorians, 1832–51* (London, Weidenfeld & Nicolson paperback, 1971), there are no index entries for 'geology', 'Lyell' or 'science'. In the subsequent volume in the same series, *Mid-Victorian Britain, 1851–75* (1971), by Geoffrey Best, there are no entries for 'evolution' or 'science'. The two references to Darwinism are fleeting and indirect. Harrison treats Malthusianism at length but neither he nor Best considers its reinforcement by biological theory, thereby providing more forceful rationalizations for its social ideology. By contrast, histories of the Victorian period written in the nineteenth century and up until *ca.* 1930 devote careful and detailed attention to evolutionism and related issues. Current literary historians are much more sensitive to the rôle of science.

[67] Arnold W. Thackray has *begun* to raise some of the relevant issues from the point of view of liberal, functionalist scholarship in the first four pages of his 'Science and Technology in the Industrial Revolution', *History of Science* 9 (1970), pp. 76–89; cf. below, note 210.

In both cases there is some domain of data, and there is a 'natural' world somewhere out there, but the interaction of the historically constrained subject with these objects is as much involved in determining what will be called a datum as it is in discovering one. Without an adequate theory to explain these interrelations, an historical account will take shape in a determinate but relatively undisciplined way. Some philosophically-inclined historians have attempted to solve this problem in the abstract, but it remains to be shown that the philosophical approach to historiography, divorced from actual historical research, will lead us very far towards a more adequate theory.

The consequence of this situation for a study on the debate of man's place in nature in nineteenth-century Britain was that when one began to see a pattern of interrelations and multiple perspectives in the writings of those who took part in the overlapping controversies, it was difficult to put the issues in a single framework. Various attempts to do so failed to produce coherence. One of them took Malthus' *Essay on the Principle of Population* as a matrix and showed that widely varying interpretations of that work were seemingly equally justified by what Malthus wrote. These interpretations— by William Paley, Thomas Chalmers, Darwin, Spencer and Wallace—were themselves interacting, so that the problem of interpretation and influence became impossibly complicated.[68] Similarly, an essay on the debate from the point of view of natural theology and the concept of progress,[69] followed later by a study of the relationship between natural theology, the periodical press and the development of scientific specialization,[70] produced a different reading of the debate. Other exercises in this vein left one feeling that research in the history of science was rather like literary criticism, where competing interpretations of a work were complementary, illuminating different aspects rather than leading progressively to a consensus[71] among scholars. If this model—or some version of it—were to be found persuasive by scholars in the history of science, they would finally cut themselves off decisively from seeing their subject as an extension of the positivist, progressive view of science itself. Let us consider a number of figures whose writings call for the use of more than one mode of interpretation.

In so far as historians of science have considered Malthus, they have concentrated on his abstract, Newtonian model of science, one which he was a pioneer in applying to man. They have also stressed his quasi-

---

[68] Young, 'Malthus and the Evolutionists', op. cit. (note 10).

[69] Young, 'The Impact of Darwin', op. cit. (note 10).

[70] Young, 'Natural Theology, Victorian Periodicals, . . .', op. cit. (note 10).

[71] See above, notes 9 and 10, and 'The Anthropology of Science', *New Humanist* **88** (July 1972), pp. 102–5.

mathematical argument on the consequences of the potential disproportion between geometrical growth of population and arithmetical development of food production. This provided a natural law purporting to explain poverty, misery, theft, famine, war and death. His concepts of 'pressure' and 'struggle' were undoubtedly fundamental in both the origins and exposition of Darwin's theory of natural selection. Darwin wrote in the Introduction to the *Origin* that his theory was 'the doctrine of Malthus, applied to the whole animal and vegetable kingdoms'.[72] Historians of science concentrate on these aspects when they give weight to Malthus at all. However, his theory was just as importantly a decisive intervention in the eighteenth-century debate on progress. Indeed, it was written in the wake of the French Revolution and in reply to the sanguine writings of Condorcet and Godwin on inevitable, unlimited social progress.[73] Similarly, his argument must be seen as an extension of Adam Smith's theory of the causes of the *Wealth of Nations* (1776), explaining the natural causes of poverty, where Smith had focussed on harmonious, progressive production of wealth.[74] Once again, Malthus' own perspective on his theory was explicitly natural theological, and the last two chapters of the first edition of his *Essay* were devoted to this topic, while his natural theological assumptions became more diffusely spread through the text in later editions. (Darwin read the 6th edition of 1826.) The political and economic debate which followed the publication of Malthus' theory raged throughout the nineteenth century and is still with us.[75] These aspects find no place in the writings of historians of science. Conversely, Darwin and evolutionism are not mentioned in the standard account of *The Malthusian Controversy*.[76]

It is well known that the nineteenth-century debate on parish relief and the Poor Laws centred around Malthus' theory. It became the fundamental touchstone of the debate on the relationship between human industry and

[72] Charles Darwin, *On the Origin of Species*, op. cit. (note 60), 1st edition (London, Murray, 1859; facsimile reprint Cambridge, Mass., Harvard, 1964; also Atheneum paperback, 1967), p. 5.

[73] Thomas Robert Malthus, *An Essay on the Principle of Population, as It Affects the Future Improvement of Society with Remarks on the Speculations of Mr. Godwin, M. Condorcet and Other Writers* (London, Johnson, 1798; facsimile reprint London, Macmillan, 1966; also Penguin paperback with an Introduction by Anthony Flew, 1970); Samuel M. Levin, 'Malthus and the Idea of Progress', *J. Hist. Ideas* 27 (1966), pp. 92–108.

[74] Elie Halévy, *The Growth of Philosophic Radicalism*, 2nd edition with corrections, trans. M. Morris (London, Faber & Faber, 1952; also paperback), pp. 225–48.

[75] Varia, *Ideas and Beliefs of the Victorians: An Historic Revaluation of the Victorian Age* (1949; reprinted New York, Dutton paperback, 1966), pp. 43, 74. See also essays and bibliographies in David V. Glass (ed.), *Introduction to Malthus* (London, Cass, 1959; also paperback).

[76] Kenneth Smith, *The Malthusian Controversy* (London, Routledge & Kegan Paul, 1951).

provision for the indigent.[77] At its own basis lay a view of nature and human nature which was deeply pessimistic and offered progress only through painful struggle, in which human inequality was taken as given, the result of God's wisdom and benevolence. This image of nature and society was carried over into the evolutionary debate, and the resulting fusion was the basis for debates on the social meaning of evolutionary theory. The line from Malthus to Darwin and on to so-called 'Social Darwinism' is unbroken and continues to the recent writings on biology and society of, for example, Morris, Ardrey, and Darlington.[78] From start to finish, this has been a reconciling approach. It has served as the basis for the secular theodicy of industrial society and depends on a class doctrine. Malthus wrote in 1798, 'If no man could hope to rise or fear to fall, in society, if industry did not bring with it its reward and idleness its punishment, the middle parts [i.e., the middle class] would not certainly be what they are now.'[79] (Thomas Chalmers echoed this doctrine in even sterner language.[80]) The use of natural law as the basis for a given view of society became a commonplace in social, political and economic theory, and the theory of evolution was employed as a new, more powerful, justification for industrial capitalism. It is no wonder that Marx and Engels wrote some of their most vehement polemics against Malthus[81] and that Marxists and other sociologists came to

[77] J. A. Banks and D. V. Glass, 'A List of Books, Pamphlets and Articles on the Population Question, published in Britain in the Period 1793 to 1880', in Glass (ed.), *Introduction to Malthus,* op. cit. (note 75), pp. 79–112; cf. G. L. Nesbitt, *Benthamite Reviewing* (New York, Columbia, 1934), pp. 58–9; Thomas Sewell, 'Malthus and the Utilitarians', *Canadian J. Econ. & Pol. Sci.* 28 (1962), pp. 268–74; J. R. Poynter, *Society and Pauperism: English Ideas on Poor Relief, 1795–1834* (London, Routledge & Kegan Paul, 1969).

[78] F. Bowen, 'Malthusianism, Darwinism and Pessimism', *North Amer. Rev.* 129 (1879), pp. 447–72; Henry George, *Progress and Poverty: an Inquiry into the Cause of Industrial Depression and of Increase of Want with Increase of Wealth . . . The Remedy* (New York, 1879; reprinted New York, Schalkenbach, 1962); Young, ' "Non-Scientific" Factors . . .', op. cit. (note 10); 'Evolutionary Biology and Ideology: Then and Now', op. cit. (note 8); 'The Human Limits of Nature', in Jonathan Miller (ed.), *The Limits of Human Nature* (New York, Dutton, in press).

[79] Malthus, *Essay,* op. cit. (note 73), Penguin edition, p. 207.

[80] Young, 'Malthus and the Evolutionists', op. cit. (note 10), pp. 119–25, 142–3.

[81] Marx to Engels, 18 June 1862: 'It is remarkable how Darwin recognises among beasts and plants his English society with its division of labour, competition, opening up of new markets, "inventions", and the Malthusian "struggle for existence".' Karl Marx and Frederick Engels, *Selected Correspondence,* 2nd edition, trans. I. Lasker (Moscow, Progress, 1965), p. 128. Engels: 'Darwin did not know what a bitter satire he wrote on mankind, and especially on his countrymen, when he showed that free competition, the struggle for existence, which the economists celebrate as the highest historical achievement, is the normal state of the *animal kingdom.* Only conscious organisation of social production, in which production and distribution are carried on in a planned way, can lift mankind above the rest of the animal world as regards the social aspect, in the same way that production in general has done this for mankind in the specifically biological aspect.'

see the union of evolutionary theory and Malthusianism (and various racist and eugenic corollaries extending to the present) as fundamental obstacles to men's believing that they could transform the existing society into a just one.[82] These aspects of the debate on man's place in nature are inseparable. To sequester the social and political debate from the scientific one is to falsify the texture of the nineteenth-century debate and to mystify oppression in the form of science.

William Paley is another writer who is given a small place in the work of historians of science on Darwinism. His *Natural Theology* (1802) and his *Evidences of Christianity* (1794) were very important in Darwin's education, and Darwin later admitted that even in overthrowing Paley's world view, he was surprised to discover how many of Paley's most basic assumptions he had retained.[83] Paley is seen as the representative of the eighteenth-century natural theology based on Design and harmony which was set aside by scientific naturalism and industrialism. His conception of natural theology was shown to be untenable in a period of growing scientific detail and finally collapsed in the *Bridgewater Treatises*, the *reductio ad absurdum* of parading the details of all the sciences *seriatim* as a cumulative series of proofs of the wisdom, goodness and benevolence of God. His was also the last plausible effort to speak of natural harmony without giving serious weight to the dimension of time and to the evidence of geology. The example with which his *Natural Theology* begins illustrates this point: finding a stone on a path implies nothing, while finding a watch implies a

---

Frederic Engels, *The Dialectics of Nature* (1873–1886), 3rd edition, trans. Clemens Dutt (Moscow, Progress, 1964), pp. 35–6. Engels to Lavrov, 12–17 November 1875: 'The whole Darwinist teaching of the struggle for existence is simply a transference from society to nature of Hobbes' doctrine of *bellum omnium contra omnes* and of the bourgeois-economic doctrine of competition, together with Malthus' theory of population. When this conjurer's trick has been performed (and I question its absolute permissibility, . . . particularly as far as the Malthusian theory is concerned), the same theories are transferred back again from organic nature to history and it is now claimed that their validity as eternal laws of human society has been proved. The puerility of this procedure is so obvious that not a word need be said about it.' *Selected Correspondence*, p. 302; cf. pp. 301–4 and *Dialectics of Nature*, pp. 312–14, and below (note 121).

[82] James Fyfe, 'Malthus and Malthusianism', *Modern Quart.* 6 (1951), pp. 200–11; J. D. Bernal, 'The Abdication of Science', ibid. 8 (1952–3), pp. 44–50; Ronald L. Meek (ed.), *Marx and Engels on Malthus: Selections from the Writings of Marx and Engels dealing with the Theories of Thomas Robert Malthus* (New York, International, 1954; also paperback); Meek's Introduction also appears as 'Malthus—Yesterday and Today', *Sci. & Soc.* 18 (1954), pp. 21–51; cf. Mary Douglas, 'Environments at Risk', *Times Lit. Suppl.* No. 3583 (30 October 1970), pp. 1273–5, where the pervasiveness of Malthusianism is touched upon.

[83] See Young, 'Malthus and the Evolutionists', op. cit. (note 10), p. 118 and 'Darwin's Metaphor', op. cit. (note 10), pp. 468–9.

watchmaker.[84] No literate person could make that simple contrast two or three decades after 1802, as the meaning of rocks became the central issue in the geological debate on the nature of God's relationship with the history of the earth. Attempts were, of course, made to integrate natural theology with the new evidence of geology, but it was no longer possible to do so on the model of a Newtonian Heavenly Clockmaker.[85]

The existing literature in the history of science fails to give sufficient weight to some of these aspects of Paley's thought, just as it fails to indicate how much of the nineteenth-century debate was conducted within the context of natural theology. Having made these points, it is important to see that historians of science reveal an almost total lack of awareness of Paley's significance in the tradition of Utilitarian ethical and social thought.[86] Going beyond this in ways which are parallel to the points made above about Malthus, there is no awareness at all of the political aspect of Paley's theodicy of harmony. Malthus went beyond Paley and set the stage for nineteenth-century rationalizations of slow change through struggle. In the area of population theory, Paley had at first considered population growth an unmixed blessing. On reading Malthus, a slight frown appeared on his brow and remained there until he could absorb pain and suffering into his theodicy and could even regard himself as an adherent to Malthus' theory.[87]

In the area of politics, he was also able to absorb pain and suffering within his generally sanguine theory. The last chapters of his *Natural Theology* (1802) were explicitly concerned with reconciling men with the status quo. He wrote, 'The *distinctions* of civil life are apt enough to be regarded as evils, by those who sit under them; but, in my opinion, with very little reason.'[88] A decade earlier, during the height of Robespierre's dictatorship and the consequent anxieties among the bourgeoisie in Britain, he had spelled out the basis for this view, in a pamphlet called 'Reasons for Contentment Addressed to the Labouring Part of the British Public':

The wisest advice that can be given is, never to allow our attention to dwell

---

[84] William Paley, *Natural Theology; or, Evidences of the Existence and Attributes of the Deity. Collected from the Appearances of Nature* (1802), new edition (London, Baynes, 1816), chs. 1–3.

[85] William Buckland's Bridgewater Treatise on *Geology and Mineralogy*, 2 vols. (London, Pickering, 1836) was of a higher standard of scientific reputability than the other works in the series, and his efforts to reconcile the evidence of geology with Scripture were *relatively* sophisticated. See Young, 'The Impact of Darwin', op. cit. (note 10), pp. 18–19 & *passim* for a discussion of the levels of debate on geology and theology.

[86] Ernest Albee, *A History of English Utilitarianism* (1901; reprinted New York, Collier paperback, 1962), ch. 9.

[87] See Young, 'Malthus and the Evolutionists', op. cit. (note 10), pp. 114–18, 140–1.

[88] Paley, *Natural Theology*, op. cit. (note 84), p. 434.

upon comparisons between our own condition and that of others, but to keep it fixed upon the duties and concerns of the condition itself.[89]

But Providence, which foresaw, which appointed, indeed, the necessity to which human affairs are subjected (and against which it were impious to complain), hath contrived, that, whilst fortunes are only for a few, the rest of mankind may be happy without them.[90]

The labour of the world is carried on by *service*, that is, by one man working under another man's direction. I take it for granted that this is the best way of conducting business, because all nations and ages have adopted it.[91]

Like Malthus, Paley's later writings contained the same reconciling doctrines but in muted form. The theme which is here mentioned explicitly—that of the basis for the hierarchical division of labour in society (with feudalistic echoes)—is justified on the combined bases of Divine Ordinance and efficiency in the writings of Adam Smith, Paley, and Malthus. In the course of the nineteenth century, its basis changes from a theological theodicy to a biological one in which the so-called 'physiological division of labour' provides a scientific guarantee of the rightness of the property and work relations of industrial society. Although it also had other roots in Continental thought (e.g. St. Simon, Comte, and German organic theories of the state), this doctrine is carried on in social and political writings up to and including the current orthodoxy.[92]

Once the apologetic and reconciling aspects of writers such as Malthus and Paley come to the fore, the ideological basis and implications of the orthodox dichotomy between science on the one hand and its social and ideological functions on the other begins to become clear. One finds that these investigations lead to the ideological foundations of the modern scientific view of the earth, living nature, man and society but that the same structures support the modern rationalizations of industrial capitalism, colonialism and imperialism. In the existing literature discussions of the

[89] William Paley, *The Works of William Paley*, new edition, 7 vols. (London, Rivington, etc., 1825), Vol. 3, pp. 315–31, at p. 318.

[90] Ibid., p. 320.

[91] Ibid., p. 324.

[92] Émile Durkheim, *Socialism and Saint-Simon* (1928), trans. C. Sattler, edited with an Introduction by A. W. Gouldner (1958; reprinted New York, Collier paperback, 1962); *The Division of Labour in Society* (1893), trans. George Simpson (1933; reprinted London, Collier-Macmillan, 1964; also Free Press paperback); F. W. Coker, *Organismic Theories of the State: Nineteenth Century Interpretations of the State as Organism or as Person*, in *Studies in History, Economics and Public Law* (1910) (Columbia University), Vol. 38, no. 2 (reprinted New York, AMS, 1967); Henry E. Barnes, 'Representative Biological Theories of Society', *Sociol. Rev.* 17 (1925), pp. 120–31, 182–95, 294–300; Cynthia E. Russett, *The Concept of Equilibrium in American Social Thought* (New Haven, Yale, 1966); Arthur Salz, 'Specialization' [Division of Labour], in Edwin R. A. Seligman (ed.) *Encyclopedia of the Social Sciences* (London, Macmillan, 1930–5), Vol. 14, pp. 279–85; *re*: 'The physiological division of labour', see below (notes 115 and 116).

scientific debate in terms of theological positions are routine, but there is almost no full-blooded attempt to include the social, political and ideological rôles played by the debate itself. These were to rationalize the existing social and political order, and to reconcile men to it. The famous controversy in the nineteenth century between science and theology was very heated indeed, and scholars have concentrated on this level of analysis. However, at another level the protagonists in that debate were in fundamental agreement. They were fighting over the best ways of rationalizing the same set of assumptions about the existing order.[93] An explicitly theological theodicy was being challenged by a secular one based on biological conceptions and the fundamental assumption of the uniformity of nature.

Once one begins to see the debate in this double perspective of science *and* ideology, it becomes necessary to keep both of its aspects constantly in the forefront and to maintain simultaneous awareness of both. It would be misleading to suggest that *they* are in tension, since they are mutually consistent and support one another. Rather, the tension lies in the mind of the scholar whose training leads him habitually to separate his subject matter into 'science' and 'contextual factors', treating one and then (if he is so inclined) the other. It is therefore very difficult indeed to refrain from treating the materials in terms of the model of 'internal' and 'external' factors, science and society.

It is very striking how we blinker ourselves and separate intellectual history from its ideological dimension. We have seen that the same Malthus who pioneered the scientific treatment of man was engaged in an important ideological task which dominated the perception of the mood of nature and society throughout the century and has remained prevalent to the present. Paley's theodicy was the best summary and popularization of an older, more harmonious, view but the theme of a higher harmony persisted in the doctrine of progress through struggle which replaced his equilibrium model. Similar accounts can be given of the writings of all the major and minor participants in the debate in the main periodicals of the early decades— *Edinburgh Review, Quarterly Review* and *Westminster Review.* Indeed, there have been political analyses of these, notably Nesbitt's study of the early

[93] Roy Porter points out that throughout this essay the *achievements* of the bourgeois revolution in the eighteenth and nineteenth centuries are ignored, a mistake which Marx never made. I am aware that in stressing a continuous tradition of rationalizing and reconciling literature and in attempting to draw attention to some fundamental continuities amid changes in capitalism, I have tremendously undervalued the advances in bourgeois rights and opportunities in industrial Britain. Marx discusses the phases of capitalism and their relationship with the division of labour in *Capital: A Critique of Political Economy* (1867), 3rd edition (1887), 3 vols. trans. S. Moore and E. Aveling (London, Lawrence & Wishart, 1970), Vol. 1, ch. 14, especially pp. 336, 348–50.

years of the *Westminster, Benthamite Reviewing*.[94] Later generations of periodicals have received similar treatment, for example, Everett's study of the *Fortnightly, The Party of Humanity*.[95] The literature on the Victorian periodicals is central to the understanding of the political meaning of the major vehicles for the conduct of the debate. Approaching the problem from this point of view leads one to find it natural that the periodicals interpreted the findings and theories of the scientists from a political perspective. Moving on to the end of the period it is also significant that the most eminent scientists, theologians, philosophers, men of letters, politicians and editors of periodicals met together in the period 1869–1880 to consider all of the aspects of the ascendancy of scientific naturalism as it bore on morality, society, and the social order. Gladstone, Walter Bagehot, John Ruskin, Alfred Lord Tennyson, Leslie Stephen, F. D. Maurice, Cardinal Manning, John Tyndall and T. H. Huxley all saw the point of their meeting together.[96] I hope that it is becoming increasingly clear why we seem unable to see the point of their Society.

The group which came together in the meetings of the Metaphysical Society shows how highly integrated the debate in the nineteenth century was. The social and intellectual milieu in which these men moved extended beyond science, philosophy and theology to include politics and literature. One of the most prominent members of this intellectual élite was George Eliot. Her milieu was that of nineteenth-century naturalism, she had close relations with Spencer and was, of course, living with Lewes. Before she met him she had developed her naturalistic philosophy under the influence of Bray, Hennell, and Combe, along with members of the positivist circle of the period. For three years she was virtual editor of the *Westminster*. Her writings integrated aspects of the prevailing naturalistic ethical, scientific and social philosophies.[97] Implicit in this integration was a reconciling political

[94] Nesbitt, op. cit. (note 77).

[95] Edwin M. Everett, *The Party of Humanity: The Fortnightly Review and Its Contributors, 1865–1874* (Chapel Hill, North Carolina, 1939).

[96] R. H. Hutton, 'The Metaphysical Society', *Nineteenth Century* 18 (1885), pp. 177–96; Alan W. Brown, *The Metaphysical Society: Victorian Minds in Crisis, 1869–1880* (New York, Columbia, 1947); cf. Young, 'Natural Theology, Victorian Periodicals. . .', op. cit. (note 10). For studies of the social, intellectual, institutional and patronage network of the scientific establishment of the period, see J. Vernon Jensen, 'The X Club: Fraternity of Victorian Scientists', *Brit. J. Hist. Sci.* 5 (1970), pp. 63–72; Roy M. MacLeod, 'The X-Club, a Social Network of Science in Late-Victorian England', *Notes & Rec. Roy. Soc.* 24 (1970), pp. 305–22.

[97] Gordon S. Haight, *George Eliot: A Biography* (London, Oxford, 1968); Bernard J. Paris, *Experiments in Life: George Eliot's Quest for Values* (Detroit, Wayne State, 1965); Anna T. Kitchel, *George Henry Lewes and George Eliot: a Review of Records* (New York, Day, 1943); Margot J. Waddell, 'Scientific Naturalism and Philosophies of Nature and Man in the Novels of George Eliot'. *Brit. J. Hist. Sci.* (in press). Miss Waddell has guided my interpretation of George Eliot's rôle in this context.

philosophy which becomes explicit in her support for the conservative radicalism of *Felix Holt*. Felix states his position in a nomination day meeting:

> But I should like to convince you that votes would never give you political power worth having while things are as they are now, and that if you go the right way to work you may get power sooner without votes. Perhaps all you who hear me are sober men, who try to learn as much of the nature of things as you can, and to be as little like fools as possible. A fool or idiot is one who expects things to happen that never can happen; . . .[98]
>
> The way to get rid of folly is to get rid of vain expectations, and of thoughts that don't agree with the nature of things.[99]

Lest this position be seen as one which George Eliot developed only in the service of fiction (an hypothesis which would miss the whole point of her intentions in writing novels), it should be added that she willingly enlarged upon its reconciling message in a didactic essay, published separately in *Blackwood's Magazine*: 'Address to Working Men by Felix Holt'.

> Well, but taking the world as it is—and this is one way we must take it when we want to find out how it can be improved—no society is made up of a single class: society stands before us like that wonderful piece of life, the human body, with all its various parts depending on one another, and with a terrible liability to get wrong because of that delicate dependence.[100]. . . Now the only safe way by which society can be steadily improved and our worst evils reduced, is not by any attempt to do away directly with the actually existing class distinctions and advantages, as if everybody could have the same sort of work, or lead the same sort of life (which none of my hearers are stupid enough to suppose), but by the turning of Class Interests into Class Functions or duties.[101]
>
> The nature of things in this world has been determined for us beforehand, . . .[102]

She again refers to 'this society of ours, this living body in which our lives are bound up'.[103] The combination of organic analogies and the reduction of social change to the uniform action of natural laws, has the effect of pure reconciliation:

> The solution comes slowly, because men collectively can only be made to embrace principles, and to act on them, by the slow stupendous teaching of the world's events.[104]

---

[98] George Eliot, *Felix Holt, The Radical* (1866), in *The Works of George Eliot*, Cabinet Edition (Edinburgh, Blackwood, n.d.), Vol. 2, ch. 30, p. 88.

[99] Ibid., p. 89; cf. Peel, *Herbert Spencer*, op. cit. (note 64), pp. 72–3.

[100] George Eliot, 'Address to Working Men by Felix Holt', *Blackwood's* **103** (1868), pp. 1–11; reprinted in Thomas Pinney (ed.), *Essays of George Eliot* (London, Routledge & Kegan Paul, 1963), pp. 415–30, at p. 420.

[101] Ibid., p. 421.

[102] Ibid., p. 422.

[103] Ibid.

[104] Ibid., p. 428.

But now, for our own part, we have seriously to consider this outside wisdom which lies in the supreme unalterable nature of things, and watch to give it a home within us and obey it.[105]

George Eliot's interpretation of the social meaning of scientific naturalism is of a piece with the ideological statements of Paley and Malthus.

The argument, as presented so far, is easy prey to the distinctions of an internalist, since he would simply argue that the aspects of Malthus, Paley, the periodical press, and certainly George Eliot, which I have discussed, are external to the history of science as he understands it. I am, in effect, lending credence to his own reasons for largely ignoring their relevance, except in contextual terms. In order to make my case stronger, we must move closer, knowing that we are many steps away from figures whose writings would touch his definition of science. Although I am profoundly out of sympathy with the idea of a continuum from contextual to internal factors, if we take that approach, it is still a matter of relative indifference where we place ourselves on it or the related one extending from allegedly pure scientists to marginal figures. It is also obvious that the case can be developed more easily on the basis of evidence from some figures than from others. After all, there must have been *some* basis for the development of the orthodox distinction between internal and external factors.

Robert Chambers' *Vestiges of the Natural History of Creation* provides a relatively easy example, both for my position and for the internalist's. That is, it is a commonplace that his science was very shaky. At the same time, his generalizations were of fundamental importance in the development of the debate on man's place in nature. He had no well-developed mechanism for explaining evolutionary change; several were mentioned, but they were not mutually consistent. He confused the general principles of scientific naturalism with the specific, then-unknown causes of evolution. He was heavily criticized for these failings by his contemporaries, notably by Herschel and by Huxley, whose review of a later edition of *Vestiges* was so vituperative that even Huxley regretted its tone.[106] Chambers did not even become aware of the scientific weaknesses of his work until he reflected on

[105] Ibid., p. 429. For further discussion of the ideological issues reflected in *Felix Holt*, see Raymond Williams, *Culture and Society, 1780–1950* (London, Chatto & Windus, 1958; reprinted with Postscript Harmondsworth, Penguin paperback, 1963), Penguin edition, pp. 112–19.

[106] Sir John F. W. Herschel, 'Address', *Report of the 15th Meeting of the British Association for the Advancement of Science*, Cambridge, 1845 (London, Murray, 1846), pp. xxvii–xliv, at pp. xlii–xliii; [T. H. Huxley], 'The Vestiges of Creation', *Brit. & For. Med. Rev.* 13 (1854), pp. 425–39; T. H. Huxley, 'On the Reception of the "Origin of Species"', in Francis Darwin (ed.), *The Life and Letters of Charles Darwin*, 3 vols. (London, Murray, 1887), Vol. 1, ch. 5, at pp. 187–9.

the almost universal criticism which it received at the hands of better-qualified scientists. He then wrote a volume, *Explanations*, and appended apologetic passages to later editions of *Vestiges*, in which he made it clear that he had not aimed or claimed to spell out the specific causes of evolution but to establish the general principles of scientific naturalism in the realms of life, mind and society.[107] In doing this he struck the keynote of contemporary scientific naturalism and drew general conclusions which the experts reached only decades later.

For present purposes, however, it is important to note that *Vestiges* was in a direct lineage from Malthus and Paley, and indeed from the *Bridgewater Treatises*, which could also be used to develop the general thesis. Chambers' final chapters were expressions of the developing theodicy of naturalism, and once again, the message was social progress through reconciliation with the laws of nature:

> To secure the immediate means of happiness it would seem to be necessary for men first to study with all care the constitution of nature, and, secondly, to accommodate themselves to that constitution, so as to obtain all the realizable advantages from acting conformably to it, and to avoid all likely evils from disregarding it.
>
> . . . we must endeavour to place ourselves, and so to act, that the arrangements which Providence has made impartially for all may be in our favour, and not against us; such are the only means by which we can obtain good and avoid evil here below.[108]
>
> It may be that, while we are committed to take our chance in a natural system of undeviating operation, and are left with apparent ruthlessness to endure the consequences of every collision into which we knowingly or unknowingly come with each law of the system, there is a system of Mercy and Grace behind the screen of nature, which is to make up for all casualties endured here, and the very largeness of which is what makes these casualties a matter of indifference to God. . . . To reconcile this to the recognised character of the Deity, it is necessary to suppose that the present system is but a part of a whole, a stage in a Great Progress, and that the Redress is in reserve.[109]
>
> Thinking of all the contingencies of this world as to be in time melted into or lost in the greater system, to which the present is only subsidiary, let us wait the end with patience, and be of good cheer.[110]

---

[107] [Robert Chambers], *Explanations: a Sequel to 'Vestiges of the Natural History of Creation'* by the author of that Work (1845), 2nd edition (London, Churchill, 1846); Robert Chambers, 'Proofs, Illustrations, Authorities, etc.', in *Vestiges*, 12th edition (1860), with an Introduction relating to the Authorship of the Work by Alexander Ireland (Edinburgh, Chambers, 1884), pp. i–lxxxii.

[108] [Chambers], *Vestiges*, 1st edition, op. cit. (note 57), pp. 380–1.

[109] Ibid., pp. 384–5.

[110] Ibid., p. 386.

Chambers has here reconciled the Malthusian conception of progress through struggle with a calm sense of Paleyan harmony. In preparing his views he relied heavily on the doctrines of phrenology, particularly the work of George Combe, whose popularizations of the doctrines of Franz Joseph Gall and J. C. Spurzheim were immensely popular in the period as a vehicle for ideas of social welfare, self-improvement and progress.[111] Combe, along with Charles Bray, also influenced the views of nature of George Eliot. If we wish to relate these doctrines directly to the theory of reconciliation to industrial capitalism, we can do so by two paths. The first is through the wide current of Calvinism running through all these works, leading us to Weber's thesis in *The Protestant Ethic and the Spirit of Capitalism*. The second path leads to the theory of which Weber's thought was in important ways a bourgeois transformation, i.e., to marxism.[112] The same year in which Chambers extolled a faith in progress and harmony based on a partially secularized theodicy embracing the entire physical and living universe, Friedrich Engels (a twenty-four-year-old German whose family owned a cotton mill in Manchester) described the wretched *Condition of the Working-Class in England in 1844*. His reaction to Malthusianism and the legislation which was enacted under its influence was very different:

Meanwhile the most open declaration of war of the bourgeoisie upon the proletariat is Malthus' Law of Population and the New Poor Law framed in accordance with it. We have already alluded several times to the theory of Malthus. We may sum up its final result in these few words, that the earth is perennially overpopulated, whence poverty, misery, distress, and immorality must prevail; that it is the lot, the eternal destiny of mankind, to exist in too great numbers, and therefore in diverse classes, of which some are rich, educated, and moral, and others more or less poor, distressed, ignorant, and immoral. Hence it follows in practice, and Malthus himself drew this conclusion, that charities and poor-rates are, properly speaking, nonsense, since they serve only to maintain, and stimulate the increase of, the surplus population whose competition crushes down wages for the employed; that the employment of the poor by the Poor Law Guardians is equally unreasonable, since only a fixed quantity of the products of labour can be consumed, and for every unemployed labourer thus furnished employment, another hitherto employed must be driven into enforced idleness, whence private undertakings suffer at cost of Poor Law industry; that, in other words, the whole problem is not how to support the surplus population, but how to restrain it as far as possible. Malthus declares in plain English that the right to live, a right previously asserted in favour of

[111] Owsei Temkin, 'Gall and the Phrenological Movement', *Bull. Hist. Med.* **21** (1947), pp. 275–321, especially pp. 307–13; Millhauser, *Just Before Darwin*, op. cit. (note 57).

[112] Max Weber, *The Protestant Ethic and the Spirit of Capitalism* (1904–5), trans. T. Parsons (London, Allen & Unwin, 1930; also paperback); cf. Young, 'Evolutionary Biology and Ideology: Then and Now', op. cit. (note 8), pp. 191, 196.

every man in the world, is nonsense. He quotes the words of a poet, that the poor man comes to the feast of Nature and finds no cover laid for him, and adds that 'she bids him begone', for he did not before his birth ask of society whether or not he is welcome. This is now the pet theory of all genuine English bourgeois, and very naturally, since it is the most specious excuse for them, and has, moreover, a good deal of truth in it under existing conditions. If, then, the problem is not to make the 'surplus population' useful, to transform it into available population, but merely to let it starve to death in the least objectionable way and to prevent its having too many children, this, of course, is simple enough, provided the surplus population perceives its own superfluousness and takes kindly to starvation.[113]

Engels argues, in short, that poverty is a political—man-made—and not a natural phenomenon. The contrast between the sweet reconciling reason of Chambers' passage in his chapter on the 'Purpose and General Condition of the Animated Creation' and the polemical agitation of the passage from Engels' chapter on 'The Attitude of the Bourgeoisie towards the Proletariat' helps to point out the two perspectives on the debate on man's place in nature which must be seen if we are to move from loose contextual references (e.g., to 'the social dimension') to a genuinely radical historiography.[114]

The continuum through Malthus, Paley and Chambers leads on to Herbert Spencer, perhaps the most influential of all the interpreters of the philosophical, ethical, social and political meaning of Victorian scientific naturalism. The case for an intimate mixture of socio-political and scientific considerations in his thought has been made out many times and hardly needs rehearsing, especially since the appearance of Peel's biography and MacRae's edition of his essays.[115] There is no need to reveal Spencer's motives as primarily socio-political, since he repeatedly makes the point himself. Indeed, he turned to phrenology, to psychology and then to biology in search of new guarantees to replace those which had been found wanting in theism and in Utilitarianism. The problem about Spencer is not that of showing that he conforms to the position being argued here. Rather, it is

[113] Frederick Engels, *The Condition of the Working Class in England in 1844, from Personal Observation and Authentic Sources* (1845), Eng. trans. (London, 1892), 2nd edition, trans. revised Institute of Marxism-Leninism, Moscow (London, Panther paperback, 1969, with an Introduction by Eric Hobsbawm), pp. 308–9.

[114] See above (note 93). In developing a general thesis about reconciling rationalizations I have undervalued the *relatively* progressive social theories and activities of nineteenth-century radicals and reformers, of whom Chambers was a notable example.

[115] Peel, *Herbert Spencer*, op. cit. (note 64); Herbert Spencer, *The Man versus the State, with Four Essays on Politics and Society*, edited with an Introduction by Donald Macrae (Harmondsworth, Penguin paperback, 1969); cf. *The Evolution of Society: Selections from Herbert Spencer's Principles of Sociology*, edited with an Introduction by Robert L. Carneiro (Chicago, 1967). For modern analogies, see J. D. Y. Peel, 'Spencer and the Neo-Evolutionists', *Sociology* 3 (1969), pp. 173–91.

to get historians to see how central his work and influence were to the nineteenth-century debate, both among scientists and the broader public. His reputation has suffered most among the leaders of thought in the period because subsequent scientists (followed dutifully by historians) have anachronistically dismissed him for holding a 'Lamarckian' theory of the mechanism of evolution. Two things should be noticed about his position. First, that it was a theory which, though embattled, was taken seriously throughout the nineteenth century and, indeed, was given increasing weight by Darwin (just as Spencer allowed an increasing rôle for natural selection). This point should lend perspective to the dismissal of Spencer as a serious figure. Second, he was unequivocal in pointing out that he attached great weight to the question of the mechanism of evolution precisely *because of* its ethical, educational, social and political consequences. Throughout his mature life he was seeking a scientific basis for a doctrine of inevitable progress which would justify his belief in an extreme form of *laissez-faire* economic and social theory.[116]

A. R. Wallace held a different theory of the mechanism of evolutionary change, one which was initially based on a fusion of ideas drawn primarily from Lyell and Malthus. When he saw that this theory had implications which were in conflict with his more fundamental belief in socialism, it was not socialism which yielded, but his belief in the Malthusian mechanism of natural selection. Important exceptions were made to his earlier belief in the total adequacy of natural selection to account for man and his mind. It should be noticed that it was not only his socialism but also his spiritualism and his belief in phrenology which he came to feel were in conflict with total adherence to natural selection. This is not the place to develop an analysis of the relationships among these influences (and their own deeper political meaning) and the biological findings which worried him. The point which I wish to make here is merely that they were all mixed up together.[117]

Similar cases can be made out for each of the significant figures in the mainstream of the evolutionary debate and the wider debate on man's place in nature. In each case—e.g., those of Buckland, Whewell, Wilberforce, Sedgwick, Powell, Chalmers, Lewes, Carpenter, Mivart—scientific, philo-

---

[116] Herbert Spencer, 'Progress: Its Law and Cause', *Westminster Rev.* 11 (1857), 445–85; 'The Social Organism' and 'The Factors of Organic Evolution', in *Essays, Scientific, Political and Speculative*, 3 vols. (London, Williams & Norgate, 1901), Vol. 1, pp. 265–307, 389–478. 'The Social Organism' is also reprinted in the Penguin edition, op. cit. (note 115). Aspects of Spencer's work are discussed in all the papers listed above (notes 8–10); on the 'physiological division of labour', see especially *Mind, Brain and Adaptation*, pp. 159, 167–9, and 'Evolutionary Biology and Ideology: Then and Now', op. cit. (note 8), pp. 184–5, 202.

[117] See above, note 61.

sophical, theological and explicitly political considerations form a closely woven network of factors in their own theories and in their considerations of the theories of others. Interpreters of evolutionary theory such as the Duke of Argyll, William Graham, Ernst Haeckel (who was influential in translation) followed the same pattern. But what of Lyell and Darwin, the real scientists whose ideas were at the heart of the debate as it occurred and whose theories dominate the secondary literature to an even greater extent? Part of the answer has been given elsewhere in research which points out the importance of theological, philosophical, and other non-positivist factors in their work.[118] In the case of Lyell, new evidence about the over-riding rôle of his concerns about man's special status has recently come to light in Wilson's edition of Lyell's hitherto unpublished *Scientific Journals on the Species Question*, and Rudwick has argued forcefully that this evidence calls for an orderly retreat on the part of those who formerly argued that Lyell could be studied along relatively internalist lines.[119] Even so, the rôle of social and political factors in the work of both Lyell and especially of Darwin, is a highly mediated one. Having got part of the way by showing the centrality of theological and philosophical questions in the origins, substance and vicissitudes of their theories, one is left with the need for a subtle and complex theory of mediations if it is to be possible to make a strong case for them as figures who should be viewed in the simultaneous perspectives of science and socio-political ideas. It is beyond doubt that their theories were central to others' reconciling and apologetic doctrines. It is also becoming increasingly clear that orthodox accounts which stress the growth of scientific naturalism as a development away from traditional theological and social doctrines, must be fundamentally reconsidered. In their place we require an interpretation which shows the deeper continuities.[120] Rather than focus on the overthrow of the relatively static theistic cosmology by a secular and progressive one, an interpretation must be worked out which stresses the development from one theodicy—in both its scientific and its social aspects—to another. The first was suitable for a relatively static and rural economy while the other was developed for a rapidly-changing and industrializing society. Although the theories of Lyell and Darwin were at the centre of the rôle of science in this change of rationalization, it may be necessary in the short run to over-emphasize the

[118] Young, 'Darwin's Metaphor', op. cit. (note 10).

[119] Leonard G. Wilson (ed.), *Sir Charles Lyell's Scientific Journals on the Species Question* (New Haven, Yale, 1970); L. G. Wilson, 'Sir Charles Lyell and the Species Question', *Amer. Scientist* 59 (1971), pp. 43–55; M. J. S. Rudwick points out the significance of these journals in his reviews in *Brit. J. Hist. Sci.* 5 (1971), pp. 408–9.

[120] See above, notes 93 and 114. Once again, this claim leaves the radicalism of the philosophic radicals and other nineteenth-century reformers unevaluated.

breadth and texture of the wider movement of which their work was but a part, however essential to it their particular contributions were. Once we have gained a broader view of the general movement of nineteenth-century naturalism, it will be a much less daunting task to place their work in it, without unduly exaggerating or minimizing their respective and related rôles.

It should be granted that the work and influence of Lyell and Darwin were less intentionally and obviously an expression of more basic socio-economic forces and structures than, for example, the work and influence of Chambers, Spencer and Wallace. Similarly, their greater scientific prestige meant that those who employed their theories for socio-political purposes could claim a sounder foundation in the nature of things—in scientific laws—for their extrapolations and generalizations. A scholar who interprets the history of science in terms of internal and external factors or some related model, would make a sharp distinction between their work and the contexts in which it developed and into which it fed. Rather like Lyell and Darwin, who withdrew from the political issues raised by their work, he would not want to get involved in that sort of thing.[121] A radical scholar would make two replies,

[121] It should be recalled that Marx had great admiration for Darwin's theory. He wrote to Lasalle in January 1861, 'Darwin's book is very important and serves me as a natural-scientific basis for the class struggle in history. One has to put up with the crude English method of development, of course'. Marx and Engels, *Selected Correspondence*, op. cit. (note 81), p. 123. It is important to bear in mind that Marx saw the conflation of Darwinism with Malthusianism—and their generalization into a single law of the 'struggle for life'—as nothing more than 'a very impressive method—for swaggering, sham-scientific, bombastic ignorance and intellectual laziness'. Ibid., p. 240; see also above (note 81). Nevertheless, Marx and Engels often made analogies between their work and Darwin's. For example, Engels did so at Marx's funeral and in his Preface to the English edition of *The Communist Manifesto*. But when Marx wrote to Darwin in 1880, asking if he could dedicate chapters 12 and 13 of the English translation of Vol. 1 of *Capital* to the great naturalist, one can imagine Darwin's feelings. Darwin's perfect reply was published in 1931: 'Dear Sir,—I thank you for your friendly letter and the enclosure. The publication of your observations on my writings, in whatever form they may appear, really does not need any consent on my part, and it would be ridiculous for me to grant my permission for something which does not require it. I should prefer the part of the volume not to be dedicated to me (although I thank you for the intended honour), as that would to a certain extent suggest my approval of the whole work, with which I am not acquainted.

'Although I am a keen advocate of freedom of opinion on all questions, it seems to me (rightly or wrongly) that direct arguments against Christianity and Theism hardly have any effect on the public; and that freedom of thought will best be promoted by that gradual enlightening of human understanding which follows the progress of science. I have therefore always avoided writing about religion and have confined myself to science. Possibly I have been too strongly influenced by the thought of the concern it might cause some members of my family if in any way I lent my support to direct attacks on religion.

'I am sorry to refuse you any request but I am old and have little strength, and the reading of proofs (as I know from present experience) imposes a heavy strain on me. (Signed) Charles Darwin.' E. Kolman, 'Marx and Darwin', *Labour Monthly* 13 (1931), pp. 702–5, at p. 702.

one on the internalist's ground and one on his own. The first is that, by the internalist's own criteria of understanding, his historiography unnecessarily blinkers his own perspective and impoverishes his understanding of the very problems which interest him. Second, a radical approach requires that the socio-political basis and its interaction between the putatively autonomous scientific results be explored in depth and detail. He must make this effort in order to understand the rôle of scientific rationality and its technological expressions (and affiliations) in maintaining the established order of society and in sustaining the false consciousness[122] which prevents men from believing that it can be transformed into a society in which the division of labour need not be hierarchical and exploitative, one in which inegalitarian structures are no longer maintained by being mystified and justified by a spurious foundation in the laws of nature. Put simply, a radical and critical historiography of science is required so that it can be seen that the present order of society is a fundamentally political, not a natural and inevitable one, and that it is therefore open to change from the struggles of men, once their consciousness is freed from the fetters of deference to a natural basis for the present social order.[123]

However, in order for this change of consciousness to occur, it will be necessary to evaluate critically the model of science and its history which considers the substance—the findings and theories—of intellectual development to be relatively autonomous and independent of socio-political determination. The cases of Lyell and Darwin are thus part of a much wider problem. They are, relatively speaking, the purest of the scientists in the Victorian debate and as such are nearer to the positions of physicists,

[122] For discussions of the concept of 'false consciousness', see Nigel Harris, *Beliefs in Society: The Problem of Ideology* (London, Watts, 1968; also Penguin paperback, 1971), Penguin edition, pp. 228 ff.; Mannheim, *Ideology and Utopia*, op. cit. (note 8), pp. 62–7, 84–7. Engels wrote in 1893, 'Ideology is a process accomplished by the so-called thinker consciously, it is true, but with false consciousness. The real motive forces impelling him remain unknown to him; otherwise it simply would not be an ideological process. Hence he imagines false or seeming motive forces'. Marx and Engels, *Selected Correspondence*, op. cit. (note 81), p. 459.

[123] Some historians of science who have read this passage have found it startlingly inconsistent with the aims and style of historical scholarship as they understand them. Some of the documents which can aid in building a bridge between their consciousness and marxist interpretations of science, scientific rationality and oppression are the following: Herbert Marcuse, *One Dimensional Man. The Ideology of Industrial Society* (London, Routledge & Kegan Paul, 1964; also Sphere paperback, 1968); Jürgen Habermas, 'Technology and Science as "Ideology" ', in his *Toward a Rational Society*, trans. Jeremy Shapiro (London, Heinemann, 1971; also paperback), ch. 6; Trent Schroyer, 'Toward a Critical Theory for Advanced Industrial Society', op. cit. (note 7), in Hans P. Dreitzel (ed.), *Recent Sociology, No. 2*, pp. 210–34, also reprinted in Fisher (ed.), *The Revival of American Socialism*, op. cit. (note 6), ch. 16; see also above (notes 3 and 8). Whether or not they will decide to cross that bridge is, of course, another matter.

chemists and mathematicians. If one is studying the writings of town planners, political scientists, sociologists, anthropologists, economists, psychologists, ethologists, physiologists, geneticists, molecular biologists, chemists, physicists, or mathematicians, one finds oneself at different points on a continuum, between writings which obviously reflect socio-political assumptions and those which do not (obviously). In studying disciplines which are near the beginning of that list, ideological assumptions appear, as it were, on the surface of the page. At the other extreme, a student of the history of physics or mathematics—or indeed of recent molecular biology— would be very incredulous if faced with an interpretation of his data which stressed ideological assumptions which appeared to play no part in the data before him.[124] Of course, there are those who defend the positivist purity of the woolliest of the social sciences and model them on the physico-chemical sciences. At the other extreme, there are those who have attempted to account for findings in so-called pure science by claiming that they are direct, unmediated expressions of economic forces in the period or, alternatively, of the psychopathology of the scientist as seen from a psychoanalytic point of view. Isaac Newton is generally considered to be the paradigm scientist of the scientific revolution, and Boris Hessen's classical essay on 'The Social and Economic Foundations of Newton's "Principia" ' is an example of the former approach,[125] while Frank Manuel's *A Portrait of Isaac Newton* is a notable example of the psychoanalytic view.[126]

What is required in the first instance, it seems to me, is a radical historiography which goes far beyond these simplistic approaches and which is based on a sufficiently flexible theory of mediations between socio-economic base and intellectual superstructure, so that it can take account of scientific developments at any point on the continuum of 'purity' of the sciences. This approach to the concept of mediation is not new. It is a revival of views which were expressed by Marx and Engels. However, the concept has degenerated in the hands of those who have reduced marxism to 'vulgar marxism'. The theory of mediation must include not only concepts for working along that continuum but must also address itself to the assumptions on which the continuum rests, that is, the paradigm of explanation of modern science which was elaborated in the sixteenth and seventeenth centuries in specific socio-political conditions and which forms the basis of the orthodox practice of scientists and of historians who study the advancing edge of objectivity.[127]

[124] These issues are given further consideration in Young, 'Evolutionary Biology and Ideology: Then and Now', op. cit. (note 8) and 'The Anthropology of Science', op. cit. (note 71).

[125] In Bukharin *et al., Science at the Cross Roads,* op. cit. (note 5), pp. 147–212.

[126] Frank E. Manuel, *A Portrait of Isaac Newton* (Cambridge, Mass., Harvard, 1968).

[127] See above, note 124.

The development of such an historiography is a fundamental desideratum both for those who have found themselves working with a relativist and contextualist approach which is not well-articulated, and for those who seek to make the historical study of science play a part in changing the world into one which is genuinely liberating, socialist and egalitarian. Of course, many historians of science will wish to continue—as one put it—'working in my own corner', either implicitly or explicitly making a contribution to the maintenance of the existing order of society. They will argue that a call for a 'radical' historiography of science is 'dragging politics into the classroom' and that the *status quo* in the subject is apolitical. Others would argue on political grounds that the present approach to the subject *is* political and that it has the *right* politics.

What are the available perspectives for a marxist historiography? What is available, and what is wrong with it?

V

Reverting once again to Needham's marxist thesis about men and their ideas being born of their time, it is important to add the corollary that changed men with changed consciousness are the product of changed times. This is as true of the present as it was of the nineteenth-century debate on man's place in nature. It was also true of Joseph Needham's development. In his essay on 'Metamorphoses of Scepticism' (1941), he reports that 'The process of socialisation of my outlook, however, really began with the General Strike in 1926 [in which he was on the wrong side] and was completed by the rise to power of Hitlerite fascism in 1933.'[128] Between these two events there occurred the Second International Congress of the History of Science and Technology in London (1931), at which the Soviet delegation put forth the version of marxist historiography of science which inspired the approaches of Needham and Bernal, and which influenced Crowther and others. Needham later wrote in 'Limiting Factors in the History of Science',

> In sum, we cannot dissociate scientific advances from the technical needs and processes of the time, and the economic structure in which all are embedded. . . . The history of science is not a mere succession of inexplicable geniuses, direct Promethean ambassadors to man from heaven. Whether a given fact would have got itself discovered by some other person than the historical discoverer had he not lived it is certainly profitless and probably meaningless to enquire. But scientific men, as Bukharin said, do not live in a vacuum; on the contrary,

---

[128] Needham, *Time: The Refreshing River*, op. cit. (note 2), p. 12.

the directions of their interests are ever conditioned by the structure of the world they live in. Further historical research will enable us to do for the great embryologists what has been well done by Hessen for Isaac Newton. . . .[129]

Similar tributes to Hessen can be found in writings of the other two major marxist historians of science in England. Bernal wrote, that 'Hessen's article on Newton . . . was for England the starting point of a new evaluation of the history of science.' This tribute appears in his *The Social Function of Science*, a work which was one of the most notable expressions of the new marxist historiography.[130] J. G. Crowther did not identify himself as a marxist to the extent that Needham and Bernal did, but he was influenced by marxist historiography. The Acknowledgements of his *The Social Relations of Science* began as follows: 'The views of B. Hessen and T. Veblen have provided much inspiration for this book.'[131] In the course of the book he gives an account of the Congress and its influence.

> Hessen gave the first concrete example of how science should be interpreted as a product of the life and tendencies of society. . . . Hessen's demonstration of the depth and range of Newton's dependence on the ideas promulgated by the epoch in which he appeared, made a profound impression on some of the younger members of the congress. It transformed the study of the history of science, and out-moded the former conceptions of the subject, which treated it as governed only by the laws of its internal logical development. Henceforth, no satisfactory history of science could be written without giving adequate attention to the dependence of science on social factors.[132]
>
> . . . The movement, of which Hessen's essay was the most stimulating expression, transformed the history of science from a minor into a major subject. It showed that a knowledge of the history of science was not only of entertaining antiquarian

---

[129] Ibid., pp. 144–5. Needham's admiration for Hessen has not diminished in the intervening forty-one years. In his 'New Foreword' to the 1971 reprint of *Science at the Cross Roads* (op. cit., note 5), he wrote, 'Perhaps the outstanding Russian contribution was that of Boris Hessen, who made a long and classical statement of the Marxist historiography of science, taking as his subject of analysis Isaac Newton.' It was 'a veritable manifesto of the Marxist form of externalism in the history of science. . . . This essay, with all its unsophisticated bluntness, had a great influence during the subsequent forty years, an influence still perhaps not yet exhausted; hence its present reprinting is to be welcomed. . . . The trumpet-blast of Hessen may therefore still have great value in orienting the minds of younger scholars towards a direction fruitful for historical analyses still to come, and may lead in the end to a deeper understanding of the mainsprings and hindrances of science in East and West, far more subtle and sophisticated than he himself could ever hope to be' (pp. viii-ix).

[130] J. D. Bernal, *The Social Function of Science* (London, Routledge & Kegan Paul, 1939; reprinted with a postscript 'After Twenty-five Years', Cambridge, Mass., M.I.T. paperback, 1967), M.I.T. edition, p. 406n.

[131] J. G. Crowther, *The Social Relations of Science* (1941), 2nd edition (London, Cresset, 1967), p. vii.

[132] Ibid., p. 431.

interest, but was essential for the solution of contemporary social problems due to the unorganized growth of a technological society.[133]

Clearly, Needham, Bernal, Crowther and others who were very inspired by the interpretation of the history of science which the Soviet delegation brought to the Congress, and their subsequent work in the field (though in the cases of the three persons mentioned it was a part-time activity, albeit a prolific avocation) continued to adhere to the model of intellectual work in science as a direct and relatively straightforward superstructural expression of the socio-economic base in a given society. This model proved useful in its time, but I do not propose to discuss it in detail.[134] It is noteworthy that its basic documents remain in print, and many of them have recently been reprinted, in particular the papers presented by the Soviet delegation to the London Congress in 1931. The reprint contains an excellent analytical introduction by P. G. Werskey 'On the Reception of *Science at the Cross Roads* in England'.[135] The revival of interest in this historiographic approach is significant, but it is basically pointless to attempt to recapture an atmosphere which played no part in the education of younger historians of science

[133] Ibid., p. 432. Crowther reports a revealing aspect of the atmosphere at the Congress which also raises current issues about the relationship between scholarship, political commitment and action: 'None of the amateur or professional students of the history of science could think of any comment for opening a discussion on the Russians' enthusiastic and exciting papers. After a pause, a twenty-year-old youth named David Guest drew attention to the significance of their views, stressing especially the historical element in all their philosophical and scientific concepts, and contrasting this with the non-historical concepts employed by Pearson and Russell in their philosophy of science. No other speaker could think of anything more to say. Guest subsequently graduated with first class honours in philosophy in Cambridge University, and was killed in Spain in 1938, fighting with the International Brigade in defence of the Republican Government.' (pp. 431–2). Cf. Bernal's remarks, quoted in *Science at the Cross Roads*, op. cit. (note 5), p. xxi.

[134] For expositions and discussions of the base-superstructure model, see George V. Plekhanov, *Fundamental Problems of Marxism* (1908), trans. Julius Katzer (London, Lawrence & Wishart, 1969), p. 8, chs. 8, 9, 11, 12, and p. 180 fn. 11; Karl Korsch, *Marxism and Philosophy* (1923), trans. Fred Halliday (London, New Left Books, 1970); V. A. Fanasyev, *Marxist Philosophy: a Popular Outline*, trans. Leo Lempert (Moscow, Progress, 1963), ch. 11, especially pp. 197–200; G. Kursanov (ed.), *Fundamentals of Dialectical Materialism*, trans. Vic Schneirson (Moscow, Progress, 1967). Wilhelm Reich was making penetrating criticisms of vulgar marxism in the 1930s. See, for example, *What is Class Consciousness?* (1933) trans. *anon.* (London, Socialist Reproduction, 1971); *The Mass Psychology of Fascism* (1933), 3rd edition, revised and enlarged, trans. Vincent R. Carfagno (New York, Farrar, Straus & Giroux, 1969; also paperback) especially chs. 1, 9; *The Sexual Revolution* (1932), 4th edition, revised, trans. Theodore P. Wolfe (London, Vision, 1969; also paperback), especially p. xvii and Part 2. For a Maoist view, see Rossana Rossanda, 'Mao's Marxism', in Ralph Miliband and J. Saville (eds.), *The Socialist Register, 1971: a Survey of Movements and Ideas* (London, Merlin paperback, 1971), pp. 53–80.

[135] Op. cit. (note 5), pp. xi–xxix.

in the present generation and which, when recovered, is clearly to be set aside. Werskey's introduction and his related studies of the period and the genre are reliable guides for those interested in this topic, but it is an *historical* topic, and I propose to discuss it only as reflected in criticisms of its limitations.[136]

Conservative and radical historians are united in the belief that the base-superstructure model, as advocated by Hessen and his followers, led too easily to crude economic reductionism.[137] This is not to say that it could not be skilfully applied to yield subtle historical work, as it undoubtedly did in the major writings of Needham and Bernal. However, in their zeal to follow Marx and Engels in deploring the practice of writing the history of the sciences 'as if they had fallen from the skies',[138] many practitioners of the marxist historiography of the period after 1931, fell into the opposite error of writing it as if it always rose directly and straightforwardly from the base, without mediation. Moreover, they did not even entertain the possibility that the metaphors of that model—base, superstructure—might need to be

[136] P. G. Werskey, 'British Scientists and "Outsider" Politics, 1931–1945', *Science Studies* 1 (1971), pp. 67–83 and the references to his other papers cited therein. Mr. Werskey has been particularly generous and helpful in suggesting that I consult many of the documents which have been drawn upon in formulating my views on marxist historiography.

[137] Ollman points out what an easy target the distinction is in *Alienation*, op. cit. (note 8), pp. 6–9. It is, frankly, difficult to recover the enthusiasm generated by Hessen's essay. He makes a gesture to the complexity of his problem: 'It would, however, be too greatly simplifying and even vulgarizing our object if we began to quote *every problem* which has been studied by one physicist or another, and every economic and technical problem which he solved . . . the above general analysis of the economic problems of the epoch would not be sufficient. We must analyse more fully Newton's epoch, the class struggles during the English Revolution, and the political, philosophic and religious theories are reflected in the minds of the contemporaries of these struggles.' (Op. cit., notes 5 and 125, p. 177). Even so, no amount of disarming qualification can prevent the reaction one feels against the simple one-to-one correlations between economic and technological problems on the one hand and scientific ideas on the other, of which the bulk of his argument consists. For example, 'The above specified problems embrace almost the whole of physics. If we compare this basic series of themes with the physical problems which we found when analysing the technical demands of transport, means of communication, industry and war, it becomes quite clear that these problems of physics were fundamentally determined by these demands. . . . We have compared the main technical and physical problems of the period with the scheme of investigations governing physics during the period we are investigating, and we come to the conclusion that the scheme of physics was mainly determined by the economic and technical tasks which the rising bourgeoisie raised to the forefront' (pp. 166–7). In the face of this approach, one can only feel sympathy and understanding on being told that in the 1940s, young Anglo-American historians of science turned to the internal history of ideas as practised by Koyré and Meyerson with a sense of relief, excitement and liberation.

[138] See Robert K. Merton, *Social Theory and Social Structure* (1949), 3rd edition (London, Collier-Macmillan, 1968), p. 587.

complemented, if not finally superseded. Hall was entirely correct when he wrote in 1963, 'In its crudest form at any rate the socio-economic interpretation of the scientific revolution as an offshoot of rising capitalism and mercantile militarism has perished without comment,'[139] A decade later it can be added that its revival is a mistake, except as a basis for developing a more subtle version of the fundamental marxist thesis. This is precisely what the most sophisticated of the historians who have been influenced by marxism have recently been doing. Eric Hobsbawm, a marxist economic historian, provides an exposition of the initial effects and of the debasement of Marx's fundamental insights.

> Yet those of us who recall our first encounters with historical materialism may still bear witness to the immense liberating force of such simple discoveries [relating ideas to economic conditions]. However, if it was thus natural, and perhaps necessary, for the initial impact of Marxism to take a simplified form, the actual selection of elements from Marx also represented an historical choice. . . . The bulk of what we regard as the Marxist influence on historiography has certainly been vulgar-Marxist. . . . It consists of the general emphasis on the economic and social factors in history which have been dominant since the end of the Second World War in all but a minority of countries (e.g. until recently West Germany and the United States), and which continues to gain ground. We must repeat that this trend, though undoubtedly in the main the product of Marxist influence, has no special connection with Marx's thought. The major impact which Marx's own specific ideas have had in history and the social sciences in general, is almost certainly that of the theory of 'basis and super-structure'; that is to say of his model of a society composed of different 'levels' which interact.[140]

At this point the historian of science begins to feel the primitive state of discussion on these issues in his own discipline, and the historian of the biological and human sciences is in a particular difficulty. He is still striving to apply certain of the most basic insights of marxism as a corrective to positivist approaches both to the history of science in general and especially to his own subject. Thus, even the vulgar marxist tradition has played an important part in attempting to transform history into a critical discipline among the social sciences. Hobsbawn writes,

> The major contribution of Marxism to this tendency in the past has been the critique of positivism, i.e. of the attempts to assimilate the study of the social sciences to that of the natural ones, or the human to the non-human. This implies the recognition of societies as systems of relations between human

---

139 Hall, 'Merton Revisited', op. cit. (note 32), p. 9.
140 Eric J. Hobsbawm, 'Karl Marx's Contribution to Historiography', in *Marx and Contemporary Scientific Thought*, op. cit. (note 8), pp. 197–211, at p. 202; cf. pp. 200–1. See also Hobsbawm, 'From Social History to the History of Society', *Daedalus* 100 (1971), pp. 20–45.

beings, of which the relations entered into for the purpose of production and reproduction are primary for Marx. It also implies the analysis of the structure and functioning of these systems as entities maintaining themselves, in their relations both with the outside environment—non-human and human—and in their internal relationships. Marxism is far from the only structural-functionalist theory of society, though it has good claims to be the first of them, but it differs from most others in two respects. First, it insists on a hierarchy of social phenomena (e.g. 'basis' and 'superstructure'), and second, on the existence within any society of internal tensions ('contradictions') which counteract the tendency of the system to maintain itself as a going concern.[141]

These points are directly germane to the analysis of the nineteenth-century debate on man's place in nature, a debate in which the modern theory which reduces the human to the non-human was being elaborated. It also points to the rôle of the debate itself in maintaining a given social order in the face of rapid socio-economic change and to the important interactions between intellectual developments and social ones. Yet just as historians of science are getting near to these insights, marxist historians in other fields are moving on, partly on the basis of a new political context and partly as a result of reading more widely in Marx's writings, some of which are newly published. Hobsbawm continues,

> Moreover, the diminishing returns on the application of vulgar-Marxist models have in recent decades led to a substantial sophistication of Marxist historio-graphy. Indeed, one of the most characteristic features of contemporary western Marxist historiography is the critique of the simple, mechanical schemata of an economic-determinist type. However, whether or not Marxist historians have advanced substantially beyond Marx, their contribution today has a new importance, because of the changes which are at present taking place in the social sciences. Whereas the major function of historical materialism in the first half-century after Engels' death was to bring history closer to the social sciences, while avoiding the oversimplifications of positivism, it is today facing the rapid historicisation of the social sciences themselves. For want of any help from academic historiography, these have increasingly begun to improvise their own —applying their own characteristic procedures to the study of the past, with results which are often technically sophisticated, but, as has been pointed out, based on models of historic change in some respects even cruder than those of the 19th century. Here the value of Marx's historical materialism is great, though it is natural that historically minded social scientists may find themselves less in need of Marx's insistence on the importance of economic and social elements in history than did the historians of the early 20th century; and conversely might find themselves more stimulated by aspects of Marx's theory which did not make a great impact on historians in the immediately post-Marxian generations.[142]

---

[141] Ibid., p. 203.
[142] Ibid., p. 209.

I appreciate that the quotation of long passages from Hobsbawm's analysis of 'Karl Marx's Contribution to Historiography' may appear confusing and that his remarks call for careful exegesis and translation into terms which apply directly to problems in the historiography of science. My aim in quoting them is to bring into sharp relief the distance between the level of debate on these issues among certain social and political historians and that among the mainstream of professional historians of science. This is not to say that no professional historians of science have addressed themselves to issues which would also interest a marxist. However they have not done so in a theoretically self-conscious way. The separation of historians of science from these issues, which Hall pointed out, had already lasted for a decade by 1963, has since grown more marked, and the incomprehensibility and apparent irrelevance of Hobsbawm's analysis can be argued to be a measure of the gap which must be bridged if a radical historiography of science which is relevant to current conditions is to be developed.[143]

It can, of course, be argued that the base-superstructure distinction is only useful to historians of science as a vehicle for freeing themselves from the restrictions of the internalist-externalist distinction and that having once freed themselves, they should lay aside the model which has led so disastrously to economic reductionism in vulgar marxism. It may be that the base-superstructure model is too encrusted with barnacles or too redolent of stale debates to be refurbished and made useful. My own current position is that the employment of a sufficiently subtle theory of mediations and interactions between socio-economic factors and intellectual life would make the base-superstructure model servicable once again, at least in an interim way, until we can develop a fully relational, totalising approach. But most historians of science who have any sympathy at all with radical historiography are still at the stage which Hobsbawm mentioned as 'first encounters' which have a tremendously liberating force, and they are tempted to be sanguine about possibilities which those who have had more experience with this model are likely to discount.

Raymond Williams made this point very forcefully in his discussion of 'Marxism and Culture' in *Culture and Society* where he was at pains to separate Marx and Engels from their vulgarizers.[144] He refers to an important letter which Engels wrote to J. Bloch in 1890:

[143] The following statement about historical studies in general is simply not true of the history of science: '. . . history is a discipline into which Marxist ideas and a Marxist approach have already penetrated very deeply, so that there is not such a gap between a Marxist historian and a non-Marxist colleague as there would be, for instance, among philosophers.' Taylor, 'Marxism and Empiricism', op. cit. (note 3), p. 229.

[144] Williams, *Culture and Society*, op. cit. (note 105), Penguin edition, ch. 6, especially pp. 271–4.

According to the materialist conception of history, the *ultimately* determining element in history is the production and reproduction of real life. More than this neither Marx nor I have ever asserted. Hence if somebody twists this into saying that the economic element is the *only* determining one, he transforms that proposition into a meaningless, abstract, senseless phrase. The economic situation is the basis, but the various elements of the super-structure—political forms of the class struggle and its results, to wit: constitu-tions established by the victorious class after a successful battle, etc., juridical forms, and even the reflexes of all these actual struggles in the brains of the participants, political, juristic, philosophical theories, religious views and their further development into systems of dogmas—also exercise their influence upon the course of the historical struggles and in many cases preponderate in deter-mining their *form*. There is an interaction of all these elements in which, amid all the endless host of accidents (that is, of things and events whose inner inter-connection is so remote or so impossible of proof that we can regard it as non-existent, as negligible) the economic movement finally asserts itself as necessary. Otherwise the application of the theory to any period of history would be easier than the solution of a simple equation of the first degree.[145]

After giving some examples and suggesting that it would be far better to read Marx's writings and his own and not to approach the problem at second-hand, Engels continues,

Marx and I are ourselves partly to blame for the fact that the younger people sometimes lay more stress on the economic side than is due to it. We had to emphasise the main principle *vis-à-vis* our adversaries, who denied it, and we had not always the time, the place or the opportunity to give their due to the other elements involved in the interaction. But when it came to presenting a section of history, that is, to making a practical application, it was a different matter and there no error was permissible. Unfortunately, however, it happens only too often that people think they have fully understood a new theory and can apply it without more ado from the moment they have assimilated its main principles, and even those not always correctly. And I cannot exempt many of the more recent 'Marxists' from this reproach, for the most amazing rubbish has been produced in this quarter, too. . . .[146]

Williams notes that this formulation is richer than that of vulgar marxism but still finds it static. Its application by marxist theorists of culture seemed to him to employ different aspects of it in varying ways 'as the need serves'.[147] His position in *Culture and Society* was very much like that of current

[145] Ibid., p. 260. I have extended the quotation and have taken it from Marx and Engels, *Selected Correspondence*, op. cit. (note 81), p. 417.

[146] Ibid., pp. 418–19. In a letter to Mehring in 1893, Engels developed the point that he and Marx had failed sufficiently to stress the process of mediation ('ways and means') and the relative autonomy of ideas and their interactions with history. It is quoted at length as an appendix to this paper. Cf. Ollman, *Alienation*, op. cit. (note 8), pp. 9–11.

[147] Williams, *Culture and Society*, op. cit. (note 105), Penguin edition, p. 266.

historians of science. His grasp of the fine texture of cultural history in nineteenth-century Britain and his determination to remain true to his materials, led him to take up a relativist and contextualist standpoint without any clear theoretical position. His own historiographic certainties were negative: he could neither adhere to the approach of idealist and élitist cultural historians nor could he allow himself to conform to the impoverishing model of vulgar marxism.

Over a decade later his writings on these issues began to reflect a much surer and more optimistic approach to historiography. In a recent essay on 'Ideas of Nature' in nineteenth-century Britain, he offered a rendering which reflects an ability to maintain simultaneously the same perspectives which were being suggested above: 'What is often being argued, it seems to me, in the idea of nature is the idea of man; and this not only generally, or in ultimate ways, but the idea of man in society, indeed the ideas of kinds of societies.'[148] The context of this passage suggests an ability on his part to unite science, the philosophy of nature and theories of man and society in a single theoretical framework.

The perspective of Raymond Williams is illuminating for another reason. Not only is he able to avoid vulgar marxism while writing from a socialist point of view, but he is also someone who increasingly shows an ability to mediate between the Old Left and the New Left. He continues to play a leading part in theoretical formulations for English-speaking socialists begun in the 1960s. It was said above that while all men are children of their time, changed men are children of changed times. The developments which Needham, Bernal and others refer to as leading them to see science and society from a socialist perspective have their parallels in developments in the 1960s in the growth of civil disobedience, in the Campaign for Nuclear Disarmament in Britain and the struggles for civil rights in America; in the growth of radical student movements, particularly in America, Germany, France, Japan, Italy, and to a lesser extent in Britain; in the world-wide protest movement against the American intervention in Southeast Asia; in the bitter disappointment over the policies of the British Labour Government; in the invasion of Czechoslovakia; in repressive political trials in Britain and America. Among committed members of the Communist Party there was an earlier reaction to the invasion of Hungary, followed by the revelations about Stalin and other repressive measures extending from treatment of minorities to the absurdities of the Lysenko

---

[148] Raymond Williams, 'Ideas of Nature', *Times Lit. Suppl.* No. 3588 (4 December 1970), 1419–21, at p. 1419. See also Williams, 'An Introduction to Reading in Culture and Society', in Fred Inglis (ed.), *Literature and Environment: Essays in Reading and Social Studies* (London, Chatto & Windus, 1971), pp. 125–39.

Affair in biology.[149] Although there are continuities between the civil disobedience movements and the more radical phase which grew up around the Vietnam War, it was in the period after 1964 that the implications of these developments for our perception of liberal institutions and liberal scholarship began to penetrate the consciousness of scholars and slowly led them to reflect about the political meaning of their work.[150] In the last

[149] See Zhores A. Medvedev, *The Rise and Fall of T. D. Lysenko*, trans. I. M. Lerner (New York, Columbia, 1969); David Joravsky, *The Lysenko Affair* (Cambridge, Mass., Harvard, 1970). Joravsky's liberal, functionalist account requires analysis from a radical perspective, one which I hope to provide in a forthcoming book on the relationship between biology, ideology and the human sciences. Cf. below (note 204) and 'Evolutionary Biology and Ideology: Then and Now', op. cit. (note 8), pp. 186–7, 204, 205.

[150] It is a hopeless task to attempt to cite a literature covering this network of events and issues. Much has not been written down, since many activists aren't into that, while much that does get written is ephemeral. For a perspective on 'The Movement' from the point of view of rock music, see R. M. Young, 'The Functions of Rock', *New Edinburgh Rev.* No. 10 (December 1970), pp. 4–14. A short list of books and collections of documents touching on a number of aspects of the development of the New Left is given below. In the same way that it is difficult for scholars who have been trained in the current orthodoxy seriously to entertain marxism, it is hard for them to grasp the perspective of the New Left and to see its direct relevance to their praxis. The challenge to state a *precise* definition of the New Left is, in the end, a mystification, since its essential nature is exploratory and experimental. The two most constant features are its libertarianism and its anti-authoritarianism. It is neo-marxist rather than marxist in the orthodox sense and involves a deep distrust of old leftist parties, their ideology and the development of the praxis of the Soviet Union. Similarly, it is very ambiguous about its relations with the claim that the industrial working class is the sole revolutionary agent. Beyond these characteristics, there are endless factions and sects, with particular orientations towards, e.g. Maoism, black liberation, women's liberation, Third World movements, ecology, etc. And beyond this, one must find one's own struggles and theoretical orientations. See, for example, Paul Jacobs and Saul Landau, *The New Radicals* (Harmondsworth, Penguin paperback, 1966); Alexander Cockburn and Robin Blackburn (eds.), *Student Power: Problems, Diagnosis, Action* (Harmondsworth, Penguin paperback, 1969); Carl Oglesby (ed.), *The New Left Reader* (New York, Grove paperback, 1969), especially Rudi Dutschke, 'On Anti-authoritarianism' and Martin Nicolaus, 'The Unknown Marx', pp. 243–53 and 84–110; Priscilla Long (ed.), *The New Left: a Collection of Essays* (Boston, Sargent paperback, 1969).

For examples of excellent critical scholarship, reinterpreting traditional issues from a New Left perspective, see Chomsky, 'Objectivity and Liberal Scholarship', op. cit. (note 6); David Horowitz, *From Yalta to Vietnam: American Foreign Policy in the Cold War* (London, MacGibbon & Kee, 1965; revised edition Harmondsworth, Penguin, 1969). Eugene D. Genovese, *The Political Economy of Slavery: Studies in the Economy and Society of the Slave South* (New York, Random House, 1965; also Vintage paperback); *The World the Slaveholders Made: Two Essays in Interpretation* (New York, Random House, 1969; also Vintage paperback).

In order to gain some sense of the 'life style' of the libertarian aspects of 'The Movement', one can get hints from reading the following: Jeff Nuttall, *Bomb Culture* (London, MacGibbon & Kee, 1968; also Paladin paperback, 1970); Theodore Roszak, *The Making of a Counter-Culture: Reflections on the Technocratic Society and Its Youthful Opposition* (London,

few years, writers (including Williams) who centred around the May Day
Manifesto group and the *New Left Review* played an important part in
bringing about a reconsideration of approaches to scholarship and political
strategy, at the same time that they set up a translation industry for bringing
the attention of others to important theoretical developments in the marxist
tradition which had been going on in relative isolation from Anglo-
American scholars. (A similar rôle was being played by C. Wright Mills in
the United States.)[151] These are the writings of Continental and émigré
ideologues extending from Rosa Luxemburg, Georg Lukács and Antonio
Gramsci,[152] to the works of the Frankfurt School—Horkheimer, Adorno,

---

Faber, 1970; also paperback); Jerry Rubin, *Do It! Scenarios of the Revolution* (New York,
Simon & Schuster paperback, 1970); David A. de Turk and A. Poulin, Jr., *The American
Folk Scene: Dimensions of the Folksong Revival* (New York, Dell paperback, 1967); Josh
Dunston, *Freedom of the Air: Song Movements of the 60's* (New York, International paperback, 1965); Jerry Hopkins, *The Rock Story* (New York, Signet paperback, 1970); Ralph J.
Gleason, *The Jefferson Airplane and the San Francisco Sound* (New York, Ballantine paperback,
1969). The rock music aspect of the movement is already (since John Lennon's epitaph,
the song 'God') an *historical* phenomenon, and young people in the New Left are currently
without even a loosely coherent perspective. It is now, arguably, a time for much more
serious consciousness-raising, reading and discussion.

Some of the ephemeral documents of the Movement are beginning to be collected and
republished. See, for example, Peter Stansill and David Z. Mairowitz (eds.), *BAMN
(By Any Means Necessary): Outlaw Manifestos and Ephemera, 1965–1970* (Harmondsworth,
Penguin paperback, 1970); Mitchell Goodman (ed.), *The Movement Toward a New America:
The Beginnings of a Long Revolution (a Collage)* (New York, Knopf paperback, 1970).

[151] C. Wright Mills, 'Letter to the New Left', *New Left Rev.*, 1961; reprinted in Long,
The New Left, op. cit. (note 150), pp. 14–26. Mills died in 1962. For a bibliography of his
writings and reviews of his work, see Irving L. Horowitz (ed.), *Power, Politics and People:
The Collected Essays of C. Wright Mills* (London, Oxford, 1963; also paperback), pp.
614–41; see also I. L. Horowitz (ed.), *The New Sociology: Essays in Social Science and Social
Theory in Honor of C. Wright Mills* (New York, Oxford, 1964; also paperback). For a
critical perspective, see Fredy Perlman, *The Incoherence of the Intellectual: C. Wright Mills'
Struggle to Unite Knowledge and Action* (Detroit, Black & Red, 1970); Irving M. Zeitlin,
'The Plain Marxism of C. Wright Mills', in Fischer (ed.), *The Revival of American Socialism*,
op. cit. (note 6), pp. 227–43.

[152] This is a *very* rapidly-growing translation industry. A selection: Mary-Alice Waters
(ed.), *Rosa Luxemburg Speaks* (New York, Pathfinder paperback, 1970); J. Peter Nettl, *Rosa
Luxemburg* (Oxford, 1966; abridged paperback edition, 1969); Georg Lukács, *History
and Class Consciousness* (1923); new edition (1967) with Preface, trans. Rodney Livingstone
(London, Merlin, 1971); 'Lukács on his Life and Work' (interview), *New Left Rev.*
No. 68 (July/August 1971), pp. 49–58; István Mészáros (ed.), *Aspects of History and Class
Consciousness* (London, Routledge & Kegan Paul, 1971); I. Mészáros, *Lukács' Concept of
Dialectic, with Biography, Bibliography and Documents* (London, Merlin paperback, 1972);
Antonio Gramsci, *The Modern Prince and Other Writings*, trans. Louis Marks (New York,
International, 1957; also paperback); *Selections from the Prison Notebooks of Antonio Gramsci*,
ed. and trans. Quintin Hoare and G. N. Smith (London, Lawrence & Wishart, 1971);
Giuseppe Fiori, *Antonio Gramsci: Life of a Revolutionary* (1965), trans. Tom Nairn (London,
New Left Books, 1970); A. Pozzolini, *Antonio Gramsci: An Introduction to His Thought*
(1968), trans. Anne F. Showstack (London, Pluto, 1970).

and especially Marcuse—followed more recently by Habermas and Schmidt[153] and on to the related works of Walter Benjamin and Lucien Goldman.[154] There are, of course, other writers whose works have played an important part in Continental debates on marxism. In particular, continuous discussion of these issues has been occurring in Yugoslavia, Czechoslovakia, Austria and Italy, but these have not on the whole been brought to the attention of scholars in England and America. The writings of Sartre and Garaudy are more accessible.[155] There is also a growing literature from the network of people who are routinely called 'structuralists'—some of whom themselves reject the designation or affiliation with others who are so labelled—Lévi-Strauss, Foucault, Althusser and Balibar.[156] Not all of these would be

[153] Göran Therborn, 'The Frankfurt School', *New Left Rev.* No. 63 (September/October 1970), pp. 65–96; Theodor W. Adorno *et al.*, *The Authoritarian Personality* (1950; reprinted New York, Norton paperback, 1969); T. W. Adorno, *Prisms,* trans. Samuel and Shierry Weber (London, Spearman, 1967); 'Sociology and Psychology', *New Left Rev.* No. 46 (November/December 1967), pp. 67–97 and No. 47 (January/February 1968), pp. 79–96. Some of the relevant writings of Marcuse are listed above (note 123) and in Young, 'Evolutionary Biology and Ideology: Then and Now', op. cit. (note 8), p. 204 and 188n, to which should be added *Five Lectures: Psychoanalysis, Politics, and Utopia,* trans. Jeremy J. Shapiro and S. M. Weber (London, Allen Lane, 1970); 'Re-examination of the Concept of Revolution', *New Left Rev.* No. 56 (July/August 1969), pp. 27–34; 'Contributions to a Phenomenology of Historical Materialism' (1928), *Telos* No. 4 (Fall 1969), pp. 3–34; Jerry Cohen, 'Critical Theory: the Philosophy of Marcuse', *New Left Rev.* No. 57 (September/October 1969), pp. 35–51; articles on Marcuse by M. Jay, R. Aronson and P. Breines in Fischer (ed.), *The Revival of American Socialism,* op. cit. (note 6), chs. 13–15; Jürgen Habermas, 'Technology and Science as "Ideology" ', op. cit. (note 123); *Knowledge and Human Interest* (1968), trans. Jeremy J. Shapiro (London, Heinemann, 1972; also paperback), especially Appendix; Göran Therborn, 'Jürgen Habermas: A New Eclectic', *New Left Rev.* No. 67 (May/June 1971), pp. 69–83; Alfred Schmidt, *The Concept of Nature in Marx* (1962), trans. Ben Fowkes (London, New Left Books, 1971). There are worthwhile articles, reviews and discussions by and about Lukács and the Frankfurt School and related marxist topics, in nearly every issue of *Telos* (see above, note 6).

[154] Walter Benjamin, *Illuminations* (1955), trans. Harry Zohn (New York, Harcourt, Brace & World, 1968; also Schocken paperback); 'The Sociology of Knowledge and the Problem of Objectivity', in L. Gross (ed.), *Sociological Theory: Inquiries and Paradigms* (New York, Harper & Row, 1967), pp. 335–57; Lucien Goldmann, *The Human Sciences and Philosophy* (1966), trans. Hayden V. White and R. Anchor (London, Cape, 1969; also paperback); Miriam Glucksmann, 'Lucien Goldmann: Humanist or Marxist?' *New Left Rev.* No. 56 (July/August 1969), pp. 49–63.

[155] Jean-Paul Sartre, *The Problem of Method* (1960), trans. Hazel E. Barnes (London, Methuen, 1963); Roger Garaudy, *Karl Marx: the Evolution of His Thought* (1964), trans. Nan Apotheker (London, Lawrence & Wishart, 1967; also paperback); *The Turning Point of Socialism,* trans. Peter and Betty Ross (London, Collins, 1970; also Fontana paperback); *The Whole Truth* (1970), trans. P. and B. Ross (London, Fontana paperback, 1971).

[156] For example, Claude Lévi-Strauss, *Structural Anthropology* (1958), trans. Claire Jacobson and B. G. Schoepf (London, Allen Lane, 1968); *The Savage Mind* (1962), trans. anon. (London, Weidenfeld & Nicolson, 1966); Michel Foucalt, *The Order of Things:*

happy to be identified as part of a marxist tradition, but it does seem fair to suggest that all of them have worked out their positions *in relation to* marxism. Some of the issues raised in this debate will be considered below. For the present I only want to suggest that historians of science could more profitably turn to this literature for stimulating issues and analogies than to the writings of traditional historiography.

In an extremely perceptive and moving essay on 'Literature and Sociology: in Memory of Lucien Goldmann', Raymond Williams provides a clear expression of the experience of finally finding someone who spoke to his intellectual and political needs. In suggesting that his sense of fellowship has great relevance for the current situation in the historiography of science, I appreciate that orthodox historians will feel that I am begging large questions about the putative differences between the history of science and cultural history. I can only say that I have begun to consider these questions elsewhere and state the interim conclusion that a consequence of the perspective being argued here is that the history of science is far more like the histories of literature, art and society than has hitherto been supposed and that to appreciate this, it is necessary to adopt an 'anthropological' point of view for considering the history of science.[157]

Williams speaks of his embarrassment in earlier decades over the conflict

---

An Archaeology of the Human Sciences, trans. *anon.*, with a new foreword (London, Tavistock, 1970). The work of Althusser has produced a very large—and polarized—literature, has inspired a new journal, *Theoretical Practice* (see above, note 6), and has increasingly shaped the editorial policy of the *New Left Review*. The Althusserian point of view is the most highly-developed Marxist critique of science and must be taken seriously, even if (especially if) it proves a far from satisfactory approach. A selection of the relevant literature: Louis Althusser, *For Marx* (1965), trans. Ben Brewster (London, Allen Lane, 1969); L. Althusser and Étienne Balibar, *Reading Capital* (1968), trans. B. Brewster (London, New Left Books, 1970); L. Althusser, *Lenin and Philosophy and Other Essays* (1966–70), trans. B. Brewster (London, New Left Books, 1971); 'Philosophy as a Revolutionary Weapon', *New Left Rev.* No. 64 (November/December 1970), 3–11; Paul Hirst, 'Althusser and Philosophy', *Theoretical Practice* No. 2 (April 1971), 16–30; special issue on 'Marxism and the Sciences', ibid., Nos. 3–4 (Autumn 1971), especially Michel Fichant, 'The Idea of a History of the Sciences', pp. 38–62; Norman Geras, 'Althusser's Marxism: An Account and an Assessment', *New Left Rev.* No. 71 (January/February 1972), pp. 57–86. For critical evaluations of Althusserism, see Andrew Levine *et al.*, 'Althusser', *Radical America*, Vol. 3, No. 5 (September 1969), 3–51; A. Levine, 'A Reading of Marx, II', ibid., Vol 4, No. 6 (August 1970), pp. 54–69 and Dale Tomich, 'Comment', ibid., pp. 69–72; Leszek Kolakowski, 'Althusser's Marx', in Milliband and Saville (eds.), *The Socialist Register, 1971*, op. cit. (note 134), pp. 111–28.

[157] See below, pp. 433–5. I have touched on these issues in 'Evolutionary Biology and Ideology: Then and Now', op. cit. (note 8), 'The Anthropology of Science', op. cit. (note 71) and 'The Human Limits of Nature', op. cit. (note 78), but the general position is very far from being satisfactorily worked out. The present essay is a further attempt to move towards a better grasp of these issues.

between the cultural élitism of the group of critics writing for *Scrutiny* and contemporary vulgar marxism.

Marxism, as then commonly understood, was weak in just the decisive area where practical criticism was strong: in its capacity to give precise and detailed and reasonably adequate accounts of actual consciousness: not just a scheme or a generalisation but actual works, full of rich and significant and specific experience. And the reason for the corresponding weakness in Marxism is not difficult to find: it lay in the received formula of base and superstructure, which in ordinary hands converted very quickly to an interpretation of superstructure as simple reflection, representation, ideological expression—simplicities which just will not survive any prolonged experience of actual works. It was the theory and practice of reductionism—the specific human experiences and acts of creation converted so quickly and mechanically into classifications which always found their ultimate reality and significance elsewhere—which in practice left the field open to anybody who could give an account of art which in its closeness and intensity at all corresponded to the real human dimension in which art works are made and valued.[158]

It was above all, as I have said, the received formula of base and superstructure which made Marxist accounts of literature and thought so often weak in practice. Yet to many people, still, this formula is near the centre of Marxism, and indicates its appropriate methodology for cultural history and criticism, and then of course for the relation between social and cultural studies. The economic base determines the social relations which determine consciousness which determines actual ideas and works. There can be endless debate about each of these terms, but unless something like that is believed, Marxism appears to have lost its most specific and challenging position.[159]

Now for my part I have always opposed the formula of base and superstructure: not primarily because of its methodological weaknesses but because of its rigid, abstract and static character. Further, from my own work on the nineteenth century, I came to view it as essentially a bourgeois formula; more specifically, a central position of utilitarian thought. I did not want to give up my sense of the commanding importance of economic activity and history. My inquiry in *Culture and Society* had begun from just that sense of a transforming change. But in theory and practice I came to believe that I had to give up, or at least to leave aside, what I knew as the Marxist tradition: to attempt to develop a theory of social totality; to see the study of culture as the study of relations between elements in a whole way of life; to find ways of studying structure, in particular works and periods, which could stay in touch with and illuminate particular art works and forms, but also forms and relations of more general social life; to replace the formula of base and superstructure with the more active idea of a field of mutually if also unevenly determining forces. That was the project of *The Long Revolution*, and it seems to me extraordinary, looking back, that I did

---

[158] Raymond Williams, 'Literature and Sociology: in Memory of Lucien Goldmann', *New Left Rev.* No. 67 (May/June 1971), pp. 3–18, at p. 9.

[159] Ibid., p. 10.

not then know the work of Lukács or Goldmann, which would have been highly relevant to it, and especially as they were working within a more conscious tradition and in less radical an isolation. I did not even then know, or had forgotten, Marx's analysis of the theory of utility, in *The German Ideology*, in which—as I now find often happens in reading and re-reading Marx—what I had felt about the reductionism now embodied in the base-superstructure formula was given a very precise historical and analytic focus.

This being so, it is easy to imagine my feelings when I discovered an active and developing Marxist theory, in the work of Lukács and Goldmann, which was exploring many of the same areas with many of the same concepts, but also with others in a quite different range. The fact that I learned simultaneously that it had been denounced as heretical, that it was a return to Left Hegelianism, left-bourgeois idealism, and so on, did not, I am afraid, detain me. If you're not in a church you're not worried about heresies; the only real interest is actual theory and practice.

What both Lukács and Goldmann had to say about reification seemed to me the real advance. For here the dominance of economic activity over all other forms of human activity, the dominance of its values over all other values, was given a precise historical explanation: that this dominance, this deformation, was the specific characteristic of capitalist society, and that in modern organized capitalism this dominance—as indeed one can observe—was increasing, so that this reification, this false objectivity, was more thoroughly penetrating every other kind of life and consciousness. The idea of totality was then a critical weapon against this precise deformation; indeed, against capitalism itself. And yet this was not idealism: an assertion of the primacy of other values. On the contrary, just as the deformation could be understood, at its roots, only by economic analysis, so the attempt to overcome and surpass it lay not in isolated witness or in separated activity but in practical work to find, to assert and to establish more human social ends in more human political means.[160]

And so we come back to Malthus again and to the development of nineteenth century naturalism, embracing first the history of the earth and then of life, mind and society in a reifying perspective of unvarying and inescapable natural law. One must transcend the prevailing historiography of science in order to understand the history of science as the history of reification: 'the process through which relations between men take on the appearance of relations between things; human society and human history, the products of man, appear not as the products of social activity but as alien and impersonal forces, laws of nature which impose themselves on humanity from without'.[161] The crucial episode in the history of reification occurred in the

[160] Ibid., pp. 10–11.

[161] Gareth Stedman Jones, 'The Marxism of the Early Lukács: an Evaluation', *New Left Rev.*, No. 70 (November/December 1971), pp. 27–64, at p. 28; cf. Peter Berger and S. Pullberg, 'Reification and the Sociological Critique of Consciousness', ibid., No. 35 (January/February 1966), pp. 56–71. The classical Neo-marxist discussion is G. Lukács, 'Reification and the Consciousness of the Proletariat', in *History and Class Consciousness*, op. cit. (note 152), pp. 83–222.

Victorian debate on man's place in nature and laid the foundations of the modern extension of the process to all aspects of thought and life. It was the nineteenth-century debate which led to the conclusion that morality and social theory could be—must be—natural sciences and that poverty and inequality and the hierarchical division of labour are natural, not political, phenomena.[162]

Of course, man is both a natural and a political being, but the reduction of the latter to the former converts the general principle of scientific naturalism into an instrument for domination which elicits deference and resignation.[163] A crucial moment in the reversal of this process is the overthrow of the views of science and its history which have led to the mystification known as reification. If we are to understand why men defer to experts about how they can and should live and to biology for the limits of human nature and society, we must understand the historical process which produced this set of abstractions and led them to become commonsense and to replace outrage, sapping men's faith in the ability to transform society for the benefit of all. The road from Adam Smith and Malthus to the present and to the critiques by radicals from Marx and Engels to Marcuse's *One Dimensional Man* is an unbroken one, and the prerequisite for a radical, critical view of science and its history is the recovery of outrage at this colossal confidence trick.[164] The advancing edge of objectivity must be replaced by a revival of radical consciousness which is developed concomitantly with the growth of radical will and action.

[162] R. M. Young, 'Darwinism and the Division of Labour', *The Listener* No. 2264 (17 Aug. 1972), pp. 202–5.

[163] See above, note 123. Marcuse argues that in the course of the development of modern science, 'The "nature of things", including that of society, was so defined as to justify repression and even suppression as perfectly rational'. 'Glorification of the natural is part of the ideology which protects an unnatural society in its struggle against liberation.' (*One Dimensional Man*, op. cit., note 123, Sphere edition, pp. 122, 187). Schroyer develops this thesis: 'Contemporary science and technology have become a new form of legitimating power and privilege. . . . We argue that the scientistic image of science has become the dominant legitimating system of advanced industrial society. . . . It is our thesis that the scientistic image of science is the fundamental false consciousness of our epoch.' ('Toward a Critical Theory for Advanced Industrial Society', op. cit. (note 123), pp. 210, 212, 213).

[164] 'Whereas Marx was able to formulate his critical theory as a critique of the purest ideological expression of equivalence exchange, i.e. classical political economy, we are forced to broaden our critique to the positivistic theory of science itself.' (Ibid., p. 213). For a useful marxist history, see Leszek Kolakowski, *Positivist Philosophy from Hume to the Vienna Circle* (1966), revised edition, trans. Norbert Guterman (Harmondsworth, Penguin, 1972), especially chs. 2–4 *re* the nineteenth century and ch. 8 *re* the political rôle of current positivism. See also his anti-Stalinist, though marxist, essays: *Marxism and Beyond: On Historical Understanding and Individual Responsibility*, trans. Jane Z. Peel (London, Pall Mall, 1968; also Paladin paperback, 1971).

The scientists and historians who were attracted by vulgar marxism still believed that science was an unequivocally progressive force, subject only to use and abuse and better or worse planning. Better planning, it was thought, was occurring in the Soviet Union.[165] A generation later, many writers of the New Left have also been reluctant to include the ideas and assumptions of science in the superstructural realm which they are criticizing. There is still a tendency to treat science as an exception, while indulging in criticisms of its abuse in the form of scientific and technological rationality. Indeed, it can be argued that some of the marxists who have been most concerned with science—the Althusserians—are indulging in a theoretical perpetuation of Stalinism.[166] Thus, with this positivism—both in its post-1930s form and its current one—has gone deference towards the model for the development of socialism which was allegedly being implemented in the Soviet Union.[167] For the earlier generation, this occurred in the period before there was general awareness of the nature of the purge trials of the

[165] From the point of view of current critiques of the dangers of, and domination by, science and technology, Bernal's *The Social Function of Science* (1938), op. cit. (note 130) is a very optimistic document. His main themes were to argue against any limitation on the development of science (He saw science *in* danger and not *as* danger.) and to identify the progress of science with the progress of society, with science increasingly serving as a model and guide for social improvement. In the last decade, Bernal's vision of science as an unequivocally progressive force has come increasingly to be seen more in terms of Frankenstein's monster. See especially chs. 14–16.

[166] See Kolakowski, 'Althusser's Marxism', op. cit. (note 156), pp. 122 ff. On the notion of marxism as object, see Young 'Evolutionary Biology and Ideology: Then and Now', op. cit. (note 8), p. 197.

[167] 'Characterized in the first place by its assertion of the radical distinction between judgments of fact and judgments of value, between external reality which is subject to "objective" laws and human activity which can at most pass moral judgments on this reality or modify it by means of technical action based on the knowledge and utilization of these objective laws, positivism corresponds to situations where the structures of society are so stable that their existence seems unaffected by the action of men who compose them and experience them. It is true that Marx had exposed at length, in *Capital*, the illusions of the fetishism of commodities, which makes economic and even historical laws appear independent of the will of men and comparable to natural laws. This warning had not, however, been enough to prevent later Marxists from falling victim to the same illusion, in so far as they lived inside a society which was relatively stable and apparently little affected by the transforming action of social classes.

'In the Bolshevik camp a situation which until 1917 was in many ways different never-theless favoured a related ideology, most clearly expressed in Lenin's *What Is To Be Done?* [Because of the low level of political consciousness of the proletariat, it could not lead a revolution]. . . . This meant that this part had to be played by the party, an organization of professional revolutionaries, whose action as the collective engineer of the revolution would implant a socialist consciousness in the working class. The Bolsheviks thus arrived at an equally positivist and objectivizing conception of society. . . .' Lucien Goldmann, 'Reflections on *History and Class Consciousness*', in Mészáros (ed.), *Aspects of History and Class Consciousness*, op. cit. (note 152), pp. 65–84, at pp. 67–8.

late 1930s, and long before many socialists came to feel that Stalinism was the truth of Bolshevism. A society which promises socialism imposed from above, without bringing the consciousness of the people into full account and without allowing theoretical and practical diversity in implementing it —a repressive approach embodied in 'democratic centralism' in the Soviet Union and imperial hegemony wherever it could be exercised—could only sanction 'socialist realism' in the arts and vulgar marxism in the critique of culture.[168] Nigel Harris points out that vulgar marxism was the obvious expression of Stalinism and that it is evident throughout Stalin's theoretical writings:

A 'material base' to society, the economic functioning of society, as almost independent of men. Ideas merely 'reflected' material reality (a point from early Lenin), and political consciousness merely 'reflected' the state of technology— put crudely, gasometers produce poetry via men. The actual lumps of the economic base—steel plants, cranes, factories and so on—seemed to possess a life of their own and to compel society to transform itself in conformity with this life. E. P. Thompson describes the relationship thus: 'Ideas are no longer seen as the medium by which men apprehend the world, reason, argue, debate, and choose; they are like the evil and wholesome smells arising from the imperialist and proletarian cooking pots.'

There is little or no interaction between the base and the rest of society, the 'superstructure', only the base dragging the reluctant superstructure along behind: 'The superstructure is created by the base precisely to serve it, to actively help it to take shape and consolidate it, to actively fight for the elimination of the old moribund base, together with its superstructure.'[169]

So the specific expressions of marxism which came to be known as 'vulgar' are themselves the essence of the very reification which Lukács and Gold-mann—following the early Marx—decried. This formulation was the predominant one which was available to admirers of socialism in action in the 1930s and after. A generation later, some theorists of the New Left are returning to the writings of the early Marx and are arguing that these are

[168] See, for example, Maurice Brinton, *The Bolsheviks and Workers' Control, 1917 to 1921: the State and Counter-Revolution* (London, Solidarity paperback, 1970); E. H. Carr, 'Revolution from Above: The Road to Collectivization', in his collection of essays, *October Revolution: Before and After* (New York, Knopf, 1969), pp. 95–109; Garaudy, *The Turning Point of Socialism*, op. cit. (note 155). For an extremely evocative recreation of the consequences of vulgar marxism and Stalinism in the arts, see Nadezhda Mandelstam, *Hope Against Hope: a Memoir*, trans. Max Hayward (London, Collins & Harvill, 1971), especially ch. 55, which begins, ' "It turns out we are part of the superstructure", M. said to me in 1922. . .' (p. 258).

[169] Harris, *Beliefs in Society*, op. cit. (note 122), p. 154. Harris is here referring to Stalin's famous essay *Concerning Marxism in Linguistics* (Moscow, Foreign Languages, 1950), in which Stalin was *liberalizing* the relationship between ideological control and scientific practice. See Joravsky, *The Lysenko Affair*, op. cit. (note 149), pp. 150 ff.

continuous with his later writings and provide the basis for a richer theory of the relationship between socio-economic factors and intellectual life.[170] They are also free from deference to the Soviet model and its intellectual expressions and are, indeed, attempting to elaborate a theory of culture which is inconsistent with the Stalinist experience. One can remain in sympathy with the struggles of the Soviet people without adhering to the theory of culture which expressed the severest limitations of that path to socialism.

If there is any life remaining in the base-superstructure distinction, it must be seen in the light of a much richer conception of the base, one which recovers the manifold aspects of men's experience as expressed in Marx's early writings. His comprehensive view of man, nature and society is now being dug out from layer after layer of interpretation from the point of view of crude economism and supported by the discovery and exegesis of such fundamental documents as the *Economic and Philosophic Manuscripts of 1844* and the *Grundrisse*.[171]

VI

Before looking further into the potentialities which are held out by New Left and related writings and which can be employed in the development of a radical historiography of science and applied to the nineteenth-century debate on man's place in nature, some attention should be given to the available alternatives to Hall's position[172] in the current bourgeois historiography of science. I shall briefly consider three positions—only briefly, because I agree in this respect with one of the main current leaders in the field, Thomas Kuhn, that passé paradigms are not really refuted but merely left behind when people come to see science in terms of a new framework. The three alternatives are Imre Lakatos' arguments for 'demarcationism' and for the history of science to be conducted as 'rational reconstruction', Thomas Kuhn's theory of 'paradigm shifts' as the process by which science

---

[170] See below, pp. 433–4.

[171] The main new sources for this approach are Karl Marx, *Economic and Philosophic Manuscripts of 1844*, trans. Martin Milligan (Moscow, Foreign Languages, 1961); Marx and Engels, *The German Ideology* (1845–46), trans. Clemens Dutt *et al.* (Moscow, Progress, 1964); Marx, *Pre-Capitalist Economic Formations* (1857–1858), trans. Jack Cohen (London, Lawrence & Wishart, 1964), which contains a portion of the *Grundrisse der Kritik der Politischen Ökonomie*, which David McLellan is in the process of bringing out in English. For introductory analyses, see McLellan, *Marx before Marxism* (London, Macmillan, 1970; revised edition Harmondsworth, Penguin paperback, 1972); *The Thought of Karl Marx: an Introduction* (London, Macmillan, 1971; also paperback).

[172] See above, pp. 356–7.

changes, and Robert Merton's Weberian view of the sociology of science. The main burden of my argument is simple: since all three of these positions depend upon a fundamental distinction between 'internal' and 'external' factors, between science and its context, they must ultimately be transcended. That is not quite all, however, since some insight might be gained from noticing the ideological work which such theories perform in maintaining the central position of scientific rationality at the expense of the possibilities of liberation.

Of the three positions, Lakatos' demarcationism and his call for 'rational reconstruction' is the easiest to deal with. It is the extreme case of interpreting the past in the light of the present, of seeing the history of science in the perspective of current orthodoxies. In his papers[173] and his public performances, Lakatos draws a sharp line between the scientists and the non-scientists. Since all the candidates for the category of 'scientist' turn out to have—as he puts it—'misbehaved' more or less often in their careers, the acceptable list tends to reduce itself to Galileo and Newton, who are easily dispensed with by relativists and contextualists who can point to Galileo's Platonism and deductivism and to Newton's preoccupations with alchemy, Biblical chronology and related aspects of his work which are integral both to his science and to his philosophy of nature. The other side of Lakatos' firm demarcation is normally filled facetiously with, for example, Marx, Hegel, Marcuse, Habermas. When critics ask where such ambigious philosopher-scientists such as Paracelsus, van Helmont, Priestley, Lyell, Darwin, Wallace, etc., or the disciplines of astrology, alchemy, physiognomy, phrenology, mesmerism, spiritualism, and so on are to be placed, confusion sets in. Lakatos' model is supposed to be falsifiable by means of the close study of the writings of scientists, conducted by reputable scholars, but when points are raised which make nonsense of a rigid demarcation between science and other factors at the centre of the scientists' work the scientists are said to have 'misbehaved' or the scholars who make the challenge turn out not to qualify as 'reputable'. At the suggestion that the rigid distinction between fact and value cannot be maintained, he says that at that point he reaches for his machine gun, while at the suggestion that his position can best be described as 'meta-methodological Stalinism', he smiles. In order to make the data of the writings of scientists conform to his model, it is necessary to jettison the fine texture of history and to re-write the theories

[173] Imre Lakatos, 'Falsification and the Methodology of Scientific Research Programmes', in Lakatos and Musgrave (eds.), *Criticism and the Growth of Knowledge*, op. cit. (note 22), pp. 91–196; 'History of Science and Its Rational Reconstructions', in Roger C. Buck and R. S. Cohen (eds.), *Boston Studies in the Philosophy of Science*, Vol. 8 (Dordrecht, Holland, Reidel, 1971), pp. 91–136.

of past scientists as though they knew what we know now or, more accurately, to stress the aspects of their work which can be tidied up in the light
of modern knowledge. Any strong evidence of intermingling cognitive and
social, constitutive and contextual, factors is relegated to the realm of
motivation or of use and abuse of truth. Lakatos grants that these are
interesting subjects, but they are unrelated to the serious study of the
history of scientific rationality itself. Like most approaches to the history
of science which begin with *logical* preoccupations in the *philosophy*
science, his development of the Popperian position proves itself both
distorting and irrelevant to the preoccupations of historians. In short, most
significant questions are begged as a result of appealing to a fundamentally
positivist conception of 'rationality'.

Thomas Kuhn's theory of *The Structure of Scientific Revolutions* must be
taken altogether more seriously, if only because it has been the single most
influential view in the historiography of science for the past decade. While
its catalytic rôle is very important, it is also necessary to take a critical view
of the reactions which it has facilitated. It has undoubtedly had an important
liberating effect on practising scientists (both natural and social) and on
students in the history and philosophy of science. It was not too sanguine to
ask, when his essay first appeared, if it heralded 'A Revolution in the
Historiography of Science'?[174] I shall argue that the intervening decade has
shown that as far as any radical political implications which might have
flowed from his work are concerned, the clear answer is 'No'. However, it
has the significant merit of having freed the self-consciousness both of
scientists and of neophytes in the study of its history and concepts from a
unilinear, cumulative, progressive conception of scientific change. His own
work—as well as that of the critics, interpreters and expositors of it—has
suffered from an important uncertainty over pursuing its sociological or its
philosophical implications. Kuhn brought sociological criteria to bear on
the alternating periods of stability and change within the scientific community and immensely sharpened our sense of the social process of conceptual change. His distinctions between 'normal' and 'revolutionary' periods
in science, and between pre-paradigm and paradigmatic states of scientific
disciplines have proved very attractive, and in some circles the term
'paradigm' has replaced those of 'concept', 'theory', and 'idea'. Kuhn would
be the last to be glad about the loose use of his key theoretical term, but he
has admitted that it is very difficult indeed to specify the units of change
which qualify as paradigm shifts.[175] Even so, scientists who read his book

[174] See above, note 22.
[175] 'Reflections on My Critics', in Lakatos and Musgrave (eds.), *Criticism and the
Growth of Knowledge*, op. cit. (note 22), pp. 231–78, at p. 234.

say that it is the only writing in the history and philosophy of science which comes near to reflecting the way science feels to those who are doing it.

At the philosophical level his theory raised fundamental questions, leading even the attentive reader to find in his writings a strong support for historical relativism. Men who think in different paradigms seem to inhabit different worlds with sharp discontinuities between them. Facts which are explained in one paradigm become less important if unexplained in another. The image of the slow accumulation of anomalies which produce a fundamental reorientation of explanatory priorities—a 'gestalt switch'—highlights the discontinuities in scientific change at the expense of leaving its conceptual—and especially its methodological—continuities unexplained. It is not surprising that the apparently relativistic implications of his theory produced considerable alarm among rigorous, traditional philosophers of science, who have argued that he is in danger of undermining the very basis of rationality in science.[176]

But it turns out that there is little to worry about. Kuhn has been unequivocal and vocal in dissociating himself from the potentially radical implications of his theory in both the sociological and epistemological realms. His comments on the reception of his theory and his 'Reflections on My Critics' have been very reassuring, and even affable.[177] But more important than this, his analysis of the relationship between 'internal' and 'external' factors has reinforced the very distinction which he appeared at one time to be bringing into question. Having produced a lucid analysis of the dilemma,[178] he has gone on to stress the internal factors at the expense of the external ones, thereby failing even to address himself to the issue of their mutual interpenetration and the deeper claim that they are constitutive of each other. On the question of the origins of Darwin's theory, for example, he considers the influence of Malthus to be 'vitally important' and challenges those who do not agree to explain the proliferation of evolutionary theories in the pre-Darwinian era. He goes on to say, however,

> Yet these speculative theories were uniformly anathema to the scientists whom Charles Darwin managed to persuade in the course of making evolutionary theory a standard ingredient of the Western intellectual heritage. What Darwin did, unlike these predecessors, was to show how evolutionary concepts should be applied to a mass of observational materials which had accumulated only

---

[176] Israel Scheffler, *Science and Subjectivity* (New York, Bobbs-Merrill paperback, 1967), especially pp. 15–19 and ch. 4.

[177] Op. cit. (note 175) Kuhn is particularly keen to dissociate himself from Paul Feyerabend's conceptual anarchism (pp. 234–5) and to rebut charges of 'irrationality', 'mob rule', and 'relativism' (pp. 259–66).

[178] Thomas S. Kuhn, 'History of Science', in David L. Sills (ed.), *International Encyclopedia of the Social Sciences* (New York, Macmillan, 1968), Vol. 14, pp. 74–83.

during the first half of the nineteenth century and were, quite independently of evolutionary ideas, already making trouble for several recognised scientific specialties. This part of the Darwin story, without which the whole cannot be understood, demands analysis of the changing state, during the decades before the *Origin of Species*, of fields like stratigraphy and paleontology, the geographical study of plant and animal distribution, and the increasing success of classificatory systems which substituted morphological resemblances for Linnaeus' parallel-isms of function. The men who, in developing natural systems of classification, first spoke of tendrils as 'aborted' leaves or who accounted for the differing number of ovaries in closely related plant species by referring to the 'adherence' in one species of organs separate in the other were not evolutionists by any means. But without their work, Darwin's *Origin* could not have achieved either its final form or its impact on the scientific and the lay public.[179]

He deplores the absence of a literature 'which has attempted to explain the emergence of Darwinism as a response to the development of *scientific* ideas or techniques' and finds irony in an analysis of the relations between Malthus and the evolutionists which does not 'make any attempt to deal with the technical issues which may have helped to shape Darwin's thought'. The irony is said to lie in the failure to make that attempt in a treatment which seeks to break down the barrier between internal and external factors by stressing the crucial rôle of Malthus' theory in the formulation of the mechanism for evolutionary change independently discovered by Darwin and Wallace.[180]

We are here involved in many layers of irony. At the simple level, to some of Darwin's contemporaries—for example Wallace, Owen, and Lewes—the so-called 'speculative' theories of evolution were far from 'uniformly anathema'. Wallace was inspired by Chambers' theory to seek a mechanism, Owen was attracted by it (but unconvinced by Darwin), and Lewes defended both. Similarly, Lyell's attempted refutation of Lamarck's theory was decisive in leading Spencer to accept and expound a version of Lamarckianism, and Spencer's evolutionary theory was accepted whole-heartedly by Hughlings Jackson, whose influence was fundamental in the development of evolutionary neurology and neurophysiology. Such examples can be greatly multiplied in support of the conclusion that it is difficult to maintain a sharp distinction between the nature and influence of the Darwin-Wallace theory and of those of the so-called speculative evolutionists.[181]

It is certainly true, as Kuhn argues, that geological, paleontological and

[179] Thomas S. Kuhn, 'The Relations between History and History of Science', *Daedalus* **100** (1971), pp. 271–304, at pp. 281–2.

[180] Ibid., pp. 301–2. Kuhn is commenting on Young, 'Malthus and the Evolutionists', op. cit. (note 10).

[181] The evidence for these assertions is spelled out in the monograph and series of papers cited in notes 9 and 10.

zoogeographical data were central to the work of both Darwin and Wallace and especially that Darwin's *Notebooks* and his published writings contain and explain 'large classes of facts'. Kuhn is also obviously right to say that without consideration of these data, the whole of the origin, development and reception of their theories cannot be understood. But the task does not merely involve a judicious mixture of so-called 'internal' and 'external' factors. The factors are indeed mixed, and Kuhn's position is an advance on the unrelenting internalism of Smith and de Beer.[182] The very sort of close analysis of the relevant empirical findings in their intricate relations with Darwin's speculations and the theories which he considered in the course of his reflections, has been undertaken by Howard Gruber in his conceptual biography of Darwin's discovery.[183] In addition, a study of the crucial rôle of Malthus' theory helps to redress the balance of historical analysis, but even a balanced view cannot overcome the deeper irony in Kuhn's approach.

His theory is based on the claim that it is a build-up of anomalous empirical findings which leads finally to a paradigm shift. That is, the accumulation of internal findings, in the social context of the internal logic of the scientific community, is the decisive factor. He relegates 'the ambient intellectual milieu' to 'the rudimentary stages of the development of a field' and, after that, confines its influence 'to the concrete technical problems with which the practitioners of that field engage'.[184] This approach, of course, misses out both the general theoretical level in a given science and, more importantly, the central rôle of socio-economic assumptions in the philosophies of nature, man and society which underlie and constrain, and in some cases determine, the content of scientific theories and even particular facts. As I have argued above, these pervasive assumptions are more obvious—i.e., less mediated— in the biological and human sciences. It is not surprising that Kuhn's background in the physical sciences has led him to develop a general theory which is hardly relevant to the features of the non-physico-chemical sciences which are most interesting to historians and which lead most directly to an ideological analysis.

Looking briefly at Kuhn's view of the influence of science on the general culture, his claims are more extreme and even astonishing: 'Science, when it affects socio-economic development at all, does so through technology.'[185] He goes on to argue that science and technology should, as a first approxi-

---

[182] See above, pp. 362–4.

[183] Howard E. Gruber, *Darwin on Man: A Psychological Study of Scientific Creativity, together with Darwin's Early and Unpublished Notebooks*, transcribed and edited by Paul H. Barrett (New York, Dutton, in press).

[184] Kuhn, 'The Relations between History and History of Science', op. cit. (note 179), p. 280.

[185] Ibid., p. 283.

mation, be treated 'as radically distinct enterprises'.[186] A consistent application of his approach in this aspect of the subject would preclude an appreciation of the rôle of science and scientific rationality in shaping the assumptions
of modern capitalistic society, extending even to its concept of rationality.
In the particular case of the nineteenth-century debate it would lead one to
declare irrelevant the whole structure of social, political, economic and
related theories which was erected on the basis of extrapolations from 'and
rationalizations of' evolutionary theory.

When the consequences of the internalist-externalist dichotomy and the
ultimate failure of Kuhnian historiography to transcend it finally become
clear, it also becomes evident that the excitement caused by his approach is,
in the end, a mystification. In the course of finding his insights liberating
because of their introduction of social factors in the process of conceptual
change, it was not noticed that he thereby excluded socio-economic factors
from the *substance* of science, and he precluded the fundamental task of
analysing the levels of the relevance of the assumptions on which they were
based.

In some respects, Kuhn's critique lends support to a contextualist approach,
but in the end he repudiates the relativist—and certainly the ideological—
affiliations which his admirers have offered. He has provided us with a sort of
internalist's contextualism concerned with a social milieu, but it is the social
context of scientists in the society of science, not in the world.[187] It is for this
reason that the similarities between Kuhn, Gillispie, Popper and Lakats are
greater than those between Kuhn and others whose research can be described
as full-bloodedly (many would still add sanguinely) contextualist and relativist. He appears to be more of an ally than he is, because he undermines
belief in one continuous rational tradition in science, but he has shown
himself to be very unwilling to allow radical implications (either philosophical
or political) to be drawn from his position. Indeed, he is not unnaturally
rather like Lakatos in expecting the history of science to conform to his
model, and hires colleagues and directs research accordingly. It should be
obvious from the argument of this paper that all of us do this more or less
self-consciously. The politically partisan nature of these activities and their
parent ideological positions must, however, be made explicit. Then we can
get on with determining whose view of science and its history is *just* and
liberating rather than whose is 'rational'. Debates about justice are frankly
concerned with conflicts of values, while those about rationality appeal to
the very concept of objectivity which is in question. (Of course, the issue

---

[186] Ibid., p. 285.
[187] See S. B. Barnes, 'Sociological Explanation and Natural Science: a Post-Kuhnian
Appraisal', *Sociological Rev. Monographs* (in press).

will not be decided in a debating chamber.[188]) Kuhn might forcefully reply that we see the world according to different paradigms of science. The rebuttal to that is that we see science through different paradigms of the *world* as it is, as it ought to be, and might be if men were freed from the mystifications of scientific rationality.

Kuhn's work is representative of the highest standards in the prevailing orthodoxy in the history of science. It is also seductive, in that it takes us to the limits of the current orthodoxy and appears to go beyond them. His own reactions to those who have attempted to step beyond the established tradition help to show that he ultimately draws back.

Robert K. Merton is the doyen of the sociology of science. His pioneering study of *Science, Technology and Society in Seventeenth-Century England* was his doctoral dissertation, was first published in 1938 as an entire issue of *Osiris*, and has recently reappeared with a new introduction in a paperbound edition.[189] His volume of essays on *Social Theory and Social Structure*—including seven chapters on the sociologies of science and of knowledge—went through three editions and nineteen printings between 1949 and 1968.[190] When Merton took up the subject of the sociology of science, it was a neglected area. Although it remains so, interest in it is growing (largely for political and ecological reasons), and most subsequent work has taken Merton's writings as a starting point for development or criticism. His is therefore the central position for assessing the sociological analysis of science. He sees his own work in the tradition of the sociology of knowledge, an approach which studies the relations among the social origins, the substance and the rôle of ideas.[191] The originator of modern analyses of the sociology of knowledge was Karl Mannheim, whose *Ideology and Utopia*

[188] See above, note 123; cf. [R. M. Young], 'Science *versus* Democracy', *Science or Society?* No. 1 (May 1971, 5 Salisbury Villas, Station Road, Cambridge), pp. 2–3; *anon.* 'Self-Management', *Bulletin of the British Society for Social Responsibility in Science* No. 15, (January/February 1972, 70 Great Russell Street, London), pp. 10–12; P. Chaulieu, 'Workers' Councils' and the Economics of a Self-Managed Society', *Solidarity Pamphlet* No. 40 (March 1972, 27 Sandringham Road, London).

[189] Robert K. Merton, *Science, Technology and Society in Seventeenth-Century England* (1938; reprinted with a new Preface and bibliography, New York: Harper paperback, 1970); cf. the review by P. M. Rattansi, 'Science and the Glory of God', *New York Review of Books* (6 May 1971), pp. 34–5 and the debate on the Merton hypothesis which was conducted for several years in *Past & Present*, beginning with H. F. Kearney, 'Puritanism, Capitalism and the Scientific Revolution', No. 28 (1964), pp. 81–101 and continuing at least until B. J. Shapiro, 'Latitudinarianism and Science in Seventeeth-Century England', ibid., No. 40 (1968), pp. 16–41; most of the relevant references are listed in Merton's 1970 bibliography (above); cf. S. F. Mason, 'Science and Religion in 17th Century England', *Past & Present* No. 3 (1953), pp. 28–44.

[190] Merton, *Social Theory and Social Structure*, op. cit. (note 138), chs. 14, 15, 17–21.

[191] Ibid., p. 585.

itself reflects a bourgeois version of Lukács' argument in *History and Class Consciousness*. The other major influence on Mannheim was Max Weber, whose *The Protestant Ethic and the Spirit of Capitalism* provides the conceptual framework for Merton's approach. Weber has been called 'the bourgeois Marx', Mannheim 'the bourgeois Lukács'.[192] Merton's intellectual debts extend further, to the Harvard 'Pareto Circle', led by Lawrence J. Henderson. This was a group explicitly concerned with providing an alternative to the growing radicalism in Cambridge, Massachusetts in the years of the Great Depression.[193] In the light of this intellectual parentage, it would be surprising if Merton's approach proved to be of much use to a radical historiography, and there are, it turns out, no surprises in store.

The criticisms to be made about Merton's approach from the point of view of the requirements of a radical historiography of science are relatively straightforward and add up to the conclusion that he never confronts the *substance* of scientific findings, theories, or assumptions about nature, man and society. He is far in advance of the traditional internalist history of ideas in addressing himself to the social and ideological context and basis of the ethos of science, but his approach to these issues is expository and analytic and is entirely lacking a critical dimension. His analyses are concerned with the rôle, the tempo and the choice of research problems of science and scientists; externalist criteria are applied to these. Yet despite his commitment to treating the sociology of science as a branch of the sociology of knowledge, he nowhere confronts the very relationships which define the domain of investigation of the sociology of knowledge. He is concerned with science on the one hand and its context on the other, and that very distinction precludes serious analysis, even according to the limited goals of the sociology of knowledge. He is aware of the central issue in the Mannheimian approach—whether or not there is any knowledge which is 'objective' or value-neutral—but he skirts it and its relativistic implications. Instead of addressing it directly, he mentions it as a fear on the part of those who oppose the sociological approach to science.[194] In the light of these features of his work, it is not surprising that he is dismissive of the marxist approach. Even in expressing sympathy for the general investigation of 'the interplay between socio-economic and scientific development' and in

[192] George Lichtheim, *The Concept of Ideology and Other Essays* (New York, Vintage paperback, 1967), pp. 3–46, at p. 35; op. cit. (notes 8, 112, 152).

[193] Barbara S. Heyl, 'The Harvard "Pareto Circle"', *J. Hist. Behav. Sci.* 4 (1968), pp. 316–34; [R. M. Young], 'Mystification in the "Scientific" Foundations of Sociology', *Science or Society?* No. 2 (June 1971), pp. 9 –11(see above, note 188); Russett, *The Concept of Equilibrium,* op. cit. (note 92); L. J. Henderson, *On the Social System: Selected Writings,* edited with an Introduction by Bernard Barber (Chicago, 1970; also paperback).

[194] Merton, *Social Theory and Social Structure,* op. cit. (note 138), p. 586.

pointing out the limits of the vulgar marxist line, he simply sidesteps the central problem and concentrates on the 'structural determinants of scientists' behaviour'.[195]

A few examples will illuminate his characteristic avoidance of the substance of science. In his work on the seventeenth century, he says, 'Much of our study will, in fact, be devoted to the isolation of some of the extra-scientific elements which strongly influenced, if they did not determine, the centering of scientific attention upon certain fields of investigation.'[196] There is no analysis of the results of tilling those fields, only of what religious, social and economic motives led men to do so. Elsewhere he mentions characteristic preoccupations such as the influence of the Protestant ethic on 'the attitudes of scientists toward their work. Discussions of the why and wherefore of science bore a point-to-point correlation with the Puritan teachings on the same subject.'[197] Similarly, for the virtuosi, 'science found its rationale in the end of all existence: glorification of God.'[198] He discusses 'a set of largely implicit assumptions which made for the ready acceptance of the scientific temper characteristic of the seventeenth and subsequent centuries.'[199] He is particularly interested in the social and theological roots of the value-free orientation of science—the scientific ethos—and their influence on the motives and personal attitudes of natural philosophers in the period, as well as their modern expressions. Finally, he directs attention to the socio-economic factors which facilitate and obstruct scientific activities in general and as directed toward individual problems.[200]

I want to make three points about Merton's work. First, as I have said, neither his interests nor his methods lead him into the substance of science: he need never read through and analyse the argument of a given scientific paper or treatise. He can skim it in order to determine its topic rather than concern himself with what is said about the topic. He seeks to identify the

[195] Ibid., pp. 661–4. In his 1938 essay (published a year earlier) Merton was less critical of vulgar marxism. He wrote, 'In the discussion of the technical and scientific problems raised by certain economic developments, I follow closely the technical analysis of Professor B. HESSEN in his provocative essay, "The Social and Economic Roots of NEWTON'S 'Principia'," . . . [op. cit., notes 125 and 5] Professor HESSEN'S procedure, if carefully checked, provides a very useful basis for determining empirically the relations between economic and scientific development. These relations are probably different in an other than capitalistic economy since the rationalization which permeates capitalism stimulates the development of scientific technology.' He elsewhere says, 'The following discussion is heavily indebted to HESSEN. . . .' (*Science, Technology and Society,* op. cit., note 189, pp. 142n, 185n; cf. pp. 206–7 and *Social Theory and Social Structure,* pp. 661, 664).
[196] Merton, *Science, Technology and Society,* p. 54.
[197] Merton, *Social Theory and Social Structure,* pp. 629–30.
[198] Ibid., pp. 634–5; cf. pp. 628–9.
[199] Ibid., p. 635.
[200] Ibid., chs. 17, 18, 20, 21.

religious affiliations of its practitioners, to determine the subject matter of papers and treatises, and to read personal and reflective works which link science—as an activity directed to particular topics—to the Puritan ethos. The internal-external dichotomy can never be transcended, since the substance of science is never considered in detail.

Second, there is no critical approach to the socio-economic, theological and ideological factors which are being related to science. These are taken as given, and there is therefore no evaluative dimension to the inquiry. Since the objectivity of scientific findings and theories is also taken as given, the exercise reduces itself to a desiccated correlation. Since a critical approach to the ideological dimensions and their relations to the substance of scientific findings and theories is the *sine qua non* of a radical historiography, Merton can provide no aid in that endeavour, except as a case study in mystification. Little purpose would be served by dwelling on this point, since excellent critiques are available of the parent tradition of Weberian analysis in Herbert Marcuse's essay on 'Industrialisation and Capitalism in the Work of Max Weber'[201] and M. D. King's devastating criticism of Merton's (and to a lesser extent Kuhn's) positivist approach and his assumptions, in a penetrating essay on 'Reason, Tradition, and the Progressiveness of Science'.[202]

The third point to be made about Merton's work has a reflexive quality which brings us back to our original problem. That is, his assumptions are functionalist. Indeed, he is probably the most sophisticated expositor of the functionalist approach in the social sciences.[203] In a separate study, I have attempted to provide a critical outline of functionalism, the relevant aspect of which for present purposes is the profoundly reactionary effect of attempting to base sociological assumptions on the model of evolutionary biology, since the concepts of adaptation, equilibrium and survival of the

---

[201] In Marcuse, *Negations: Essays in Critical Theory*, trans. Jeremy J. Shapiro (Boston, Beacon, 1968; also paperback), pp. 201–26; cf. Norman Birnbaum, 'Conflicting Interpretations of the Rise of Capitalism: Marx and Weber', *Brit. J. Sociol.* 4 (1953), 125–41; Alvin W. Gouldner, *The Coming Crisis in Western Sociology* (London, Heinemann, 1971; also paperback), pp. 179–80.

[202] *History and Theory* 10 (1971), pp. 3–32; cf. King, 'Science and the Professional Dilemma', in Julius Gould (ed.), *Penguin Social Science Survey* (Harmondsworth, Penguin paperback, 1968), pp. 34–73. There is a useful criticism of Merton's adaptation to American functionalist norms in J. G. Crowther, *Science in Modern Society* (London, Cresset, 1967), ch. 47. See also S. B. Barnes and R. G. A. Dolby, 'The Scientific Ethos: a Deviant Viewpoint', *Archiv. Europ. Sociol.* 11 (1970), pp. 3–25.

[203] See Merton, *Social Theory and Social Structure*, op. cit. (note 138), Part I, 'On Theoretical Sociology' (Reprinted as a Free Press paperback, 1967). See also N. J. Demerath III and R. A. Peterson (eds.), *System, Change, and Conflict: a Reader on Contemporary Sociological Theory and the Debate Over Functionalism* (London, Collier-Macmillan, 1967).

system are taken as given and value-neutral.[204] In the context of the problem of studying the nineteenth-century debate on man's place in nature, it is particularly ironic that one of the main reconciling rationalisations which was erected *on the basis of* evolutionary theory could be used as an approach for the critical study of the origins and ideological rôle *of* evolutionary theory. The use of an approach which assumes the validity of biological analogies to evaluate critically the rôle of biological analogies seems rather unpromising.

Once again, let us recall the distance between the position of Lakatos, Kuhn and Merton on the one hand and the orthodoxy in the historiography of science reflected in Hall's article, 'Merton Revisited', on the other. In order to gain a clear perspective on the work of each of them, it would be necessary to examine in detail the meaning of their being children of their times according to the theory of mediation which is being called for here. In the case of Merton, we have a number of analyses of the intellectual tradition of functionalism and its particular social basis at Harvard in the 1930s from the research of Barbara Heyl.[205] This would have to be complemented by a detailed knowledge of Merton's personal history. Such studies would play a useful rôle in the development of a radical historiography, but they do not accurately reflect the gap between that and the current consciousness of the profession of the history of science. For the great majority of those doing research in the 1960s, the current view is still expressed by Hall's analysis. How do he and the positions he defends reflect *their* times?

The decade which Hall reviews was one of avoidance of overt political agitation. It was the period which lay between the era of witch-hunting of Communists (the Army-McCarthy hearings were in 1953) and the emergence of the student movement (the Berkeley protests and the Gulf of Tonkin Resolution occurred in 1964). This atmosphere was reflected, although less intensely, in other NATO countries. In the relevant period, Hall worked in Cambridge, Indiana and London. That period was also the era of the conservatives' claim that we had reached 'the end of ideology'.[206] Hall's thesis not only fails to find promise in Merton's approach, but in rejecting it, he equates it with externalism. He is thereby unable to make a clear distinction between idealist explanations in the light of the Puritan ethos and other arguments about the intellectual superstructure which relate it directly to economic and social factors. That is, in the period in which no-one was

[204] In a book tentatively entitled *Ideology and the Human Sciences*, to be published by Longmans and Doubleday; cf. above (note 149).

[205] Op. cit., note 193.

[206] I have discussed this issue in 'Evolutionary Biology and Ideology: Then and Now', op. cit. (note 8), pp. 182, 198–201; see also below (note 215).

confronting marxism, Hall was allowing the Weberian-Mertonian tradition to stand for all sorts of socio-economic interpretations. For example, his analysis of Merton's work is interrupted by comments on the marxist historian S. Lilley, without any indication that there is a fundamental distinction between their approaches: Lilley asks a question which would not occur to Weber or Merton—'What is Puritanism a mediation *of*?'[207]

Hall's fastidious reaction to the assumptions of Merton's thesis clearly shows the strong disinclination which the advocates of the internal history of scientific ideas feel toward approaching science in even contextualist terms. They would presumably be even less inclined to welcome a step beyond contextualism—from the ethos and activities of science and scientists to their substantive findings and theories. If Weber was a bourgeois Marx, and Merton a functionalization and Americanization of that position, and if Hall finds *this* too political and social, how can one expect any sympathetic understanding of marxism itself? However, it would be wrong to conclude that Hall's position is merely one of false consciousness.[208] On the contrary, he is candid about the basis of internalism as seen from his own point of view:

> One issue between the externalist and the internalist interpretation is this: was the beginning of modern science the outstanding feature of early modern civilisation, or must it yield in importance to others, such as the Reformation or the development of capitalism? Before 1940 most general historians and many historians of science would have adopted the latter position; since 1940 nearly all historians have adopted the former one. Why this change should have come about is not hard to imagine.
>
> By this I do not mean to suggest that the problems raised by the sociologists of science are obsolete; on the contrary, as some scientists like J. D. Bernal have been saying for a long time and many more are saying now, they are immensely real and direct at this moment. Consequently the historical evolution of this situation is of historical significance too, and I believe we shall return to its consideration when a certain revulsion from the treatment of scientists as puppets has been overcome, when (if ever) we are less guiltily involved in the situation ourselves so that we can review it without passion, and when a fresh approach has been worked out. This will not, I imagine, take the form so much of a fusion between two opposite positions in the manner of the Hegelian dialectic, as the demarcation of their respective fields of application with some degree of accuracy. There may also develop a socio-techno-economic historiography whose study will be the gradual transformation of society by science and not (as too often in the past) the rapid transformation of science by society. All this will require a fine analysis, a scrupulous drawing of distinctions and a

---

[207] Hall, 'Merton Revisited', op. cit. (note 32), p. 6.
[208] See above (note 122) and Lichtheim, *The Concept of Ideology*, op. cit. (note 192), pp. 15, 18–22, 41, 43–6.

careful avoidance (except under strict controls) of evidence drawn from subjective, propagandist and programmatic sources. A true sociology of science will deal with what actually happened and could happen, not with what men thought might happen or should happen.[209]

It would be difficult to imagine a better expression of the position of a humane, liberal scholar, speaking with complete sincerity. Hall suggests a socio-political basis for the trend towards internalist historiography of ideas: the abuse of science and scientists since the early days of World War II and the continuation of this in the atomic arms race. Remove these abuses, and the students of science will be able to return to distinguishing clearly and carefully the sphere of science from that of society in a dispassionate, scrupulous, objective way. All trace of subjective, political or ideological bias must be kept out (except under strict controls). Lastly, a true sociology of science should be concerned with the facts, not with the aims or purposes of men.[210]

Hall's explanation should be contrasted on the one hand with J. G. Crowther's review of it and on the other with the criteria of a radical historiography. Crowther devotes two short chapters to it: 'A New Scholasticism' and 'External and Internal Influences on Research'.[211] Once again, we should recall the epigraph: Crowther is a child of the same times that inspired Needham and Bernal, and one would therefore expect his interpretation to err on the side of direct correlation between socio-political and intellectual developments. It should not be thought, however, that the earlier marxist historiography was simply mistaken: it was merely woefully incomplete. Crowther does not appeal to liberal revulsion but to political and economic self-interest on the part of historians of science:

The social relations of science, which have such an intense bearing on contemporary politics, are liable to become controversial and disturbing. It is easy to understand the attraction of other aspects of the history of science,

---

[209] Hall, 'Merton Revisited', op. cit. (note 32), pp. 14–15.

[210] The potential fruitfulness of reconciliation and further development of the liberal and functionalist perspectives of Hall and Merton are canvassed by Arnold W. Thackray in an admirably clear review of many of the issues considered in the present essay: 'Science: Has Its Present Past a Future?', in Roger H. Stuewer (ed.), *Historical and Philosophical Perspectives of Science* (Minneapolis, Minnesota, 1970), pp. 112–27. Thackray argues for a '*via media* which avoids the extreme formulations' of either pure history of ideas or of vulgar marxism (p. 124). His position is eclectic (p. 126). He has begun to apply his approach in 'The Industrial Revolution and the Image of Science', in Everett Mendelsohn and A. W. Thackray (eds.), *Science and Values: Patterns of Tradition and Change* (in press). The conference at which this paper was read showed clearly that Mertonianism is alive and well.

[211] Crowther, *Science in Modern Society*, op. cit. (note 202), chs. 45–46.

which are less liable to invoke controversy, and offer the promise of a long period of undisturbed study.[212]

Nor are the foundations which disburse research funds

disposed to support researches which might attract public attention in a period of political excitement. They do not favour researches which might throw an unfavourable light on how the capital, with which the foundations are endowed, was originally accumulated. In recent times, this has often been the result of the exploitation of science and invention.

Hence there is a tendency to support researches on non-controversial, non-social and non-political subjects, and the young historian of science is under pressure to choose subjects which appear to have as little relevance as possible to contemporary events, and engage in the solitary explorations of scholasticism.[213]

He explains the small volume of work on the social relations of science since 1940 and more particularly since 1953, as

the result of a long-range natural protective action, by dominant interests that do not wish to have the social and political implications of their scientific policy comprehensively investigated.

They prefer that historians of science should withdraw into the socially dis-embodied history of scientific ideas. This would tend to establish the notion that science exists without any obligations to society.

So, since 1940, the historians of science have given less and less aid to the solution of the problem of science in its relation to society. The effect of this is to strengthen the traditional conservative theory of the dominant interests, and widen and harden the ancient fissure between the intellectual and the social life. A ruling conservative ideology is left more firmly than ever in control of the new scientific powers.[214]

It is a typical vulgar marxist exaggeration to attribute self-conscious policies and intentions to the ruling élite and to young historians of science. I cannot speak for the grant-giving foundations, but as one who began research in the period, I can only say that there was in us an unanalysed sense that communism was bad but no conscious juxtaposition of marxism with the history of science. We were not keeping our heads down: it simply never occurred to us to select socially relevant problems or to approach them from a Mertonian, much less a marxist, perspective. False consciousness is not self-consciousness: it is highly mediated by the social and intellectual context. But, as we have noted, changed men with changed consciousness are the product of changed times which react upon the interpretation of science in society. And that brings us back to the New Left.

[212] Ibid., p. 288.
[213] Ibid., pp. 289–90.
[214] Ibid., pp. 290–1; cf. pp. xiv–xvii.

VII

It would be gratifying if—having attempted to work our way through the mystifications of the internal history of ideas, vulgar marxism, and the historiographic approaches of Lakatos, Kuhn and Merton—we could turn with confidence to the writings of current marxist historians for guidance. Instead, we find that many marxists are as deferential to science as their bourgeois antagonists, and it becomes necessary to pick one's way with great care through a whole body of literature for suggestions and hints. The New Left has been very successful indeed in bringing about the end of 'the end of ideology' in intellectual life—or, rather, in unmasking it.[215] The marxist views of intellectual production, the role of the intelligentsia and of the universities have, in varying degrees, been revived by New Left writers. They have also produced a critique which grants a potentially significant revolutionary rôle to students and intellectuals and have pointed out their long-term interests in making common cause with the workers against the ruling élites. But they stop short at the door of the scientific laboratory.[216]

An excellent example of this comes from an acrimonious debate between two socialist historians, Perry Anderson and E. P. Thompson. Anderson is the editor of the *New Left Review*, the leading intellectual periodical of the New Left. Although many would argue that its policy leaves a great deal to be desired in the realm of political agitation and praxis (especially with respect to the relations between the rôle of intellectuals and the struggles of other workers), its rôle in consciousness-raising has been considerable. Partially in collaboration with Tom Nairn, Anderson wrote a series of

[215] See above (note 206); S. W. Rousseas and J. Farganis, 'American Politics and the End of Ideology', in Horowitz (ed.), *The New Sociology*, op. cit. (note 151), pp. 268–89; Christopher Lasch, 'The Cultural Cold War: a Short History of the Congress of Cultural Freedom' in his collection of essays, *The Agony of the American Left* (New York, Random House, 1968; also Vintage paperback), ch. 3; Donald C. Hodges, 'The End of "The End of Ideology" ', *Amer. J. Econ. & Sociol.* **26** (1967), pp. 135–46; and (wait for it) Alasdair MacIntyre, 'The End of Ideology and the End of the End of Ideology', in his collection of essays, *Against the Self-Images of the Age* (London, Duckworth, 1971), ch. 1. For a more general, albeit liberal, discussion of the rôle of intellectuals, see Philip Rieff (ed.), *On Intellectuals) Theoretical Studies. Case Studies* (New York, Doubleday, 1969; also Anchor paperback), especially J. P. Nettl, 'Ideas, Intellectuals, and Structures of Dissent', Anchor edition, pp. 57–134.

[216] In Britain there is a growing movement centering around BSSRS (see above, note 188). The American radical movement of Scientists and Engineers for Social and Political Action (SESPA) publishes the bi-monthly, *Science for the People* (9 Walden Street, Jamaica Plain, Mass. 02130). These activities at the level of agitation must be complemented by a recognition of the unity of science and its metaphysics with politics; see Marcuse, *One Dimensional Man*, op. cit. (note 123), ch. 9, especially pp. 181–6.

articles which covered a wide range of issues which had—and have—to be
faced by the British Left. Edward Thompson is an ex-member of the British
Communist Party (post-Hungary) and the author of a distinguished and
massive study of *The Making of the English Working Class* (1963), a model of
marxist analysis of the fine texture of a historical development.[217] It was
from the depths of his detailed scholarship that Thompson made a highly
polemical critique of Anderson and Nairn's theoretical endeavours, especially
their sweeping generalizations.[218] It is not central to my present purpose to
attempt to adjudicate their respective positions, although the issues which
they were debating are basic ones which must be faced in any serious marxist
analysis.[219] Rather, I want to focus on their treatment of my own topic.

The first point is that Anderson and Nairn simply do not mention Darwin
in their critique of the Victorian bourgeoisie in the nineteenth century.[220]
In this they have much in common with other intellectual, political and
social historians of the period. But in pointing out this omission, Thompson
gives an assessment of Darwin and evolutionism which is simplistic in the
extreme and is difficult to credit to such a meticulous and incisive scholar.
Darwin is represented as the pure empirical scientist, the quintessential
example of the best in the British tradition. Thompson notices the basis of
the facts which Darwin considers in the 'culture of agrarian capitalists, who
had spent decades in empirical horticulture and stock-breeding'.[221] It is true
that Darwin relied very heavily on data from the world of domestic stock-
breeding, yet Thompson's account utterly fails to notice Darwin's place in
the intellectual tradition of ideological debate on the philosophies of nature,
man and society. At this deeper level it is absurd to suggest that Darwin was
a pure scientist and to contrast him sharply with Huxley, who is represented
as a pure ideologue.[222]

Marx and Engels saw Darwin's rôle and context very clearly; how could
Thompson have forgotten it? When he turns to the evolutionary debate, he
reveals almost total ignorance:

[217] Op. cit. (note 65); cf. his masterly integration of data from varying levels of analysis,
'The Moral Economy of the English Crowd in the Eighteenth Century', *Past & Present*
No. 50 (1971), pp. 76–136.
[218] E. P. Thompson, 'The Peculiarities of the English', in Ralph Miliband and J.
Saville (eds.), *The Socialist Register, 1965* (London, Merlin paperback, 1965), pp. 311–62.
The articles of Anderson and Nairn are cited in Thompson's footnotes.
[219] The debate ended with a withering critique of the respective positions, and terms
of reference of Thompson and Anderson: Nicos Poulantzas, 'Marxist Political Theory in
Great Britain', *New Left Rev.* 43 (May/June 1967), pp. 57–74.
[220] Anderson concedes this in his reply, 'Socialism and Pseudo-Empiricism', *New
Left Rev.* No. 35 (January/February 1966), pp. 2–42, at pp. 17–18.
[221] Thompson, 'Peculiarities of the English', p. 335.
[222] Ibid.

There should have been more crisis than there was, more of a parting of the ideological heavens. The intellectuals should have signalled their commitments; signed manifestos; identified their allegiances in the reviews. The fact that there was comparatively little of this may be accounted for by the fact that Darwin addressed a protestant and post-Baconian public, which had long assumed that if God was at issue with a respectable Fact (or if a dogma was at odds with a man's conscience) it was the former which must give way.[223]

The simplest way of dealing with this and Thompson's other pronouncements on Darwin would be—as a first approximation—to place 'not' at the relevant place in each sentence; the result would be closer to the situation in the period than what he does say.

It would be helpful if Perry Anderson's (even more) polemical reply simply corrected Thompson's crudities, but it only compounds them. He reveals no better appreciation of one of the most vehement debates among all levels of society in the period—and especially in manifestoes and reviews. Anderson merely (and erroneously) notes that the issue had been settled earlier—in the Enlightenment.[224] This only adds to the absurdities, since at one level the debate was very heated indeed, while at another it is clear that evolutionism was *not* opposed to theism. It only shifted the level of analysis of a theistic approach: identifying the laws of nature with the laws of God. On the question of whether or not there was a crisis, the writers in the Victorian periodicals studied by Ellegård in his book on the reception of Darwin's theory of evolution in the British periodical press, 1859–72, would have been surprised to be told this: the mere listing of the periodicals covers fifteen pages, while a comparable list of pamphlets and books would be very extensive indeed. Even the most unsatisfactory accounts of the debate acknowledge the tremendous controversy which evolutionism engendered: in particular, the dramatic confrontation between Bishop Samuel Wilberforce and T. H. Huxley at the Oxford meeting of the British Association in 1860 is the incident in the Victorian crisis of faith which is perhaps the most widely-known of all.[225]

[223] Ibid.
[224] Anderson, 'Socialism and Pseudo-Empiricism', section on 'The Rôle of Religion and Science', pp. 17–23, especially p. 20.
[225] These points are nearly all commonplaces in the literature on the Darwinian debate, except for the claim that the debate was conducted almost wholly within a theistic context. See the papers cited above (note 10) and the literature cited therein, especially Ellegard, *Darwin and the General Reader*, op. cit. (note 50). The Wilberforce-Huxley confrontation is discussed as part of a clichéd account in Young, 'The Impact of Darwin', *op. cit.* (note 10), p. 19. There are numerous other points of detail and substance on which both Thompson and Anderson show surprising and disheartening ignorance of the rôle and scope of the nineteenth-century debate and the issues of science and scientific naturalism.

But Thompson and Anderson share a far more misleading and fundamentally unmarxist assumption which Anderson puts unequivocally as 'the
basic ontological difference between the natural and the human sciences'. He
concludes that 'it follows, of course, that the modern natural sciences are
relatively (not, of course, absolutely) asocial in character. They partake of a
"natural objectivity", which is precisely that of the structure of their object.
Darwinism is no exception.'[226] For both Thompson and Anderson, Darwin
was a pure scientist. Anderson makes a sharp distinction between Darwin as
a scientist and the ideological uses to which his theory was put:

> Darwin's discoveries were not, of themselves, ideological: it was their *use* which
> was—and about this he [Thompson] says nothing at all. Yet Darwinism is
> probably the most dramatic case history of a scientific theory giving immediate
> birth to a social ideology. No other scientific discovery was ever as rapidly
> 'politicized' as this. 'The survival of the fittest' and 'the natural law of selection'
> became a ruthless celebration of Victorian racism and imperialism: these axioms
> provided a benison for class society, and a mystique for militarism. They did so
> in the name of a *natural* destiny inscribed in the course of *things*.[227]

Anderson is certainly right to relate the evolutionary movement to the
social and political philosophies which employed 'Darwinism' as a rationalization.[228] But, unlike Marx and Engels, he fails to see the continuity between
Utilitarianism, the wide movement which included Darwinian and other
evolutionary theories, and post-Darwinian social and political rationalizations. It is an unbroken tradition, aspects of which have been discussed
above, and which must be seen in a single, totalizing framework embracing
science and society and the interplay between them. The distinction between
the constitutive aspects of science and the contextual factors which play upon
it must be broken down. One can only say '*Et tu*' to two of Anderson's
concluding remarks: Thomson's 'tribute to Darwin . . . over-simplifies the
question of the social character of natural science. . . .' 'He has leapt into
controversies about which he knows little; he has allowed literary flourishes
to get the upper hand over sober accuracy.'[229]

226 Anderson, 'Socialism and Pseudo-Empiricism', p. 19.

227 Ibid., p. 20.

228 There is a very large literature on 'Social Darwinism', while the use of biological
theory as a rationalization of social and political theories is very nearly ubiquitous. The
following sources review much of the literature; Richard Hofstadtev, *Social Darwinism
in American Thought* (1944), 2nd edition (Boston, Beacon paperback, 1955); Philip P.
Wiener, *Evolution and the Founders of Pragmatism* (Cambridge, Mass., Harvard, 1949;
also Harper paperback, 1965); Bernard Semmel, *Imperialism and Social Reform: English
Social-Imperial Thought, 1895–1914* (London, Allen & Unwin, 1960; also Anchor paperback,
1968); Russett, *The Concept of Equilibrium,* op. cit. (note 92); Demerath and Peterson (eds,).
System, Change, and Conflict, op. cit. (note 203); Peel, 'Spencer and the Neo-Evolutionists',
op. cit. (note 115); cf. above (note 204).

229 Anderson, 'Socialism and Pseudo-Empiricism', pp. 40, 41.

It would be misleading to suggest that Anderson's position over Darwinism is an aberrant mistake. On the contrary, it is consistent with his whole approach to ideology. In his otherwise valuable critique of the ideology of British intellectual life, 'Components of the National culture',[230] he explicitly excludes the natural sciences and the creative arts from his analysis and suggests that

> the dose of 'objectivity' in the natural sciences and 'subjectivity' in art is symmetrically greater than either in the social sciences . . ., and they therefore have correspondingly more mediated relationships to the social structure. They do not, in other words, directly provide our basic concepts of man and society— the natural sciences because they forge concepts for the understanding of nature, not *society*, and art because it deals with man and society, but does not provide us with their *concepts*.[231]

I gather that Anderson has changed his views since writing this, but since the change is in the direction of the Althusserian position—one which intensifies the problems being discussed here—the point being made here remains.[232]

It should be clear by now that the approach I am advocating implies that the 'more mediated relationships' of the natural sciences to the social structure should become the central concern of a radical historiography of science. Once again, it should be stressed that this is not a new suggestion but a revival of the original marxist position, one which has been vitiated by vulgar marxism. Anderson's claims are very ironic in the present, since the natural and especially the biological sciences *are*—both directly and analogically—providing basic concepts for the interpretation of man and society. Until we understand the central rôle of scientific rationality in the network of issues which sustains the hierarchical division of labour, there is an irreconcilable conflict between the sources of the potential liberation of man from struggle with nature and with his fellow man on the one hand and genuine democracy on the other. The undertaking of this analysis is being blocked by the distinction which Anderson wants to maintain in both the Darwinian case and in the analysis of current intellectual life. The

[230] Perry Anderson, 'Components of the National Culture', *New Left Rev.* No. 50 (July/August 1968), pp. 3–57; reprinted in Cockburn and Blackburn (eds.), *Student Power*, op. cit. (note 150), pp. 214–84. Reverting for a moment to the autobiographical mode of section II of this essay, I find it ironic that a friend gave me a copy of Anderson's article in 1968, stressing its significance for my work. I did not see the point of reading it until December 1970, in the midst of the British Home Secretary's (ultimately successful) campaign to deport Rudi Dutschke.

[231] Ibid., p.5; cf. Young, 'Evolutionary Biology and Ideology: Then and Now', op. cit. (note 8), pp. 192–3.

[232] See above, note 156.

same approach should be applied both to Darwin and to the present. As Marx said,

> Upon the different forms of property, upon the social conditions of existence, rises an entire superstructure of distinct and peculiarly formed sentiments, illusions, modes of thought and views of life. The entire class forms and creates them out of its material foundations and out of the corresponding social relations.[233]

It is particularly ironic that Anderson should—in a marxist argument—preclude Darwin from ideological analysis, since one can buy a whole volume of writings by Marx and Engels arguing against the Malthusian basis of Darwinism and the whole philosophy of nature and man which this putatively unideological Darwin crowned with scientific respectability.[234] Where are we to draw the line between the sources, content and influence of Darwinism? Anderson should be able to see that the applicability of ideological analysis to the nineteenth-century debate on man's place in nature is not as contentious as it would be, for example, in physics, where the mediations are far more complex and subtle. Ideological analysis was originally applied to the examination of social ideas in terms of their social location—the class interests of those who produce them and benefit from them, along with the general conditions that affect their nature, distribution and acceptance.[235] The point about evolutionary theory is that it is the central conception *linking man and social theory to natural science*. The very existence of evolutionary theory and its general acceptance raises the question of whether sharp distinctions can be made between those disciplines which are amenable to ideological analysis and those which are not or are qualitatively less so.

In this form the question highlights the pivotal position of biology between the natural and the human sciences. In the present connection, however, the problem arises in another form. The distinction between the inescapability of ideology in the liberal arts—including history—and the alleged objectivity of the natural sciences places the historian *of* science in an absurd position. All of these problems come together if one is an (1) historian of (2) science whose interest is (3) evolutionary theory (4) in the period when the putative boundaries between science and matters to do with human nature were the (5) subject of intense public debate. Even if one wanted to argue for sharp distinctions between the physico-chemical

---

[233] Karl Marx and Frederick Engels, *Selected Works* in one volume (London, Lawrence and Wishart paperback, 1970), p. 117.

[234] Meek (ed.), *Marx and Engels on Malthus*, op. cit. (note 82).

[235] Charles Frankel, 'Theory and Practice in Marx's Thought', in *Marx and Contemporary Scientific Thought*, op. cit. (note 8), pp. 20–32, at p. 27.

and other sciences, the history of biological and human sciences in the nineteenth century could not confidently be excluded from the domain of ideological analysis on the most restricted definition of that domain. My own position is that ideological analysis knows no boundaries, but even those who do not grant this position should be able to see its applicability to the nineteenth-century debate on man's place in nature.

If we cannot turn directly to otherwise sophisticated and subtle writers of the New Left, where can we look for guidance and solidarity in the task of developing a radical, libertarian socialist approach to science and its history? There are grave limitations to the usefulness of the writings of Thompson and Anderson on the matters discussed above, but other aspects of their work are important and useful. Similarly, as we have seen, the historiographic writings of Hobsbawm and Williams are directly relevant. There are other helpful current writers who will be mentioned below, but surely the first task is to return to Marx:

> The ideas of the ruling class are in every epoch the ruling ideas: i.e., the class which is the ruling *material* force of society, is at the same time its ruling *intellectual* force. The class which has the material means of production at its disposal, has control at the same time over the means of mental production, so that thereby, generally speaking, the ideas of those who lack the means of mental production are subject to it. The ruling ideas are nothing more than the ideal expression of the dominant material relationships, the dominant material relationships grasped as ideas; hence of the relationships which make the one class the ruling one, therefore, the ideas of its dominance. The individuals composing the ruling class possess among other things consciousness, and therefore think. Insofar, therefore, as they rule as a class and determine the extent and compass of an epoch, it is self-evident that they do this in its whole range, hence among other things rule also as thinkers, as producers of ideas, and regulate the production and distribution of the ideas of their age: thus their ideas are the ruling ideas of the epoch. . . . If now in considering the course of history we detach the ideas of the ruling class from the ruling class itself and attribute to them an independent existence, if we confine ourselves to saying that these or those ideas were dominant at a given time, without bothering ourselves about the conditions of production and the producers of these ideas, if we thus ignore the individuals and world conditions which are the source of the ideas, we can say, for instance, that during the time that the aristocracy was dominant, the concepts honour, loyalty, etc., were dominant, during the dominance of the bourgeoisie the concepts of freedom, equality, etc. The ruling class on the whole imagines this to be so. This conception of history, which is common to all historians, particularly since the eighteenth century, will necessarily come up against the phenomenon that increasingly abstract ideas hold sway, i.e. ideas which increasingly take on the form of universality. For each new class which puts itself in the place of one ruling before it, is compelled, merely in order to carry through its aim, to represent its interest as the common interest of all the

C.P.H.S.—15

members of society, that is, expressed in ideal form: it has to give its ideas the form of universality, and represent them as the only rational, universally valid ones.[236]

Although Marx is here talking about non-scientific ideas, it is the general form of his thesis which should provide the perspective for a radical interpretation of science and its history in any epoch. In particular, the development of the philosophies of nature, man and society which run through the debate of man's place in nature in nineteenth-century Britain needs to be interpreted in the light of general trends from a theory suitable for a pastoral, agrarian, aristocratic world to one which reflects a competitive, urban, industrial one. In the same period the view of God changed from a natural theology of harmony in nature and society (with direct appeals for explanation of their order), to a Deity identified with the self-acting laws of nature. The latter were laws of progress through struggle, in which the inequalities of society were not justified by a Divinely-ordained social status. Rather, they were based on a biological one of the hierarchical division of labour which, in turn, depended on the universal law of 'the *physiological* division of labour'.[237] Science did not replace God: God became identified with the laws of nature. Adam Smith, Paley, Malthus, Darwin, Chambers, Spencer, the 'Social Darwinists' and the emergence of functionalism, pragmatism, psycho-analysis and numerous other theorists and their schools can then be seen as part of a continuous development. That development was the substitution of one form of rationalization of the hierarchical relations among men, for another—from the projection of natural theology to the reification of man and society through biologism. Belief in the mystical union of 'order and progress'[238] altered in the course of the debate from natural theology, in alliance with the mechanical psychology of associationism, to biological evolutionism in alliance with an organic version of associationism.[239] In the course of this period there was a change from the

[236] Marx and Engels, *The German Ideology,* op. cit. (note 171), pp. 61–2.

[237] See above, notes 92, 116, 162.

[238] This slogan found its way from Comtism and its British disciples and part-allies (e.g. J. S. Mill, G. H. Lewes, and Herbert Spencer) and went as far afield as the rationalizations of the Mexican científicos, the legend on the flag of Brazil, and Oriental westernizers. See, for example, W. M. Simon, *European Positivism in the Nineteenth Century* (Ithaca, Cornell, 1963); John Womack, Jr., *Zapata and the Mexican Revolution* (London, Thames & Hudson, 1969; also Penguin paperback, 1972), Thames & Hudson ed., pp. 10, 218, etc.; J. M. Dunn, *Modern Revolutions,* op. cit. (note 39), p. 49; Benjamin Schwartz, *In Search of Wealth and Power: Yen Fu and the West* (Cambridge, Mass., Harvard, 1964; also Harper paperback, 1969), *passim.*; D. W. Y. Kwok, *Scientism in Chinese Thought, 1900–1950* (New Haven, Yale, 1965).

[239] Young, 'Association of Ideas', op. cit. (note 10); cf. above (note 204).

theodicy of sufficient reason of Adam Smith and William Paley to that of Herbert Spencer: cosmic evolution. The secularization which occurred in the period was really a new context for belief in progressive harmony based on 'the nature of things'. Looking further, it is worth considering whether or not the new technological revolution in our own time is merely an extension of the same historical development—based on organic analogies but embodied in scientific management, management science, cybernetics, systems analysis, operational research, and the world of computers.[240]

This view of the debate extending from the Newtonian conception of society in Adam Smith to the present, is based on a general notion of history and epistemology: the processes by which conceptions of nature come to be defined is fundamentally the same as those by which conceptions of society are developed. The traditional distinction between genetic and analytic accounts in philosophy and science should be softened so as to mesh with the weaker use of that distinction in interpersonal and social interpretations. Similarly, the whole distinction between the content and validity of an idea and its context should also be considerably softened. Nothing is ultimately contextual: all is constitutive, which is another way of saying that all

---

[240] This is, of course, the argument of Marcuse, Habermas and Schroyer (see above, notes 123, 153). However, the analysis can be developed in much greater detail and scope, with explicit links between evolutionism and particular industrial rationalizations and techniques. A short-list of relevant sources: Raymond Aron, 'Development Theory and Evolutionist Philosophy', in his *The Industrial Society: Three Essays on Ideology and Development* (London, Weidenfeld & Nicolson, 1967), pp. 48–91; Samuel Haber, *Efficiency and Uplift: Scientific Management in the Progressive Era, 1890–1920* (Chicago, 1964); Horace B. Drury, *Scientific Management: a History and Criticism* (1915), 3rd edition, *Studies in History, Economics and Public Law* (Columbia, 1922; reprinted New York, AMS, 1968); Loren Baritz, *The Servants of Power: A History of the Use of Social Science in American Industry* (Middletown, Conn., Wesleyan, 1960); Elton Mayo, *The Human Problems of an Industrial Civilization* (New York, Macmillan, 1933); *The Social Problems of an Industrial Civilization* (1945; London, Routledge & Kegan Paul, 1949); A. Tillet *et al.* (eds.), *Management Thinkers* (Harmondsworth, Penguin paperback, 1970); Denis Pym (ed.), *Industrial Society: Social Sciences in Management* (Harmondsworth, Penguin paperback, 1968); Roger Williams, *Politics and Technology* (London, Macmillan paperback, 1971); Michael Bosquet, 'The "Prison Factory"', *New Left Rev.* 73 (May/June 1972), pp. 23–34. Biologistic and scientistic rationalizations of the hierarchical division of labour are not a capitalistic monopoly. See, for example, V. I. Lenin, 'The Immediate Tasks of the Soviet Government' (1918), in *Selected Works* in one volume (London, Lawrence & Wishart, 1969, pp. 400–31, especially p. 417; cf. p. 450. Lenin was not so enthusiastic about Taylor in 1914. See 'The Taylor System—Man's Enslavement by the Machine', in V. I. Lenin, *On Workers' Control and the Nationalisation of Industry*, trans. *anon.* (Moscow, Progress, 1970), pp. 15–17. For modern versions, see V. G. Afanasyev, 'The CPSU and the Theory and Practice of Scientific Management of Society' and E. A. Arab-Ogly, 'Scientific and Technological Revolution and Social Progress', in *Development of Revolutionary Theory by the CPSU*, trans. David Skvirsky (Moscow, Progress, 1971), pp. 237–57, and 361–79.

relationships are dialectical.[241] Of course external nature exists, but all attempts to know it—to qualify or quantify it in any way—are inescapably mediated through human consciousness, and consciousness is a socio-political and ideological mediator. The nature of consciousness is inconsistent with any positivist view of coming to know external nature. Things exist independent of man, but man (i.e., human praxis) is the *measure* of all things. An exclusively contemplative relationship to nature or to man is out of the question except as a result of the praxis of those whose interests are served by such a posture. It is in our practical behaviour—encountering, suffering, struggling and co-operating—that we come to know ourselves, one another and things.[242] If we are to continue to attempt to employ the base-superstructure distinction in any form, and, *a fortiori*, if we are to move beyond it, our perception of the base must be enriched so as to include an epistemology of struggle, not a passive, contemplative attitude towards nature and man—or indeed towards the study of their history.

From Marx and the framework he provides, we can move on to Lukács, whose analysis of reification provides tools for looking more closely at the ways in which science has been used for the purpose of reconciling men to the *status quo*. In *History and Class Consciousness* he developed with great care

[241] 'Marx admitted no absolute division between nature and society, and hence no fundamental methodological distinction between the natural sciences and historical science. As he wrote in the *German Ideology*: "We know only a single science, the science of history. History can be contemplated from two sides, it can be divided into the history of nature and the history of mankind. However the two sides are not to be divided off; as long as men exist the history of nature and the history of men are mutually conditioned." An "opposition between nature and history" is created by the ideologists in that they exclude from history the productive relation of men to nature' (Schmidt, *The Concept of Nature in Marx*, op. cit., note 153, p. 49). 'But if nature and society are internally related (Marx explicitly denies nature and history are "two separate things"), an examination of any aspect of either involves one immediately with aspects of the other' (Ollman, *Alienation*, op. cit., note 8, p. 53).

[242] Schmidt writes, 'Marx's polemic against Feuerbach in the *German Ideology* is an absolutely classic demonstration of the point that the natural sciences, a main source of materialist assertions, provided no immediate consciousness of natural reality at all, because man's relation to reality is not primarily theoretical but practical and modificatory. In their field of vision, their methodology, even in the content of what they regard as matter, the natural sciences are socially determined' (*The Concept of Nature in Marx*, pp. 32–3). Marx said, 'Here, as everywhere, the identity of nature and man appears in such a way that the restricted relation of men to nature determines their restricted relation to one another, and their restricted relation to one another determine men's restricted relation to nature, just because nature is as yet hardly modified historically; . . .' (*The German Ideology*, op. cit., note 171, p. 42). 'Human thought in general, and therefore scientific thought, which is a particular aspect of it, are closely related to human conduct and to the effects man has on the surrounding world' (Goldmann, *The Human Sciences and Philosophy*, op. cit., note 154, p. 26).

and subtlety the ways in which the categories of science were applied to economic and social relations in ways which led to fatalism.[243] One of the major themes in the book is that 'nature is a social category'.[244] Most of his analysis is devoted to the ways in which society is objectified. For example,

> What is important is to recognise clearly that all human relations (viewed as the objects of social activity) assume increasingly the objective forms of the abstract elements of the conceptual systems of natural science and of the abstract substrata of the laws of nature.[245]
> The view that things as they appear can be accounted for by 'natural laws' of society is, according to Marx, both the highpoint and the 'insuperable barrier' of bourgeois thought.[246]

Although the majority of his examples are drawn from the ways in which classical economics employed the model of natural science in the process of reification, Lukács' argument can be extended from economic and social laws to the scientific—especially the biological—laws on which these

[243] 'Lukács critique of science was aimed at the contemplative position which he claimed it implied. To regard society as governed by scientific laws, was, according to Lukács, to take a reflective attitude to it, instead of intervening actively to change it and thereby transcend its laws' (Therborn, 'The Frankfurt School', op. cit., note 153, p. 75; cf. pp. 82–3).

[244] Lukács, *History and Class Consciousness*, op. cit. (note 152), p. 130.

[245] Ibid., p. 131.

[246] Ibid., p. 174. There are numerous passages in Lukács' analysis which are extremely apposite to the basic line of argument of the present essay, e.g., pp. 11, 14, 19, 23, 38, 49, 54, 70, 101, 135, 157, 178, 181, 231, 237, 240–1, 245, 314, 334. Lukács argued that the ideology of science blinds men's vision to the realities of their own existence in the social world. 'It does this by inculcating the illusion, or what Marx called the "false consciousness", that existing social arrangements are governed by immutable laws, very much like those which prevail in the processes of the physical world and, like them, beyond the power of man to change. To anyone so indoctrinated—and this, *ceteris partibus* may comprise all classes of the population—all social processes take on an illusory or "reified" appearance in the sense that they come to be regarded as having an "objective", external reality of their own, as though they were something other than the activities of members of the same society in their relations with one another. This, according to Lukács, is the clue to man's split personality as a member of modern society; what he thinks he does in that capacity (his consciousness) bears no relation to what he actually does (his existence). If anything, the accomplishments of science tend to nourish the very social irrationality already fostered by the reified structure of its parent ideology. By isolating the facts of the empirical world for specialized study, for example, the social sciences have to disregard the organic unity which alone gives them meaning and, by treating them as hard and fast data, they convert what are essentially potentialities into finalities' (Leopold Labedz, 'Relativism and Class Consciousness: Georg Lukács', in L. Labedz (ed.), *Revisionism: Essays on the History of Marxist Ideas* (London, Allen & Unwin, 1962), pp. 142–65, at pp. 158–9; cf. p. 164).

extrapolations are based.[247] At one point he appears to provide a general warrant for extending ideological analysis from scientized (or reified) social relations to science itself:

> Nature is a social category. That is to say, whatever is held to be natural at any given stage of social development, however this nature is related to man and whatever form his involvement with it takes, i.e. nature's form, its content, its range and its objectivity are all socially conditioned.[248]

---

[247] 'It was a great and important historical advance when the Enlighteners of the eighteenth century started to investigate the natural conditions surrounding social development and attempted to apply the categories and results of the natural sciences directly to the knowledge of society. Naturally, this gave rise to much that was perverse and unhistorical, but in the struggle with the traditional theological conception of history it signified a very considerable advance at the time. It was quite different in the second half of the nineteenth century. If historians or sociologists now attempted to make Darwinism, for example, the immediate basis of an understanding of historical development, this could only lead to a perversion and distortion of historical connections. Darwinism becomes an abstract phrase and the old reactionary Malthus normally appears as its sociological "core". In the course of later development the rhetorical application of Darwinism to history becomes a straightforward apology for the brutal dominion of capital. Capitalist competition is swollen into a metaphysical history-dissolving mystique by the "eternal law" of the struggle for existence' (G. Lukács, *The Historical Novel* (1937), trans. Hannah and Stanley Mitchell (London, Merlin, 1962; reprinted Harmondsworth, Penguin paperback, 1969), Penguin edition, p. 207).

[248] Lukács, *History and Class Consciousness*, op. cit. (note 152), p. 234. This passage has not been allowed to pass without comment. Given the tremendous weight of traditional dualism and empiricism, it is very difficult indeed to see Lukács' analysis in fully relational, dialectical terms. It is thus easy to conclude that Lukács has here crossed over into idealism. Schmidt objects specifically to the passage: 'Lukács pointed correctly to the socio-historical conditioning of all natural consciousness as also of phenomenal nature itself. But in Marx nature is not *merely* a social category. [Lukács did not say that it was—RMY]. It cannot be totally [again—RMY] dissolved into the historical process of its appropriation in respect of form, content, extent and objectivity. If nature is a social category, the inverted statement that society is a category of nature is equally valid [Yes]. Although nature and its laws subsist independently of all human consciousness and will for the materialist Marx, it is only possible to formulate and apply statements about nature with the help of social categories' (*The Concept of Nature in Marx*, op. cit., note 153, p. 70). He elsewhere says of Lukács that 'he dissolves nature, both in form and content, into the social forms of its appropriation' (p. 96, cf. p. 228 fn. 148). If Schmidt, a younger member of the Frankfurt school, can express this view, it is not surprising that it is put far more strongly in Stedman Jones' Althusserian essay 'The Marxism of the Early Lukács: an Evaluation', op. cit. (note 161), sections 3 and 4—'The Assault on Science' and 'Science and Class Struggle', pp. 44–64. He refers to Lukács' 'romantic, antiscientific thematic', raises the crucial issue and gives an answer which sets clear limits to ideological analysis of the methods and substance of science: 'Historical materialism can theorise the significance of scientific activity as is social practice and can formulate the specific social and historical conditions in which new sciences have emerged: but it does not thereby arbitrate their validity or their *scientificity*. To believe otherwise is to conflate the social

But Lukács does not himself extend the analysis to nature itself in *History and Class Consciousness* and indeed Gramsci complained of this:

> It seems that Lukács asserts that one can only speak the dialectic for the history of man but not for nature. He may be right and he may be wrong. If his assertion presupposes a dualism between nature and man he is wrong, because he falls into a view of nature proper to religion and Greco-Christian philosophy and also into idealism, which in reality does not manage to unite men and nature and relate them together other than verbally. But if human history should be conceived also as the history of nature (also through the history of science), how can the dialectic be separated from nature?[249]

This fundamental issue has remained unresolved in Marxist thought. On the one hand, one of the bases of marxism is the reaction against converting men into things, leading marxists to posit a distinction between their dialectical view of man-and-nature and a reductive, positivist conception of nature, while on the other hand, dualist ontologies are anathema both because of their reconciling rôle in the history of thought and because they lead to various forms of idealism. The only alternative to dualism would seem to be to treat science and nature in an anthropological perspective, thereby fully extending ideological analysis to the history of science and its metaphysical assumptions. Aspects of this critique can be developed by combining the analyses of A. N. Whitehead and E. A. Burtt with ones which have been influenced by Lukács, especially that of Herbert Marcuse in *One Dimensional Man*.[250] I understand that this sort of extension is exactly what Lukács was doing in his *Ontologie*, the work which he was completing when he died.[251]

The conception of 'an anthropological perspective' mentioned above might easily be misleading, since much current anthropology is based on the very mystifying functionalist conception which one is attempting to transcend. Thus, the term 'anthropological' must be defined with care. I am using it to refer to two sorts of writing. First, that of the early Marx, who was

---

bearers of a science with its substantive contents; the materialist history of a science with its epistemology' (p. 62). It is precisely this conflation——yielding a double perspective—which I am advocating and which I believe Lukács saw. For further criticism of Lukács as idealist, see Poulantzas, 'Marxist Political Theory in Great Britain', op. cit. (note 219), pp. 61–3.

[249] Gramsci, 'Critical Notes on an Attempt at a Popular Presentation of Marxism by Bukharin', in *The Modern Prince and Other Writings*, op. cit. (note 152), p. 109.

[250] See above, note 28; Marcuse, *One Dimensional Man*, op. cit. (note 123), Sphere edition, chs. 6 and 9, especially pp. 180–6; cf. pp. 115–19.

[251] My sources for this information are members of Lukács family and scholars who have been doing research on Lukács, Rudi Dutschke and John Fekete, who have had extensive discussions about Lukács' research in the period before he died in June 1971; cf. 'Lukács on His Life and Work', op. cit. (note 152), pp. 51–2.

unequivocally approaching nature and society in terms of human values.[252] I
agree with those who argue that there is no 'epistemological break' between
these and his later work. The writings of Lukács, Goldmann, Mészáros and
Ollman fall clearly within this approach to Marx, while those of Avineri,
Schmidt and Lefebvre are helpful.[253] The second sense in which I am using the
term 'anthropological' is to refer to the approaches of two social anthropologists
whose work seems to me to be adaptable for our purposes in developing a
richer theory of mediation, one which may lead ultimately to a totalizing
perspective. The first is Robin Horton, whose anthropological study of the
affinities between 'African Traditional Thought and Western Science' helps
us to see the relativity of ways of 'ordering the world'.[254] He does this
without considering that the various ways of doing this are greater or lesser
approximations to the indubitably 'objective' way of western science. The
comparative method which he employs can be seen as the social scientists'
version of the experimental method in science. But the worries which his
ideas have caused among British philosophers would seen to imply that his
challenge is promising.[255] Horton does not appear to be developing this line
of enquiry further into the analysis of western concepts of science and
rationality. This is unfortunate, but the issues are being pursued with
considerable subtlety and depth by Mary Douglas. Beginning with her
study of pollution and taboo in *Purity and Danger* she has, with increasing
boldness and imagination, applied the approach of the anthropologist to the
economic, technological and scientific cosmologies of her own culture.
Professor Douglas is more interested in the texture of the mediation of social
arrangements in the conventions of society than she is in questions about
power or about what these highly-ordered conventions are mediations *of*.
Even so, her analyses are excellent models for our own efforts, and she has

[252] See above, note 171.
[253] The conception of an 'epistemological break' is Althusserian (see above, note 156).
In addition to the writings of Lukács, Goldmann, Lefebvre, Schmidt, and Ollman cited
above (notes 3, 8, 152–4, 167), see István Mészáros, *Marx's Theory of Alienation* (London,
Merlin, 1970; also paperback); 'Alienation and the Necessity of Social Control', in
Miliband and Saville (eds.), *Socialist Register, 1971*, op. cit. (note 134), pp. 1–20; Shlomo
Avineri, *The Social and Political Thought of Karl Marx* (Cambridge, 1968; also paperback);
for a review of Avineri's book which shows starkly what is at issue between neo-marxism
and Bolshevism, see David Fernbach, 'Avineri's View of Marx', *New Left Rev.* No. 56
(July/August 1969), pp. 62–8; Henri Lefebvre, *Dialectical Materialism* (1940), trans. John
Sturrock (London, Cape paperback, 1968) .
[254] Robin Horton, 'African Traditional Thought and Western Science', *Africa* 37
(1967), pp. 50–71, 155–87.
[255] Bryan R. Wilson, *Rationality* (Oxford, Blackwell, 1970), includes essays on the
issues raised by the relativism of the social sciences and an abbreviated version of
Horton's essay (which should be read in the full version).

suggested—however little she may wish to set in train politically radical thinking—ways of freeing ideological analysis from the restrictions which which have kept it away from the domain of natural science. Her recent essay on 'Environments at Risk' and her Inaugural Lecture, 'In the Nature of Things', offer important encouragement and guidance for explicitly political investigations of scientific cosmologies.[256] It may be that Horton and Douglas were only—by analogy—extending the relativism of social anthropology to our own society, but they have provided very significant tools for a radical analysis of science.

I can point out one other writer whose work promises to play an important part in the development of a radical critique of science and its history: Alfred Sohn-Rethel. He has taken on the daunting task of relating the problem of the origins and rôle of abstract—particularly scientific—thought to the basic problem of the separation between mental and manual labour, i.e., to one of the fundamental bulwarks of hierarchical, anti-democratic societies. He has also integrated this analysis with a critique of the ideology of science as applied in industrial processes. These analyses are contained in a book and a number of papers which are not widely-known in Britain and America and are almost completely unknown to historians of science.[257]

Any impression which may be forming, that a long analysis is by now degenerating into a reading list, would be accurate. In formulating a radical historiography of science there is much of existing and earlier approaches to be overcome and little, as yet, available which has been sufficiently worked out as a new approach. But it should not be thought that this is a bad position for a marxist to find himself in:

> The materialist doctrine that men are the products of circumstances and up-
> bringing, and that, therefore, changed men are products of other circumstances
> and changed upbringing, forgets that it is men that change circumstances and
> that the educator himself needs educating.[258]

---

[256] Mary Douglas, *Purity and Danger: An Analysis of the Concepts of Pollution and Taboo* (London, Routledge & Kegan Paul, 1966; also Penguin paperback, 1970); 'Environments at Risk', op. cit. (note 82); 'In the Nature of Things', *New Society* (9 December 1971), pp. 1133–1138. Professor Douglas' most recent enquiries into conventions and boundaries are as yet unpublished.

[257] Alfred Sohn-Rethel, *Geistige und körperliche Arbeit: zur Theorie der gesellschaftlichen Synthesis* (Frankfurt, Suhrkamp Verlag, 1970; also paperback); 'Mental and Manual Labour in Marx', in Paul Walton and Stuart Hall (eds.), *Situating Marx* (London, Human Context, 1972; also paperback); 'The Dual Economics of Transition' (in press).

[258] Marx and Engels, *Selected Works*, op. cit. (note 233), p. 28 ('Theses on Feuerbach, III').

VIII

I have tried to suggest that we set aside the 'internalist-externalist' dichotomy in the historiography of science and that we consider going beyond the marxist 'base-superstructure' model to a far richer and more subtle theory of mediations, moving toward a theory of totality. The base-superstructure distinction is overlaid with decades of vulgarization, and it should ultimately be superseded after its strengths and weaknesses have been better understood by students of science and its history. For by then it will have served its purpose as a bridge between the crude polarity of 'idealism *versus* materialism', and a genuinely totalizing, relational theory of man, nature and society.[259] I am not yet able to gain a clear picture of what that would mean.

In attempting to take a more radical approach to the nineteenth-century debate on man's place in nature, however, it is abundantly clear that the scientific and ideological perspectives should be held in constant tension and that in considering man, God, nature and society, no one of these topics should be considered except in its relations with the other three. Relativism and contextualism are useful as first approximations, but they are ultimately useless to socialists unless they are subsumed under a strong ideological approach. In the nineteenth century the boundaries between men and nature were in dispute. On the whole, nature won, which means that reification won. It is still winning, but some radicals are trying to push back the boundaries of reifying scientism as far as they can, and a critical study of the development of the models which underlie reifying rationalizations may be of service to them as they begin to place science in history—the history of men and events.

That leads me back to the book which occasioned this essay. The title one chooses for a book involves a mixture of issues—its domain, the author's perspective, commercial appeal, and so on. For several years I was sorry that T. H. Huxley had usurped the obvious one in his collection of essays on comparative structure and function: *Man's Place in Nature*.[260] The problem

---

[259] For discussions of the concept of totality, see Sartre, *The Problem of Method*, op. cit. (note 155), pp. 25 ff. and ch. 2: 'The Problem of Mediations and Auxiliary Disciplines'; Goldmann, *The Human Sciences and Philosophy*, op. cit. (note 154), p. 127, etc.; Lefebvre, *The Sociology of Marx*, op. cit. (note 3); Mészárus, *Lukács' Concept Dialectic*, op. cit. (note 152), ch. 6; Althusser and Balibar, *Reading Capital*, op. cit. (note 156). Ollman's *Alienation*, op. cit. (note 8), approaches the whole of Marx's work from the point of view of the philosophy of 'internal relations' and is a very helpful guide towards a totalizing perspective.

[260] Thomas H. Huxley, *Man's Place in Nature* (1863), *Collected Essays* (London, Macmillan, 1894), Vol. 7, chs. 1–3 (reprinted Ann Arbor, Michigan paperback, 1959).

was made much worse when I saw that one of the seminal figures in pheno-menological social theory, Max Scheler, had used the same title in 1928.[261] But, of course, in coupling the contextualist and relativist approach with a strong ideological perspective and a radical commitment, the problem has solved itself, and the solution becomes obvious: *Nature's Place in Man*.

[261] Max Scheler, *Man's Place in Nature* (1928), trans. Hans Meyerhoff (Boston, Beacon, 1961; also Noonday paperback, 1962); 'On the Positivistic Philosophy of the History of Knowledge and Its Law of Three Stages' and 'The Sociology of Knowledge: Formal Problems', in Curtis and Petras (eds.), *The Sociology of Knowledge*, op. cit. (note 8), pp. 161–9, 170–86; cf. pp. 16–18 *re* Scheler's opposition to Marx.

Appendix

(See above, note 146)

Frederick Engels from London to F. Mehring in Berlin, 14 July 1893: '. . . one more point is lacking, which, however, Marx and I always failed to stress enough in our writings and in regard to which we are all equally guilty. That is to say, we all laid, and *were bound* to lay, the main emphasis, in the first place, on the *derivation* of political, juridical and other ideological notions, and of actions arising through the medium of these notions, from basic economic facts. But in so doing we neglected the formal side—the ways and means by which these notions, etc., come about—for the sake of the content. This has given our adversaries a welcome opportunity for misunderstandings and distortions. . . .

'Ideology is a process accomplished by the so-called thinker consciously, it is true, but with a false consciousness. The real motive forces impelling him remain unknown to him; otherwise it simply would not be an ideological process. Hence he imagines false or seeming motive forces. Because it is a process of thought he derives its form as well as its content from pure thought, either his own or that of his predecessors. He works with mere thought material, which he accepts without examination as the product of thought, and does not investigate further for a more remote source independent of thought; indeed it is a matter of course to him, because, as all action is *mediated* by thought, it appears to him to be ultimately *based* upon thought.

'The historical ideologist (historical is here simply meant to comprise the political, juridical, philosophical, theological—in short, all the spheres belonging to *society* and not only to nature) thus possesses in every sphere of science material which has formed itself independently out of the thought of previous generations and has gone through its own independent course of development in the brains of these successive generations. True, external facts belonging to one or another sphere may have exercised a codetermining influence on this development, but the tacit presupposition is that these facts themselves are also only the fruits of a process of thought, and so we still remain within that realm of mere thought, which apparently has successfully digested even the hardest facts.

'It is above all this semblance of an independent history of state constitutions, of systems of law, of ideological conceptions in every separate domain that dazzles most people. If Luther and Calvin "overcome" the official Catholic religion or Hegel "overcomes" Fichte and Kant or Rousseau with his republican *Contrat social* indirectly "overcomes" the constitutional Montesquieu, this is a process which remains within theology, philosophy or political science, represents a stage in the history of these particular spheres of thought and never passes beyond the sphere of thought. And since the bourgeois illusion of the eternity and finality of capitalist production has been added as well, even the overcoming of the mercantilists by the physiocrats and Adam Smith is accounted as a sheer victory of thought; not as the reflection in thought of changed economic facts but as the finally achieved correct understanding of actual conditions subsisting always and everywhere—in fact, if Richard Coeur-de-Lion and Philip Augustus had introduced free trade instead of getting mixed up in the crusades we should have been spared five hundred years of misery and stupidity.

'This aspect of the matter, which I can only indicate here, we have all, I think, neglected more than it deserves. It is the old story: form is always neglected at first for content. As I say, I have done that too and the mistake has always struck me only later. So I am not only far from reproaching you with this in any way—as the older of the guilty parties I certainly have no right to do so; on the contrary. But I would like all the same to draw your attention to this point for the future.

'Hanging together with this is the fatuous notion of the ideologists that because we deny an independent historical development to the various ideological spheres which play a part in history we also deny them any *effect upon history*. The basis of this is the common undialectical conception of cause and effect as rigidly opposite poles, the total disregarding of interaction. These gentlemen often almost deliberately forget that once an historic element has been brought into the world by other, ultimately economic causes, it reacts, can react on its environment and even on the causes that have given rise to it. . . .' (Marx and Engels, *Selected Correspondence*, op. cit., note 81, pp. 459–60.)

XX

FROM 'ENCHYME' TO 'CYTO-SKELETON':

THE DEVELOPMENT OF IDEAS ON THE CHEMICAL ORGANIZATION OF LIVING MATTER*

Mikuláš Teich

For the biochemist the problem of organic form is ultimately unavoidable.
Joseph Needham, 'Chemical Aspects of Morphogenetic Fields', 1937.†

Although the last few years have seen the publication of several studies dealing with various aspects of the development of the chemistry of life, no comprehensive history of biochemistry has been produced since the book by Lieben which was published thirty-seven years ago and has remained available only in German.[1] Some of the more recent work covering broader specialized areas, such as nutrition[2] or cell respiration,[3] is very valuable. These books, shorter articles, and studies and occasional historical reviews by professional biochemists[4] demonstrate that there is a need for a professional approach to the history of biochemistry which should take its legitimate place in the history of science and medicine.

* I prepared this article while I held a Visiting Research Fellowship supported by the Wellcome Trust in the University of Oxford for which I am deeply thankful. I wish also to express my appreciation of the hospitality shown to me by the Principal and Fellows of Brasenose College during this time.

† In J. Needham and D. E. Green (eds.), *Perspectives in Biochemistry* (Cambridge, 1937), p. 66.

[1] F. Lieben, *Geschichte der physiologischen Chemie* (Leipzig-Wien, 1935). Reprinted in 1970.

[2] E. V. McCollum, *A History of Nutrition* (Boston, Houghton, Mifflin, 1957).

[3] D. Keilin, *The History of Cell Respiration and Cytochrome* (Cambridge, 1966).

[4] J. Needham (ed.), *The Chemistry of Life: Eight Lectures on the History of Biochemistry* (Cambridge, 1970); T. W. Goodwin (ed.), *British Biochemistry Past and Present* (London and New York, 1970).

Periodization of Modern Biochemistry

For some time I have been engaged in work on the history of biochemistry
with the aim of filling some of the gaps in our historical knowledge of biochemistry. My first concern was to establish when modern biochemistry was
born and to draw a broad outline of its growth. Biochemistry matured
into its modern form at the close of the nineteenth century. There were
three main stages of development:[5]

(1) a gradual separation of the science of chemistry of life from the still
    unified body of chemistry into organic chemistry (c.1800–c.1840);
(2) linking of organic chemistry with physiology to form physiological
    chemistry (c.1840–c.1880);
(3) the transformation of physiological chemistry into biochemistry which
    began around 1880.

By its separation from physiology, physiological chemistry was transformed into modern biochemistry the growth of which can be also divided
so far into three phases; the most significant lines of investigation can be
characterized approximately as[6, 7]

(1) the rise of enzyme chemistry (c. 1880–c. 1920);
(2) the development of dynamic biochemistry on the basis of enzyme
    chemistry (c. 1920–c. 1940);
(3) studies of structural and dynamic relations in biochemical changes which
    began to intensify around 1940.

The exploration of the origins and the phasing of modern biochemistry
was essential because they provided the necessary framework for the
programme I have embarked on, that is, to place the development of modern
biochemistry within the broader context of biological history and especially
to consider its relations to the great generalizations, the cell theory and the
theory of evolution.

[5] M. Teich, 'On the Historical Foundations of Modern Biochemistry', *Clio Medica* 1
(1965), pp. 41–57; a slightly shortened version of this article appeared in J. Needham
(ed.), op. cit. 4, pp. 171–91.
[6] M. Teich, 'The History of Modern Biochemistry: The First Phase (1880–1920)',
Actes du XIᵉ Congrès international d'histoire des sciences Varsovie-Cracovie 24–31 Août 1965
(Wrocław-Varsovie-Cracovie, 1968), v, pp. 223–6.
[7] M. Teich, 'The History of Modern Biochemistry: The Second Phase: c.1920–1940/45',
XIIᵉ Congrès international d'histoire des sciences Paris 1968, Actes (Paris, 1971), viii, pp. 199–
203.

'Internal' or 'External' History of Biochemistry : the Case of Fermentation

From the very beginning of the studies it became clear that it was impossible to present the development of biochemistry merely as the history of the chemistry of life in itself. Like his colleagues in other fields, but more so because of the state of the discipline, the historian of biochemistry has often to choose a partial approach, that is, to select and concentrate on a topic which covers only a relatively small portion of the problems the biochemists had sought to solve in the past. This is why much of the earlier historical writing on biochemistry has often been concerned with particular aspects of the subject, such as the history of proteins, the synthesis of urea, experimental methodology and so on, presented largely as an 'internal' history of ideas and techniques.

The division of research into an 'internal' and 'external' history of science is useful, but cannot be taken as absolute. A picture of the past mechanically reconstructed on the basis of the two histories will inevitably be distorted. Real history is an integral process, not divided into 'internal' and 'external' compartments, but governed by a multitude of circumstances, deriving from and belonging inseparably to spheres inside and outside science.

Take the example of fermentation, probably the earliest biochemical process made use of by man in daily life. Assumptions about fermentation and explanations of its nature, already important in the history of natural, technical, and medical, knowledge before chemistry became a science at the end of the eighteenth century, played an even more decisive rôle afterwards. Indeed, Lavoisier formulated and applied the principle of the conservation of matter (one of the corner-stones of modern chemistry) in a discussion of alcoholic fermentation, which he considered one of the most extraordinary operations in chemistry. He was interested in the change of sugar into alcohol and carbon dioxide. 'To solve these two questions' he writes 'it is necessary to be previously acquainted with the analysis of the fermentable substance and of the products of the fermentation.' This is immediately followed by the formulation of the principle of the conservation of matter: 'We may lay it down as an incontestible axiom, that, in all the operations of art and nature, nothing is created; an equal quantity of matter exists both before and after the experiment; the quality and quantity of the element remain precisely the same; and nothing takes place beyond changes and modifications in the combination of these elements. Upon this principle the whole art of performing chemical experiments depends. We must always suppose an exact equality between the elements of the body examined and those of the products of its analysis.' After demonstrating in detail the results

of his analytical determinations and calculations, Lavoisier reiterated the
basic idea of his theoretical and experimental approach to the study of
chemical changes: 'We may consider the substances submitted to fermentation, and the products resulting from that operation, as forming an algebraic
equation; and, by successively supposing each of the elements in this equation
unknown, we can calculate their values in succession, and thus verify our
experiment by calculation, and our calculation by experiment reciprocally.
I have often successfully employed this method for correcting the first results
of my experiments, and to direct me in the proper road for repeating them
to advantage.'[8] The chapter on alcoholic fermentation in Lavoisier's classic
Traité elémentaire de chimie (1789) is even more extraordinary because it contained an accurate (very nearly) conclusion about the chemical relations
between sugar, alcohol and carbon dioxide, based on erroneous analytical
figures. Nevertheless, as Arthur Harden put it: 'From this point commences
the modern study of the problem.'[9]

The central position of the idea of metabolism in the development of
biochemistry cannot be doubted. Yet there is still little appreciation of the
background leading to the formulation of this concept by Theodor Schwann
in his classic book on the cell (about which more is to follow). It was the
culmination of a series of brilliant investigations, achieved within an
unbelievably short time (1835–1839), embracing the examination of the
tension-length relationship in muscle; the study of digestion including the
discovery of pepsin; the experimental enquiry into the nature of putrefaction
and the demonstration of the improbability of spontaneous generation; the
elucidation of the connection between the growth of yeast globules and the
production of alcohol during fermentation. By putting together his results
on digestion, on alcoholic fermentation, and on the cell, Schwann assumed
that the study of fermentation held the key to the understanding of cell-life.
It was in this connection that he introduced the idea of metabolic change in
the cell.

After 1839 there was a perceptible shift in the scientific output of Schwann,
who, although comparatively young, did not keep up his previous exertions.
It has now become clear that this was primarily due to the confrontation of
the scientific interpretation of the phenomena of life and of his religious
beliefs. Schwann began his great active phase of scientific activity assuming,
as his pioneering experiments in muscular physiology showed, that biological processes could be completely reduced to physics and chemistry.
Perhaps the discovery that alcoholic fermentation was due to a vital activity
contributed to Schwann's subsequent decision to examine in more detail the

[8] A.-L. Lavoisier, *Elements of Chemistry* (New York, 1965), pp. 130–1, 140.
[9] A. Harden, *Alcoholic Fermentation* (London, 1911), p. 3.

relations of scientific knowledge of living phenomena and the teachings of the Catholic Church, which he accepted. Beginning with the cell he intended to explore all fundamental aspects of life, including the brain and the nature of consciousness, culminating in enquiry into the position of man as a biological and moral being, but he never finished it. For many years he used to enter his thoughts into a Notebook or he wrote them up into short sections, but he refrained from publishing practically any of it. Judging from the account given by Marcel Florkin,[10] who had seen the hand-written material, there is little doubt that Schwann abandoned his erstwhile material-ist outlook and, instead, endeavoured to integrate scientific findings within the framework of Christian beliefs.

In the long run Schwann's ideas regarding the fundamental importance of the study of fermentation for the understanding of the chemistry of life have been fully confirmed. However, before that chemists like Berzelius, and especially Liebig, rejected completely the idea that fermentation had anything to do with life. Liebig maintained that it was a purely chemical process, related to putrefaction and decay of those organic bodies which had already ceased to live. Whether fermentation constituted a pure chemical reaction in which the yeast ferment decomposed sugar into alcohol or had something to do with the life of yeasts became the subject of a prolonged discussion, often known as the mechanist-vitalist controversy. Liebig and Pasteur were perhaps the most prominent participants in the later stages of the contro-versy, but there were others, and it became important not only because of the scientific and philosophical issues involved. It is possible, for example, to draw attention to Pasteur's findings (from 1857) that besides alcohol and carbon dioxide, alcoholic fermentation yielded also succinic acid, glycerol and other products, or Emil Christian Hansen's work on pure yeasts in the Carlsberg laboratory (1882). The scientific discoveries of Pasteur and Hansen brought about far-reaching technical and economic consequences in the production of wine, beer and spirits.

The disclosure of cell-free fermentation by Eduard Buchner (1897) furnishes another case of the difficulty of separating the 'internal' and 'external' aspects of scientific development. It found itself at the nodal point where science, medicine,[11] philosophy, technology and industry met. The demonstration of fermentation outside living cells, proving that both

[10] M. Florkin, *Naissance et déviation de la théorie cellulaire dans l'oeuvre de Théodore Schwann* (Paris, 1960). See also R. Watermann, *Theodor Schwann Leben und Werk* (Düsseldorf, 1960).

[11] In his substantial contribution, dealing with the background to Buchner's discovery (which appeared after my article was ready for press), Robert Kohler, *J. Hist. Biol.* 4 (1971), pp. 35–61, quite rightly draws the attention to the often neglected immunological setting of the extraction of zymase from a cell-free yeast juice and emphasizes that 'there was no real continuity between the earlier controversies over fermentation and Buchner's

Liebig and Pasteur had been partly right and partly wrong, indicated that the mechanist and vitalist approaches were not as widely apart from each other, as it had seemed. Indeed, it could be argued on the basis of what they wrote, that the 'mechanist' Liebig adopted at times a 'vitalist' position and the 'vitalist' Pasteur a 'mechanist' one. Following Buchner's discovery the systematic study of ferments (enzymes) and their rôle in cellular chemistry was taken up. It brought to fruition the ideas which had been in the mind of Schwann and of later investigators, among them prominently the wine-merchant *cum* scientist Moritz Traube, who very clearly perceived that the understanding of the chemistry of living processes depended on an accurate theory of fermentation.[12]

Some of the more important developments in early enzyme chemistry originated in the work of physical chemists on the environment in which biochemical changes were taking place. Among others, it resulted in the recognition of the phosphate buffer system by A. Fernbach and L. Hubert (1900),[13] the ancestry of which can be traced to the understanding of the rôle of phosphates in agriculture, reinforced by the idea of the constancy of the internal environment, pointed out by Claude Bernard as the condition for normal activities in the living organism.

Much of research along these lines was carried out in the laboratories of the Carlsberg Foundation in Copenhagen, established by the brewer J. C. Jacobsen in 1876, for fundamental studies of chemical and physiological problems, related to brewing.[14] It was here that Hansen demonstrated the obnoxious effects of 'wild' yeast, which eventually led Jacobsen to introduce important changes in brewing practice because pure yeast cultures began to be used in his breweries. There is no doubt that in research conducted in the chemical section of the Carlsberg laboratory, first headed by Johannes Kjeldahl, and after his death by S. P. L. Sørensen, relations between science

---

discovery'. To establish a sharp dividing line between continuity and discontinuity and its nature in the development of scientific knowledge is not always easy. However, there is more than 'a historical mythology' which places Buchner's discovery within the context of the Liebig-Pasteur controversy. As Dr. Kohler himself points out, Eduard Buchner worked in Nägeli's laboratory on alcoholic fermentation and published a paper on it in 1886. Then he continues: 'Nägeli had been involved in the dispute over fermentation and upheld a version of Liebig's chemical theory in his book, *Theorie der Gährung,* which appeared in 1879. In the heated debate over the rôle of oxygen, Nägeli also opposed Pasteur's claim that yeast fermented less actively with air present than without. Eduard Buchner, too, criticized Pasteur's experimental procedure, but his own carefully controlled experiments confirmed Pasteur's claim: fermentation *per cell* was greater in the absence of air.'

[12] M. Traube, *Gesammelte Abhandlungen* (Berlin, 1899).

[13] A. Fernbach, L. Hubert, *Comptes rendus des séances de l'Académie des sciences,* 181 (1900), pp. 293–5. Cf. also F. Szabadváry, *History of Analytical Chemistry* (Oxford-London, 1966), pp. 363–4.

[14] J. Pedersen, *The Carlsberg Foundation* (Copenhagen, 1956).

and the brewing industry existed, though not necessarily on a linear basis. Sørensen's name is indissolubly linked with his studies of the influence of the hydrogen ion concentration on enzyme reactions, which he had likened to temperature effects and in the course of which he had developed the immensely useful concept of the hydrogen ion exponent, pH (1909).[15] Although the work was not ostensibly undertaken in direct response to needs of brewing, the available evidence suggests that the connections were not as remote as they would seem. The mediating link can be found in the studies of some other Carlsberg workers, more specifically related to scientific problems of brewing, and referred to by Sørensen in his highly seminal paper. Thus Fr. Weis was concerned with enzymes during germination of the barley. He published a paper (1903) in which, analogously to Fernbach and Hubert, he described a marked effect of the phosphate buffer on proteolytic enzymes from malt.[16] At that time, according to Sørensen, two other Carlsberg workers, Petersen and Sollied, confirmed these findings in relation to proteolytic enzymes from yeast. In the light of Sørensen's work on the hydrogen ion concentration the physical-chemical nature of the constancy maintained by buffer solutions found its explanation. Owing as it did to what could be put in the service of the fermentation industries, Sørensen's contribution acquired a much wider significance because it provided the key for the understanding and control of those chemical, biological, and industrial, processes which require a specific optimum range of acidity and alkalinity.

The point of these lines is not to present a thorough historical account of fermentation but to illustrate some of the multiform reciprocal links, which make the historical study of the phenomena of fermentation as much a part of the history of biochemistry (and other sciences), as a part of the technological and economic history of several industries. Another feature of the history of fermentation is that it was intimately connected with endeavours to answer the perennial question on the origin of living matter and the nature of the processes underlying it. Over periods of time the theoretical conceptions worked out 'inside' science to answer this problem reflected variously 'outside' philosophic, ideological and indeed political attitudes.[17]

[15] S. P. L. Sørensen, *Biochem. Ztschr.* 21 (1909), pp. 130–304.

[16] Fr. Weis, *Comptes rendus des travaux du Laboratoire de Carlsberg*, 5 (1903), 203–4, 283.

[17] Pasteur's work on beer was motivated by patriotic considerations arising out of France's defeat by Germany. Cf. L. Pasteur: 'L'idée de ces recherches m'a été inspirée par nos malheurs. Je les ai entreprises aussitôt après la guerre de 1870 et pursuivies sans relâche depuis cette époque, avec la résolution de les mener assez loin pour marquer d'un progrès durable une industrie dans laquelle l'Allemagne nous est supérieure.' *Etudes sur la bière* (Paris, 1876), vii.

Equally, it is not possible in the following sections of the article, devoted to a rather specialized enquiry into concepts on the chemical organization of living matter from about 1840 up to 1940, to bring out the totality of relations which, at any stage, were involved in their formulation. It is an attempt to describe some of the historical relations of morphology and biochemistry starting from the work of Theodor Schwann and Jan Evangelista Purkyně up to the pursuits of (Sir) Rudolph Peters and Joseph Needham. It will be particularly concerned with:

(a) the concepts of the cell and protoplasm from the morphological and chemical view;
(b) the experimental work on protoplasm contractility, tissue oxidation, and the idea of the 'giant' molecule;
(c) enzyme activity and cell structure;
(d) attempts to establish a link between the dynamic side of the changes in the cell and its structure.

Seemingly, the concepts followed an 'internal' line of development. But short references to vital staining, to the twin study of enzymes and vitamins, to investigations of the structure of fibres (not to speak of fermentation chemistry) will at least indicate that 'external' elements, such as the manufacture of dyes and drugs, the needs of war and dietetics, or the efforts to replace natural materials by man-made plastics and fibres, were integrally implicated in the thinking about them.

The Cell Theory and the Theory of Cells of Schwann

Schwann published the first results of his findings, dealing with the analogies in the structure and growth of plants and animals, in three articles in 1838.[18] They were then incorporated into the first and second sections of the book *Mikroskopische Untersuchungen über die Uebereinstimmung in der Struktur und dem Wachsthum der Thiere und Pflanzen* which were published in two parts in 1838 and contained Schwann's cell theory. In addition to this Schwann produced also a theory of cells forming the third section of the book, which bears 1839 as the publishing date.[19]

[18] Th. Schwann, *Neue Notizen aus dem Gebiete der Natuer und Heilkunde* (L. F. Froriep and R. Froriep), v (1838), pp. 33–6, 225–9; vi (1838), pp. 21–3.
[19] Th. Schwann, *Mikroskopische Untersuchungen über die Uebereinstimmung in der Struktur und dem Wachsthum der Thiere und Pflanzen* (Berlin, 1839). An English translation *Microscopical Researches into the Accordance in the Structure and Growth of Animals and Plants*, appeared 1847 in London. Further citations will refer to the English version.

There had been others before Schwann who groped towards a cell theory but none of them had gone further and formulated a theory of cells.[20] It is the combination of both which makes Schwann's work a landmark in the history of morphology and biochemistry. This comes out clearly, for instance, when the achievements of Schwann and Purkyně are compared. Purkyně perhaps more than anybody else before Schwann came nearest to the formulation of a comparative theory of microscopic structures in plants and animals together with a rudimentary theory of their formation. But he recognized the merit of Schwann's attempt to work out a generalization in a field where others (including himself) had tried before and had not succeeded.[21]

The usual interpretation of Schwann's contribution to the establishment of the cell theory is that he introduced the notion of the cell as the basic structural unit of all plant and animal tissues. Schwann himself saw his achievement in a different light. He stressed that the main feature of *his* cell theory consisted in the recognition of the formation of cells to be the *common* principle in the development of plants and animals.[22]

As earlier and contemporary microscopists had done, Schwann accepted the cell to be the most widespread elementary part (*Elementarteil*) of plants but not of animals. The elementary parts of animals, according to Schwann, were diverse; apart from cells, tubes and globules, they consisted mostly of fibres. In fact, the aim of his famous treatise was to show that whatever the function of the elementary parts, their origin was the same. Because of this, the elementary parts of organisms ceased to be formations existing side by side, without any mutual relations. The cell became just as much a morphological as a physiological concept.

Deeply convinced that teleological explanations of living processes should be eschewed as much as possible Schwann outlined a theory of cells. In the particular case of living activity he was concerned with, that is, the formation of cells, he perceived at that time, at any rate, no difficulty in visualizing the natural steps from chemistry to morphogenesis. 'As the elementary materials of organic nature', he wrote, 'are not different from those of the inorganic kingdom, the source of the organic phenomena can only reside in another

[20] For a comprehensive factual treatment of the history of the cell theory, see J. R. Baker, *Quart. Journ. Microscop. Sc.*, **89** (1948), pp. 103–25; **90** (1949), pp. 87–108, 331; **93** (1952), 157–90; **94** (1953), pp. 407–40; **96** (1955), 449–81; cf. also A. Hughes, *A History of Cytology* (London and New York, 1959); M. Florkin, op. cit. (10). These publications contain references to some of the older literature.

[21] See review of Schwann's book by J. E. Purkyně (Purkinje), *Opera selecta* (Prague, 1948), pp. 116–20.

[22] Cf. Schwann's postscript where he defended his priority against the claims of Valentin, op. cit. (19), pp. 219–20.

combination of these materials, whether it be in a peculiar mode of union of the elementary atoms to form atoms of the second order, or in the arrangement of these conglomerate molecules when forming either the separate morphological elementary parts of organisms, or an entire organism.'[23]

Schwann linked the formation of cells directly to cell-life and in doing so he differentiated between the *plastic* (morphological) and *metabolic* processes. He associated the morphological features of cell formation, for instance, the building up of fibres, with presumably physical interaction at molecular level, whereas the metabolic phenomena he conceived of as chemical changes involving either the constituents inside the cell, or the exchange of materials between the cell and its surrounding medium. Schwann considered the structureless substance in the interior of the cell and the intercellular substance surrounding the cell to be essentially the same. Nevertheless, he distinguished between the cell contents and the external *cytoblastema*, and because of this, he ascribed to the cell membrane a major active rôle in the transport of materials into and out of the cell. It is interesting that he questioned whether the processes in the passage across the membrane were not electrochemical, similar to those taking place in a galvanic pile. Finally, he discussed the possibility that the formation of the elementary parts of organisms and their growth could involve crystallization of the structureless substance, capable at the same time of imbibition.

Although scientific progress undermined Schwann's particular view on cytogenesis, which he owed to Schleiden,[24] he was proved right in visualizing the chemistry, morphogenesis and physiology of the cell as interrelated aspects of the same process. This appears clearly from the way he linked up his studies on fermentation with his work on the development of cells. 'I could not avoid bringing forward fermentation as an example', he wrote in a long footnote, 'because it is the best-known illustration of the operation of the cells, and the simplest representation of the process which is repeated in each cell of the living body.'[25] After describing some experiments showing that fermentation was associated with living activity he concluded: 'The foregoing inquiry into the process by which organized bodies are formed, may perhaps, however, serve in some measure to recommend this theory of fermentation to the attention of chemists.'[26] It is evident that Schwann was well aware of the connection between the chemistry and the physiology of the cell, whereas Liebig hardly ever spoke of it. Not surprising is it then that they held opposite views on the nature of fermentation.

[23] Schwann, op. cit. (19), pp. 190–1.
[24] M. Schleiden, *Arch. Anat. Physiol. wiss. Med.* (Müller), (1838), pp. 137–76.
[25] Schwann, op. cit. (19), p. 197.
[26] Ibid., p. 198.

Despite a common start cell morphology and cell chemistry for a long time developed side by side more or less independently of each other. The study of protoplasm comprising empirical observations and theoretical generalizations became the dividing and unifying area where, eventually, morphology and chemistry of life had to take stock of each other.

Protoplasm and the Changing Concept of the Cell

It appears that the term 'protoplasm' was first used by Purkyně on 16 January 1839 in his talk to the Silesian Patriotic Society in Breslau.[27] The theme of the talk was clearly related to the writings by Schwann on the analogies in the structural elements of the animal and plant organism which made their appearance during the previous year. But how did he come to employ this term? Various suggestions have been advanced, pointing out the fact that Purkyně, originally destined for holy orders, must have become familiar with the word *protoplastus* used in old Catholic hymns for Adam. Also the word *plasma* occurs in old Latin church texts.[28]

Be that as it may, the concept of protoplasm developed undoubtedly in connection with previously held views that life was associated with a special kind of *substance* which Buffon called *matière vivante*[29] and G. R. Treviranus *lebende Materie, Lebensstoff,* or *Lebensmaterie,*[30] i.e., living matter. There was also the concept of the formative ground substances, capable of transformation in stages into mature plants and animals which goes back to the work of C. F. Wolff (1759).[31]

With all these ideas Purkyně and his collaborators in Breslau were thoroughly familiar in the thirties of the last century, when they were engaged in the systematic microscopic examination of animal tissues. Purkyně was among the first to draw attention to the analogies and dissimilarities between the microscopic structure of plants and animals. From his earlier embryological work—he discovered among others the *vesicula germinativa* or the nucleus in the avian egg—he was only too well aware that the development of organisms from the embryonic into the adult state took

[27] *Opera selecta*, pp. 114–15. Breslau is now in Poland and known under its Polish name Wrocław.

[28] F. K. Studnička, *Protoplasma*, 27 (1936–37), pp. 619–25.

[29] Buffon, *Histoire naturelle, générale et particulière* II (2nd edition, Paris, 1750), pp. 30–40.

[30] G. R. Treviranus, *Biologie oder Philosophie der lebenden Natur für Naturforscher und Aerzte* II (Göttingen, 1803), pp. 311, 403–4.

[31] C. F. Wolff, *Theoria generationis* (editio nova, aucta et emendata, Hallae ad Salam, 1774).

place in stages. At first (1834) he seemed to have employed the already known botanical term *cambium* both for the plant and animal rudimentary living substance. His close collaborator Valentin spoke (1835) of *Urstoff* or *Urmasse* (primordial substance or mass) to which the cytoblastema of Schwann was closely related.[32]

Like Schwann, Purkyně did not approach the structure of tissues from a purely morphological point of view. Purkyně and Valentin had already demonstrated in their work on ciliary motion, that morphological physiology could achieve remarkable results in the elucidation of fundamental biological processes.[33] The same broad outlook is revealed in Purkyně's work on the structure of gastric glands which cannot be separated from his chemical and physiological investigations of what then was called 'artificial digestion', following closely in the footsteps of Eberle, J. Müller and Schwann.[34]

It was on 19 September 1837, at the Annual Meeting of the German Naturalists and Physicians, which, as it happened, took place in Prague, that Purkyně reported on his work on the structure of gastric glands and artificial digestion. He employed the term *enchyme* for the contents of 'granules', and it was the granular appearance under the microscope of the liver, spleen, thymus, thyroid, kidney, pancreas and other organs which made him state that as animals consisted of 'granules' (with a central nucleus), so plants were composed of 'granules or cells'. The comparative study of tissue enchyme led Purkyně not only to express the idea of a morphological analogy in the microscopic structure of plants and animals but also the hope that their biochemical and physiological features would be studied. He clearly stated that each plant cell has its *vita propria,* and that it forms its own specific contents by absorbing suitable substances from, and discharging them into, the general sap. He visualized a similar process of formation and dissolution of animal enchymes. At the end of his communication he suggested that their examination would contribute to the understanding of embryonic development, pathogenic processes and *pseudoplasmas.*

It is not clear what Purkyně meant by the term *pseudoplasma* but it appears that he believed in the existence of different forms of 'plasms' or living matter, in various stages of development and differentiation. Thus he conceived 'cambium' and 'protoplasm' to be analogical embryonic living materials of plants and animals. 'Enchyme' was the living matter of mature glands and other animal tissues, not exclusively granular in nature. For Purkyně granules were only one of the basic organized states of living

---

[32] Studnička, op. cit. (28), p. 621.

[33] M. Teich, *British J. Hist. Sc.* **5** (1970), pp. 168–77.

[34] V. Kruta, *Physiologia Bohemoslovenica,* **7** (1958), 1–8; *Československá gastroenterologie a výživa,* **21** (1967), pp. 1–12.

matter, and fibres could be another. All this was linked to the view that the state of living matter oscillated between the fluid and the solid. Proto-plasmatic granules, i.e. animal embryonic cells, were jelly-like, neither solid nor fluid, and this state seemed to persist in animals when they became mature. On the other hand, with plants the permeation of the fluid and the solid states was of a short duration, limited to early stages of their existence. In the course of the development of the plant, due to the separation of the fluid from the solid, true plant cells and then vessels were eventually formed.[35]

Whereas Purkyně conceived of protoplasm as the embryonic living material of animals, von Mohl introduced it in 1846 into botany to denote the living substance which gave rise to structures like the cell nucleus and cell membrane, in fact to the new cell. This remarkable physiological capability of the jelly-like substance to form new cells impressed von Mohl so much that he considered Schleiden's term for it, *Schleim* (mucilage), inadequate and suggested, therefore, that it should be replaced by *proto-plasm*.[36] Whether von Mohl arrived at this term independently of Purkyně, or made himself familiar with it, after having read about it in Reichert's review of advances in microscopical anatomy in *Müllers Archiv* (1841) is difficult to decide.[37]

The term protoplasm was reintroduced into zoology by R. Remak when he was engaged in his fundamental studies of egg cleavage. In 1852 he equated the yolk to the protoplasm of the egg cell.[38] Three years later, after discussing the varying meanings of 'yolk' in the literature, and its rôle in holoblastic and meroblastic cleavage, he made it clear that it was necessary to differentiate between protoplasm or *zooplasm* as living matter and yolk as food material.[39]

However, a certain time elapsed before the notion gained ground that life could not be considered independently of matter and that protoplasm was, in fact, a complex body, or to use T. H. Huxley's characterization, the 'physical basis or matter of life' (1868).[40]

Most of the work on the cell in the forties and fifties was carried out by botanists. They distinguished between cell sap and cell contents, the

[35] *Opera selecta*, pp. 109–11.
[36] H. v. Mohl, *Ann. Mag. Nat. Hist.*, xviii (1846), pp. 2–3.
[37] C. Reichert, *Arch. Anat. Physiol. wiss. Med.* (Müller), (1841), clxiii; Studnička, op. cit. (28), pp. 621–2.
[38] R. Remak, *Arch. Anat. Phys. wiss. Med.* (Müller), (1852), p. 50.
[39] R. Remak, *Untersuchungen über die Entwicklung der Wirbelthiere* (Berlin, 1855), pp. 81–2.
[40] T. H. Huxley, *Collected Essays* I (London, 1894), pp. 130–1. For a valuable summary of the mechanist-vitalist debate on the nature of protoplasm in England between Thomas Henry Huxley and Lionel Smith Beale see G. L. Geison, *Isis* 60 (1969), pp. 273–92.

importance of which began to be recognized in the life of the vegetable cell as compared with the membrane. One of the consequences of this was that the older concept of the vegetable cell as a mere bounded cavity or space began to be undermined. Comparisons were drawn and similarities discovered in the vegetable cellular material and *sarcode*, the jelly-like contractile transparent nitrogenous living substance which constituted, according to Dujardin (1835), the bodies of lower animals.[41]

M. Schultze (1861) suggested that this term should be dropped and replaced by the word protoplasm which began to be used for cell substance. Because animal cells did not seem to be bounded, yet in general they did not coalesce and retained their integrity, Schultze was led at the same time to formulate a new definition of the animal cell. He declared that it constituted nothing but a lump or blob of nucleated protoplasm.[42]

Schultze was not the only person who voiced his doubts about the adequacy of the cell picture, essentially derived from Schwann. This can be gathered from E. Brücke's critical remarks which were published in the same year as Schultze's article. Brücke raised the question whether cells constituted the 'elementary organisms' as they appeared to do, or could be subdivided into still lower living units. He was concerned with the findings that the cell was usually not a space enclosed by a membrane, that its contents were not fluid, and that the presence of the nucleus could not always be confirmed.

Brücke considered the cell contents to be the most important part of the cell, and it was in connection with this that he made an interesting observation. He asserted that the application of the concepts of 'solid' and 'liquid', as known in physics, to the state of the cell substance, had no validity. The organic compound molecules contained in living cells arranged themselves in such a manner that the resulting complex structure deserved the special name of 'organization'.[43]

The emergence of the central importance of the viscous cell substance changed within two decades the concept of the cell so that its outline became, in the true sense of the word, blurred. The cell became, as it were, submerged into protoplasm, a complex form of matter, endowed with the fundamental attribute of contractility.

[41] F. Dujardin, *Annales sc. natur. (Zoologie)*, Ser. 2, 4 (1835), pp. 343–77.

[42] M. Schultze, *Arch. Anat. Phys. wiss. Med.* (Reichert, Du Bois Reymond), 1861, pp. 1–27.

[43] E. Brücke, *Sitzb. K. Akad. Wiss.* (Vienna), xliv, 1861, 2. Abth. H. vi–x, pp. 381–406.

Contractility, Respiration and the 'Giant Molecule'[44]

It was widely believed that protoplasmic contractility could usefully be studied in muscle. It should be remembered that Schultze's paper was concerned in the first place with the genesis of the muscle fibre and its structure. Following out this historically, great significance attaches to Kühne's efforts to find out something more specific about protoplasm and muscle contractility in terms of chemistry (1864). Employing a 10 per cent NaCl solution he obtained from muscle a protein compound which he named myosin. As it turned out the discovery of this substance became of signal consequence in the history of knowledge of muscle contraction.[45]

Only three years later the existence of myosin was taken into consideration in the explanation of the nature of chemical processes in muscle.[46] Ever since the discovery of oxygen, its rôle in living activities, muscle contraction not excepted, had been considered crucial. When it became known that muscle could contract in the absence of oxygen supply, L. Hermann proposed that the oxygen needed for contraction was already present in the muscle. According to this view it was firmly bound to, and part of, a complex energy-generating substance, called 'inogen', which underwent continuous decomposition and resynthesis in the muscle tubes and plasm. The decomposition was slow and spontaneous in the resting muscle but was instantaneous during exertion.

To account for the known facts in muscular activity Hermann maintained that inogen was really a carbon-nitrogen precursor which during contraction decomposed into carbon dioxide, lactic acid and myosin. He had more difficulty in showing what happened during restitution. There was already some doubt whether Liebig was right in suggesting that during muscular exertion proteins were metabolized. Voit drew attention to the known fact that people inhabiting mountainous regions, when out working, consumed

[44] This section deals with some aspects of the history of 'giant' molecules which, according to J. Needham, 'now requires detailed historical research', *Notes and Records of the Royal Society of London*, **17** (1962), p. 158.

[45] W. Kühne, *Untersuchungen über das Protoplasma und die Contractilität* (Leipzig, 1864). Cf. also A. Szent-Györgyi: 'I felt I had now enough experience for attacking some more complex biological process, which could lead me close to the understanding of life. I chose muscle contraction. With its violent physical, chemical, and dimensional changes, muscle is an ideal material to study. If one embarks on such a new field one usually does not know where to begin. There is one thing one can always do, and this I did: repeat the work of old masters. I repeated what W. Kühne did, a hundred years earlier. I extracted myosin with strong potassium chloride (KCl) and kept my eyes open.' *Ann. Rev. Bioch.* **32** (1963), 9.

[46] L. Hermann, *Untersuchungen über den Stoffwechsel der Muskeln, ausgehend vom Gaswechsel derselben* (Berlin, 1867).

mostly fat bacon. Traube pointed out that bees were capable of continuous flight, although they were fed on sugar, and suggested that during contraction oxygen was first taken up by a ferment-carrier and then transferred to a non-nitrogenous substrate. Finally, Fick and Wislicenus confirmed that protein metabolism could not account for the energy expenditure needed to perform a mountain ascent. Knowing these circumstances Hermann eventually concluded that myosin was not directly involved in muscular activity. With some hesitation he followed the previous suggestion of Traube and developed the idea that myosin acted as a ferment in combining oxygen with the hitherto undiscovered nitrogenous body to produce the complicated inogen substance. In fact in this process myosin acted as an organic catalyst in a cycle of chemical changes.[47]

The inogen hypothesis was condemned outright as an example of unproductive speculative thinking ever since W. M. Fletcher and F. G. Hopkins demonstrated that there could be no common precursor for carbon dioxide and lactic acid and that it depended on conditions whether they were produced simultaneously in the absence of oxygen supply (1907).[48] It would be superfluous to restate the signal contribution of Fletcher and Hopkins to the study of muscle contraction. As has been pointed out many times, the modern era of work on the nature of muscle action had its beginning with their investigation. Yet the hard fact remains that the validity of the inogen hypothesis was accepted for more than thirty years, and this makes it difficult to ignore, at least from a historian's point of view. Could it be that despite the just criticism which was meted out to it, the inogen hypothesis reflected some fundamental characteristics of muscle contraction? Before tackling this question let us turn to contemporary attempts to study and explain protoplasmic activities other than contractility.

Inogen was the first in a series of 'giant' molecules which just at the turn of the century were thought of as fundamental components of protoplasm, responsible for life itself. If inogen was conceived of in the course of the elucidation of protoplasmic contractility, the 'giant' or 'living protein' molecule was evolved in the examination of the site of respiration.

From 1865 onwards E. Pflüger began to be interested in physiological combustion. After investigating the life functions of frogs deprived of oxygen supply he concluded that respiration was cellular in nature (1875). With Pflüger the cell again assumed a central position; it was synonymous with living matter, and respiration its basic property.

[47] Ibid., pp. 61–2, 64–9, 80–2, 91–2; L. Hermann, *Grundriss der Physiologie des Menschen* (5th edition, Berlin, 1874), pp. 231–4; *Elements of Human Physiology* (trans. A. Gamgee) (London, 1875), pp. 253–6.

[48] W. M. Fletcher and F. G. Hopkins, *Proc. Roy. Soc.* (B), **89** (1915–17), pp. 444–67. Croonian Lecture, held 9 December 1915, MS received 22 November 1916.

In order to explain the survival of frogs under anaerobic conditions he supposed, as Hermann had before him, the existence of intramolecular oxygen bound to protein molecules in living cells. In order to differentiate this from the 'dead' reserve protein found in egg white or plant seeds, Pflüger considered the protein with the intramolecular oxygen as 'living'. The production of carbon dioxide by respiring cells was explained by him as due to intramolecular heat, bringing about the decomposition of the living protein molecules into carbon dioxide, water and cyan groups. At the same time, from the breakdown products protoplasm, i.e., living protein molecules, was constantly regenerated. In connection with the consideration of how protoplasmic growth occurred Pflüger suggested that living protein molecules could undergo constant polymerization and form 'giant' molecules. He thought it feasible that living masses, for instance the whole nervous system, consisted of a single 'giant' protein molecule.[49] As we know the existence of that kind of protein molecule and its transformation has not been confirmed.

Pflüger was greatly praised later on for his decisive contributions to knowledge of the respiratory process. But he was also sharply criticized for his speculations on the nature of the processes in living cells, associating them with the decomposition and restitution of the unstable living protein. What has not always been pointed out is that the praise and criticism apply to interrelated sides of the same work of Pflüger, which included also a theory of the origin of life (with the cyanogen radicals playing a major rôle), still discussed by serious students in the field.[50]

In Pflüger's time his authority carried great weight, and he influenced many investigators, among them P. Ehrlich, who, accepting Pflüger's views on tissue oxidation, was interested in obtaining a deeper insight into the local distribution of the oxidation and reduction processes in the cell. In 1885 he put forward for consideration the idea that the protoplasmic giant molecule envisaged by Pflüger consisted of a central portion, i.e. a nucleus, and side-chains. Whereas the chemical structure of the nucleus was, according to Ehrlich, related to specific cell functions, the side-chains were connected with the vital activities of the cell. The atomic groups of some side-chains acted as sites for the taking on of oxygen and releasing it, and the groupings in the other side-chains attracted molecules which were to be oxidized.[51]

Max Verworn attempted to unite the main features of the ideas of Pflüger and Ehrlich. In 1895, Verworn criticized Pflüger's concept that the living

[49] E. Pflüger, *Arch. Ges. Physiol.*, (Pflüger), 10 (1875), pp. 251–367, 641–4.

[50] A. I. Oparin, *The Origin of Life on the Earth* (3rd edition, London, 1957), pp. 82–4; J. D. Bernal, *The Origin of Life* (London, 1967), p. 225.

[51] P. Ehrlich, *Collected Papers* I (London and New York, 1956), pp. 436–7.

molecule must be protein. He pointed out that no protein was known to be as unstable as the assumed living protein. Instead, he proposed that the living protein molecule should be replaced by a 'biogen' molecule the chemical structure of which Verworn imagined to be built up on the lines advanced by Ehrlich. The biogen molecule, as set out in 1903, was a very complex carbon compound with a benzene ring as the nucleus for a number of side-chains. These contained on the one hand nitrogen and perhaps iron atoms and functioned as receptors of oxygen. On the other hand there were also present atomic groups (e.g. aldehydic) serving as the substrate for the oxidative dissociation of the biogen molecule.[52]

The giant molecules were not, as is often assumed, thought of as units of living matter. Hermann was primarily concerned with the explanation of muscle contractility, Pflüger with tissue oxidation and the origin of life, while Verworn looked for a general metabolic mechanism. Biogens were not so much units of living matter as carriers of life, undergoing metabolic changes within protoplasm. Protoplasm was not conceived of as a unitary substance either chemically or morphologically. The cell was the only form of living matter capable of continuous existence.[53]

That the cell was *the* elementary organism was by no means universally accepted at the turn of the century. This critical attitude towards the cell theory led M. Heidenhain to propose his 'protomere' theory. He introduced the term 'living mass' (*lebendige Masse*) to cover various forms of living matter, of which the cell was only one. All forms of living mass were composed of smallest living units—protomeres.

The protomere had a chemical basis but was not a chemical molecule. Chemical processes underlying the so-called lower functions, i.e., basic metabolism and general increase of the mass of protoplasm were visualized as not having, in the first place, a spatial orientation. The spatial orientation came into play when the protomeres, as the carriers of the so-called higher functions (contractility, impulse conduction, secretion, resorption and specific growth), arranged themselves into living matter spatially organized.[54]

Heidenhain proposed a division of labour among the investigators of living matter. Biochemistry should study the structure and dynamic processes taking place within the protomeres. Microscopy should investigate the architecture of living matter. Dynamic physiology should take up the treatment of functions like contractility, impulse conduction, etc. deriving from the spatial orientations of the protomeres.[55]

[52] M. Verworn, *Die Biogenhypothese* (Jena, 1903), p. 69.

[53] M. Verworn, *General Physiology* (trans. 2nd German edition, 1897) (London, 1899), pp. 80, 484; *Die Biogenhypothese*, p. 102.

[54] M. Heidenhain, *Plasma und Zelle* I (Jena, 1907), pp. 498–9.

[55] Ibid., p. 500.

Enzymic Activity and Cellular Structure

Let us again turn our attention briefly to the cell. As already indicated, among the first to enquire into the oxidation-reduction processes inside the cell was Ehrlich. He was also interested in the processes of dyeing and followed keenly the theoretical and practical sides of this, developed mainly in Germany; he combined both interests and discovered vital staining. It was based on observations that certain coal tar dyes were reduced to their 'leuco-bases' when brought into contact with living cells. Thus were laid the foundations of the methylene blue method, developed later by Torsten Thunberg, for the study of cell respiration.[56]

Without following up the matter it becomes apparent that Ehrlich's work formed a component part of the history of synthetic dyes and drugs and their manufacture. This in turn cannot be separated from the economic, social and political movement which had given rise to, and marked, the growth of Germany in the second half of the nineteenth century. We had already occasion to get acquainted with some of the multifarious antecedents and consequences of the disclosure of cell-free fermentation by Buchner. Thus to account for two of the scientific advances which had fundamentally shaped the further course of modern biochemistry it is not enough to place them merely in their scientific frame.

The demonstration of cell-free fermentation showed in a new light the rôle of intracellular enzymes in living processes. Enzymes rang the death-knell for biogens, but Verworn battled hard to save the biogen hypothesis of metabolic change. In the course of his analysis he made some acute observations, envisaging, for instance, that the enzyme was also a substrate.[57] He noted that in many respects enzymes were supposed to operate in a similar way to biogens. Hence those who believed that there were points of contact between both conceptions could consider the biogen hypothesis as an improved form of the enzyme hypothesis. Verworn himself, however,

---

[56] Ehrlich, op. cit. (51), p. 438 f. Among the dyes he used Ehrlich did not mention methylene blue. Thunberg pointed out that the first description of this dye in connection with cell respiration studies was given by H. Dreser, and that Ehrlich used it later under the influence of Dreser's publication. Cf. *Skand. Arch. Physiol.* **35** (1918), p. 193; *Quart. Rev. Biol.* v (1930), pp. 322–7. There is no doubt that Dreser thought on similar lines to Ehrlich and independently of him published the first description of the use of methylene blue in respiration experiments. See H. Dreser, *Ztschr. f. Biol.* **21** (1885), pp. 41–66. After reading Dreser's communication Ehrlich published in the same year two papers on methylene blue with a critical discussion of some of Dreser's results. In a footnote to the first paper Ehrlich emphasized that he knew about methylene blue before he had read Dreser's article and that he had already lectured on it on 18 December 1884. Cf. *Collected Papers* I, pp. 497–508.

[57] M. Verworn, op. cit. (52), p. 15.

considered this approach an idle one. The enzyme hypothesis was no less hypothetical than the biogen hypothesis. But whereas the biogen hypothesis assumed that metabolic changes could be all explained from the changes of one compound, the enzyme hypothesis 'requires a great number of various enzymes in each cell, acting in a coordinated manner and each performing only its special function'.[58] In conclusion Verworn maintained that where there was a conflict between the principles of singularity (i.e. biogen) and plurality (i.e. enzymes), the rules of scientific research demanded that the first should be given priority.

The discovery of zymase was of fundamental importance because it paved the way towards a conception of a chemically organized living matter and in this way also to a new sub-microscopic picture of protoplasm. Perhaps nobody formulated these ideas so early as F. Hofmeister[59] and F. G. Hopkins.[60]

For both of them, the colloidal ferments were the most important instruments the cell possessed for the performance of ordered chemical reactions. They were both very critical of explanations of living processes based on the decomposition and restitution of this or the other living molecule. They thought of cells as highly differentiated but integrated systems. But there were also certain differences, perhaps not so much of principle, as of emphasis.

Hofmeister was very concerned to justify that there was nothing inherently improbable in the suggestion of a multitude of chemical reactions simultaneously taking place and involving about 2×10^{13} water, protein, lipoid, crystalloid and salt molecules in a space of, say, $8000\mu^3$. This was the assumed volume of a liver cell on the basis that it was a cube whose side measured 20μ. Hofmeister likened the cell to a physico-chemical machine, potentially self-regulated by the reversible nature of enzyme catalyzed reactions. From Hofmeister's description the possibility of a co-ordinated sub-microscopical physico-chemical world underlying the visible morphological structures in the cell began to emerge.

Hopkins did not imply as clearly as Hofmeister, at any rate before the First World War, that the gap between the biochemical and morphological fields could be bridged. For him the living cell was not so much a physico-chemical machine as a colloidal system of phases coexisting in a dynamic equilibrium. His own work on muscle chemistry and vitamins made him

---

[58] Ibid., p. 113.

[59] F. Hofmeister, *Naturwissenschaftliche Rundschau*, 16 (1901), pp. 581–3, 600–2, 612–14; *Schriften der wissenschaftlichen Gesellschaft in Strassburg*, 17 (1912); *Ztschr. Morph. Anthrop.*, 18 (1914), pp. 717–24.

[60] F. G. Hopkins, *Rep. Brit. Ass.* (1913), pp. 652–68.

realize that the chemical events within this system were associated with reactions between relatively small molecules which could be elucidated by current experimental methods.

What Hofmeister and Hopkins were really facing was the problem of the relations of enzymic activity and cell structure, a question of some importance, for instance, in the history of cell respiration, yet relatively little discussed.

Some of the most important contributions to our knowledge of cell respiration derived from T. Thunberg who began his series of extensive studies early in this century. In the course of his work he developed a method for the study of reducing systems in tissue preparations by placing them in special vacuum tubes together with methylene blue. The method, which was an enlargement of the work by Ehrlich and Dreser in the eighties of the last century, led to the demonstration of dehydrogenases, and furthermore, in conjunction with Wieland's concept of oxidation of organic compounds by removal of hydrogen, to the hydrogen-activation theory. The question whether the action of dehydrogenases was affected by cellular structure Thunberg answered in the negative.[61]

By contrast, O. Warburg, the originator of the rival oxygen-activation theory, stressed the rôle of the structural elements in enzymic activity. Warburg came into the field before the First World War when he noticed a reduction of respiration in cytolysed and mechanically disrupted cells. For him this was evidence that cell respiration was connected in some way with cell organization. He pursued his studies further, and enzymes as such did not at first play a great rôle for him in the explanation of the respiratory process. He attributed much significance to cellular iron and assumed that respiration was due to iron surface catalysis. Later Warburg modified his views by attaching the greatest importance in the activation of molecular oxygen to an iron-containing organic catalyst, related to the blood pigment, the so-called *Atmungsferment* (respiratory ferment).[62] The controversy between the followers of the rival theories and the way it was resolved, is a part of history which has been treated before and is of no immediate concern to us here.[63]

Compared with the stir created by this issue the problem of the relations of enzymic activity and cell structure left the biochemists on the whole

[61] T. Thunberg, *Ergeb. Physiol.* (1911), 330; **39** (1937), pp. 76–116; *Skand. Arch. Physiol.*, **35** (1918), pp. 163–95; **40** (1920), pp. 1–91; *Quart. Rev. Biol.*, v (1930), pp. 318–47.

[62] O. Warburg, *Arch. Ges. Physiol.* (Pflüger), **145** (1919), pp. 277–9; O. Warburg, O. Meyerhof, ibid., **148** (1912), p. 295; O. Warburg, *Biochem. Ztschr.*, **119** (1921), pp. 135, 152–3; *Jhrb. ges. Physiol. exp. Path.*, i (1923), p. 143.

[63] See D. Keilin, op. cit. (3), also M. Dixon, 'The History of Enzymes', in J. Needham (ed.), op. cit. (4), pp. 30–7.

fairly cool. Only a few paid any attention to it, among them J. H. Quastel, who together with some colleagues examined bacterial systems and concluded that the activity of dehydrogenases was dependent on the integrity of the cell. This idea did not find much favour with Thunberg, who believed Quastel went astray because he worked with bacteria. 'In the case of such cells,' he wrote, 'it is tempting to look upon the solid structure as essential for functions which in other cells are not necessarily bound to a structure. If we work on larger cells and on tissues of softer consistency we are not so easily tempted to attach such a high importance to the cell structure for the dehydrogenation processes.'[64]

In 1935 H. A. Krebs, in a paper dealing with deamination, discussed in some detail relations of enzymic activity and cell structure. He found that kidney and liver possessed two deaminating enzymic systems which differed in the way they behaved on extraction of the ground tissue. As a pupil of Warburg, Krebs was familiar with his teacher's investigations on the effect of dilution on the respiration of laked erythrocytes, and he decided to examine the process of extraction in detail. He showed that *l*-amino acid deaminase, responsible for the deamination of the naturally occurring *l*-series, was inactivated not by grinding or disruption of cell, but by dilution. 'These results lead to a distinction between enzymes or systems,' wrote Krebs, 'which act independently of the amount of fluid in which they are dissolved or suspended and systems which act only within a small range of concentration and are destroyed if the medium is diluted. The vast majority of the common enzymes belong to the first group. The second group comprises those reactions which have been considered as being bound up with the structure of the living cell, i.e., the bulk of oxidations and fermentations.'[65]

Related to this work were the experiments by J. Yudkin, who confirmed two years later that the cause of reduced activity of the bacterial dehydrogenases on lysis was in some way linked with the structure of the cell.[66]

When in 1939 I. M. Korr reviewed the field he found it in a rather vague state. Korr asserted that in practice not enough attention was really paid to the fact that chemical reactions were taking place not in a pure buffer solution but in a complex microheterogeneous environment. Further, the relations of chemical and morphological processes in the cell, according to Korr were reciprocal. Korr wrote under the influence of the 'co-ordinative biochemistry' outlook of R. A. Peters, and the 'cyto-skeleton' idea of P.

[64] J. H. Quastel, *Trans. Farad. Soc.*, **26** (1930), p. 857; T. Thunberg, *Quart. Rev. Biol.*, v (1930), p. 333.
[65] H. A. Krebs, *Biochem. J.*, **29** (1935), pp. 1639–43.
[66] J. Yudkin, ibid., **31** (1937), pp. 1065–8.

Wintrebert, F. Vlès and J. Needham, who urged the need for considering much more closely the relations of the dynamic side of cell processes and the sub-microscopical features of cell architecture.[67]

Co-ordinative Biochemistry

Peters started his research career with J. Barcroft before the First World War, on the nature of the union of oxygen with haemoglobin. His first doubts on the purely physical explanation of adsorption arose then and they were reinforced when he became familiar, in the later stages of the war, with the effects of war gases on living tissues, which he was asked to examine. He began to look for a chemical or physico-chemical approach to permeability and adsorption. He found enough support in the work of Hardy, Harkins Langmuir and Adam on mono-molecular films of organic compounds at interfaces to press for a new understanding of integrated cell activity. Peters formulated his theoretical views for the first time in his Harben Lectures of 1929 in the USA. He distinguished two types of surfaces, the external surface presented by the cell to its environment, and internal surfaces or interfaces within the cell, and suggested that 'a cell surface whether internal or external may be made of molecules so anchored that they constitute a chemical mosaic. This conception makes co-ordinative biochemistry possible.'[68]

A year later, in a discussion on the structure of living matter arranged by the Faraday Society, Peters returned to his challenge of the current biochemical doctrine. Referring to 'normal architecture of the cell' and the 'dynamic equilibrium' as the two notions often employed to characterize life Peters said—

> The view which is presented here differs from most others in the stress which is laid upon architecture. Its keynote is the complete (or nearly complete) structural organization of the cell. I believe this to be organized not only in respect to its grosser parts such as the nucleus, but also in regard to the actual chemical molecules of which it consists. Owing to the microheterogeneous nature of the system, surface effects take precedence over ordinary statistical, mass-action relationships, and become in the ultimate limit responsible for the integration of the whole and therefore the direction of the activities. It is believed that the directive effect of the internal surfaces is displayed predominantly by an organized network of protein molecules forming a three-dimensional mosaic extending throughout the cell. The enzymes would form part of this

[67] E. M. Korr, *Cold Spring Harbor Symp. Quant. Biol.*, vii (1939), pp. 74–93.
[68] R. A. Peters, *Biochemical Lesion and Lethal Synthesis* (Oxford, 1963), p. 221.

structure their activity being largely controlled by the mosaic. This conception of living matter reconciles some difficulties; for instance independent chemical reactions could proceed simultaneously in various parts of the cell, but at the same time the mosaic could not react as a whole to stimuli transmitted from the cell surface.[69]

Although Hopkins accepted, in general, the notion of a 'geography of the cell' he was not very explicit about what should be understood by 'chemical organization' of the cell.[70] Peters critically developed Hopkins' views on the colloidal conditions in which the dynamic equilibrium in the cell came to be established by bringing enzyme and surface chemistry together. However important enzymes were, Peters could not see his way to consider them as primarily responsible for the integrity of the cell. He made it clear that he was concerned with sub-microscopic cell organization, within which biochemical cycles such as the lactic acid-glycogen interconversion could operate.[71]

It can hardly be said that Peters' communication was welcomed by all participants. One of the leading physical chemists present, F. G. Donnan, felt that the expression 'chemical architecture of the cell' used by Peters was too vague. Viscosity experiments on living protoplasm did not offer support for structural conceptions. If the expression stood for molecular chain orientation then this view, although not experimentally confirmed, was not original.[72] A much sharper attack came from one of the researchers working in the Cambridge laboratory, B. Woolf. He did not think that sub-visible structures had to be postulated for chemical reactions to proceed in the cell. Buchner with the yeast press juice and Meyerhof with the cold water extract of muscle had shown that enzymes could operate in simple systems. This line of study should be continued by concentrating not only on the substrates of enzyme activity but also on the intermediate stages of metabolism. Peters' supposed invisible structure in protoplasm could hardly be tested in experimental practice and thus the suggestion was not very helpful for the

[69] R. A. Peters, *Trans. Far. Soc.*, 26 (1930), pp. 797–8.

[70] Cf. J. S. D. Bacon: 'With the advantage of a further thirty-five years' study of cellular structure we must be careful in interpreting his (i.e. Hopkin's—M.T.) statements. In particular it is difficult to understand exactly what he meant by "chemical organization", and how he would have reacted to what has now been discovered about the fine structure of cell organelles. Undoubtedly, he visualized intracellular catalysts as being "organized" to the extent that some were fixed to membranes, but it would seem that by chemical organization he meant the linking of individual enzymes through the extreme specificity of their catalytic reaction. His assertion was in part a response to those who denied that an *in vitro* study of intracellular enzymes could explain how they acted in the highly-structured cell system.' W. Bartley, H. L. Kornberg, J. R. Quayle (eds.), *Essays in Cell Metabolism* (Hans Krebs Dedicatory Volume) (London, 1970), pp. 61–2.

[71] R. A. Peters, *Trans. Far. Soc.*, 26 (1930), pp. 797–806.

[72] F. G. Donnan, ibid., pp. 815–16.

advance of biochemical knowledge.[73] Among those who came to the support of Peters, however, was J. Needham, who believed that the speakers had tended to take a rather crude view of what Peters was trying to convey. Needham reminded the audience that certain results obtained by ultraviolet spectrophotometry of sea-urchin eggs upon cytolysis showed a difference in absorption protein curves, and asked: 'Does this not indicate in the living cell the proteins are associated with the other constituents in some radically different way from that which occurs in mixtures of compounds which are not organized?'[74]

Before we turn our attention to Needham's 'cyto-skeleton', an idea closely related to the co-ordinative conception, let us briefly follow its further handling by Peters. He reaffirmed his position in 1937, but this time he was able to discuss the hypothesis, much more than before, in the light of experimental evidence provided by himself and other investigators.[75]

Peters was able, especially, to draw upon his researches on vitamin B_1, originally begun in 1922 in Cambridge, and then continued in Oxford. Investigations of the brain of pigeons fed upon polished rice led to the establishment of the co-enzymic rôle of vitamin B_1 in the oxidation of pyruvate. The vitamin B_1 deficiency studies gradually developed into an area where nutrition and intracellular chemistry met pharmacology and pathology. Peters showed that disturbances in enzyme activity resulted in a 'biochemical lesion' and 'lethal synthesis', injuring or bringing the life of the cell to an end. The theoretical and experimental work of Peters demonstrated the value of the dynamic and structural approaches to cell organization and, in fact, proved that they had to be interconnected.[76]

The 'Cyto-Skeleton'

We referred already to J. Needham's concern with problems of cellular organization when he came to the support of Peters at the Faraday Society discussion on the structure of living matter. As a pupil of Hopkins, Needham had learned from his teacher that the cell was a complex heterogenous system, endowed with a 'chemical geography', in which reactions between simple molecules were catalyzed by colloidal enzymes. Needham's deeper interest in cellular organization was aroused by listening to Peters' lectures

[73] B. Woolf, ibid., pp. 816–17.

[74] J. Needham, ibid., p. 819.

[75] R. A. Peters, in J. Needham and D. E. Green (eds.), *Perspectives in Biochemistry*, (Cambridge, 1937), pp. 36–44.

[76] R. A. Peters, op. cit. (68), pp. 54, 174, 180, 202.

on 'physico-chemical cytology' which in turn owed so much to W. B. Hardy's stress on the need for obtaining more solid information on the colloidal state.[77] As a result of these influences Needham matured into a dedicated practitioner of physical biochemistry, especially in the fields of embryology and morphogenesis, which he made his own. For him the problem of biological organization was fundamental, and he believed it was within the possibilities of experimental science to take up the age-old questions and answer them meaningfully, if not exhaustively. It was for this reason that he opposed vitalism because he thought it inhibitory to further advance.

This then, is the message of Needham's *Order and Life*, a book published in 1936 on the basis of his Terry Lectures at Yale University, delivered there the year before. Six years after Peters' first formulation, Needham discussed and developed the conception of co-ordinative biochemistry by which he meant 'the extension of morphology into biochemistry and the bridging of the gulf between the so-called sciences of matter and the so-called sciences of form'.[78] The thirty-two years which elapsed between the first and second editions have not impaired one of the main arguments presented in the book, namely that 'biochemistry and morphology should, then, blend into each other instead of existing, as they tend to do, on each side of an enigmatic barrier'.[79] As one reads the account one sees more distinctly that the young Cambridge biochemist, in the light of knowledge existing in the mid-thirties, re-stated much more clearly ideas which Hofmeister had earlier proclaimed.

Needham reviewed the evidence for the thesis that morphological patterns were chemically conditioned and found it in diverse results and fields, such as changes in respiration upon cytolysis, relations between cellular structure and oxidation mechanism, the co-ordination of enzyme systems in muscular work and elsewhere. Above all, he showed a remarkable grasp of new knowledge of properties of large molecules derived from centrifugation, flow birefringence and X-ray studies of complex substances of biological origin. Possibilities were opening at last to give some substance to the idea previously expressed long before by investigators like Schwann, that morphogenesis was a process akin to crystallization. It became apparent that for the understanding of physiological processes such as muscle contraction knowledge of molecular organization was no less important than that of molecular reactions. This had important repercussions for the tackling of the fundamental problem of protoplasmic organization.

---

[77] W. B. Hardy, *Collected Scientific Papers* (Cambridge, 1936).
[78] J. Needham, *Order and Life* (2nd edition, Cambridge, Mass., 1968), p. 138.
[79] Ibid., p. 139.

Many histologists and embryologists had tended to accept, since the last quarter of the previous century, one or the other 'histological' descriptions of protoplasmic structure such as granular, filar, fibrous, reticular, alveolar, foam-like and so on. It was believed that whatever the pattern, the structure constituted an elastic but fairly rigid framework. These views persisted well into the third and fourth decade of this century as shown, for instance, in the term 'cytosquelette' introduced by Wintrebert (1931). Needham thought that it could prove useful to describe a markedly different situation where molecular events within the cell were responsible for morphological, dynamic, patterns and brought in the concept of 'cyto-skeleton' or 'cell-skeleton' to cover it, as also the 'co-ordinative' model of Peters. The cell-skeleton was believed to be a fluid lattice with large molecules playing an important part in protoplasmic organization.[80] Did it mean that the once rejected 'giant' molecules were, in fact, being resurrected?

Although the impact of the giant molecule concept upon biological thought cannot be totally discounted, the history of the study of large molecules in the twentieth century can be traced rather to Nägeli's hypothesis, stated in the forties of the last century, that starch grains and cellular fibres are composed of sub-microscopic elongated crystalline particles, which he called 'micelles'. Not long ago J. S. Wilkie undertook a detailed study of Nägeli's work and concluded that Nägeli's views were progressively rejected in the period from 1882 to 1920.[81]

From the early 1920s, however, the study of large 'polymeric' molecules such as cellulose, starch, rubber, and proteins, began to receive a strong impulse from industry. There were not a few factories in existence for the manufacture of materials which were, in effect, only modifications of natural high polymers. Due to the endeavours of W. N. Haworth, H. Staudinger, W. H. Carothers, K. H. Mayer, H. Mark, T. Svedberg and others the study of large molecules advanced fairly rapidly.[82] An important rôle in the bringing together of biological and technical studies was the work on textile fibres by W. T. Astbury, J. B. Speakman and others. It is interesting to read in Astbury's book *Fundamentals of Fibre Structure* (1933) what he had to say about Nägeli's work: 'The moral of this romance of research, and of many similar stories in the history of science, is again that of the profound unity underlying all natural phenomena. We need not suppose for a moment that Nägeli, as he pored over his plant cells and starch grains ever gave a thought to the welfare of textile industry, but the fact remains that his

[80] Ibid., pp. 149 f.
[81] J. S. Wilkie, *Annals of Science*, 16 (1960), p. 209.
[82] H. Mark, *Impact of science on society*, xviii (1968) (1), pp. 27–33.

problems were in essence our problems, and therefore the gifts of his genius were as much for industry as for biology.'[83]

It was Astbury's work which led Needham to the aphorism that 'biology is largely the study of fibres'. The realization that proteins were crystalline fibres provided a basis for picturing the cell-skeleton as a web-like structure, composed of chain-like molecules which was continuously built up and broken down. There could be no doubt that the concept of the cell-skeleton signified a substantial swing from the position of Hopkins, the great advocate of small molecules, who had little use for large molecules in cell-life. Needham recognized the change when he wrote: 'In general, it may be said that in biochemistry the interest has in recent years been slowly shifting from chemical reactions as such to the chemical, indeed "organic", structure of the great molecules. How these great molecules come to be there, how they are formed and how replaced, brings us always back to the realm of morphology. But for the unification of biochemistry and morphology, which every reflective biologist has at heart, this shift of interest is an excellent thing.'[84] It speaks both for the teacher and the pupil that these sentences appeared in Needham's contribution to the *Hopkins-Festschrift* of which he was also one of the editors.

All the same, plausible and attractive direct evidence for the existence of the cyto-skeleton was hard to come by. Needham admitted this in his Terry Lectures and also later in *Biochemistry and Morphogenesis* (1942), which included an extended version of the material and argument presented in *Order and Life*. He was encouraged in his belief, for instance, by the findings in the thirties of Fauré-Fremiet on thixotropy and Pfeiffer on cytoplasmic anomalous viscosity.[85] By the time the book was published, Needham and his co-workers had just finished struggling with the problem by publishing three extensive papers on the anomalous viscosity of protein solutions.[86]

Needham assumed that the conception of a cyto-skeleton composed of

[83] W. T. Astbury, *Fundamentals of Fibre Structure* (London, 1933), pp. 100–1.

[84] J. Needham, op. cit. (75), p. 78.

[85] J. Needham, *Biochemistry and Morphogenesis* (Cambridge, 1942), pp. 659–61.

[86] I. A. S. C. Lawrence, J. Needham, S.-C. Shen, *J. Gen. Physiol.*, 27 (1943–4), pp. 201–32.

    II. A. S. C. Lawrence, M. Miall, J. Needham, S.-C. Shen, *J. Gen. Physiol.*, 27 (1943–4), pp. 233–71.

    III. M. Dainty, A. Kleinzeller, A. S. C. Lawrence, M. Miall, J. Needham, D. M. Needham, S.-C. Shen, *J. Gen. Physiol.*, 27 (1943–4), pp. 355–99.

The work was begun in 1940 by Joseph Needham, Shih-Chang Shen and Dorothy M. Needham, with A. S. C. Lawrence as rheological adviser. Arnošt Kleinzeller and Margaret Miall joined the group in 1941 and Mary Dainty in 1942. Cf. *J. Gen. Physiol.*, 27 (1934–44), p. 355. There were also short reports in *Nature*, 146 (1940), pp. 104–5; 147 (1941), pp. 766–8; 150 (1942), pp. 46–9.

micellar fibrils could be tested by following morphogenetic problems, as for instance the morphological changes in the developing neural tube of the amphibian embryo which involved a great lengthening of the cuboidal ectodermal cells. Could it be that this elongation had something to do with the presence of anisometric particles in the embryonic stages of the organism? From simultaneous measurements of flow-birefringence and anomalous viscosity of protein fractions obtained not only at the neurula stage, but also from unfertilized and fertilized eggs they found a certain justification for this assumption—

> The fibrillar micellar 'cyto-skeleton' is proving difficult to demonstrate precisely . . . our contribution to the question lies in the fact that in the amphibian embryo there is a protein or a group of proteins in the total euglobin class which spreads instantaneously into a surface film having the property of anomalous flow. Its molecules therefore readily pass into the fibrillar stage. . . . The union of all these facts into a coherent picture of morphological change at neurulation is, however, a matter for the future.[87]

It is characteristic of Needham to perceive possible connections between seemingly widely separated areas and to explore them. In 1939 the biochemical world was startled by the discovery of W. A. Engelhardt and M. N. Ljubimova that the protein myosin behaved as the enzyme adenosine triphosphatase.[88] For the first time in the history of biochemistry and physiology a direct relation between microstructure and chemical action was experimentally established. It is not without interest that as a result of the new suggestion some of the older ideas that myosin was an organic catalyst and also a means of transforming chemical into mechanical energy, though not confirmed in the form proposed by Traube and Hermann, were to a certain degree vindicated.

The discovery of the enzymic property of the contractile component of the muscle is usually considered as the start of a new era in muscle chemistry. Among the first to enlarge the subject on lines similar to those of the Soviet investigators was the Needham group, especially Dorothy Needham. Muscle chemistry was her primary interest, and she had already many important achievements to her credit, such as the observation, fundamental to the understanding of muscle glycolysis, that the oxidation of phosphor-glyceraldehyde to phosphor-glyceric acid was coupled with adenosine-triphosphate synthesis.[89] One might have thought that the study of myosin properties

[87] *J. Gen. Physiol.*, **27** (1943–4), p. 231.

[88] W. A. Engelhardt, M. N. Ljubimova, *Nature*, **144** (1939), pp. 668–9; cf. also W. A. Engelhardt, *Yale Jl Biol. Med.*, **15** (1942–3), pp. 21–38 (originally published 1941 in Russian).

[89] D. M. Needham, R. K. Pillai, *Biochem. J.*, **31** (2) (1937), pp. 1837–51. See also Dorothy Needham's *Machina Carnis The Biochemistry of Muscular Contraction in its Historical Development* (Cambridge, 1971).

could hardly have had much to do with chemical embryology. Still, the in the thirties of Fauré-Fremiet on thixotropy and Pfeiffer on cytoplasmic experiments were ostensibly designed for the study of the effect of adenosine-triphosphate on myosin by measuring the changes in the double refraction of flow and viscosity of myosin solutions, combining the von Muralt-Edsall demonstration of myosin flow-birefringence[90] with the Engelhardt-Ljubimova discovery of the enzymatic property of myosin. The findings of the Needham group confirmed the contractile enzyme hypothesis of the Soviet investigators in so far as adenosine-triphosphate caused a *reversible* decrease in the axial ratio of the myosin particles, and they considered several possible explanations for this, e.g. an actual contraction of the molecule, or a sliding effect of complex micelles. At that time the important protein actin, discovered contemporaneously by Szent-Györgyi and his school in Hungary, was not known in England, but it is now clear that the solutions in the Cambridge co-axial viscosimeter must have consisted largely of actomyosin. This would have dissociated to give the separate molecules of each protein. As is now known they polymerize to form filaments in the intact muscle, and contraction occurs by a mechanism in which one set of these slides along another.

For J. Needham, however, the implication of the study went beyond muscle contraction because he thought that contractility was associated with many embryonic processes. He wondered whether inductor substances did not behave as contractile enzymes in a fashion similar to myosin and whether such transformations were not at the root of histogenesis.

At a conference on 'Explanation in Biology' held in June 1968 in the United States, one of the participants, S. Grobstein, said—

> It is well known that complex macromolecular patterns can self-assemble from dissociated molecules; that is, what we have been talking about at the cellular level also occurs at the molecular one. Native collagen, for example, can be solubilized in dilute acid so as to completely lose its fibrous character. On return to neutrality, the still intact collagen monomers reassemble into fibres and these can be shown to have a characteristic ultrastructural banding pattern whose detail can be altered by the conditions of the reassembly.[91]

Employing the terminology of information theory as applied to genetics he posed the question whether there could be 'a useful formal approach to the problem of the conversion of property from one level to those of another?'[92] What Grobstein had in mind was a transition like this: micro-

[90] A. v. Muralt, J. T. Edsall, *J. Biol. Chem.*, 89 (1930), pp. 315–50, 351–86.
[91] C. Grobstein, *J. Hist. Biol.*, 2 (1969), p. 202.
[92] Ibid., p. 203.

molecule→macromolecule→polymer→ultrastructural array→cell→tissue→ organ→organism.

He was answered appropriately by W. Coleman who referring to *Order and Life* said—

> Needham's prescience included a full discussion of structural chemistry. This was not, of course, the chemistry of DNA, but it was the chemistry which then seemed alone able to handle problems of development.[93]

Indeed, as Needham began to see it in the early 1940s, nature (and society) consisted of a series of dialectically connected levels of organization and the cyto-skeleton embodied one of these transitional levels or 'envelopes'.[94]

Later combined biochemical and electron-microscopical explorations of the cell, especially in the 1950s, which are outside the period considered in this article, have fully justified the earlier ideas of Needham and others on the chemical organization of living matter. Perhaps the following passage from a recent well-known textbook on cell biology illustrates adequately this connection between the scientific past and the present—

> Most workers agreed that although the ground substance appeared optically homogeneous in the light microscope, it must nevertheless contain a sub-microscopic skeleton responsible for its elastic properties. These early workers conceived of a *cytoskeleton* of highly elongated particles interacting with each other to form a 'brush heap' or gel. As recently as 1950 Francis Crick wrote about protoplasm being like 'mother's sewing basket' filled with spools of thread, buttons, and knitting needles in untidy array. Until recently it looked as if no 'knitting needles' or cytoskeleton of elongated particles was likely to be found in the ground substance except for the very obvious fibrillar organization of special cells such as striated muscle cells or special cellular structures

---

[93] W. Coleman, ibid., p. 218.

[94] Compare what J. Needham wrote in his Introductory Essay 'Metamorphoses of Scepticism' in the third collection of his essays and addresses—'We cannot consider nature otherwise than as a series of levels of organisation, a series of dialectical syntheses. From ultimate physical particle to atom, from atom to molecule, from molecule to colloidal aggregate, from aggregate to living cell, from cell to organ, from organ to body, from animal body to social organisation, the series of organisational levels is complete. Nothing but energy (as we now call matter and motion) and the levels of organisation (or the stabilised dialectical syntheses) at different levels have been required for the building of our world. The consequences of this point of view are boundless. Social evolution is continuous with biological evolution, and the higher stages of social organisation, embodied in advanced ethics and in socialism, are not a pious hope based on optimistic ideas about human nature, but the necessary consequence of all foregoing evolution. We are in the midst of the dialectical process, which is not likely to stop at the bidding of those who sit, like Canute, with their feet in the water forbidding the flood of the tide.' *Time, the Refreshing River (Essays and Addresses, 1932–1942)* (London, 1943), p. 15. See also pp. 31, 122 f., 233 f.

such as the spindle, the centrioles or the cilia. In the last few years, however, electron microscopists have begun to discover *filaments* and *microtubules* in a large variety of cell types, and it now seems clear that these elongated rods are a universal component of the ground substance.[95]

Conclusion

Historical writing on biochemistry has too often been concerned with relatively isolated aspects of the subject. It is curious that the place of biochemistry within the broader context of biological history has been little examined. Thus the development of biochemistry in its relation to the cell theory has been hardly considered.

This brief survey has shown the essential rôle of the historical relations of cell morphology and cell chemistry in the advancement of contacts between biology and chemistry and, therefore, in the development of biochemistry from about 1840. At times differing in approach and the use of techniques, the cross-fertilization of the two fields eventually produced the linking of the structural and dynamic outlooks in the study of the chemical organization of living matter. Although characteristic of so much of the biochemistry of our days this trend was by no means so apparent before 1940. Thus relations of enzymic activity and all structure were examined only sporadically. This was historically conditioned because biochemists were, understandably, preoccupied with work on pure enzymes. But as H. A. Krebs pointed out the—

purification of enzymes may modify their properties and that it is therefore essential to investigate the behaviour of enzymes not only in the purified state but also in their natural environment. The study of crude enzymes may bring to light important characteristics which may escape attention in the examination of the pure enzymes. This follows from the consideration that living cells are systems whereby the whole is more than the sum of its components. The integration of the parts to a unit involves an arrangement where the component parts influence each other's behaviour. In the terms of chemistry this interlocking implies that cell constituents modify each other's chemical reactivity an interplay which is an essential part of the regulatory mechanisms. It distinguishes a complex unit from a complex mixture.[96]

---

[95] A. G. Loewy and Ph. Siekevitz, *Cell Structure and Function* (2nd edition) (London, 1970), p. 55.
[96] H. A. Krebs, in M. Kasha, B. Pullman (eds.), *Horizons in Biochemistry* (Albert Szent-Györgyi Dedicatory Volume) (New York and London, 1962), p. 291.

This is in line with the warnings of N. W. Pirie[97] and A. Tiselius[98] who have emphasized that the biochemist has to be concerned just as much with the 'impure', the 'non-uniform', the 'dissimilar', and the 'heterogeneous' as he is with the 'pure', the 'uniform', the 'identical' and the 'homogeneous'.

When R. A. Peters proposed in 1929 and 1930 to take into account the inner architecture of the cell on the basis of a 'co-ordinative biochemical' approach, his views were apparently not acceptable to the pure colloid chemists, though the more thoughtful biochemists began to appreciate the necessity of linking biological form and chemistry together as reflected in the ideas of the cell micro-morphology put forward by J. Needham.

The story is complex because in the actual development of biochemistry, as indicated by the history of fermentation chemistry, vital staining, or the knowledge of the nature of fibres, the 'external 'and the 'internal' factors do not operate separately. The study of the history of biochemistry has to bear in mind the social and philosophical aspects just as it has to take into consideration the relations of structural and dynamic approaches against the background of important generalizations such as the cell theory.[99] In this way, perhaps, a rational comprehension and presentation of the intellectual and technical difficulties in the development of modern biochemistry and its place in history can be attained.

---

[97] N. W. Pirie, *British J. Phil. Sc.*, **2** (1951–2), 275–7; *Arch. Biochem. Biophys. Suppl.*, **1** (1962), 21–9.

[98] A. Tiselius, *Ann. Rev. Biochem.*, **37** (1968), 4, 14–15.

[99] The same argument applies also to the theory of evolution. Not before dynamic biochemistry became well established did the relevance of evolutionary concepts for the understanding of some of the universal biochemical processes such as the citric acid cycle begin to be appreciated.

BIBLIOGRAPHY OF JOSEPH NEEDHAM

Books

Science, Religion, and Reality (ed., with a contribution), Sheldon, London, 1925; Macmillan, New York, 1928. 2nd ed. Braziller, New York, 1955.

Man a Machine, Kegan Paul, London, 1927; Norton, New York, 1928.

The Sceptical Biologist (essays and addresses), Chatto & Windus, London, 1929; Norton, New York, 1930.

Chemical Embryology (3 vols.), Cambridge Univ. Press, 1931; repr. Hafner, New York, 1963.

The Great Amphibium (essays and addresses), SCM Press, London, 1932.

A History of Embryology, Cambridge Univ. Press, 1934. 2nd ed. 1959. Russian tr. by A. V. Yodinci & V. P. Karpova, Moscow, 1947.

Order and Life (Terry Lectures at Yale University), Yale Univ. Press, 1935. Italian tr. by M. Aloisi, Einaudi, Torino, 1946. Repr. M.I.T. Press, Cambridge, Mass., 1968.

Christianity and the Social Revolution (ed. [with J. Lewis *et al*.], with a contribution), Gollancz, London, 1935.

Adventures Before Birth (tr. from the French of Jean Rostand), Gollancz, London, 1936.

Perspectives in Biochemistry (F. G. Hopkins Presentation Volume; ed. [with D. E. Green], with a contribution), Cambridge Univ. Press, 1937. Repr. 1938.

Background to Modern Science (lectures arranged by the Cambridge Univ. History of Science Committee), ed. [with W. Pagel]), Cambridge Univ. Press and Macmillan, New York, 1938.

The Levellers and the English Revolution (under ps. Henry Holorenshaw), Gollancz, London, 1939. Russian tr. by S. M. Raskinoi & B. F. Semenov, Moscow, 1947. Italian tr. by C. de Cugis & G. Mori, in 'Saggi sulla Rivoluzione Inglese del 1640', ed. C. Hill, Feltrinelli, Milano, 1957. American ed., Fertig, New York, 1971.

Science in Soviet Russia (ed. [with J. Sykes Davies]), Watts, London, 1942.

Biochemistry and Morphogenesis, Cambridge Univ. Press, 1942. Repr. 1950, and (with new introduction), 1966.

The Teachers of Nations, Addresses and Essays in Commemoration of John Amos Komensky (Comenius) (ed.), Cambridge Univ. Press, 1942.

Time, the Refreshing River (essays and addresses), Allen & Unwin, London, 1943. Repr. 1948.

History is on Our Side (essays and addresses), Allen & Unwin, London; Macmillan, New York, 1945.

Chinese Science (album of pictures taken during the Second World War), Pilot Press, London, 1945.

Science Outpost (papers from the Sino-British Science Cooperation Office), Pilot Press, London, 1948. Chinese tr. by Hsü Hsien-Kung & Liu Chien-Khang, Shanghai, 1947; also, in 2 vols., by Chang I-Tsun, Thaipei, 1952, 1955. Japanese tr. in preparation, Heibonsha, Tokyo.

Hopkins and Biochemistry (F. G. Hopkins Memorial Volume; ed. [with E. Baldwin]), Heffer, Cambridge, 1949.

Science and Civilisation in China (7 vols. in 11 or more parts), Cambridge Univ. Press, 1954– . In collaboration with Wang Ling, Lu Gwei-Djen, Kenneth Robinson, Ho Ping-Yü, Lo Jung-Pang, Nathan Sivin, Ch'ien Ts'un-Hsün, Ohta Eizō, Huang Jen-Yü and others. Chinese tr. by Huang Wên-Shan *et al.* 1972– ; Japanese tr. in preparation, Shisakusha, Tokyo.

The Development of Iron and Steel Technology in China (Second Dickinson Lecture), Heffer, Cambridge (for the Newcomen Society), 1958. Repr. 1965.

Heavenly Clockwork [with Wang Ling & Derek de S. Price], Cambridge Univ. Press (for the Antiquarian Horological Society), 1960.
Within the Four Seas (essays and addresses), Allen & Unwin, London, 1970.
The Grand Titration (essays and addresses), Allen & Unwin, London, 1970. Italian tr., Mulino, Bologna, in the press; Japanese tr. by Hashimoto Keizō, in the press.
Clerks and Craftsmen in China and the West (collected lectures and addresses), Cambridge Univ. Press, 1970. Japanese tr. by Ushiyama Teruo *et al.*, Shobō Shinsha, Tokyo, in the press.
The Chemistry of Life (lectures arranged by the Cambridge Univ. History of Science Committee) (ed., with an introduction), Cambridge Univ. Press, 1970.
La Tradition Scientifique Chinoise (essays and addresses), Hermann, Paris, in the press.

Journal Publications and Occasional Papers

(excluding those in Biochemistry and Experimental Embryology and Morphology)

'The Philosophical Basis of Biochemistry', *Monist*, 1925, **35**, 27.
'Recent Developments in Biochemistry', *Outlook*, 1926, **58** (no. 1486), 184.
'Lucretius Redivivus; the Hope of a Chemical Psychology', *Psyche*, 1927 (no 27), 3.
Appendix, 'Biochemistry and Mental Phenomena', in *The Creator Spirit* (Hulsean Lectures), by C. E. Raven, Hopkinson, London, 1927.
Materialism and Religion, Benn, London, 1929.
'Philosophy and Embryology: Prolegomena to a Quantitative Science of Development (I), Physico-Chemical Embryology', *Monist*, 1930, **40**, 193.
'Laudian Marxism', *Criterion*, 1932, **12** (no. 46), 56.
'Biology and Mr. Huxley' (review of *Brave New World*), *Scrutiny*, 1932, **1** (no. 1), 76.
'Biology (Today and Tomorrow)', art, in *Science Today and Tomorrow* (Morley College Lectures), ed. E. V. Hubback; Williams & Norgate, London, 1932.
'Molly Dancing in East Anglia' [with A. L. Peck], *Journ. English Folk Dance and Song Soc.*, 1933 (3rd Ser.), **1** (no. 2), 79.
'Limiting Factors in the Advancement of Science as Observed in the History of Embryology', *Yale Journal of Biology and Medicine*, 1935, **8**, 1.
'The Geographical Distribution of English Ceremonial Dance Traditions', *Journ. English Folk Dance and Song Soc.*, 1936 (3rd Ser.), **3** (no. 1), 1.
'The Dances of Podhale (Poland)', *Journ. English Folk Dance and Song Soc.*, 1937 (3rd Ser.), **3** (no. 2), 117.
'A Discussion of Religion' [with C. Lamont], *Science & Society*, 1937, **1** (no. 4), 487.
'Christianity and Communism', *Modern Churchman* (Conference Number), 1937.
'Integrative Levels; A Revaluation of the Idea of Progress' (Herbert Spencer Lecture, Oxford Univ.), Oxford, 1937. Also in *Modern Quarterly*, 1937, **1**, 3.
'The Rise and Fall of Western European Science', *Manufacturing Chemist*, 1938, **9** (no. 2).
'Science Technology and Society in Seventeenth Century England' (review of R. K. Merton), *Science & Society*, 1938, **2** (no. 4), 566.
Foreword to *Biology and Marxism* by Marcel Prenant, tr. C. D. Greaves, Lawrence & Wishart, London, 1938.
'The Springtime of Science', *Chemical Practitioner*, 1939, **12** (part 2), 17.
'Voices from the English Revolution' (under ps. Henry Holorenshaw), *Modern Quarterly*, 1939, **2** (no. 1), 35.
'The Nazi Attack on International Science', *Biology*, 1941, **6** (no. 3), 107.
'Biological Science in the U.S.S.R.', *Nature*, 1941, **148**, 362.

'Matter, Form, Evolution and Us', *World Review*, 1941 (Nov.), **15**; collected in *This Changing World*, ed. J. R. M. Brumwell, Routledge, London, 1944, p. 27.

'Chung-Kuo chih Kho-Hsüeh yu Wên-Hua (Science in Chinese Culture)' (in Chinese), address at the Annual Conference of the Science Society of China at Mei-than, Kweichow, 1944.

'Report of the First Year's Working of the Sino-British Science Cooperation Bureau', Chungking, 1944.

'On Science and Social Change', *Science & Society*, 1946, **10** (no. 3), 225.

Report of the Second and Third Years' Working of the Sino-British Science Cooperation Bureau, Chungking, 1946.

Science and Unesco; International Scientific Cooperation—Tasks and Functions of the Secretariat's Division of Natural Sciences, Pilot Press, London, 1946 (also in French tr.).

Science and Society in Ancient China (Conway Memorial Lecture), Watts, London, 1947.

'The Chinese Contribution to Science and Technology', from *Reflections on Our Age* (Unesco Lectures, 1946), ed. D. Hardman & S. Spender, Allan Wingate, London, 1948.

Science and International Relations (Robert Boyle Lecture), Blackwell, Oxford, 1948.

'The Ballad of Mêng Chiang Nü Weeping at the Great Wall' (tr. with Liao Hung-Ying), *Sinologica*, 1948, **1** (no. 3).

The Liaison Work of Unesco's Field Science Cooperation Offices, Paris, 1948; (also in French, Spanish, Chinese, Arabic and Hindi trns.).

'The Unity of Science: Asia's Indispensable Contribution', *Asian Horizon*, 1949, p. 55.

'L'Unité de la Science: L'Apport indispensable de l'Asie', *Archives Internationales d'Histoire des Sciences*, 1949, **7**, 563.

Introduction to *Contemporary Chinese Woodcuts* [with Hetta Empson and Zderek Hrdlicka], Fore & Collet, London, 1950.

Human Law and the Laws of Nature in China and the West (Hobhouse Memorial Lecture), Oxford University Press, 1951. Japanese tr. *Shisō*, 1965–6.

'Natural Law in China and Europe; Parts I and II', *Journal of the History of Ideas*, 1951, **12**, 3.

'A Contribution to the History of Chinese Dietetics' [with Lu Gwei-Djen], *Isis*, 1951, **42**, 13.

'The History of Science and Technology in India and South East Asia', *Nature*, 1951, **168**, 64. A preview and review of 'Symposium on the History of the Sciences in South Asia', *Proc. National Institute of Sciences of India*, 1952, **18**.

'Biochemical Aspects of Form and Growth', art. in *Aspects of Form*, ed. L. L. Whyte; Lund Humphries, London, 1951, p. 77. 2nd ed. 1968.

'Ancient and Mediaeval Chinese Thought on Evolution' [with Donald Leslie], *Proc. National Institute of Sciences of India* (New Delhi), 1952, **7**.

'Chinese Science Revisited', *Nature*, 1953, **171**, 237 and 283.

'Relations between China and the West in the History of Science and Technology', *Actes du Septième Congrès International d'Histoire des Sciences,* Jerusalem, 1953.

'The Pattern of Nature-Mysticism and Empiricism in the Philosophy of Science: Third-Century B.C. China, Tenth-Century A.D. Arabia, and Seventeenth-Century A.D. Europe', art. in *Science, Medicine and History*, Singer Presentation Volume, ed. E. A. Underwood, Oxford University Press, 1953.

'Thoughts on the Social Relations of Science and Technology in China', *Centaurus*, 1953, **3**, 40.

'Le Dialogue Europe-Asie', *Comprendre*, 1954 (no. 12); *Synthèses*, 1958, **143**, 91. Japanese tr. *Gendai Shisō* (sp. no.), 1957, p. 121.

'Prospection Géobotanique en Chine Médiévale', *Journal d'Agriculture Tropicale et de Botanique Appliquée*, 1954, **1** (nos. 5–5), 143.

'Horner's Method in Chinese Mathematics: Its Origins in the Root-Extraction Procedures of the Han Dynasty' [with Wang Ling], *T'oung Pao*, 1955, **43,** 345.

'The Peking Observatory in A.D. 1280 and the Development of the Equatorial Mounting', from *Vistas in Astronomy* (Stratton Presentation Volume; ed. A. Beer), Vol. 1, Pergamon, London and New York, 1955.

'L'Asie et l'Europe devant les problèmes de la science et de la technique', *Europe-Chine*, 1955, **116,** 24.

'Remarks on the History of Iron and Steel Technology in China', in 'Actes du Colloque International, Le Fer à travers les Ages', *Bull. Fac. Lett. Univ. Nancy*, 1956, **16,** 93.

'Chinese Astronomical Clockwork' [with Wang Ling & Derek de S. Price], *Nature*, 1956, **177,** 600. Chinese tr. by Hsi Tsê-Tsung, *Kho Hsüeh Thung Pao*, 1956 (no. 6), p. 100. Also in *Actes du VIIIᵉ Congrès International d'Histoire des Sciences*, Florence, 1956, p. 325.

'Iron and Steel Production in Ancient and Mediaeval China' (abstract of Second Dickinson Lecture), *Transactions of the Newcomen Society*, 1956–57, **30,** 141.

'Mathematics and Science in China and the West', *Science & Society*, 1956, **20,** 320.

'The Dialogue of Europe and Asia', *United Asia*, 1956, **8** (no. 5), 1. Sinhalese tr. by M. Wickramasinghe, Colombo, 1960.

' "Spiked" Comets in Ancient China' [with Arthur Beer & Ho Ping-Yü], *Observatory*, 1957, **77,** 137.

'Les Mathématiques et les Sciences en Chine et dans l'Occident', *Pensée*, 1957, **75,** 3.

Review of 'Structure de la Médecine Chinoise' by P. Huard, *Discovery*, 1957, p. 490.

'Les Sciences en Chine Mediévale' [with A. Haudricourt], in *Histoire Générale des Sciences*, P.U.F., Paris, vol. 1, 477.

'Asien und Europa im Spiegel wissenschaftlicher und technischer Probleme', *Geist und Zeit*, 1957, **2,** 35.

'The Translation of Old Chinese Scientific and Technical Texts', from *Aspects of Translation*, ed. A. H. Smith. Secker & Warburg, London, 1958. Also in *Babel*, 4 (no. 1).

'Il dialogo tra l'Europa e l'Asia', *Ulisse*, 1958, **5,** 643.

Chinese Astronomy and the Jesuit Mission: An Encounter of Cultures, China Society, London, 1958.

'Les Sciences en Extrême-Orient du XVIᵉ au XVIIIJ siècle' [with J. Chesneaux], from *Histoire Générale des Sciences (de 1450 a 1800)*, vol. 2, 681. P.U.F., Paris, 1958.

'Wheels and Gear-Wheels in Ancient China', in *Actes du IXᵉ Congrès International d'Histoire des Sciences*, Barcelona, 1958.

'An Archaeological Study-tour in China, 1958', *Antiquity*, 1959, **33,** 113.

'The Missing Link in Horological History: A Chinese Contribution' (Wilkins Lecture), *Proc. Roy. Soc. A*, 1959, **250,** 147.

'The Dialogue of Europe and Asia' (in Bengali, tr. Krishna Dhav), *Bharat-Chin*, 1959, 1 (no. 3), 3.

'Automata', from *Enciclopedia Universale dell'Arte*, Istituto per la Collaborazione Culturale, Venice and Rome, 1959, vol. 2.

'The Laboratory Equipment of the Early Mediaeval Chinese Alchemists', [with Ho Ping-Yü], *Ambix*, 1959, 7 (no. 2), 58.

'An Early Mediaeval Chinese Alchemical Text on Aqueous Solutions' [with Ts'ao T'ien-Ch'in & Ho Ping-Yü], *Ambix*, 1959, 7 (no. 3), 122.

'Ondes et particules dans la pensée scientifique chinoise' [with Kenneth Robinson], *Sciences*, 1959 (no. 4), p. 65.

'Elixir Poisoning in Mediaeval China' [with Ho Ping-Yü], *Janus*, 1959, **48,** 15.

'Theories of Categories in Early Mediaeval Chinese Alchemy' [with Ho Ping-Yü,] *Journal of the Warburg and Courtauld Institutes*, 1959, **22** (nos. 3–4), 173.

'The Wheelwright's Art in Ancient China; I, The Invention of "Dishing" ' [with Lu Gwei-Djen & Raphael A. Salaman], *Physis*, 1959, **1,** 103.

'The Wheelwright's Art in Ancient China; II, Scenes in the Workshop' [with Lu Gwei-Djen & Raphael A. Salaman], *Physis*, 1959, 1, 196.

Review of *The Phenomenon of Man,* by P. Teilhard de Chardin, *New Statesman,* 1959 (7 Nov.).

'Science and Society in Ancient China', *Mainstream*, 1960, 13 (no. 7), 7.

'Les Contributions Chinoises à l'Art de Gouverner les Navires', in *Actes du cinquième Colloque International d'Histoire Maritime,* S.E.V.P.E.N. 1960 (1966); *Scientia,* 1961 (French and English tr.).

'Efficient Equine Harness; The Chinese Inventions' [with Lu Gwei-Djen], *Physis,* 1960, 2, 121.

'The Past in China's Present', *Centennial Review,* 1960, 4 (no. 2), 145.

Classical Chinese Contributions to Mechanical Engineering (Earl Grey Lecture, Univ. of Newcastle), 1961.

'The Chinese Contribution to the Development of the Mariner's Compass', *Scientia,* July 1961.

'The Earliest Snow-Crystal Observations' [with Lu Gwei-Djen], *Weather,* 1961, 16 (no. 10), 319.

'Human Law and the Laws of Nature', from *Technology, Science and Art; Common Ground,* Hatfield College of Technology, 1961.

'Aeronautics in Ancient China', *Shell Aviation News,* 1961 (no. 279), 2; (no. 280), 15.

'Hygiene and Preventive Medicine in Ancient China' [with Lu Gwei-Djen], *Journal of the History of Medicine and Allied Sciences,* 1962, 17, 429. Abridgement in *Health Education Journal,* September 1959.

'Christianity and the Asian Cultures', *Theology,* 1962, 65 (no. 593), 180.

'Frederick Gowland Hopkins' (Royal Society Centenary Lecture), *Perspectives in Biology and Medicine,* 1962, 6 no. 1), 2. Also in *Notes and Records of the Royal Society of London,* 1962, 17 (no. 2), 117.

'Astronomy in Classical China', *Quarterly Journal Royal Astron. Soc.,* 1962, 3, 87.

'Du Passé Culturel, Social et Philosophique Chinois dans ses Rapports avec la Chine contemporaine', *Comprendre,* 1962 (nos. 21–22, 23–24), 261.

'The Pre-Natal History of the Steam-Engine', *Transactions of the Newcomen Society,* 1962–63, 35, 3.

'China and the Origin of (Qualifying) Examinations in Medicine' [with Lu Gwei-Djen], *Proc. Roy. Soc. Med.,* 1963, 56 (no. 1), 1.

'Grandeurs et Faiblesses de la Tradition Scientifique Chinoise', *Pensée,* 1963 (no. 111).

'China's Philosophical and Scientific Traditions', *Cambridge Opinion,* 1963, 36, 11.

Review of *Mediaeval Technology and Social Change* by Lynn White, *Isis,* 1963, 54, 418.

'Poverties and Triumphs of the Chinese Scientific Tradition', in *Scientific Change* (Report of History of Science Symposium, Oxford, 1961), ed. A. C. Crombie, Heinemann, London, 1963.

'China and the Invention of the Pound-Lock', *Transactions of the Newcomen Society,* 1963–64, 36, 85.

'Mediaeval Preparations of Urinary Steroid Hormones' [with Lu Gwei-Djen], *Medical History,* 1964, 7, 101. Abridged in *Nature,* 1963, 200 (no. 4911), 1047.

'Science and Society in China and the West', *Science Progress,* 1964, 52 (no. 205), 50.

'Science and China's Influence on the World', art. in *The Legacy of China* (ed. R. Dawson), Oxford, 1964.

'Chinese Priorities in Cast-Iron Metallurgy', *Technology and Culture,* 1964, 5 (no. 3), 398.

'Glories and Defects of the Chinese Scientific and Technical Traditions', art. in *Neue Beiträge zur Geschichte der alten Welt* (ed. E. C. Welskopf), vol. 1. Akademie-Verlag, Berlin, 1964.

'Science and Society in East and West', *Science & Society,* 1964 (no. 4), p. 385. In Bernal

Presentation Volume, *The Science of Science*, ed. M. Goldsmith & A. McKay. Souvenir, London, 1964, repr. Penguin, London, 1966.

'An 8th-Century Meridian Line: I-Hsing's Chain of Gnomons and the Pre-history of the Metric System' [with A. Beer, Ho Ping-Yü, Lu Gwei-Djen, E. G. Pulleyblank and G. I. Thompson]. *Vistas in Astronomy*, 1964, **4**, 3.

'Understanding the Past is the Key to the Future', *Far East Trade and Development*, 1965.

'A Korean Astronomical Screen of the Mid-Eighteenth Century from the Royal Palace of the Yi Dynasty (Choson Kingdom, 1392 to 1910)' [with Lu Gwei-Djen], *Physis*, 1966, **8**, 137.

Foreword to *Window on Shanghai; letters from China 1965–7* by Sophia Knight, Deutsch, London, 1967.

'The Dialogue between Asia and Europe', art. in *The Glass Curtain between Asia and Europe*, ed. Raghaven Iyer; Oxford, 1965, p. 279. Germ. tr. by M. von Schön & G. Mehling, Callwey, München, 1968.

'A Further Note on Efficient Equine Harness; The Chinese Inventions' [with Lu Gwei-Djen], *Physis*, 1965, **7**, 70.

Time and Eastern Man (Henry Myers Lecture, 1964), Royal Anthropological Institute, London, 1965. Also as 'Time and Knowledge in China and the West', art. in *The Voices of Time*, ed. J. T. Fraser; Braziller, New York, 1966, p. 92.

'Organiser Phenomena after Four Decades; a Retrospect and Prospect'. (Introduction for the reprint of *Biochemistry and Morphogenesis*) 1966. Also as contribution to the J. B. S. Haldane Memorial Volume, *Haldane and Modern Biology*, ed. K. R. Dronamraju, Johns Hopkins Press, Baltimore, 1968.

'The Optick Artists of Chiangsu' [with Lu Gwei-Djen], *Proceedings of the Royal Microscopical Society*, 1966, **1**, Part **2**, 59 (abstract). In *Studies in the Social History of China and South East Asia* (Purcell Memorial Volume, ed. J. Ch'en & N. Tarling), Cambridge Univ. Press, 1970. Also in *Proceedings of the Royal Microscopical Society*, 1967, **2**, 113.

'Proto-Endocrinology in Mediaeval China' [with Lu Gwei-Djen], *Japanese Studies in the History of Science*, 1966 (no. 5), p. 150.

'Naturvidenskab og samfund i øst og vest', *Dansk Udsyn*, 1966, **2**, 155. (Danish tr. of 'Science and Society in East and West'.)

'Magnetic Declination in Mediaeval China' [with P. J. Smith], *Nature*, 1967, **214** (no. 5094), 1213.

'The Roles of Europe and China in the Evolution of Oecumenical Science', *Advancement of Science*, 1967, **24** (no. 119), 83. Also in *Journal of Asian History*, 1967, **1**, 1.

'Records of Diseases in Ancient China' [with Lu Gwei-Djen], art. in *Diseases in Antiquity*, ed. D. Brothwell & A. T. Sandison; Thomas Springfield, Illinois, 1967.

'Skin Colour in Chinese Thought', note in *Race*, 1968, 249.

'Sex Hormones in the Middle Ages [with Lu Gwei-Djen], *Endeavour*, 1968, **27** (no. 102), 130.

'The Development of Botanical Taxonomy in Chinese Culture' in *Actes du XIIe Congrès International d'Histoire des Sciences*, Paris, 1968, p. 127.

'The Voyage of Surgery', *Guy's Hospital Reports*, 1968, **117**, 139.

'The Esculentist Movement in Mediaeval Chinese Botany; Studies on Wild (Emergency) Food Plants' [with Lu Gwei-Djen], *Archives Internationales d'Histoire des Sciences*, 1969 (no. 84–85), 225.

Foreword to *The Question Mark; the End of Homo Sapiens*, by Hugh Montefiore, Collins, London, 1969.

'Artisans et Alchimistes en Chine et dans le Monde Hellénistique', *Pénsee*, 1970 (no. 152).

'China and the West', in *China and the West; Mankind Evolving*, ed. A. Dyson & Bernard Towers, Garnstone, London (for the Teilhard de Chardin Association), 1970.

The Refiner's Fire; the Enigma of Alchemy in East and West (Second Bernal lecture), Birkbeck College, London, 1971.

'Desmond Bernal; a Personal Recollection', *Cambridge Review*, 1971, **93**, 33.

Hand and Brain in China [with Joan Robinson, Edgar Snow & T. Raper], Anglo-Chinese Educational Institute, London, 1971.

Foreword to *Science at the Cross-Roads* (Papers of the Soviet Delegation, by N. I. Bukharin, B. Hessen *et al.*, from the Second International Congress of the History of Science and Technology, Kniga, London, 1931); repr. Cass, London, 1971, ed. R. M. McLeod, with Introduction by P. G. Werskey.

'Mao and the Dark Aspects', *Theoria to Theory*, 1971, **5**, 154.

'Do the Rivers Pay Court to the Sea? The Unity of Science in East and West', *Theoria to Theory*, 1971, **5**, 68.

'The Coming of Ardent Water' [with Lu Gwei-Djen & Dorothy M. Needham], *Ambix*, 1972, **19**, 69.

'Altes China-junges Europa', *Die Waage,* 1972, **11**, 97.

'Ancient Chinese Oecology and Plant Geography; the Case of the *Chü* and the *Chih*', art. in *Balazs Memorial Volume*, ed. F. Aubin, 1973.

'A Chinese Puzzle; Eighth or Eighteenth?', art. in *Pagel Presentation Volume,* ed. A. G. Debus, 1973.

INDEX